An Economy Based on Carbon Dioxide and Water

Michele Aresta · Iftekhar Karimi ·
Sibudjing Kawi

Editors

An Economy Based on Carbon Dioxide and Water

Potential of Large Scale Carbon Dioxide Utilization

 Springer

Editors
Michele Aresta
IC^{2}R srl, Lab H124, Tecnopolis
Valenzano, Italy

Iftekhar Karimi
Department of Chemical and Biomolecular
Engineering
National University of Singapore
Singapore, Singapore

Sibudjing Kawi
Department of Chemical and Biomolecular
Engineering
National University of Singapore
Singapore, Singapore

ISBN 978-3-030-15867-5   ISBN 978-3-030-15868-2   (eBook)
https://doi.org/10.1007/978-3-030-15868-2

Library of Congress Control Number: 2019936294

This Springer imprint is published by the registered company Springer Nature Switzerland AG
The registered company address is: Gewerbestrasse 11, 6330 Cham, Switzerland

Preface

The Carbon Dioxide Problem

Surface (biomass) and sub-surface (fossil-C such as coal, oil, gas) C-based assets have been, and will be for long time yet, used as source of energy and goods by humans. The use of all such materials causes the formation of CO_2, which is emitted into the atmosphere. The intensity of the emission of CO_2 has continuously grown and parallels the growth of both the population and average standard of life. (Figs. 1 and 2).

This has caused the continuous accumulation of atmospheric CO_2 which has reached 408 ppm these days with respect to 275 ppm of the preindustrial era. The atmospheric level of CO_2 is the "warning light" for the "health" of our planet. A correlation exists between the consumption of energy—the CO_2 emission—the accumulation in the atmosphere and the increase of the planet temperature (Figs. 3 and 4) that may cause non-return catastrophic events. Whether CO_2 is the direct actor or is an "indicator" of the impact caused by human activities on the atmosphere, is in question. As a matter of fact, humans are using C-based resources in a "highly inefficient" way: the efficiency of conversion of chemical energy (fossils or biomass) into other forms of energy (electrical, thermal, mechanical, etc.) ranges around 27–35%. This means that 73–65% of the original chemical energy is released to the atmosphere in the form of heat, often at high temperature, causing its direct heating. As the emission of CO_2 is related to the amount of C-based fuels burned or goods used, it becomes an easy "witness" of the impact humans are causing on Earth (even if not the direct cause). However, there is a general fear that continuing to accumulate CO_2 into the atmosphere (or continuing with an inefficient use of natural resources) may increase the temperature of our planet. Although CO_2 is a greenhouse gas (GHG), its atmospheric concentration is much lower than that of other GHGs, e.g. water vapour. Is the atmospheric CO_2 that causes planet heating or the released heat or both? A question rises: Supposed that we capture all the produced CO_2, but continue to heat the atmosphere with inefficient conversion of chemical energy (C-based energy sources) into other forms of energy, shall we "cool down" our planet? The most effective solution would be to reduce the use of C-based fuels (that will imply the reduction of CO_2 emission) and the release of

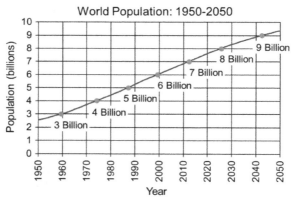

Source: U.S. Census Bureau, International Data Base, August 2016 Update.

Fig. 1 Growth of world population

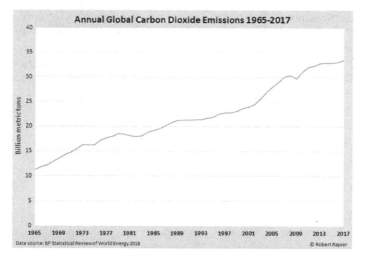

Fig. 2 CO_2 emission

heat to the atmosphere. This will require a great effort in terms of more efficient technologies in the use of primary C-based sources of energy, covering the entire chain of production of electric energy, heating systems, industrial plants and transport sector: our whole life! This will take long time and demand huge economic means. As for now, the policy is to reduce the emission of CO_2 into the atmosphere, will or will not such practice be the solution to the climate change (CC), shifting from fossil-C (that has produced a net increase of the atmospheric CO_2-level) to renewable-C and to perennial C-free energy sources.

How to avoid CO_2, thus?

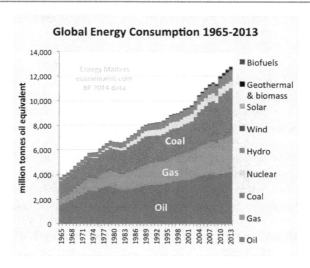

Fig. 3 Energy consumption

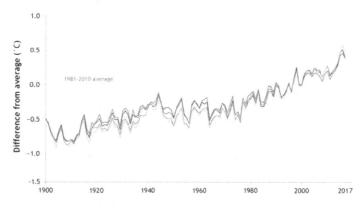

Fig. 4 Increase of average temperature

The agreed target (ONU 2018 and previous International Agreements) is to stay below 2° rise of the average temperature of our planet with respect to 1990. We are already 1.5° above; therefore, we must urgently slow down the use of fossil carbon in order to stay within the target. Defossilization of energy is the necessary action, and the proposal is to advance in such direction in next years. From now to 2050, fossil-C should slowly but steadily be replaced by *quasi*-zero-C emission energy sources. Alternatives are the use of solar, wind, geothermal, hydro-energy sources (SWGH, perennial energy sources). The question is whether or not such sources will provide energy with the requested density and intensity for powering industrial sites and other intensive energy users (megalopolis). (see P. Moriarty, D. Honnery,

What is the global potential for renewable energy? Renewable and Sustainable Energy Review, 2012, 16, 244–252). On the other hand, the use of biomass will cover a minor part of the energy needs limited by the regeneration rate. The production rate of CO_2 is much higher (around 6–15 g_{CO2} m^{-2} s^{-1} from the combustion of several kinds of wood-see Tran, H.C.; White, R.H. *Fire and Materials,* **1992,** *16,* 197–206.) than the fixation of CO_2 by photosynthesis (the best values are observed in microalgae (see Adamczyk, M.; Lasek, J.; Skawinska, A. *Appl Biochem Biotechnol,* **2016,** *179,* 1248–1261), better fixing agents than terrestrial plants, that are in the range 0.1–1.7 g_{CO2} l^{-1} day^{-1} or 0.00057–0.0098 g_{CO2} m^{-2} s^{-1}). Therefore, the combustion rate is from 1000 to 10,000 times faster than the C-fixation by photosynthetic microorganisms and microalgae. Will nuclear come back? This is not excluded, if safe production of electricity will be developed. Our future looks quite problematic and to what extent energy-poor countries in Asia, America and Africa will rise their standard of life is an open question, as it is not easy to state whether Northern America, Europe, Japan will have all to restate their habitudes. Shall we necessarily go towards a global lower energy consumption? When this will occur? In the meanwhile, attention is devoted to reduce the CO_2 emission into the atmosphere.

Measures for the Reduction of the CO_2 Atmospheric Level

The technologies for reducing the use of fossil-C and the emission of CO_2 into the atmosphere can be categorized as reported below.

- *Higher efficiency in the conversion of chemical energy (fossil-C and biomass) into other forms of energy*: thermal, electrical, mechanical, etc. As said above, the average efficiency ranges today around 27–35%; therefore, 73–65% of the original chemical energy is released to the atmosphere in the form of heat, causing its direct heating. Improving the efficiency of the actual energy system is possible using both upstream and downstream technologies, but the economic cost is often quite high. In practice, power stations, industrial processes, transportation means should all be restyled and made adequate to the target of reducing the loss of energy in the form of heat. As a lower profile strategy, new power and industrial plants should all be built according to new efficiency criteria (e.g. all power plants should be built as IGCC), with much higher CAPEX. Heat could be recovered (when and where it makes sense) and used in a range of diverse low-temperature applications (in part this is already done, even if with rare-spot applications). The efficiency of primary energy conversion could be increased by 10–20 points, according to the different sectors. An improvement of the actual system is, thus, possible but it will take long time and demand a lot of economic resources.
- *Efficiency in the utilization of the different forms of energy, avoiding misuses.* This is the easiest route to C-saving and should be implemented by each single person. The implementation of virtuous practices can produce reduction of energy used, and thus produced, with reduction of conversion of fossil-C and CO_2 emission with a possible target of 5–10%.

- *Decarbonization of energy, by using perennial primary sources*, alternative to fossil fuels, such as SWHG. Such change is ongoing, and solar-, wind- and geo-power are being efficiently used, even if still at a moderate global intensity (5–8 % of global consumption) and leopard-spot distribution. It must be emphasized that perennial sources have a different availability on our planet and regional solutions must be implemented. Conversely, fossil-C (coal, oil, gas) is easily transportable and can reach any point on our planet: this is a plus in favour of existing fossil-C-based technologies. The moment being, anyway it is worth to emphasize that virtuous countries exist, which produce large amounts of their energy from SWHG (https://www.climatecouncil.org.au/11-countries-leading-the-charge-on-renewable-energy/) as reported in Table 1.
  It is worth to emphasize that if the energy sector can be decarbonized, the chemical, polymer, pharmaceutical, cosmetic, materials industry will not. Therefore, C-based goods will continue to exist and most likely the use of fossil-C will be continued for long time yet in the future. As a matter of fact, the international programme 205050 foresees 50% reduction of emissions with respect to 1990 (for USA with respect to 2005) for EU and G20 countries by 2050. Afterwards, the reduction should continue at a rate of 1.5% per year.
- *Use of renewable carbon, such as biomass.* Renewable carbon has a different abundance on Earth and can find a different utilization. Biomass (land and aquatic) already finds application in several fields: from energy to chemicals and materials production. Because of the relative abundance/scarcity, even biomass has an odd usability and its transport is not energetically and economically efficient. Even in this case, fossil-C is a winning option for its higher energy density.

Table 2 says that liquid fuels are the best options for energy storage and transportation. Therefore, the conversion of electricity produced *via* SWGH into liquid fuels by combining CO_2 reduction and water oxidation is a convenient way for storing and transporting energy.

Table 1 Use of energy sources alternative to fossil-C in some virtuous countries

Country	Actual use of SWHG for electricity generation	Perspective use of SWHG
Sweden		100% 2050
Costa Rica	99% (2015)	C-neutral 2021
Nicaragua	54%	90% 2021
Scotland	97% of household electricity	
Germany	36.1+ % average with some >85% spot	
Uruguay	95%	
Denmark	42%	100% (2050)
China	1000 coal mines closed in 2018	
Morocco		50% 2020
Kenya	51% (2015)	70% (2020)

Table 2 Energy density of a variety of energy carriers

Energy carrier	MJ/kg
Diesel	36
Gasoline	34
Coke	30
Brown coal	18
Methanol	17
Bio-oil from algae	13
H_2(liquid, 20 MPa)	9
H_2(compressed)	2
Battery (lead)	0.33
Battery (lithium)	2.8

Capture of produced CO_2 from point-continuous sources or from the atmosphere. Capture can be applied either to point-continuous-fixed sources (industries or power plants) or even everywhere if the source is the atmosphere. If practised at reasonable economic and energetic costs, capture will make available large amounts of CO_2 that can be either disposed (CCS) or converted into chemicals, materials and fuels (CCU). The separation of CO_2 from flue gases from power stations or industrial processes (Table 3) requires an amount of energy, which depends on the composition of the flue gas and the use one wishes to make of separated CO_2 that drives its purity. The separation is a process that requires energy, and in the actual energy system implies the use of fossil-C. It has been calculated by DoE-USA that recovering CO_2 and disposing it in land reservoirs 40 km away from the power station where it is generated (a particularly rare and favourable case) consumes 25% of the produced electric energy. Such amount increases with the distance of shipping and the nature of disposal site: it may reach 40%+. Consequently, the capture and housing of CO_2 expand the extraction of fossil-C because additional energy is necessary for such end-of-pipe operation. Modern power generation technologies (such as IGCC) proceed to the decarbonization of fossil fuels before combustion. This means that fossil fuels are converted into CO_2 and H_2 and the latter is used for clean energy generation, but not zero-C emission as CO_2 is anyway produced even if in the precombustion phase.

Table 3 CO_2 emission from industrial sectors

Industrial sector	Mt_{CO2}/y produced
Cement	>1000
Oil refineries	850–900
Iron and steel	ca. 900
Fermentation	>200
Ethene and other petrochemical processes	155–300
Ammonia	160
LNG sweetening	25–30
Ethene oxide	10–15

In the short term, it may make sense to recover CO_2 from industrial sources that may provide more pure CO_2 than flue gases from power stations. Table 3 shows that ca. 3 500 Mt/y of quite pure CO_2 is available, which covers ca. 10% of the total emission. The advantage of recovering–converting CO_2 on-site brings to a closed loop. Obviously, this approach is very proficient for sites where several industrial activities are present and an efficient recovery of CO_2 is possible: clustering process is a way to optimize the energy and waste system. And this can be an interesting option for the future of our industry and the energy sector.

Recovery from the atmosphere (medium–long term) has the advantage of not requiring an industrial site or a power station as source. Interestingly, the atmosphere can provide also water vapour: this means that co-processing atmosphere-sourced CO_2 and water vapour will make possible to produce Syngas and thereafter energy-rich products (fuels). Such practice is very suggestive of Nature and can be implemented everywhere on our planet, making possible a local production of necessary fuels and chemicals completely decoupled from the existence in situ of natural fossil resources or emitters, with great benefit for non-industrialized countries which are not even rich of fossil-C.

An Economy Based on CO_2 and Water: it is a *vision* today. Will science and technology be able to make it a reality? Yes, but the correct conditions for the use of perennial energies must be developed and investment in research is necessary at the correct level, with the integrated cooperation of academy, industry and governments.

It is an investment in the future of humankind and our planet.

This book makes the point on where we are and where we have to go for exploiting such option.

After presenting the potential and bottlenecks of large-scale CCU (Chap. 1), the capture (Chap. 2) and the technological applications (Chap. 3), various aspects of utilization are discussed in detail, namely carbonation of basic natural or industrial matrices (Chap. 4), conversion into energy-rich products (Chaps. 5 and 6), electrochemical and photo-electrochemical conversion (Chap. 7) and Plasma technologies (Chap. 8). Bio-based routes are discussed in Chaps. 9–11, highlighting the integration of biotechnologies and catalysis. Chapter 12 makes a techno-economic and energetic analysis of selected CO_2-based processes. At the end, the perspective use of CCU is presented.

Michele Aresta
IC^2R srl, Lab H124, Tecnopolis
Valenzano, Italy

Contents

Large Scale Utilization of Carbon Dioxide: From Its Reaction with Energy Rich Chemicals to (Co)-processing with Water to Afford Energy Rich Products. Opportunities and Barriers

1

Michele Aresta and Francesco Nocito

Abstract

This chapter makes the analysis of the possible routes for large scale CO_2 utilization (CCU). Processes that convert CO_2 into chemicals, materials and fuels are discussed, as they are part of the strategy for reducing the CO_2 emission into the atmosphere. Technical uses of CO_2, which do not imply its chemical conversion, are discussed in Chap. 3, while mineralization and carbonation reactions for the production of inorganic materials are treated in Chap. 4. Here, the catalytic synthesis of organic products with a market close to, or higher than, 1 Mt/year is discussed, presenting the state of the art and barriers to full exploitation. Minor applications are summarized, without a detailed analysis as their contribution to CO_2 reduction is low, even if they can favour the development of a sustainable chemical industry with reduction of the environmental impact. Energy products (C1 and Cn molecules) are discussed for some peculiar aspects in this chapter, as their catalytic production will be extensively presented in following chapters where the potential of using CO_2 *and water* as source of fuels is analysed for its many possible applications setting actual limits and future perspectives. A comparison of Carbon Capture and Storage-CCS and CCU is made, highlighting the pros and cons of each technology.

M. Aresta (✉)
IC²R srl, Lab H124, Tecnopolis, via Casamassima km 3, 70010 Valenzano, Italy
e-mail: michele.aresta@uniba.it

F. Nocito
Department of Chemistry and CIRCC, University of Bari, 70126 Bari, Italy

© Springer Nature Switzerland AG 2019
M. Aresta et al. (eds.), *An Economy Based on Carbon Dioxide and Water*,
https://doi.org/10.1007/978-3-030-15868-2_1

1

1.1   Introduction

The capture of CO_2 from power plants and industrial processes is a way to avoid that it enters the atmosphere, as announced in the Preface. The concentration of CO_2 in flue gases from power stations, that represent the largest point-source, averages 14%: it can be lower, depending on the quality of the fuel used. Industrial processes, as detailed in the Preface, emit CO_2 at concentrations that can reach 90%+ in fermentation units. CO_2 can also be recovered directly from the atmosphere, the most abundant source of carbon dioxide on our planet and the most diluted (408 ppm, [1]). Technical aspects of capture of CO_2 from the various sources are discussed in Chap. 2. Once captured, CO_2 can be either disposed or used. Disposal of CO_2 is not discussed in this book: only a brief comparison with the utilization option is made, highlighting pros and cons of both. In this chapter, an analysis of the various options of CO_2 conversion is made, highlighting the potential of such technology to reduce the atmospheric level of CO_2. Various categories of reactions are presented, according to their energetic content. The conversion into chemicals and polymers [2] is discussed in detail, while the conversion into C1 or Cn energy products is analysed from a general point of view highlighting the importance of the use of CO_2 and water in their synthesis, while their chemistry is discussed in ad hoc chapters of this book.

1.2   CCS Versus CCU: The Amount Issue

As said above, one of the approaches to control the atmospheric CO_2 level is its capture from point-continuous sources or from the atmosphere (see Chap. 2). Once captured, CO_2 can be either disposed (CCS) or used (CCU). Table 1.1 makes a comparison of the two approaches, highlighting the pros and cons. One of the main discussion topics when comparing CCS and CCU, is the amount of CO_2 that can be disposed. Besides, the energy penalties and the residence time are key issues. The two alternative strategies are quite different, as shown in Table 1.1.

We shall not discuss CCS in this paper, will devote instead our attention to the effective and efficient CO_2 conversion into chemicals, materials and fuels, targeting avoidance of large volumes of CO_2. We shall also make an exam of the potential impact on the chemical industry, and eventually on the energy sector, of a full exploitation of technologies based on CO_2 utilization. Mineralization of CO_2, i.e. its conversion into inorganic carbonates is discussed in Chap. 4. Benefits of CO_2 utilization in technological uses in which it is used as technical fluid but not converted into other chemicals, are discussed in Chap. 3.

1.3 CO_2 Conversion Options: The Energy Issue

The chemical utilization of CO_2 can be categorized into three classes [3], as shown below.

Table 1.1 CCS and CCU at glance

CCS strength	Issues	Entry	CCU strength	Issues
TRL = 8–9		TRL	**TRL = 4–9**	
Injection technologies are well known	Sites are not ubiquitary	Ubiquity	A CO_2 conversion-plant can be built in any place	When used, compounds give back CO_2
Some 5 Mt/year are disposed; sites may be characterized by a different permanence	Presently, disposal is mainly for EOR. Doubts exist about the permanence and diffusion in new sites	Amount used permanence	220 Mt/y are used. Some processes are old of 150 y (urea, aspirine, inorganic carbonates)	Only polymers may have a sequestration time of decades. CCU is not for storing CO_2 but for cycling-C
Transportation is practiced on a large scale in industrial areas	Safety of pipelines in civil areas is a bottleneck	Safety	CO_2 is not toxic. It represents the ideal substitute of phosgene or CO	Concentration above 10% volume in air may cause asphyxia.
CCS is for long time disposal. Better knowledge of disposal sites may increase safety	CCS is apt to linear economy and produces additional CO_2 during storage	Expected innovation	CCU is circular economy: it is ideally used in a solar-wind-geo (SWG) energy-frame	Value chains and clustering-processes may optimize CCU
CCS has a sense in a fossil-C based energy frame. It makes sense in the very short time and will loose power in future	CCS fails to be implemented in a frame of large scale SWG exploitation	Fossil-C versus Perennial SWG energy frame	CCU can turn on air-CO_2 without fossil-C extraction: emitted CO_2 can be recycled into chemicals, materials, fuels continuously	Formation of CO_2 is much faster than some fixation processes
CCS is said to be able to store all CO_2 generated in the combustion of C-based materials on the earth	Disposal of CO_2 is not zero CO_2-generation. The real potential must be verified	Perspective amount of CO_2 avoided	7–10 Gt/year of CO_2 can be converted into a variety of compounds and fuels, in a frame of SWG exploitation. CCU is the future	Integration of technologies and system approach is necessary

(continued)

Table 1.1 (continued)

CCS strength	Issues	Entry	CCU strength	Issues
CCS needs energy for shipping and housing CO_2. Energy cost depends on the distance btw source-disposal	A loss of usable energy is the effect: from 25 to 50 +%	Energy requirement	Reactions of CO_2 with electron-rich species are exoergonic. Reduction reactions are endoergonic	Making fuels needs hydrogen that must come from water
Popular acceptance of CCS has been randomly enquired. Attitude of people must be deciphered through larger enquiries	A lot of information must be delivered to common people before their response to accept/reject CCS can be sound	Popular acceptance	CO_2 is in the air and used by Nature for making million products: this is under the eyes of everybody. The industrial exploitation needs to be illustrated	Information still needs to be spread around for clarifying basic concepts and avoid confusion (CO vs. CO_2!)
As mentioned above, the storage technology is well known with spent gas-wells. The injection technology is well known by oil-gas industry. The knowledge of new sites must be developed to high level before use	Disposal sites require still a lot of investigation for knowing their long-term storage, leakage and diffusion. This is a prerequisite for safety	Research needed	The knowledge of the reactivity of CO_2 under different conditions is quite good, even if new catalysts can be discovered and new processes that may allow to implement a larger scale conversion	Co-processing of CO_2 and water under solar irradiation is a must for the future avoiding hydrogen production

Capture is a prerequisite for both technologies, the purity of CO_2 may be distinct for CCS and CCU

   A. Conversion into chemicals in which the $-CO_2-$ entity is preserved by incorporation of the entire moiety into a new molecule without C–C bond formation. In such case, the oxidation state of the C-atom remains equal to +4. Such processes may be low in energy as, in general, the CO_2 molecule reacts with electron rich co-reagents such as amines-RR′R″N, or O-containing species: O^{2-}, OH^-, OR^-, or similar systems. Exoergonic carbonation of inorganic bases belongs to this Class.

B. Carboxylation of organic substrates.

Carboxylation reactions, in which a C–C bond os formed and the oxidation state of C goes down to +3 from +4 can be categorized as: (i) reactions in which CO_2 reacts with energy rich molecules such as olefins, alkynes, dienes (conjugated and cumulated) to afford specialty chemicals such as lactones, pyrones, esters, among others; (ii) insertion of CO_2 into C–H bonds with formation of acids. The former (see Sect. 9) occurs under quite mild conditions and can be exploited in a energy system based on fossil-C, the latter (1.1) are quite common in Nature and are of great interest for Industry.

$$R-H + CO_2 \rightarrow R - CO_2H \tag{1.1}$$

Today, the introduction of a carboxylic moiety $-CO_2-$ into an organic substrate occurs via "oxidation" of an alkyl group (in general a methyl, $-CH_3$) (1.2a) or partial destruction of an aromatic ring (1.2b) or even by hydrolysis of a cyanide moiety (1.3).

$$R-CH_3 \overset{oxidant}{\rightarrow} R-CO_2H \tag{1.2a}$$

$$\tag{1.2b}$$

$$\tag{1.2a}$$

$$R-CN + 2H_2O \rightarrow R-CO_2H + NH_3 \tag{1.3}$$

Reactions (1.2a), (1.2b) are often characterized by low selectivity and, thus, by a low Carbon Utilization Fraction (CUF). In reaction (1.2b) two atoms of carbon are lost as CO_2 over ten. The direct carboxylation would avoid such loss as reaction (1.1) has a CFU = 1. Reaction (1.3) uses toxic species. Cyanides are produced through complex routes with high E-factor (ratio of waste to target useful product). The bottleneck in (1.1) is the activation of the C–H bond. Thermal processes seem less suited than photochemical [4] ones to such end. In fact, recently several examples of photochemical reactions can be found in the literature [5], that are overwelming thermal examples.

C. Conversion into chemicals in which the C-atom has a lower oxidation state than +3: from +2 to −4. The energy content of the resulting chemical depends on a series of factors, among which the oxidation state of the C-atom of the species into which CO_2 is converted, the number of C–H bond formed, the number of C–C bond formed. Table 1.2 shows the thermodynamic properties of some molecules derived from CO_2 upon hydrogenation. It is evident that there is not a (linear) relationship between the change of the oxidation state with respect to CO_2 (Column 3) and the change of Free Gibbs energy (Column 7) moving from CO_2 to CH_4. Acetylene has

peculiar properties and methanol and ethene (in the two latter the oxidation state is
−2) have different energy content.

Considering a series of compounds having similar structural properties (for
example compounds #7–9 in Table 1.2 in which the C atom has a sp^3 geometry), it
is possible to find a correlation between the difference in $\Delta G°$ between the reduction
product and original CO_2 (1.4) and the variation of the oxidation state of the C-atom
(Fig. 1.1).

$$\Delta G = \Delta G_{f(newmoleculeperC-atom)} - \Delta G°_{f(CO_2)} \qquad (1.4)$$

In all such conversions two actors play a key role: energy and hydrogen. Both must
be delivered to the reactive system for the conversion of CO_2 may occur and cannot
be sourced from fossil-C: energy must come from perennial sources (SWHG) and
hydrogen from waste biomass and sea- or process- waters, in order to not aggravate
the drinking-water world problem.

This is the limiting factor for the conversion of CO_2 into energy products, with
recycling of large masses of C in the form of CO_2.

1.4 CCU and Resources Conservation

The utilization of CO_2 as building block or source of carbon means recycling carbon,
mimicking Nature. This is a virtuous practice that reduces the transfer of fossil-C to the
atmosphere, supports the sustainability of the chemical industry (and transport sector)
and preserves fossil-C resources for next generations. The consumption of fossil-C
today ranges around 11.5 [7] −12.5 Gtoe/year [8] with a forecast of reaching 25 Gtoe
in 2035. In a Business as Usual (BAU) this means that the CO_2 emission will rise from
37 to over 70 Gt/year! The daunting figure is that the estimated fossil-C reserves (coal,
oil and gas) are today quantified at 941 Gtoe. According to the BP forecast [9], at the
actual rate of consumption of fossil-C, humanity have crude oil available until year
2066, natural gas until year 2068, and coal until year 2169. Such figures are impressive! We hopefully rely on the fact that the knowledge we have of the real reserves of
fossil-C is not absolutely certain and complete and, thus, on the fact that new
coal-oil-gas sites can be discovered or deeper sites not reached so far can be exploited.
With a perspective availability of fossil-C of ca. 150 years, recycling carbon (equivalent to use renewable carbon) becomes a compulsory practice for saving natural
resources. Therefore, using (atmospheric) CO_2 for making chemicals and fuels, as
Nature does, is an ethical problem, before being scientific or technical. However, one
could ask at what cost recycling can be done and whether or not recycling carbon is
sustainable.

Table 1.2 Thermodynamic properties [6] of some molecules derived from CO_2 upon hydrogenation

Compound	Formal oxidation state of C	$\Delta_{Ox} = n_{ox,X} - n_{ox,CO_2}$	$\Delta H°$ (kJ/mol)	$\Delta S°$ (J/mol)	$\Delta G°$ (kJ/mol)	$\Delta G°$ (kJ/mol C)	#
CO_2	+4	0	−393.51	213.6	−457.2	−457.2	1
CO	+2	−2	−110.53	197.6	−169.4	−169.4	2
CH_2O	0	−4					3
C_2H_2	−1	−5	226.7	200.8	166.9	88.45	4
CH_3OH	−2	−6	−238.7	126.8	−276.5	−276.5	5
C_2H_4	−2	−6	52.3	219.5	−13.1	−6.55	6
C_3H_8	−2.7	−6.7	−103.8	269.9	−184.2	−61.4	7
C_2H_6	−3	−7	−84.7	229.5	−153	−76.5	8
CH_4	−4	−8	−74.8	186.2	−130.3	−130.3	9

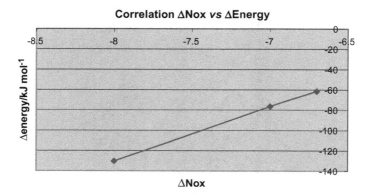

Fig. 1.1 Correlation $\Delta N_{ox}/\Delta$Energy for some reduced forms of CO_2

Fig. 1.2 Sources of energy and goods in the human history: past and future

Figure 1.2 presents the history of energy and goods sources for humans (food is not considered). Until ca. 1750, biomass was practically the only real source of energy and goods (materials, fibers). With the industrial revolution and the exploitation of coal-mines first, and oil- and gas-fields later, fossil carbon has increased its role as source of energy and goods competing and winning over biomass. Today, we are moving back to the use of SWGH and renewable-C, aiming at reducing the use of fossil-C. Perennial sources and biomass may become the main sources of energy and goods in 30 years from now, with a significant reduction of the extraction of fossil-C. For this change may occur at a significant extent, it is necessary that economic and environmental costs of innovative processes are affordable. As a matter of fact, the solar(wind)-H_2 price should be close to that of reforming-H_2 (ca. 1.3–1.5 US\$/kg): today it is some three times higher. The US Department of Energy has specified that the solar-to-hydrogen commercialization target is a energy conversion efficiency preferably >10% to compete with gasoline [10]. Presumably, a comparable solar-to-fuel conversion cost is necessary for photocatalytic or photovoltaic (PV) plus electrocatalytic CO_2 reduction may become a commercial reality. Installed PV is increasing at a high rate: from 51 GW in 2015, to 305 in 2017, 969 in 2025 [11]. By 2040, it is foreseen that the worldwide installed power of PV will rise to over 3000 GW and the cost of PV-materials will decrease to a level that will make PV able to produce low cost electrons and H_2, comparable with reforming-H_2.

1.5 CCU and Innovation

A key role CCU can play in the industrial sectors is *innovation*. The use of CO_2 may introduce new synthetic methodologies, more direct than those on stream today. Innovation in synthesis means an overall reduction of: reaction steps, use of fossil-C based materials, waste production, process energy, separation energy. The tangible benefits with CCU are: safer working conditions, raw material diversification, reduction of carbon footprint, lower environmental impact (reduction of not only GHG emission, but also reduced burden on other categories such as human toxicity, soil and water toxicology, air pollution, acidification, etc.), lower overall industrial waste production (with subsequent lower CO_2 emission in waste treatment) and less toxic, resource saving. Innovative processes based on CO_2 are those which while reducing the C-footprint, do not increase the impact on other environmental categories. This is a very important point: targeting only CO_2 reduction without taking care of emissions which can negatively impact other environmental categories is not an acceptable solution. Therefore, the LCA of new processes should take into account at least five impacts categories, namely: Climate Change Potential-CCP, Human toxicity, Soil ecotoxicity, Air Pollution, Water pollution. Reducing CCP, while increasing the other categories does not represent an acceptable solution, because for reducing the latter impacts energy will be used and this will cause CO_2 emission. Claiming inexistent benefits does not support CCU. Moreover, we believe that soon the way LCA will be carried out [12] will increase its complexity, moving from single product-process to cluster of processes and cascades of products, including waste management.

  CO_2 is the alternative to either phosgene ($COCl_2$, a toxic species, $LC_{50} = 3$ ppm) [13] or CO (poisonous gas) [14]. In both cases, the use of CO_2 will generate safer conditions for workers, with less risk and a positive impact on Capital Expenditure (CAPEX), as a lower investment in safety will be necessary, assuring longer life to plants with lower Operative Expenditure (OPEX) (less corrosion of plants). Today, safe conditions when using phosgene require the insulation of a plant in a dome connected to air treatment plants: a procedure that has strong impact on both CAPEX and OPEX. The great advantage of phosgene is its high reactivity that allows reactions to be carried out at room temperature, or nearby. Such high reactivity has an energetic and environmental cost, linked to the production of energy intensive and strongly impactant CO and Cl_2 used for making phosgene [15]. Phosgene is banned in several countries today and cannot be transported: it can only be produced and used in situ. Such limitations are making the use of phosgene strongly problematic in some geographic areas and demand for effective substitutions. CO_2 is the ideal candidate and the new synthetic methodologies are really welcome and meet the industrial interest.

CO is used in the synthesis of methanol, in FT processes, in carbonylation reactions at the rate of ca 173 GWth/year in 2017 and an estimate of 239.5 GWToe/year in 2023 [16] and its use is expected to continue to grow. Substituting CO_2 to CO may mean the use of up to 50% excess hydrogen (methanol synthesis) that is possible only if H_2 produced from water using perennial energy sources is used: the use of fossil-H_2 is a non sense as maximization of H_2 production (1.5a) is based on WGS reaction (1.5b) that releases CO_2.

$$C + H_2O \rightarrow CO + H_2 \tag{1.5a}$$

$$CO + H_2O \rightarrow H_2 + CO_2 \tag{1.5b}$$

Therefore, while the substitution of $COCl_2$ with CO_2 would be immediately possible, supposed we have the right technologies, the substitution of CO will take place in coming years when PV-H_2 will have a large production and a cost comparable to reforming-H_2.

The above are just two examples more will be discussed in coming paragraphs. Such innovative methodologies also address the raw material diversification target. The use of CO_2, as said above, allows the substitution of fossil-C, implementing the reduction of carbon footprint of industrial synthetic methodologies and lowering the environmental impact of the chemical industry.

The utilization of CO_2 even lowers the overall industrial waste production by implementing more direct syntheses. Such new processes produce less toxic waste because toxic reagents are avoided. (vide infra) All above benefits, coupled to resource saving, make the CO_2 chemistry much wished at the industrial level.

A point of interest is that CO_2 is produced in several industrial processes (Table 3 in Preface) that deliver purer streams than power plants. Therefore, the use of industrial CO_2 can be convenient also because it is produced at chemical sites where it would be easily converted, saving shipping costs. As a matter of fact, clustering chemical processes can favour CO_2 recovery and use in situ, maximizing the use of resources and minimizing energy expenditure (see Sect. 1.8).

1.6 CCU and Sustainability of the Chemical and Polymer Industry

It is worth to emphasize that if the *energy industry can be decarbonized, the chemical and polymer industry can not.* In our everyday life we mainly (85–90%) use products based on chemicals and materials produced via catalytic processes, with a minor share of natural products. In any case, all products are C-based. Even if bio-based products will be promoted and will increase their share on the market, carbon will always be around: we can imagine change its source, not to substitute it. All biomass is produced from CO_2, therefore our future will increase the use of CO_2, either directly by developing CO_2-based processes or by letting Nature fix CO_2 and using catalysis for converting biomass [17]. In this paragraph a few cases of reagent/raw material substitution will be discussed, relevant to making more sustainable the chemical industry shifting from fossil-C to CO_2. We shall not enter into the detailed mechanistic studies, this is not the scope of this paper. More specialized books ([18] and references therein) and papers ([19] and references therein) can be found in the scientific literature which deal with such aspect. Here we shall highlight the aspects that make the CO_2-based chemistry economically and environmentally appealing, while empowering sustainable processes. Energy products such as methane, methanol, long-chain hydrocarbons will not be treated in detail. They will be discussed in Sect. 8. The key aspects, which will be discussed here are relevant to the use of CO_2 as *building-block* substituting fossil-C. Only chemicals with a market close or higher than 1 Mt/year will be considered. When necessary, if hydrogen will be needed, it is intended that it will be produced from water by electrolysis using SWGH energies as primary sources.

HCO_2H, formic acid, has a market of ca. 800 kt/year and an annual growth estimated at 4.95% by 2027 [20] with a key utilization in sectors shown in Fig. 1.3. Its use as H_2-carrier/storage (vide infra) may increase its market to tens Mt/year. Formic acid is made through routes based on CO, produced via reforming of coal or methane. Its direct synthesis from CO_2 and H_2 (1.6a) is highly wished and attempted since long time. Such reaction has some thermodynamic limitations ($\Delta G° = + 6$ kcal/mol)

$$H_{2g} + CO_{2g} \rightarrow HCO_2H_1 \tag{1.6a}$$

$$H_{2g} + CO_{2g} + RR'NH \rightarrow RR'H_2{}^{+-}HCO_2 \tag{1.6b}$$

$$RR'NH_2{}^{+-}HCO_2 + HX \rightarrow HCO_2H + RR'NH_2{}^{+-}X \tag{1.7}$$

due to the negative entropic contribution (two gases are converted into a liquid) which can be won by carrying out the reaction in presence of amines, HNRR' in organic solvents [with production of ammonium formates, $HCO_2{}^-$ $^+H_2NRR'$, (1.6b)] or inorganic bases, MOH in water (with formation of HCO_2M). In all such cases the reaction has a negative $\Delta G°$ (ca. −30 kcal/mol), but salts (ammonium or Group 1 metals) more than the free formic acid are formed. The production of free

Fig. 1.3 Global formic acid market, by application, 2016 (%)

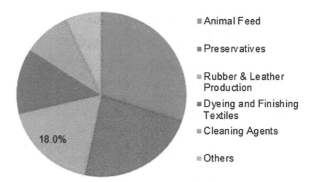

- Animal Feed
- Preservatives
- Rubber & Leather Production
- Dyeing and Finishing Textiles
- Cleaning Agents
- Others

HCO_2H requires the reaction of the salts with inorganic acids with formation of equimolar amounts of inorganic waste (1.7). On the other hand, some catalysts have been developed so far (1.6b) for the synthesis of formates which are almost ready for industrial application [21]. Attempts to shift from salts to the free acid (use of SC-CO_2 [22] or water as reaction solvents [23]) are undergoing. Should they succeed, the direct synthesis of formic acid from CO_2 and water (as source of H_2) will be ready for exploitation.

Carboxylic acids, R-CO_2H are synthetized either through routes based on CO (acetic acid, CH_3CO_2H, ca. 6 Mt/year) or via oxidative processes or hydrolysis of cyanides (1.2a)–(1.2b) [19]. The two latter routes are not very efficient for their low CUF and have a large environmental impact. The direct synthesis based on CO_2 (1.8) is highly wished, but demands the C–H activation, a not easy process.

$$R(Ar)-H + CO_2 \rightarrow R(Ar)-CO_2H \qquad (1.8)$$

Process innovation is necessary in this field and the use of radiation chemistry more than thermal processes may represent a solution [5, 6]. The production of long chain carboxylic acids from hydrocarbons and CO_2 would represent an interesting route to biodegradable surfactants with great environmental benefit. Surfactants have a market of several Mt/year and are expected to grow at a rate of 3.54%/year [24]. Direct carboxylations are quite appealing for the high level of innovation and the environmental benefits due to quasi-zero organic waste production and significant reduction of CO_2 emission.

Acrylic acids market [25] is built up by several derivatives, such as: Acrylic Esters (Methyl Acrylate, Ethyl Acrylate, Butyl Acrylate, Ethyhexyl Acrylate), Acrylic Polymers (Acrylic Elastomers, Super Absorbent Polymers, Water Treatment Polymers, Ammonium Polyacrylate, Cyanopolyacrylate) and End Use (Diapers, Surface Coatings, Adhesives & Sealants, Plastic Additives, Water Treatment). It amounts at ca. 6.6 Mt/year (2017) with a perspective annual growth of 7.6%/year and an estimated global market of 8.2 Mt by 2020. The production of acrylic acid is today based on the non-selective oxidation of propane to propene and of the latter to

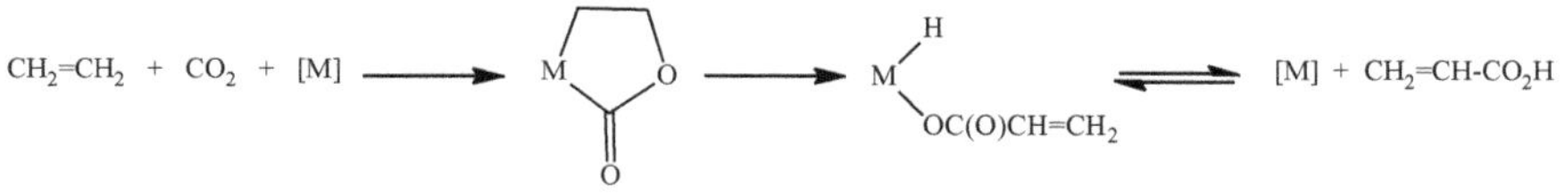

Scheme 1.1 Pathway to the formation of acrylic acid from ethene and CO_2

acrylic acid, (1.9) a route that has a low CUF and produces a lot of organic waste and CO_2 due to the difficulty of driving selective oxidation processes.

$$CH_3CH_2CH_3 \rightarrow CH_3CH=CH_2 \rightarrow HO_2CCH=CH_2 \qquad (1.9)$$

The direct reaction of ethene with CO_2 (1.10) is feasible, even if it has several thermodynamic

$$CH_2=CH_2 + CO_2 \rightarrow CH_2=CH-CO_2H \qquad (1.10)$$

and kinetic bottlenecks. The coupling is promoted by metal (Ni, Mo, W, Pd) complexes in a low or zero oxidation state with production of a metallacycle (Scheme 1.1) which is then converted into a hydrido acrylate. Reductive elimination should give back the catalyst and acrylic acid.

The reductive elimination (Scheme 1.1) is opposed by the addition of the free acid to the metal centre and is driven by the M–H and M–O bond energy. For long time Ni, Mo and W have been used as promoters in an attempt to produce the free acid [26]. Recent studies have shown that using Pd [27] an easy elimination is possible at room temperature when the alkyl esters are formed and a TON of ca. 10 was reached. Even more recently, with Ni(0) as catalyst using a base (M'OH) [28] metal acrylate (M'O$_2$CCH=CH$_2$) and water are eliminated from the hydridoacrylate (TON of ca. 10), reproducing the catalyst [M]. The elimination of the acrylate moiety is, thus, made easier by the use of the Group 1 bases, with the metal cation M'^+ that acts as a shuttle for the acrylate anion. Pd systems [29] have reached a TON of 100. The situation is evolving rapidly and there is hope that this synthetic route will have a possible exploitation in the short term with large reduction of the environmental impact. Moreover, as ethene can be made from CO_2, via electrolysis (see Chap. 7), all carbons of the acrylic moiety could be derived from CO_2.

Linear (Fig. 1.4a) *and cyclic carbonates* (Fig. 1.4b) find application in several fields: reagents, solvents, monomers for polymers, are the most relevant. Their free market is not very large (a few hundreds kt/year) as they are in general produced for captive use.

The utilization of CO_2 in this field has a great potential, and the technology is at high TRL for cyclic carbonates, but not yet mature for linear ones.

Linear carbonates can be produced by direct carboxylation of alcohols (1.11). Bottlenecks are the

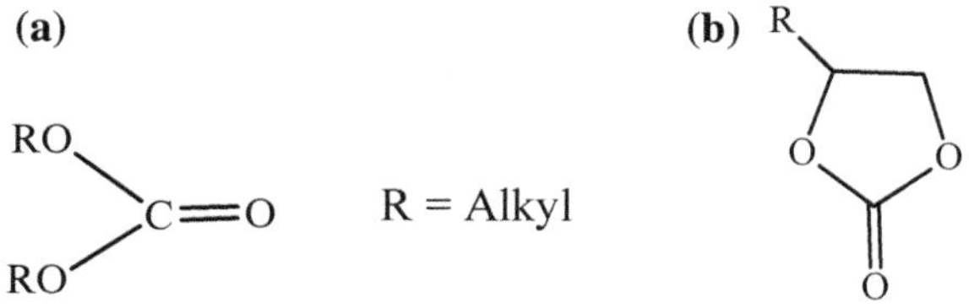

(a)   **(b)**

$$RO \diagdown \atop RO \diagup C{=}O \qquad R = Alkyl$$

Fig. 1.4 Linear (**a**) and cyclic (**b**) carbonates

$$2ROH + CO_2 = (RO)_2CO + H_2O \qquad (1.11)$$

reaction thermodynamics (the low equilibrium concentration makes that the produced carbonates cannot be efficiently distilled out) and kinetics. Reaction (1.11) is a slightly endergonic equilibrium [30] favoured by low temperature and water elimination. Active catalysts are required that may drive the reaction at a temperature as much close to 300 K. The risk is that such active catalysts may promote parallel reactions, with increase of the complexity of the reaction mixture that causes energy expenditure (and CO_2 emission) in the post reaction separation process [31]. Therefore, a deep study of catalysts is necessary and DFT studies can help to identify the best species [32]. Water elimination has been attempted using inorganic or organic water traps and membranes [33]. The latter option avoids the need to regenerate the starting water-traps, but needs development of new materials for avoiding the pressure swing (reaction-separation) in order to efficiently separate water. In fact, under high pressure and temperature existing membranes may loose their separation capacity, so that the water separation must be carried out under low pressure. Temperature and pressure swing or high expenditure of energy in the separation of products are not in favour of CO_2 emission reduction and must be kept under strict control.

Cyclic carbonates are produced mainly by direct carboxylation of epoxides [34] (1.12) or carboxylation of di(poly)ols [35] (1.13). The former route suffers the scarce availability of epoxides made with hydrogenperoxide (market of ca. 600 kt/year, much lower than the market of carbonates) and their high cost. An alternative route is the oxidative carboxylation of olefins [36] (1.14) that would avoid the synthesis of epoxides, a synthetic strategy prevented so far by the two-oxygen addition to the olefin double bond, that causes low yield of carbonates and complex reaction mixtures which demand energy for separation.

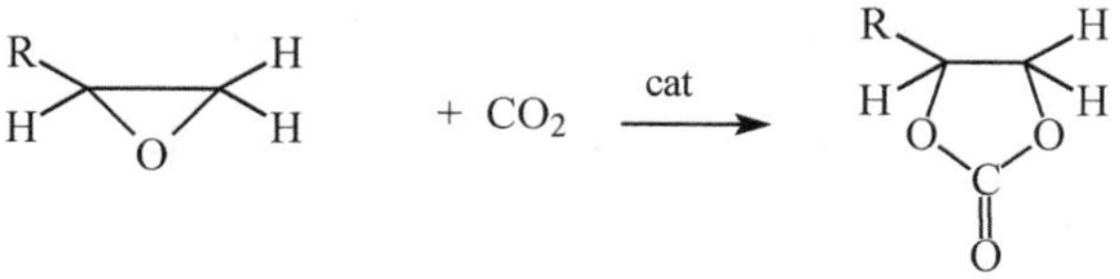

$$(1.12)$$

$$\text{HOCH}_2\text{CHCH}_2\text{OH} \quad \xrightarrow{\text{CO}_2} \quad \quad\quad (1.13)$$

$$\underset{\text{OH}}{|}$$

$$\text{RCH}_2{=}\text{CH}_2 \ + \ \text{CO}_2 \ + \ 1/_2\,\text{O}_2 \quad \longrightarrow \quad\quad\quad (1.14)$$

A way to win such drawbacks in the synthesis of organic carbonates is the use of urea as a substitute of CO_2 (1.15). Urea is made from CO_2 and NH_3 and in the synthesis of carbonates releases NH_3 that can be captured and re-used. Reaction (1.15) has no thermodynamic limitation and can be run to completion [37]. The convenience of the catalytic carbonation of methanol with urea using the reactive distillation technique has been demonstrated at a demo scale [38].

$$2\text{ROH} + \text{H}_2\text{NCONH}_2 \rightarrow (\text{RO})_2\text{CO} + 2\text{NH}_3 \quad\quad\quad (1.15)$$

Even the synthesis of cyclic carbonates can take advantage from the reaction of diols with urea (1.16)

$$(1.16)$$

$$\text{HOCH}_2\text{CH}_2\text{OH} \ + \ \text{H}_2\text{NCONH}_2 \ \rightarrow \quad\quad + \ 2\text{NH}_3$$

This area is of great industrial interest. The carboxylation with CO_2 has been developed to TRL around 5, while the reaction with urea has been developed to TRL 7 with some spot demonstration plants. Both have a great potential and require investment and Academy-Industry interactive collaboration in order to move up to higher TRL and large commercial application.

Carbamic acid ($RNH\text{-}CO_2H$) is a labile species that dissociates to the free amine (RNH_2) and CO_2, unless it is stabilized by dimerization [39]. Conversely, carbamic acid esters are stable entities used in agrochemistry as fungicides, erbicides, pesticides or in cosmetic industry and other applications.

In principle they can be produced by reacting amines (primary or secondary aliphatic amines) with CO_2 and an alkylating agent (1.17) in presence of a base (the latter can be the amine itself).

$$RR'NH + R''X + CO_2 + B \rightarrow RR'N-CO_2R'' + BH^{+-}X \qquad (1.17)$$

Reaction (1.17) is a two step process consisting of a preliminary reaction of the amine with CO_2 to afford the ammonium carbamate (1.18) and the subsequent alkylation of the carbamate (1.19).

$$2RR'NH + CO_2 \rightarrow RR'N-CO_2^-NH_2RR'^+ \qquad (1.18)$$

$$RR'N-CO_2^{-+}NH_2RR + R''X \rightarrow RR'N-CO_2R'' + NH_2RR'^{+-}X \qquad (1.19)$$

The bottleneck of such reaction is the alkylation of the amine (1.20) that can be prevented by

$$RR'N-CO_2^{-+}NH_2RR' + R''X \rightarrow RR'NR'' + CO_2 + NH_2RR'^{+-}X \qquad (1.20)$$

using a suitable Crown-Ether (CE) (1.21) that reduces the interaction of the metal cation (either Group 1 metal or ammonium) with the carbamate anion:

$$RR'N-CO_2^{-+}(NH_2RR')CE + R''X \rightarrow RR'N-CO_2R'' + (NH_2RR'.CE)^{+-}X$$
$$(1.21)$$

A good alternative to alkyl halides is the use of aliphatic carbonates (1.22) that can be prepared from CO_2 and alcohols (see above).

$$RR'NH + (R''O)_2CO \rightarrow RR'N-CO_2R'' + R''OH \qquad (1.22)$$

Ammonium carbamates are the products of reaction with CO_2 of primary or secondary aliphatic amines [see (1.18)] used as chemical sorbents for CO_2 in its capture from flue gases and other gas mixtures, an application disussed in Chap. 2.

Di-carbamates of primary amines find an industrial application as source of di-isocyanates (1.23), (1.24). The carbamation of aromatic amines requires active and selective catalysts [43]. Bio-mimetic mixed anhydrides [40–42], structurally analogous to the enzyme carboxyphosphate (Fig. 1.5) active in the elimination of ammonia in living organisms, are effective and selective catalysts in such reaction.

$$(1.23)$$

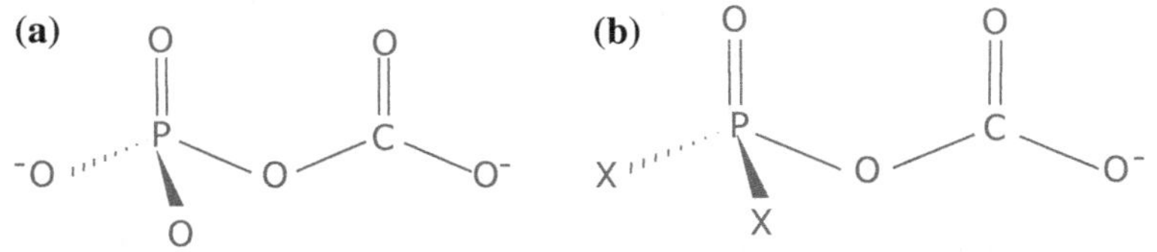

Fig. 1.5 The enzyme carboxyphosphate (**a**) and its mimetic mixed-anhydride (**b**)

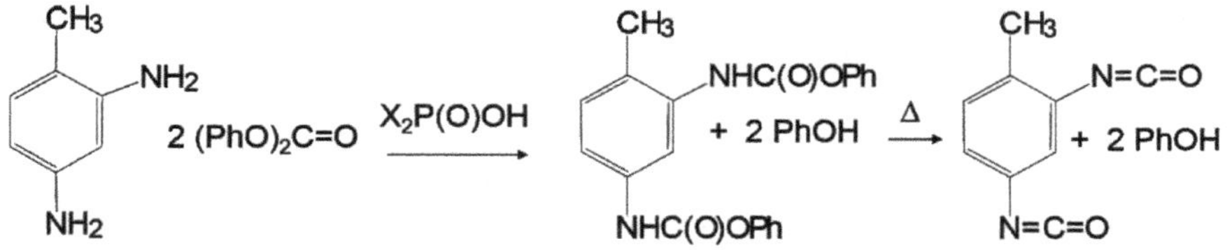

$$(1.24)$$

Isocyanates are used as co-monomers with diols $[HOCH_2(CH_2)_nCH_2OH, n = 1+]$ for the production of polyurethanes that have a market of several Mt/year and an estimated 79 BUS$ value at 2021 (53 BUS$ in 2014), used in a variety of industrial sectors from automotive to agriculture, personal care, fillers, adhesives, sealants etc. [44]. These days, the tendency is to shift from fossil-C based monomers to bio-sourced molecules, using polyols and amines derived from biomass. Such move has a key role for the sustainability of the chemical and polymer Industry and opens the way to an extensive carbon recycling, including the exploitation of CCU.

The *"methanol to olefin-MTO"* and *"methanol to propene-MTP"* processes [45] are the route to C2 (ethene) and C3 olefins (propene) with "all-carbons derived from CO_2". Therefore, large scale polymers such as polyethene-PET (market of 150 Mt/year in 2017 [46]), and polypropene-PP (market of 73.7 Mt/year in 2017 [47]) could be entirely made from CO_2 with an utilization of some 880 Mt/year. Even if the MTO and MTP [48] processes are well known since long time [49], catalysts must be improved in both the production of methanol [19] and its further conversion via dimethylether-DME. The key point is the selectivity, as at the moment, in both cases, a variety of olefins are obtained, from C2 to C8 with different abundance. Such large distribution rises the issue of cost for separation (economic and energetic) and consequent CO_2 emission attached to the energy consumed. The electro-reduction of CO_2 to ethene has reached high conversion and selectivity using Cu-electrodes [50] in water, that may open to a potential exploitation.

This field has a great interest as polymers have a large market and can become a "chemical storage" of CO_2. Their lifetime depends on the use we make: they could last even for a few decades. Another interesting approach is the direct copolymerization of olefins and CO_2, a route not so much investigated so far, but of great industrial relevance.

C–C reductive coupling of CO_2 to afford di- (ethene glycol, $HO–CH_2CH_2–OH$, EG) and poly-ols (glycerol and others) is made possible by electrochemical reduction [51], even if at very low efficiency (0.15%), using ionic liquids coated electrodes. Toshiba have improved the technology up to 0.5% selectivity in 2016. There is an interest in developing such process, considered the large market of EG (expected to reach 22 Mt/year in 2020 [52]). EG is a co-monomer in the production of polymers (polyesters and polyols) and its production from CO_2 would convert ca. 100 Mt/year of CO_2 into long living materials. Such route can also be used for the synthesis of other C2 molecules, such as glyoxylic acid (World Market (WM) 140 kt/year), glycolic acid (WM 100 kt/year), acetic acid (WM 15 Mt/year). Photo- and photoelectro-processes would be of paramount interest for developing such synthetic routes.

Polymers "all-C" derived from CO_2. As mentioned above, polymers containing CO_2 can be considered a way to store CO_2 even if not with the same permanence possible for underground storage. In addition to synthesize all units from CO_2 as described above, another foreseeable approach is possible, that is the integration of the use of biomass and synthetic chemistry/catalysis. In fact, diols electrochemically derived from CO_2 can be reacted with biomass derived dicarboxylic acids for the synthesis of polyesters, supplanting products derived from fossil-C. An example is the synthesis of polyfuranoates (1.25). They can substitute polyethene terephtalates, with much benefit.

$$(1.25)$$

The examples above exemplify low energy processes based on CO_2 which can be exploited in the short-medium term, even within a energy system based on fossil-C.

1.7   Integration of Technologies: Biotechnologies and Catalysis

Integration of biotechnology and catalysis would boost the conversion of CO_2 into added value products or energy vectors. Such approach encompasses three key aspects: (i) let Nature fix CO_2 and use catalysis for biomass conversion; (ii) enhanced biological CO_2 fixation using chemical technologies; (iii) one (catalysis or biotechnology) makes what the other (biotechnology or catalysis) cannot do.

A point to emphasize is that biosystems (whatever they are) produce what they need: with some exceptions, they are not selective towards a single product, instead they accumulate inorganics taken from the external world or produce a variety of organic products often of comparable abundance: carbohydrates, lipids, proteins, enzymes, other organic chemicals. Such high product entropy is not in the direction of the stringent rules of CCU. The energy cost of separation if often very high and causes the emission of large volumes of CO_2. Therefore, it is necessary to increase the selectivity towards the wanted product (e.g., carbohydrates or lipids). This requires either a genetic manipulation or physical stress, including the adaptation of biosystems to live in non-natural conditions. Biotechnology and catalysis are perpendicular each other, in the sense that the former is a horizontal (extensive) technology, while the latter is a vertical (intensive) technology. Several issues need to be mastered for the optimal coupling of the two technologies, trying to make biotechnology more intensive and selective towards a target product. The peculiar aspects biotechnology and catalysis can target [53] are described in the below categories.

(a) Production of aquatic biomass under "enhanced" (high CO_2 concentration or using other C-sources such as carbonates and hydrogencarbonates) or "stressed" (N- or P-starvation) conditions, under sunlight or solar-simulated light. Such approach is particularly adaptable to microalgae, which proteins, lipids or carbohydrates content can be in part adapted to their end-use (see Chap. 11).

(b) Bioelectrochemical systems (BES), where enzymes or whole microorganisms use CO_2 and PV-electrolytic hydrogen for the production of chemicals or fuels (C1 and Cn). (see Chap. 9).

(c) CO_2 fermentation with non-fossil hydrogen as energy carrier for the production of raw materials for fine-chemicals.

(d) Man-made photosynthesis, for co-processing CO_2 and water into chemicals or fuels with the aid of a (bio)-catalyst. Use of photochemical systems is of particular interest.

(e) Conversion of cellulose, an abundant non-edible terrestrial biomass. Both chemical technologies and enzymatic processes can afford intermediates (oligo-saccharides, monomeric polyols C6 and C5) that are platform molecules for the catalytic production of added value chemicals.

(f) Conversion of oils extracted from land seed-plants or aquatic biomass into chemicals used in the polymer (new dicarboxylic acids), cosmetic and agro-industry (see Chap. 11).

Noteworthy, all the accelerated CO_2 conversions listed above use non-fossil, perennial energy sources. The integration of biotechnology and catalysis may attain a better exploitation of CO_2 and water. Noticeably, biosystems require water as growth medium in perfect adherence to the strategic view of this book. An aspect of paramount interest is the use of biosystems, either directly or via redox mediators [54], in CO_2 reduction, as in (1.26), coupled with the oxidative degradation of

$$CO_2 + 8H^+ + 8e^- = CH_4 + 2H_2O \qquad (1.26)$$

organic waste, stopping the latter reaction at the level of useful products, obviously avoiding CO_2 production [55]. An attracting opportunity linked to the use of BES is the fact that they can offset overpotentials, reducing energetic costs and most likely improving selectivity. As an example, the thermodynamic potential for the methanation of CO_2 is -0.244 V (vs. SHE) at pH = 7 (Table 1.3), but it may be affected by quite large over-potentials in normal electrochemical systems that make it more energetically requiring. The use of a microbial cathode could offset over-voltage. Microbial fuel cells (Scheme 1.2) are an old concept [56] with new targets.

Electrochemically driven bioprocesses can convert CO_2 into C1 molecules such as formate [57], or into C2 molecules such as acetate [58], and other species [59].

In BES/PBES either enzymes (eventually supported, inglobated, encapsulated) or the whole microorganism can be used, both with pros and cons. Enzymes avoid the growth of biomass, but whole microorganisms have higher resistance than isolated enzymes, which stability, on the other hand, depends on the bacteria from which they are extracted. This is the case, for example, of *Formate Dehydrogenase* (*FDH*) which is very air sensitive if extracted from anaerobic *Syntrophobacter fumaroxidans* [60], and more stable when extracted from *Candida boidinii* (CbsFDH) that has a commercial use, i.e., for the regeneration of NADH from its

Table 1.3 Reduction potentials of CO_2

	$E°$ (V) versus SHE at pH = 7 in water
$CO_2 + e^- \rightarrow CO_2^-$	-1.9 (-2.1 in organic solvents)
$CO_2 + 2H^+ + 2e^- \rightarrow HCOOH$	-0.61
$CO_2 + 2H^+ + 2e^- \rightarrow CO + H_2O$	-0.52
$2CO_2 + 12H^+ + 12e^- \rightarrow C_2H_4 + 4H_2O$	-0.34
$CO_2 + 4H^+ + 4e^- \rightarrow HCHO + H_2O$	-0.51
$CO_2 + 6H^+ + 6e^- \rightarrow CH_3OH + H_2O$	-0.38
$CO_2 + 8H^+ + 8e^- \rightarrow CH_4 + 2H_2O$	-0.24
$2H^+ + 2e^- \rightarrow H_2$	-0.42

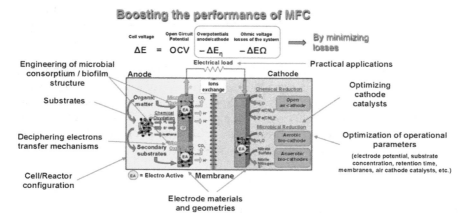

Scheme 1.2 A microbial fuel cell at work

oxidized form NAD$^+$. *Carbon Monoxide Dehydrogenases (CODH)* catalyse the two-electron reduction of CO_2 to CO. Both the formation of CO and HCO_2H are the start for the conversion of CO_2 into organics. It is worth to emphasize that enzymes avoid radical-reactions, which are initiated with the $1e^-$ transfer to CO_2 (-1.9 V vs. SHE at pH $= 7$, Table 1.3) and drive its reduction towards the $2e^-$-path lower in energy (Table 1.3) and based on proton-coupled-electron-transfers (PCET). Both *CODH and FDH* have been supported on graphite and shown to behave as reversible electrocatalysts [61], inspiring manmade photosynthetic systems. Interestingly, the use of *FDH* with *Formaldehyde dehydrogenase, $F_{al}DH$* and *Alcohol dehydrogenase, ADH,* brings to the reduction of CO_2 into CH_3OH in water at ambient conditions [62]. Each of the three enzymes uses NADH as PCET and behaves as a $2e^-$ reducing agent. The drawback of such appealing system is the formation of NAD$^+$ that must be reduced back to NADH for keeping low the economic cost of methanol. The backconversion of NAD$^+$ into NADH can be carried out through enzymatic, chemical, electrochemical, photocatalytic routes. The latter [62–64] using solar light-active heterogeneous TiO_2-modified with $[CrF_5(H_2O)]_2$ or Fe@ZnS nanoparticles [62] coupled to a Rh(III)-complex ($[CpRh (bpy)Cl]^+$) for facilitating the electron and hydride transfer from the H-donor (water or a water-glycerol solution) to NAD$^+$ produces 100–1000 mol of CH_3OH, per mol of NADH, paving the way to a practical application, anticipating solar energy utilization for hybrid CO_2 reduction [65].

While BES and PBES systems rise great expectation for their potential, several drawbacks and bottlenecks exist to their exploitation. High CAPEX and OPEX, the intrinsic kinetics of biosystems, lifetime of enzymes and devices are barriers to their full implementation. Discovering new low cost materials, engineering faster enzymes, developing hybrid biotechnologies are key steps to a better use of such interesting option that needs continuous feed of energy, rising the issue of discontinuous availability of perennial energy sources and requiring a clever engineering of energy supply with best mix of different primary sources or stored energies.

1.8 CCU, Clustering of Processes and the Energy Sector

As we have already discussed, the conversion of CO_2 into energy products requires energy and hydrogen, both not originated from fossil-C. Such stringent conditions make that at glance CCU can give only a limited contribution to the energy sector. However, if in an energy frame based on fossil-C it is possible to use CO_2 as building block for chemicals and polymers, it does not make sense to convert CO_2 into fuels. Therefore, only if perennial energy sources are used to power the process and hydrogen is derived from water it makes sense to convert CO_2 into energy products for some selected sectors, such as avio-, navy-, road-transport where electric motors (directly or indirectly) powered by solar energy cannot be conveniently used. While the detailed analysis of the state of the art on the conversion of

CO_2 into C1 and Cn energy-rich products is made in Chaps. 5 and 6, respectively, here some key elements will be discussed for clarifying the frame in which such option can really contribute to recycling-carbon and reducing CO_2 emission.

The key point to keep in mind is that most likely the existing industry organization will be revolutionized in future for avoiding dispersion of processes in separated sites. Clustering of processes and diverse activities will play a key role in order to optimize the utilization of raw materials and minimize waste production and CO_2 emission. Such approach can result particularly useful when the efficient use of resources and materials is targeted and when renewable-C utilization is operated. For us CO_2 is synonym of renewable carbon: indeed, it can be recovered and cycled incessantly, as Nature does since ever. Clustering of processes is a strategic approach to the efficient use of resources and valorisation of "waste" streams. Approaching the conversion of CO_2 via "value-chains", more than single reactions, will give a new system perspective to CCU. Scheme 1.3 is an example of integration of processes for the production of chemicals and fuels.

The integration of the above network of reactions with that reported in Scheme 1.4 opens new perspectives. In a system in which processes are clustered, "waste" H_2 can be more easily cycled into a new process for CO_2 reduction opening to a full range of products. This is the case of the use of CO_2 (combined or not with oxygen) as dehydrogenating agent (Scheme 1.4, left part, reactions (1.1)–(1.3) towards aliphatic hydrocarbon moieties (coupling of methane or dehydrogenation of propane to propene or ethylbenzene to styrene, a process that is finding industrial exploitation with some demo-plants in Korea and China) with production of C–C bonds or C=C bonds and hydrogen. Such application has been found useful for the production of C2 moieties from methane or olefins from alkyl moieties. The released hydrogen could be used for the reduction of CO_2 to useful products in situ (see Chap. 7 in [18a]).

In paragraphs below, a quick sight to the production of selected energy-rich molecules will be given, highlighting bottlenecks and perspectives, while the deeper analysis of catalytic systems will be developed in Chaps. 5 and 6.

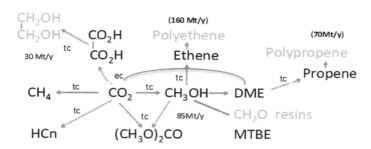

Scheme 1.3 Value-chains for CO_2 utilization (tc: thermal reactions; ec: electrochemical processes; figures in parentheses give the actual market of products made from fossil-C: in future could be made from CO_2)

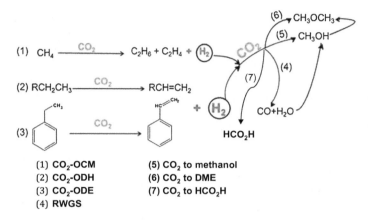

Scheme 1.4 Utilization of CO_2 as dehydrogenating agent and the use of produced H_2 for CO_2 reduction

Carbon Monoxide, CO

Metal-centres greatly favour the deoxygenation of CO_2 to CO. $\eta^2_{C,O}$-Co-ordinated CO_2 is able to transfer one oxygen atom to an oxophile intra- (1.27a) or inter-molecularly (1.27b) at 300–360 K [66], while gas-phase deoxygenation to afford oxygen-atoms (1.28) occurs at high temperature (ca. 1670–1825 K) or using energetic radiations ($\lambda < 160$ nm) [67].

$$LnM(CO_2) \rightarrow LnMO + CO \qquad (1.27a)$$

$$LnM(CO_2) + Sub \rightarrow LnM(CO) + Sub{=}O \qquad (1.27b)$$

$$CO_2 \rightarrow CO + {''}O{''} \qquad (1.28)$$

Reaction (1.28) can be promoted by metal oxides able to release and take-up oxygen [68] (1.29a), (1.29b). Coupling the CO_2 reduction to CO with water-splitting, (1.29c) besides oxygen, Syngas (CO + H_2) is produced that can be used for the production of gasoline and diesel (1.29d) [69]. Water vapour and CO_2 can be both recovered from the atmosphere, and the fuel produced is labelled "Diesel from Air".

$$MO_x + Energy \rightarrow MO_{x-1} + 1/2O_2 \qquad (1.29a)$$

$$MO_{x-1} + CO_2 \rightarrow MO_x + CO \qquad (1.29b)$$

$$MO_{x-1} + H_2O \rightarrow MO_x + H_2 \qquad (1.29c)$$

$$xH_2 + yCO \rightarrow Liquid\ fuels \qquad (1.29d)$$

Reactions (1.29a)–(1.29d) are a clear and precise demonstration of the objectives of this Book: co-processing of CO_2 and water, getting close to Nature. The photo-chemical [70] reduction of CO_2 using solar light was described by Nobel Laureate J. M. Lehn in 1983 using Re- and Ru-complexes [71]. Lehn and others also showed that CO, formates and H_2 are coproduced [72] with some selectivity issues. The photo-electrochemical reduction of CO_2 is discussed in detail for its various aspects in Chap. 7. "Solar chemistry" is of paramount importance today as co-processing of CO_2 and water using solar radiations is a way to the production of chemicals and fuels from non-fossil-C. Particular interest have risen bimetallic systems such as SnO_2-coated Cu nanoparticles which, depending on the thickness of the tin oxide layer afford more selelctively CO (0.8 nm; >93% at −0.7 V_{VHE}) [73] or formates (1.8 nm). Nanowires made of "SnO_2 on CuO" coupled with a GaInP/GaInAs/Ge photovoltaic cell, are characterized by an interesting solar-to-CO conversion efficiency of 13.4% [74]. Using perennial energy for coprocessing CO_2 and water is of paramount importance for our future and for extended C-recycling [19, 75]. Catalysts selectivity towards the formation of CO is a key issue to solve.

Methanol-CH_3OH has a worldwide market of the order of 85 Mt/year [76], and has several applications, from fuel to solvent and as a feedstock (Scheme 1.5) for the synthesis of several chemicals [77] including Cn hydrocarbons (methanol-to-gasoline route) or unsaturated hydrocarbons [78] (methanol to olefins-MTO or methanol to propene-MTP routes). It is currently produced from Syngas (1.30a) at 20 MPa and 573 K using copper-zinc oxide/chromium oxide catalysts [79] or from methane that is expanding its world-market (890 Mt liquefied in 2016, 879 in 2017 [80]).

$$CO + 2H_2 \rightarrow CH_3OH \qquad (1.30a)$$

$$CO_2 + 3H_2 \rightarrow CH_3OH + H_2O \quad \Delta H°_{298\,K} = -49.5\,kJ\,mol^{-1} \qquad (1.30b)$$

Noteworthy, since the 1970s CO_2 is left as additive to CO in the synthesis of methanol for a more complete use of hydrogen. Even if the partial hydrogenation of CO_2 is exothermic (1.30b), in order to overcome activation barriers, a temperature higher than 513 K is necessary [81] and active and selective catalysts are necessary for avoiding the formation of co-products.

Dimethyl ether, an intermediate in the synthesis of methyl acetate, dimethyl acetate and light olefins or a fuel for substituting diesel (40+% better cetane number), is produced at a rate of ca. 5 Mt/year (2016) [82] upon methanol dehy-dration, but the single step technology based on CO_2 hydrogenation is becoming more and more popular (1.31). The latter technology, reduces CAPEX and OPEX since only one reactor is required [83].

$$2CO_2 + 6H_2 \rightarrow CH_3OCH_3 + 3H_2O \quad \Delta H°_{298K} = -122.2\,kJ\,mol^{-1} \qquad (1.31)$$

Methanation of CO_2, (1.32), known as the Sabatier-Senderens reaction [84], is of great interest for both the automotive industry, as it represents a way to recycle carbon, and for biogas and natural gas valorisation, possibly without preliminary separation of the two C1 molecules [85]. Although such reaction takes place in milder conditions than the synthesis of methanol, several problems linked to heat elimination (the reaction is strongly exothermic), to catalyst substitution (Ni is used), engineering and formulation, still exist and need to be solved [86]. Nano-sized catalysts, dispersed on supports or as monoblocks appear to be possible solutions.

$$CO_2 + 4H_2 \rightarrow CH_4 + 2H_2O \quad \Delta H = -165\,\text{kJ mol}^{-1} \quad (1.32)$$

Cn hydrocarbons and higher olefins. The production of Cn hydrocarbons by direct hydrogenation

$$nCO + 2nH_2 = -(CH_2)_{n-} + nH_2O \quad \Delta H = -166\,\text{kJ mol}^{-1} \quad (1.33)$$

of CO_2 (1.33) is very appealing for producing fuels for the transport sectors, including avio and navy, while recycling large quantities of C. Challenges are the difficulty to drive the selectivity towards a single species. Iron-nanoparticle@carbon-nanotube (Fe@CNT) seems to have quite good potential [87] to drive the reaction under quasi C-neutral conditions even within the actual energy system frame. Indium-modified zeolites [88] give high yield (>78%) of Cn hydrocarbons with low methane selectivity. *Cn olefins* can be produced (54% light olefins with >50% conversion CO_2) adding K^+ as co-catalyst [87, 89].

Noteworthy, it does make sense to implement such reactions even today if excess-hydrogen is used. Flared-H_2 amounts at ca. 8 Gm^3/year, as it represents an average 5.54% of the total flared gases (150–179 Gm^3) which cause the emission of over 450 Mt/year CO_2 [90]. Once renewable, low cost hydrogen will be available the conversion of CO_2 into Cn hydrocarbons or olefins may become an option economically and energetically viable, possibly clustering processes and operations for an efficient use of materials and even waste gases, with much benefit for the environment and our society.

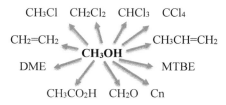

Scheme 1.5 Methanol as feedstock for the chemical industry

1.9 Miscellanea

CO_2 can be used for the synthesis of several compounds containing either the carboxylic $-CO_2-$ or the reduced carbonylic $-CO-$ moiety. Examples of the former reactions are the synthesis of lactones, pyrones or esters. Particular attention has been devoted to the reaction of alkynes and dienes (cumulated and conjugated) with CO_2 [91]. In all such reactions, key issues are the lifetime of the catalyst and the selectivity towards a given product. High TRL (6–7) has been reached in the conversion of butadiene into the lactone [92] shown in Scheme 1.6, while the conversion of allene (Scheme 1.6b) [93] or alkynes (Scheme 1.6c) [94] still has not been developed to an exploitation level.

Several other reactions of CO_2 with a variety of agents (Scheme 1.7) can be found in the literature, but they are more at a "curiosity" than at an application level and need much investigation for showing their viability in synthetic chemistry.

In all reactions in Scheme 1.7, and similar synthetic procedures, key issues are: the life and cost of the catalyst, the use of expensive (energetically and economically) co-reagents often lost after the first cycle, the yield and selectivity towards the target product, the cost (energetic and economic) of isolation of the pure target product. All above reactions seem to be more suited for replacing expensive processes for the synthesis of high added value products with a niche market (a few kt/year) more than large scale ($\gg 1$ Mt/year) chemicals. The benefit is in cleaning the chemical industry as they can reduce the environmental impact by producing less waste and consuming

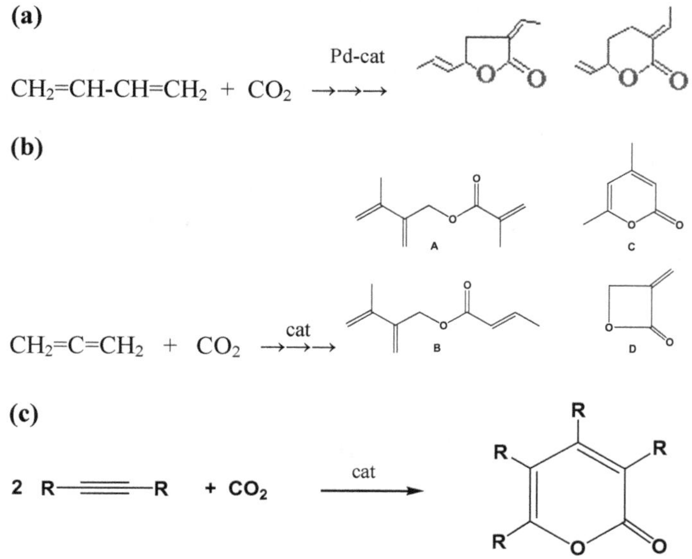

(a)

$$CH_2=CH\text{-}CH=CH_2 \ + \ CO_2 \ \xrightarrow{\text{Pd-cat}} \to \to$$

(b)

$$CH_2=C=CH_2 \ + \ CO_2 \ \xrightarrow{\text{cat}} \to \to$$

(c)

$$2 \ R\text{—}\!\!\equiv\!\!\text{—}R \ + \ CO_2 \ \xrightarrow{\text{cat}}$$

Scheme 1.6 Telomerization of dienes (butadiene or allene) and alkynes with carbon dioxide

Scheme 1.7 Conversions of CO_2 into added value products

$$\text{(i) } RNH_2 + CO_2 + H_2 \rightarrow RHNC\overset{\displaystyle O}{\underset{\displaystyle H}{\big\|}} + H_2O$$

$$\text{(ii) } 3R_3SiH + CO_2 \rightarrow CH_3OSiR_3 + R_3SiOSiR_3$$

(iii)

(iv)

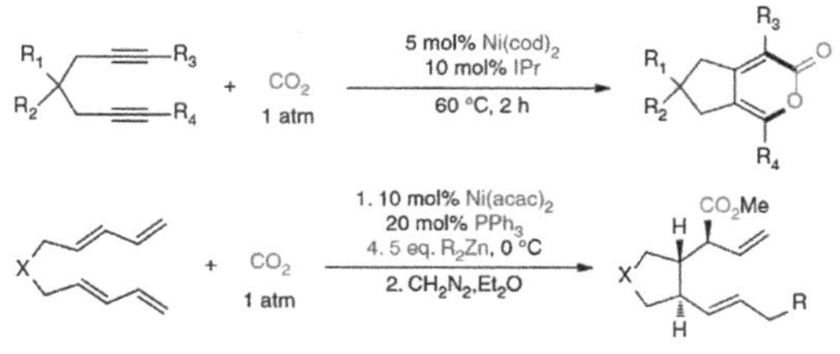

less energy. Therefore, it is of paramount importance that the reactions are quantitative and very selective and do not use reagents that have high cost and are characterized by high environmental impact and energy use.

1.10 Conclusions

An economy based on CO_2 and water: is, thus, this possible? And at what extent? We have considered in this chapter mainly those processes based on the use of CO_2 as building block, that means processes in which the entire entity is conserved in a new chemical (carboxylates, carbamates, carbonates, etc.). The processes relevant to such chemistry do not need hydrogen, but convert CO_2 into added value products having short-, medium-, long-life. The TRL level varies from 4 to 9 with the process considered, in all cases bottlenecks are well known and remedies, opening to a good perspective in the exploitation of discussed technologies in the short-medium term. Nevertheless, the market of the products is not very high. We can expect that from actual 200 Mt/year CO_2 used we can move to over 350–400 Mt/year within 2030. Apparently, this is not a great contribution to controlling the CO_2 atmospheric level. Nevertheless, we must consider that what is of interest is the *avoided* CO_2 more than the *used* CO_2. If we assume an average avoided/used ratio of 2.8, considering the benefit of technology innovation, then we can conclude that close to 1 Gt/year of CO_2 will be avoided by 2030: a more significant contribution. All processes described recycle carbon and avoid the extraction of fossil-C contributing to resource conservation and avoiding CO_2 accumulation in the atmosphere. They refer to the production of chemicals and materials that have a market some 15 times lower than fuels. In following chapters, several energy products will be examined, either C1 or Cn molecules. All of them need hydrogen: this must be derived from water, without causing drinking-water shortage for humans. In particular, with the availability of large volumes of PV-H_2 at a cost close to reforming-H_2 the conversion of CO_2 into fuels may grow to higher levels, (2–3.5 Gt/year). Co-processing CO_2 and water may also serve to produce large-scale chemicals other than fuels. Of interest is the case of the synthesis of C2 and C3 olefins from CO_2 via electrolysis that would contribute to rise the amount of CO_2 used to ca. 600 Mt/year in the medium term. Technologies discussed in next chapters, together with the utilization discussed above, demonstrate the potential of utilization of CO_2 as building block for chemicals and materials or source of carbon for fuels. A study carried out by the Catalyst Group [17] shows that combining the potential of all technologies it will be possible by 2040 avoid some $7-9\,Gt_{CO_2}/year$: this is a very interesting target that will contribute to CC control.

References

1. https://www.CO2.earth/
2. Aresta M (2010) Carbon dioxide as chemical feedstock. Wiley-WCH
3. Aresta M, Quaranta E, Tommasi I, Giannoccaro P, Ciccarese A (1995) Enzymatic versus chemical carbon dioxide utilization. Part I. The role of metal centres in carboxylation reactions. Gazz Chim Ital 125(11):509–538
4. (a) Baran T, Wojtyla S, Dibenedetto A, Aresta M, Macyk W (2015) Zinc sulfide functionalized with ruthenium nanoparticles for photocatalytic reduction of CO_2. Appl Catal B Env 178:170–176. (b) Aresta M, Dibenedetto A, Baran T, Wojtyla S, Macyk W (2015) Solar energy utilization in the direct photocarboxylation of 2,3-dihydrofuran using CO_2. Faraday Discuss 183:413–427. (c) Dibenedetto A, Zhang J, Trochowski M, Angelini A, Macyk W, Aresta M (2017) Photocatalytic carboxylation of CH bonds promoted by popped graphene oxide (PGO) either bare or loaded with CuO. J CO_2 Utilz 20:97–104
5. Liang Y-F, Steinbock R, Yang L, Ackermann L (2018) Continuous visible light-photo-flow approach for manganese-catalyzed (het)arene C–H arylation. Angew Chem Int 57:10625–10629
6. Helmenstine AM (2018) Heat of formation or standard enthalpy of formation table. Thought Co. https://www.thoughtco.com/common-compound-heat-of-formation-table-609253
7. Rotman D, MIT Technol Rev. https://www.technologyreview.com/s/425489/a-future-of-fossil-fuels/
8. Abas N, Kalair A, Khan H (2015) Review of fossil fuels and future energy technologies. Futures 69:31–49
9. https://knoema.com/infographics/smsfgud/bp-world-reserves-of-fossil-fuels
10. (a) Goto Y, Wang Q (2018) A particulate photocatalyst water-splitting panel for large-scale solar hydrogen generation. Joule. Accepted https://doi.org/10.1016/j.joule.2017.12.009. (b) Pinaud BA, Benck JD, Seitz LC, Forman AJ, Chen Z, Deutsch TG, James BD, Baum KN, Baum GN, Ardo S (2013) Technical and economic feasibility of centralized facilities for solar hydrogen production via photocatalysis and photoelectrochemistry. Energy Env Sci 6(7):1983–2002
11. https://www.power-technology.com/comment/global-pv-capacity-expected-reach-969gw-2025/
12. (a) Aresta M, Galatola M (2001) Life cycle analysis applied to the assessment of the environmental impact of alternative synthetic processes. J Cleaner Prod 7:181–193. (b) Aresta M, Caroppo A, Dibenedetto A, Narracci M (2002) Life cycle assessment (LCA) applied to the synthesis of methanol. Comparison of the use of syngas with the use of CO_2 and dihydrogen produced from renewables. In: Maroto-Valer M (ed) Envrironmental challenges and greenhouse gas control for fossil fuel utilization in the 21st century. Kluwer Academic, Plenum Publishers, New York. (c) Artz J, Müller TE, Thenert K, Kleinekorte J, Meys R, Sternberg A, Bardow A, Leitner W (2018) Sustainable conversion of carbon dioxide: an integrated review of catalysis and life cycle assessment. Chem Rev 118(2):434–504

13. Aresta M, Quaranta E (1997) Carbon dioxide: a substitute for phosgene. Chem Tech 27(3): 32–40
14. https://www.uptodate.com/contents/carbon-monoxide-poisoning
15. Rupesh S, Muraleedharan C, Arun P (2016) Exergy and energy analyses of Syngas production from different biomasses through air-steaming gasification. Front Energy, pp 1–13
16. (a) https://www.mordorintelligence.com/industry-reports/global-synthesis-gas-Syngas-market-industry. (b) https://www.reportlinker.com/syngas-reports
17. Aresta M, Dibenedetto A, LN He (2012) Analysis of demand for captured CO_2 and products from CO_2 conversion. A report exclusively for members of the carbon dioxide capture and conversion CO_2–CC programme of the catalyst group resources (TCGR)
18. (a) Aresta M, Dibenedetto A, Quaranta E (2016) Reaction mechanisms in carbon dioxide conversion. Springer. (b) Aresta M, Nobile CF, Albano VG, Forni E, Manassero M (1975) New nickel-carbon dioxide complex: synthesis, properties, and crystallographic characterization of (carbon dioxide)-bis(tricyclohexylphosphine)nickel. J Chem Soc Chem Comm 15:636–637
19. Aresta M, Nocito F, Dibenedetto A (2018) What catalysis can do for boosting carbon dioxide utilization. Adv Catal 62:49–110
20. https://www.marketresearchfuture.com/reports/formic-acid-market-1132
21. Tanaka R, Yamashita M, Nozaki K (2009) Catalytic hydrogenation of carbon dioxide using Ir (III)–Pincer complexes. J Am Chem Soc 131(40):14168–14169
22. Wesselbaum S, Hintermaier U, Leitner W (2012) Continuous-flow hydrogenation of carbon dioxide to pure formic acid using an integrated scCO$_2$ process with immobilized catalyst and base. Angew Chem Int Ed 51:8585–8588
23. Moret S, Dyson P, Laurenczy G (2014) Direct synthesis of formic acid from carbon dioxide by hydrogenation in acidic media. Nat Commun 5:1–7
24. https://www.mordorintelligence.com/industry-reports/global-market-for-surfactants-industry
25. https://www.alliedmarketresearch.com/acrylic-acid-market
26. Alvarez R, Carmona E, Galindo A, Gutierrez E, Marin JM, Monge A, Poveda ML, Ruiz C, Savariault JM (1989) Formation of carboxylate complexes from the reactions of CO_2 with ethylene complexes of molybdenum and tungsten. X-ray and neutron diffraction studies. Organomet 8(10):2430–2439
27. Aresta M, Pastore C, Giannoccaro P, Kovacs G, Dibenedetto A, Papai I (2007) Evidence for spontaneous release of acrylates from a transition-metal complex upon coupling ethene or propene with a carboxylic moiety or CO_2. Chem Eur J 13(32):9028–9034
28. Lejkowski ML, Lindner R, Kageyama T, Bodizs GE, Plessow PN, Mueller IM, Schaefer A, Rominger F, Hofmann P, Futter C, Schunck SA, Limbach M (2012) The first catalytic synthesis of an acrylate from CO_2 and an alkene—a rational approach. Chem Eur J 18 (44):14017–14025
29. (a) Wang X, Wang H, Sun Y (2017) Synthesis of acrylic acid derivatives from CO_2 and ethylene. Chem 3:211–228. (b) Li Y, Liu Z, Cheng R, Liu B (2018) Mechanistic aspects of acrylic acid formation from CO_2–ethylene coupling over palladium- and Nickel-based catalysts. ChemCatChem 10(6):1420–1430
30. See for example Chapter 6 in Aresta M, Dibenedetto A, Quaranta E (2016) Reaction mechanisms for carbon dioxide conversion. Springer
31. Aresta M, Dibenedetto A, Dutta A (2017) Energy issues in the utilization of CO_2 in the synthesis of chemicals: the case of the direct carboxylation of alcohols to dialkyl-carbonates. Cat Today 281:345–351
32. Aresta M, Dibenedetto A, Angelini A, Papai I (2015) Reaction mechanisms in the direct carboxylation of alcohols for the synthesis of acyclic carbonates. Top Catal 58(1):2–14
33. Dibenedetto A, Aresta M, Angelini A, Etiraj J, Aresta BM (2012) Synthesis characterization and use of NbV/CeIV-mixed oxides in the direct carboxylation of ethanol by using pervaporation membranes for water removal. Chem A Eur J 18(33):10524–10534

34. (a) See Ref. [30]. (b) Della Monica F, Buonerba A, Capacchione C (2018) Adv Synth Catal https://doi.org/10.1002/adsc.201801281

35. Aresta M, Dibenedetto A, Nocito F, Pastore C (2006) A study on the carboxylation of glycerol to glycerol carbonate with carbon dioxide: the role of the catalyst, solvent and reaction conditions. J Mol Cat 257(1–2):149–153

36. (a) Aresta M, Quaranta E, Ciccarese A, (1987) Direct synthesis of 1,3-benzodioxol-2-one from styrene, dioxygen and carbon dioxide promoted by Rh(I). J Mol Cat 41:355–359. (b) Dibenedetto A, Aresta M, Distaso M, Pastore C, Venezia AM, Liu C-J, Zhang M (2008) High throughput experiment approach to the oxidation of propene to propene oxide with transition metal oxides as O-donors. Catal Today 137:44–51

37. Angelini A, Dibenedetto A, Curulla-Ferre D, Aresta M (2015) Synthesis of diethylcarbonate by ethanolysis of urea catalysed by heterogeneous mixed oxides. RSC Adv 5(107):88401–88408

38. Wang M, Wang H, Zhao N, Sun Y (2007) High-yield synthesis of dimethyl carbonate from urea and methanol using a catalytic distillation process. Ind Eng Chem Res 46(9):2683–2687

39. Aresta M, Ballivet-Tkatchenko D, Belli-Dell'Amico D, Bonnet MC, Boschi D, Calderazzo F, Faure R, Labella L, Marchetti F (2000) Isolation and structural determination of two derivatives of the elusive carbamic acid. RSC Chem Commun 13:1099–1100

40. Aresta M, Dibenedetto A, Quaranta E (1998) Reaction of aromatic diamines with diphenylcarbonate catalyzed by phosphorous acids: a new clean synthetic route to mono- and dicarbamates. Tetrahedron 54(46):14145–14156

41. Aresta M, Bosetti A, Quaranta E (1996) Procedimento per la produzione di carbammati aromatici. Ital Pat, Appl, p 002202

42. Aresta M, Dibenedetto A, Quaranta E (1999) Selective carbomethoxylation of aromatic diamines: with mixed carbonic acid diesters in the presence of phosphorous acids. Green Chem 1(5):237–242

43. (a) Aresta M, Dibenedetto A (2002) Mixed anhydrides: key intermediates in carbamates forming processes of industrial interest. Chem A Eur J 8(3):685–690. (b) Aresta M, Dibenedetto A (2002) Development of environmentally friendly synthese: use of enzymes and biomimetic systems for the direct carboxylation of organic substrates. Rev Mol Biotechnol 90:113–128

44. https://www.grandviewresearch.com/industry-analysis/polyurethane-pu-market

45. https://www.engineering-airliquide.com/methanol-and-derivatives

46. http://blogs.platts.com/2017/09/07/infographic-whats-store-global-polyethylene-polypropylene-2027/

47. https://www.plasticsinsight.com/resin-intelligence/resin-prices/polypropylene/

48. http://www.chemengonline.com/methanol-to-propylene-technology/

49. Inui T, Phatanasri S, Matsuda H (1990) Highly selective synthesis of ethene from methanol on a novel nickel-silicoaluminophosphate catalyst. JCS Chem Comm 3:205–206

50. Peng Y, Wu T, Sun L, Nsanzimana JMV, Fisher AC, Wang X (2017) Selective electrochemical reduction of CO_2 to ethylene on nanopores-modified copper electrodes in aqueous solution. ACS Appl Mater Interfaces 9(38):32782–32789

51. Tamura J, Ono A, Sugano Y, Huang C, Nishizawa H, Mikoshiba S (2015) Electrochemical reduction of CO_2 to ethylene glycol on imidazolium ion-terminated self-assembly monolayer-modified Au electrodes in an aqueous solution. Phys Chem Chem Phys 17(39):26072–26078

52. http://www.grandviewresearch.com/industry-analysis/ethylene-glycols-industry

53. Aresta M, Dibenedetto A (2019) Beyond fractionation in microalgae utilization. In: Pires J, Goncalves AL (eds) Bioenergy with carbon capture and storage. Elsevier, ISBN 9780128162293

54. Rabaey K, Rozendal RA (2010) Microbial electrosynthesis—revisiting the electrical route for microbial production. Nat Rev Microbiol 8(10):706–716

55. Sugnaux M, Happe M, Cachelin CP, Gasperini A, Blatter M, Fischer F (2017) Cathode deposits favor methane generation in microbial electrolysis cell. Chem Eng J 324:228–236
56. (a) Potter MC (1911) Electrical effects accompanying the decomposition of organic compounds. Proc R Soc B 84(571):260–276. (b) Santoro C, Arbizzani C, Erable B, Ieropoulos IJ (2017) Microbial fuel cells: from fundamentals to applications. A review. Power Sources 356:225–244
57. Tremblay PL, Zhang T (2015) Electrifying microbes for the production of chemicals. Front. Microbiol 6:201–205
58. Xafenias N, Mapelli V (2014) Performance and bacterial enrichment of bioelectrochemical systems during methane and acetate production. Int J Hydrogen Energy 39(36):21864–21875
59. El Mekawy A, Hegab HM, Mohanakrishna G, Bulut M, Pant D (2016) Technological advances in CO_2 conversion electro-biorefinery: a step toward commercialization. Biores Technol 215:357–370
60. de Bok FAM, Hagedoorn PL, Silva PJ, Hagen WR, Schiltz E, Fritsche K, Stams AJM (2003) Two W-containing formate dehydrogenase (CO_2 reductases) involved in syntrophic propionate oxidation by Syntrophobacter fumaroxidans. Eur J Biochem 270:2476–2485
61. Reda T, Plugge CM, Abram NJ, Hirst J (2008) Reversible interconversion of carbon dioxide and formate by an electroactive enzyme. Proc Natl Acad Sci USA 105(31):10654–10658
62. Aresta M, Dibenedetto A, Baran T, Angelini A, Labuz P, Macyk W (2014) An integrated photocatalytic/enzymatic system for the reduction of CO_2 to methanol in bioglycerol–water. Beilst J Org Chem 10:2556–2565
63. Schlager S, Dibenedetto A, Aresta M, Apaydin DH, Dumitru LM, Neugebauer H, Sariciftci NS (2017) Biocatalytic and bioelectrocatalytic approaches for the reduction of carbon dioxide using enzymes. Energy Technol 5(6):812–821
64. Aresta M, Dibenedetto A, Macyk W (2015) Hybrid (enzymatic and photocatalytic) systems for CO_2—water coprocessing to afford energy-rich molecules. In: Rozhkova E, Katsuhiko A (eds) From molecules to materials, pathways to artificial photosynthesis. Springer, pp 149–169
65. Angelini A, Aresta M, Dibenedetto A, Baran T, Macyk W (2015) IP 0001419035, fotocatalizzatori per la riduzione nel visibile di NAD^+ a NADH in un processo ibrido chemi-enzimatico di riduzione di CO_2 a metanolo
66. Aresta M, Dibenedetto A, Quaranta E (2016) State of the art and perspectives in catalytic processes for CO_2 conversion into chemicals and fuels: the distinctive contribution of chemical catalysis and biotechnology. J Catal 343:2–45
67. Marxer D, Furler P, Tacacs M, Steinfeld A (2017) Solar thermochemical splitting of CO_2 into separate streams of CO and O_2 with high selectivity, stability, conversion, and efficiency. Energy Environ Sci 10(5):1142–1149
68. (a) Bork A H, Kubicek M, Struzik M, Rupp J LM (2015) Perovskite $La_{0.6}Sr_{0.4}Cr_{1-x}Co_xO_{3-\delta}$ solid solutions for solar-thermochemical fuel production: strategies to lower the operation temperature. J Mater Chem 3(30):15546–15557. (b) Rao CNR, Dey S (2015) Generation of H_2 and CO by solar thermochemical splitting of H_2O and CO_2 by employing metal oxides. J Solid State Chem 242(2):107–115
69. Mostrou S, Buchel R, Pratsinis SE, van Bokhoven JA (2017) Improving the ceria-mediated water and carbon dioxide splitting through the addition of chromium. Appl Catal A Gen 537:40–49
70. Amatore C, Savéant JM (1981) Mechanism and kinetic characteristics of the electrochemical reduction of carbon dioxide in media of low proton availability. J Am Chem Soc 103 (17):5021–5023
71. Hawecker J, Lehn JM, Ziessel R (1983) Efficient photochemical reduction of CO_2 to CO by visible light irradiation of systems containing $Re(bipy)(CO)_3$ X or $Ru(bipy)_3^{2+}$–Co^{2+} combinations as homogeneous catalysts. J Chem Soc Chem Commun 9:536–538
72. (a) Hori Y, Wakebe H, Tsukamoto T, Koga O (1994) Electrocatalytic process of CO selectivity in electrochemical reduction of CO_2 at metal electrodes in aqueous media.

Electrochim Acta 39(11–12):1833–1839. (b) Hori Y (2008) Electrochemical CO_2 reduction on metal electrodes. In: Vayenas (ed) Modern aspects of electrochemistry, vol 42, 3rd edn., pp 89–189

73. Li Q, Fu J, Zhu W, Chen Z, Shen B, Wu L, Xi Z, Wang T, Lu G, Zhu JJ, Sun S (2017) Tuning Sn-catalysis for electrochemical reduction of CO_2 to CO via the core/shell Cu/SnO_2 structure. J Amer Chem Soc 139(12):4290–4293

74. Schreier M, Héroguel F, Steier L, Ahmad S, Luterbacher JS, Mayer MT, Luo J, Graetzel M (2017) Solar conversion of CO_2 to CO using earth-abundant electrocatalysts prepared by atomic layer modification of CuO. Nat Energy 2(7):17087–17096

75. Zhang W, Hu Y, Ma L, Zhu G, Wang Y, Xue X, Chen R, Yang S, Jin Z (2017) Progress and perspective of electrocatalytic CO_2 reduction for renewable carbonaceous fuels and chemicals. Adv Sci 5(1):1700275–1700279

76. https://www.technology.matthey.com/article/61/3/172–182/

77. Olah GA (2013) Towards oil independence through renewable methanol chemistry. Angew Chem Int Ed 52(1):104–107

78. Tian P, Wei Y, Ye M, Liu Z (2015) Methanol to olefins (MTO): from fundamentals to commercialization. ACS Catal 5(3):1922–1938

79. (a) Raudaskoski R, Turpeinen E, Lenkkeri R, Pongrácz E, Keiski RL (2009) Catalytic activation of CO_2: use of secondary CO_2 for the production of synthesis gas and for methanol synthesis over copper-based zirconia-containing catalysts. Catal Today 144(3–4):318-323. (b) Yang C, Ma Z, Zhao N, Wei W, Hu T, Sun Y (2006) Methanol synthesis from CO_2-rich syngas over a ZrO_2 doped CuZnO catalyst. Catal Today 115(1–4):222–227

80. https://www.igu.org/sites/default/files/103419-World_IGU_Report_no%20crops.pdf

81. Ma J, Sun NN, Zhang XL, Zhao N, Mao FK, Wei W, Sun YH (2009) A short review of catalysis for CO_2 conversion. Catal Today 148:221–223

82. http://www.methanol.org/wp-content/uploads/2016/06/DME-An-Emerging-Global-Fuel-FS.pd

83. Frusteri F, Cordaro M, Cannilla C, Bonura G (2015) Multifunctionality of $Cu–ZnO–ZrO_2$/ H-ZSM5 catalysts for the one-step CO_2-to-DME hydrogenation reaction. Appl Catal B: Env 162:57–65

84. Sabatier P, Senderens JB (1902) New synthesis of methane. J Chem Soc 82:333

85. Jürgensen L, Ehimen EA, Born J, Holm-Nielsen JB (2015) Dynamic biogas upgrading based on the Sabatier process: thermodynamic and dynamic process simulation. Bioresour Technol 178:323–329

86. (a) Stangeland K, Kalai D, Li H, Yu Z (2017) CO_2 methanation: the effect of catalysts and reaction conditions. Energy Proc 105:2022–2027. (b) Brooks KP, Hu J, Zhu H, Kee RJ (2007) Methanation of carbon dioxide by hydrogen reduction using the sabatier process in microchannel reactors. Chem Eng Sci 62:1161–1170. (c) Kirchner J, Katharina J, Henry A, Kureti LS (2018) Methanation of CO_2 on iron based catalysts. Appl Catal B: Env 223:47-59. (d) Visconti, CG (2010) Reactor for exothermic or endothermic catalytic reaction WO2010/130399 & Visconti, CG (2014) Multi-structured reactor made of monolithic adjacent thermoconductive bodies for chemical processes with a high heat exchange WO2014/102350

87. Mattia D, Jones M D, O'Byrne J P, Griffiths OG, Owen RE, Sackville E, McManus M, Plucinski P (2015) Towards Carbon-Neutral CO_2 Conversion to Hydrocarbons Chem-SusChem 8(23):4064–4072

88. Gao P, Li S, Bu X, Sun Y (2017) Direct conversion of CO_2 into liquid fuels with high selectivity over a bifunctional catalyst. Nat Chem 9(10):1019–1024

89. (a) Satthawong R, Koizumi N, Song C, Prasassarakich P (2013) Bimetallic Fe–Co catalysts for CO_2 hydrogenation to higher hydrocarbons. JCOU 3–4:102–106. (b) Visconti CG, Martinelli M, Falbo L, Infantes-Molina A, Lietti L, Forzatti P, Iaquaniello G, Palo E, Picutti B, Brignoli F (2017) CO_2 hydrogenation to lower olefins on a high surface area K-promoted bulk Fe-catalyst. Appl Catal B 200:530–542

90. Emam EA (2015) Gas Flaring in industry: an overview. Pet Coal 57(5):532–555

91. Aresta M, Dibenedetto A, Quaranta E (2016) Reaction mechanisms in carbon dioxide conversion. Springer (Chapter 5)
92. (a) Pitter S, Dinjus E (1997) Phosphinoalkyl nitriles as hemilabile ligands: new aspects in the homogeneous catalytic coupling of CO_2 and 1,3-butadiene. J Mol Catal A Chem 125:39–45. (b) Behr A, Henze H (2011) Use of carbon dioxide in chemical syntheses via a lactone intermediate. Green Chem 13:25–39
93. (a) Doring A, Jolly PM (1980) The palladium catalysed reaction of carbon dioxide with allene, Tetrahedron Lett, 21:3021–3024. (b) Aresta M, Ciccarese A, Quaranta E (1985) Head to head and head to tail coupling of allene and co-condensation with carbon dioxide promoted by 1,2-bis(diphenylphosphimo)ethane(η^6-tetraphenylborate) rhodium. C1 Mol Chem 1:283-295. (c) North M (2011) Synthesis of b,g-unsaturated acids from allenes and carbon dioxide. Angew Chem Int Ed 48:4104–4105. (d) Aresta M, Dibenedetto A, Papai I, Schubert G (2002) Unprecedented formal [2 + 2] addition of allene to CO_2 promoted by $[RhCl(C_2H_4)(P^iPr_3)]_2$: direct synthesis of four membered lactone α-methylene-β-oxiethanone. The intermediacy of $[RhH_2Cl\ (P^iPr_3)]_2$: theoretical aspects and experiments. Inorg Chim Acta 334:294–300
94. (a) Hoberg H, Schaefer D, Buchart G (1982) Oxalanickelacyclopenten-drivate, ein neur typ vielseitig verwendbarer synthone. J Organomet Chem 228:C21–C24. (b) Albano P, Aresta M (1980) Some catalytic properties of Rh(diphos)(h^6-BPh$_4$). J Organomet Chem 190:243–246

Capture of CO_2 from Concentrated Sources and the Atmosphere

Xiaoxing Wang and Chunshan Song

Abstract

As the rapid rise of the atmospheric CO_2 concentration has aroused increasing concern worldwide on the global climate change, the research activities in CO_2 capture both from the concentrated CO_2 sources and the atmosphere have grown significantly. The amine based solid sorbents exhibited great promise in the near-future application for CO_2 capture owing to their advantages including high CO_2 capacity even at extremely low CO_2 concentration (e.g., 400 ppm), excellent CO_2 sorption selectivity, no need for moisture pre-removal (moisture even shows promotion effect), lower energy consumption, less corrosion and easy handling compared to liquid amine. Among them, PEI-based sorbents have been considered as one of most promising candidates and have been extensively studied. Great progress has been made in the past two decades. Hence, in this review, we summarize the recent advances with supported PEI sorbents for CO_2 capture, with an emphasis on (1) sorbent material development including the effects of support and polymer structure; (2) CO_2 sorption mechanism; (3) CO_2 sorption kinetics, (4) sorbent deactivation, and (5) practical implementation of PEI-based sorbent materials. At last, the remaining problems and challenges that need to be addressed to improve the competitiveness of sorbent-based capture

X. Wang · C. Song (✉)
PSU-DUT Joint Center for Energy Research, EMS Energy Institute, Pennsylvania State University, 209 Academic Projects Building, University Park, PA 16802, USA
e-mail: csong@psu.edu

X. Wang
e-mail: xxwang@psu.edu

C. Song
Departments of Energy & Mineral Engineering and of Chemical Engineering, Pennsylvania State University, 209 Academic Projects Building, University Park, PA 16802, USA

© Springer Nature Switzerland AG 2019
M. Aresta et al. (eds.), *An Economy Based on Carbon Dioxide and Water*,
https://doi.org/10.1007/978-3-030-15868-2_2

technologies are discussed. Through the current review, we expect it will not only offer a summary on the recent progress on the supported PEI sorbents, but also provide possible links between fundamental studies and practical applications.

2.1 Introduction

The emission of greenhouse gases into atmosphere has caused worldwide concern, particularly carbon dioxide (CO_2), which has been identified as the major reason for global climate change [1]. As of 2018, the atmospheric concentration of CO_2 has permanently reached over 400 ppm, increased rapidly from the ~ 280 ppm level before the industrial revolution [1, 2], while it changed only slightly over a thousand years from 280 ppm in 1000 to 295 ppm in 1900 based on Antarctica ice core data [2]. Without control, it is projected that the atmospheric CO_2 concentration could surpass 550 ppm level by 2050, causing significant impact on global climate and global human nutrition as well [3]. The 2015 United Nations Climate Changed Conference came to an agreement that the global temperature target is to limit the temperature increase within 2 °C [4], which stressed out the urgency of reducing CO_2 emissions [2]. The global awareness and concerns on the continuing emissions of large quantity of CO_2 has led to increased efforts to developing strategies for mitigating CO_2 emissions, both from concentrated sources such as fossil-fuel-fired power plants, oil refineries, steel industries, cement plants, etc. and from air [1, 4–9]. Among them, carbon capture, utilization and storage (CCUS) has been widely accepted as a promising option to mitigate CO_2 emissions.

Combustion of fossil fuels is the largest emission source accounting for over 80% of total CO_2 emissions [9–11]. One of the major anthropogenic sources of CO_2 emission is the fossil fuel-fired power plants. In the United States, about 30, 21 and 26% of total US greenhouse gas (GHG) emissions come from fossil fuel-fired power plants, industrial and transportation, respectively [12]. The CO_2 emissions from transport sector is also projected to double by 2050 due to a strong increase in demand for cars in developing countries and aviation [11]. CO_2 capture from concentrated sources thus plays a crucial role in maintaining or slowing the increase of the atmospheric CO_2 concentration [8]. Recently, another strategy called negative emissions has been proposed, which removes CO_2 directly from the atmosphere or "direct air capture" (DAC) [8, 9]. The combination of both strategies, i.e., CO_2 capture from both the concentration sources and from atmospheric may be the key to meet the target that limits global warming below two degrees Celsius.

The current benchmark and most mature capture technology is the aqueous alkanolamine based technology [13–16]. However, this process is highly energy intensive due to high water heating requirement during the CO_2 recovery step and it increases the cost by 25–40%, which limits the widespread of this technology [17]. Consequently, development of CO_2 capture/separation technologies with minimal energy and economic penalty has become a major global challenge [4, 5, 18]. Among them, adsorption is considered as a promising technology for efficient capture of CO_2 from gas streams. Various adsorbents such as carbon materials [19, 20], pillared clays [21, 22], metal oxides [23], molecular sieves [24, 25], zeolites [26, 27], silica gel [28, 29], carbon nanotubes [30, 31] and metal-organic framework (MOFs) [32–34] have been investigated. However, the use of these adsorbents for CO_2 capture is limited due to their low capacity and selectivity to CO_2.

Within the last two decades, amine-functionalized solid sorbents have attracted great attention for CO_2 capture from gas streams including flue gases and the atmosphere [35], as it offers much lower parasitic energy requirement due to the lower heat capacities of solids compared to water. So far, there are mainly three popular approaches reported in literature for the preparation of the amine functionalized solid sorbents: (1) in situ polymerization of reactive amine monomers inside the porous material and co-condensation of amine containing compounds with silica precursor during mesoporous material preparation [36–40]; (2) post-synthesis grafting or covalently bonding of amine compounds, mainly aminosilanes, such as (3-aminopropyl)trimethoxysilane, (3-aminopropyl)triethoxysilane, as examples, on the silica support surface [28, 41–45]; and (3) wet impregnation method to immobilize amine polymer or compounds including polyethylenimine (PEI), tetraethylenepentamine (TEPA), diethanolamine (DEA), and dendrimers [46–49]. to the nanoporous materials [50–59].

Many reviews on CO_2 capture using amine-functionalized sorbents and other type of adsorbents have been published for CO_2 capture from flue gas and the air [5, 8, 16, 61–68]. PEI is a polymer with repeating unites composed of one amine group and two carbon aliphatic CH_2CH_2 spacers. Because of its richness of amino groups in a more condensed form, relatively high nitrogen percentage in the molecule, and suitable molecular weight for thermal stability [69, 70], PEI-based sorbents have been considered as one of the most promising candidates for CO_2 capture both from

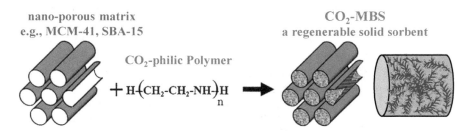

Scheme 2.1 Concept of "molecular basket" sorbents (MBS) for CO_2 capture. Adapted from literature [60]

the concentrated CO_2 sources and the ambient air. Since the supported PEI adsorbents, the so-called "molecular basket" sorbents, the concept of which is depicted in Scheme 2.1, was first reported for CO_2 capture from flue gas by Xu et al. [59], it has received increasing attention in the research community and some great progress has been made since then. In this work, we will summarize the recent advances with supported PEI sorbents for CO_2 capture, with emphasis placed on (1) sorbent material development including the effects of support and polymer structure; (2) CO_2 sorption mechanism; (3) CO_2 sorption kinetics, (4) sorbent deactivation, and (5) practical implementation of PEI-based sorbent materials.

2.2 Development Progress of Supported PEI Sorbents for CO_2 Capture

2.2.1 Effect of Support

In 2002, Song's group at Penn State [59] first reported the MCM-41 supported PEI sorbent termed as "molecular basket sorbent" (MBS) for CO_2 capture, showing a CO_2 capacity of 133 mg/g at 75 °C. MBS showed high CO_2 capacity and selectivity, good regenerability and stability [50], positive effect of moisture on sorption capacity [51], less corrosion and less energy consumption in regeneration compared to the amine-scrubbing process [52], and were thus regarded as a promising alternative for CO_2 capture. This work received a significant attention worldwide and triggered subsequent studies on amine-impregnated composite sorbents for CO_2 capture. Up to now, various support materials including MCM-41, MCM-48, SBA-15, SBA-12, SBA-16, KIT-6, HMS, MCF and activated carbons to name a few have been explored for CO_2 capture [53–58].

Table 2.1 summaries the CO_2 capture capacities of supported PEI sorbents with various porous support and various PEI type from the concentrated CO_2 stream and the atmosphere at different temperatures. The data shows that the textural structure of the support undoubtedly plays an important role in the CO_2 capture performance of the supported PEI sorbents.

Son et al. [25] applied different types of mesoporous silica including MCM-41, MCM-48, SBA-15, SBA-16 and KIT-6 for PEI impregnation and systematically studied their CO_2 sorption performance. They found that the CO_2 sorption capacity and kinetics were mainly influenced by the pore diameter of the support. The highest CO_2 capacity of 135 mg/g was obtained over the 3D structured KIT-6 with the largest pores. Sayari et al. [71, 74] reported that both the pore size and length could affect CO_2 sorption over the supported PEI sorbent. The CO_2 capacity of 210 mg/g was obtained over the pore-expanded MCM-41 supported PEI sorbent [71]. Generally, higher PEI loading gives a better CO_2 sorption capacity. A support with larger pore volume is thus beneficial as it could hold more PEI, offering higher potential to achieve higher CO_2 capacity. Using a mesocellular silica foam (MCF) with large pore volume (2.5 cm^3/g) and large pore size (31 nm), many researchers obtained a higher CO_2 capacity than both MCM-41 and SBA-15 as the

Table 2.1 Summary of CO$_2$ capture capacities over supported PEI sorbents

Support	Polyethylenimine type	Temp. (°C)	CO$_2$ conc. (%)	CO$_2$ capacity (mg/g)		References
				Dry	Humid	
MCM-41	PEI[a]	75	100	133	–	[50]
PE-MCM-41	PEI (Mn ~423)	75	100	210	–	[71]
MCM-41 (as-syn)	PEI (Mn ~423)	75	100	237	–	[69]
MCM-36	PEI (Mw ~800)	25	100	12	–	[72]
SBA-15	PEI[a]	75	100	70	–	[73]
SBA-15	PEI (Mn ~423)	75	100	173	–	[74]
SBA-15	PEI (branched, Mn ~600)	25	100	82	–	[75]
SBA-15	PEI (branched, Mn ~800)	75	100	90	–	[76]
KIT-6	PEI (linear, Mn ~600)	75	100	135	–	[25]
KIT-6	PEI (Mw ~800)	105	100	132	–	[77]
HMS	PEI (linear, Mn ~600)	75	100	184	–	[58]
HMS	PEI (Mn ~423)	75	100	180	–	[78]
Mesocellular silica foam (MCF)	PEI (Mn ~423)	75	100	201	–	[78]
TUD-1	PEI (Mn ~423)	75	100	116	130	[79]
Mesocellular silica foam (MCF)	PEI (branched, Mn ~600)	25	100	101	–	[80]
MCF	PEI (Mn ~10,000)	85	100	176	–	[81]
Silica foam with ultra-large pores	PEI (Mn ~423)	75	100	255	–	[48]
Hierarchical silica monolith	PEI (linear, Mn ~600)	75	100	210	260	[82]
mm-sized sphere silica foam	PEI (linear, Mw ~800)	75	100	188	–	[83]
Precipitated silica	PEI (branched, Mw ~25,000)	70	100	130	–	[84]
Precipitated silica	PEI (branched, Mw ~800)	70	100	147	–	
Precipitated silica	PEI (linear, Mn ~423)	70	100	173	–	

(continued)

Table 2.1 (continued)

Support	Polyethylenimine type	Temp. (°C)	CO_2 conc. (%)	CO_2 capacity (mg/g) Dry	CO_2 capacity (mg/g) Humid	References
CARiACT G10 silica	PEI (Mn ~423)	80	100	123	–	[85]
CARiACTVRG10	PEI (Mn ~423)	80	100	166	–	[86]
Precipitated silica	PEI (branched, Mw ~600)	105	100	186	–	[87]
Precipitated silica	PEI (branched, Mw ~800)	105	100	202	–	[88]
Silica gel bead	PEI[a]	50	100	51	–	[89]
Silica gel	PEI (linear, Mn ~423)	75	100	138	–	[53]
Mesoporous multilamellar silica	PEI (Mw ~600)	75	100	179	219	[90]
Trimodal nanoporous silica	PEI (Mw ~275)	75	100	172	–	[91]
Resin HP20	PEI (Mn ~600)	75	100	160	–	[92]
Carbon black	PEI (linear, Mn ~423)	75	100	135	–	[57, 93]
Mesoporous carbon	PEI (branched, Mn ~600)	30	100	205	–	[94]
Fumed silica (Aerosil 380)	PEI (linear, Mw ~5000)	70	95	155	–	[95]
Mesocellular silica foam (as-syn.)	PEI (Mn ~600)	70	67	169	–	[96]
Mesoporous silica nanotube	PEI (branched, Mw ~800)	85	60	121	–	[97]
Mesocellular silica foam	PEI (branched, Mw ~1200)	105	50	151	–	[98]
Mesocellular siliceous foam	PEI (branched, Mw ~1200)	75	50	110	131	[99]
MCM-41	PEI (branched, Mn ~600)	75	15	89	131	[51]
MCM-41	PEI (Mw ~25,000)	40	15	25	–	[100]
SBA-15	PEI (linear, Mn ~423)	75	15	105	–	[101]
SBA-15	PEI (linear, Mn ~423)	75	15	140	–	[52]
Mesocellular silica foam	PEI (linear, Mn ~423)	75	15	152	–	[102]
Hierarchically porous silica	PEI (Mw ~600)	75	15	159	–	[103]

(continued)

Table 2.1 (continued)

Support	Polyethylenimine type	Temp. (°C)	CO$_2$ conc. (%)	CO$_2$ capacity (mg/g) Dry	CO$_2$ capacity (mg/g) Humid	References
Core-shell 5A@mesoporous silica	PEI (Mw ~600)	25	15	72	222	[104]
Mesocellular silica foam	PEI[a]	75	11	146	–	[105]
Mesocellular silica	PEI (Mn ~600)	45	10	55	–	[106]
Mesocellular cellular foam	PEI (branched, Mn ~800)	60	10	150	193	[107]
Mesoporous silica capsules	PEI (Mn ~423)	75	10	196	245	[108]
Commercial silica (Grade Q-10)	PEI (Mw ~800)	75	10	102	–	[109]
Mesoporous alumina	PEI (Mw ~800)	25	10	86	–	[110]
Fumed silica	PEI (branched, Mw ~25,000)	25	400 ppm	52	78	[70]
Fumed silica (Aerosil 380)	PEI (linear, Mw ~25,000)	25	400 ppm	53	–	[95]
SBA-15	PEI (Mw ~800)	25	400 ppm	76	–	[110]
Mesoporous alumina	PEI (Mw ~800)	25	400 ppm	46	–	[110]
Mesocellular silica foam	PEI (linear, Mw ~2500)	25	400 ppm	46	–	[111]
Mesoporous silica	PEI (branched, Mw ~1800)	20	400 ppm	–	73	[112]
Resin HP20	PEI (Mn ~600)	25	400 ppm	99	–	[92]
HNTs	PEI[a]	25	400 ppm	55	–	[113]

[a]Not specified in the paper

support. For example, a capacity of 151 mg/g was reported by Subagyono et al. [98, 99]. The CO_2 capacity as high as 201 mg/g was also reported by Wang et al. over the MCF supported PEI sorbent [78]. Qi et al. [48] used a silica foam with ultra large pores. The PEI loading reached as high as 83 wt%, the CO_2 capacity of which reached 255 mg/g at 75 °C.

The support with hierarchical porous structure has also been studied and proved to be a good candidate. Ahn et al. [82] impregnated PEI onto a hierarchical silica monolith. At PEI loading of 65 wt%, the sorbent exhibited a CO_2 capacity of 210 mg/g at 75 °C. The authors attributed the high capacity to the improved mass transport because the interconnected macropores offer less diffusive resistance and the mesopores help the dispersion of PEI [82]. Qi et al. [108] synthesized meso-porous silica capsules with egg-shell structure and loaded PEI over these silica capsules for CO_2 capture. The sorbents showed extraordinary capture capacity up to 245 mg/g under simulated flue gas conditions (pre-humidified 10% CO_2). Larger particle size, higher interior void volume and thinner mesoporous shell thickness all improved the CO_2 capacity of the sorbents [108].

Yue et al. [47, 69] first used the as-synthesized mesoporous silica as a support for PEI impregnation. Without removing the surfactant, the loaded PEI could be dispersed in the micelles of the support, forming a web-like structure to capture CO_2, so that a high CO_2 capacity of 237 mg/g was achieved over this type of sorbent materials. One advantage of this method is that it can save energy and time in sorbent preparation and reduce the production of pollutants as it eliminates the surfactant removal step. However, the presence of surfactant can also limit the maximum amount of PEI loaded into the pores [91]. In addition, the CO_2 sorption kinetics over as-synthesized silica supported amine was slower, especially at a high amine loading [91].

Ma et al. developed a series of carbon supported PEI sorbents for CO_2 capture and obtained the CO_2 capacities comparable to MCM-41 and SBA-15 supported PEI sorbents [57, 93]. They studied the relationship between textural structure of carbon supports and sorption capacity of PEI-loaded carbon-based sorbents, and found the total pore volume, especially the mesopore plus macropore volume, plays a crucial role in determining the CO_2 sorption capacity, while the surface area is not a significant factor [93].

Besides the surface area, pore volume and pore diameters, pore structure and particle size of the support are also reported to be an important factor to the CO_2 sorption capacity of the supported PEI sorbents. Wang et al. [78] examined the three-dimensional (3D) mesoporous materials including MCF, MSU-J and HMS as the support in comparison with MCM-41 and SBA-15. They concluded that the large pore size and unique 3D pore structure can facilitate the CO_2 diffusion and mass transfer and offer more accessible sorption sites. Zhang et al. [53] loaded PEI on a commercial silica gel with different particle size and found that smaller particle size is better. Olah et al. [70, 84] pointed out that the support particles with more outer surface can help the dispersion of amine, allowing more accessible sites for CO_2 than those packed within the pores. Zhang et al. [114] studied the influence of silica supports including SBA-15, TUD-1 and fumed silica HS-5 on the sorption capacity of molecular basket sorbent for CO_2 capture. Among them, the fumed silica based MBS showed the highest CO_2 sorption capacity at all PEI loadings (20, 30, 40 wt%), which was attributed to its unique 3D interstitial pore structure. The fume silica supported PEI sorbents also showed a good performance for CO_2 capture directly from the air. At room temperature, the CO_2 capacity was 52 and 78 mg/g under dry and humid conditions, respectively [70].

2.2.2 Effect of PEI-Type

As shown in Table 2.1, the properties of PEI type can also influence of the CO_2 sorption capacity of the supported PEI sorbents. Generally, linear PEI with low molecular weight is better than the branched PEI with high molecular weight in terms of CO_2 capacity. Considering the fact that the CO_2 sorption ability decreases with primary amine > secondary amine > tertiary amine, and tertiary does not react with CO_2 under dry condition, the decrease in CO_2 uptake with increasing PEI molecular weight can be attributed to the decrease of primary amine content and the increase of tertiary amine content in PEI molecules [88, 115]. Since primary amine interacts stronger with CO_2, another type of amine, polyallylamine (PAA), which contains only primary amino groups with linear bone structure, was also studied in comparison with PEI [111, 116–118]. Figure 2.1 compares the CO_2 capacity of the supported linear PEI, branched PEI and PAA sorbents at different sorption temperatures [118]. The PAA sorbent showed a comparable CO_2 capacity to the PEI sorbent. In addition, all the CO_2 capacities increased with the increase of

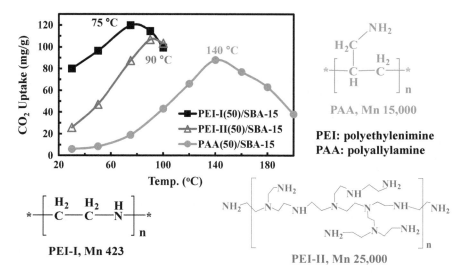

Fig. 2.1 CO_2 capacity as a function of sorption temperature over PEI-I(50)/SBA-15, PEI-II(50)/SBA-15 and PAA(50)/SBA-15 sorbents. Adapted from literature [118]

temperature, which can be explained by the sorption mechanism described in the Sect. 2.3. The optimum temperature was 75, 90 and 140 °C for PEI-I(50)/SBA-15, PEI-II(50)/SBA-15 and PAA(50)/SBA-15, respectively, which may be owing to the increased hydrogen bonding and rigid structure from linear PEI, to branched PEI and further to PAA [118].

2.2.3 Effect of Additives

Through the CO_2 sorption mechanism study (Sect. 2.3), it is understood that increasing the accessible sites per unit of the sorbent and improving the mass transfer rate (i.e., reducing the CO_2 diffusion barrier) can enhance the CO_2 sorption capacity of PEI loaded sorbents and improve the amine efficiency. One strategy is adding an additive to PEI polymer, such as polyethylene glycol (PEG), which has been widely studied. Xu et al. [50] first reported an increase in CO_2 capacity of MCM-41 supported PEI sorbent with the addition PEG at the weight ratio of 2:3 from 68.7 to 77.1 mg/g. Snape et al. [119] loaded PEI and PEG on fly ash derived carbon with altering PEI loading at a fixed PEG content and found that the addition of PEG increased the CO_2 uptake. Olah's group studied the effect of PEG on silica nanoparticles supported PEI sorbents [84, 120]. They reported that the addition of PEG enhanced the isothermal desorption of CO_2 [84] and improved the amine efficiency up to 62% [120]. Zhang et al. [114] studied the effect of PEG addition on the CO_2 capacity of SBA-15, TUD-1 and HS-5 supported PEI sorbents at varied PEI loading and sorption temperature. The results showed that adding PEG can significantly enhance CO_2 capacity and improve the amine efficiency of the sorbents, which was attributed to the inter-molecular interaction between PEI and PEG, alleviating the diffusion barrier with bulk PEI-plug layer [114]. Similar conclusion was also reported by Jones et al. [121].

Other additives such as surfactants have also been studied to promote the CO_2 sorption performance of supported PEI sorbents. Wang et al. [94, 122] explored a wide range of surfactants to promote the hierarchical silica monolith and mesoporous carbons supported PEI sorbents, as shown in Fig. 2.2. All of the additives increased the CO_2 capacity of the sorbent. The surfactant promoted sorbents showed the CO_2 capacity as high as 142 mg/g at 30 °C, and the amine utilization was over 50%, much higher than the supported PEI at this low temperature. The addition of surfactant also promoted the sorption kinetics and the regeneration performance. The authors claimed the CO_2 neutral polymer surfactant served as a diffusion promoter into PEI films to create extra CO_2 diffusion channels, which could transfer CO_2 into the deeper PEI film and contribute to the higher capacity of the sorbent [122]. Same results were observed over the mesoporous carbon supported PEI sorbents with the addition of surfactants [94].

Choi et al. [123, 124] synthesized nanoporous silica supported PEI sorbents with 1,2-epoxybutane (EB) modification, as depicted in Fig. 2.3c. Although the total CO_2 capacity decreased with the addition of EB, the sorbent exhibited a good working capacity of 97 mg/g (Fig. 2.3a) and significantly improved the long-term

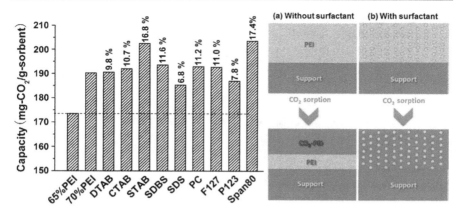

Fig. 2.2 CO_2 capacity of supported PEI sorbents with and without the addition of 5% surfactants along with the proposed scheme for the role of surfactant. Adapted from literature [122]

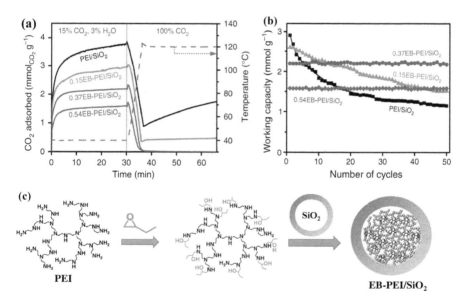

Fig. 2.3 **a** CO_2 sorption-desorption profiles of the EB-PEI/SiO₂ sorbent in a TSA cycle. **b** CO_2 working capacities of sorbents plotted over the number of TSA cycles, and **c** Scheme for preparation of EB-PEI/SiO₂ sorbents. Adapted from literature [123]

stability in a temperature swing adsorption (TSA) process (100% CO_2 at 120 °C, Fig. 2.3b). It also reduced the heat of CO_2 sorption, facilitated CO_2 desorption (>99%), and improved the oxidation-resistance of the supported PEI sorbent.

Apart from organics, the use of inorganics has also been reported to improve the CO_2 capture performance of supported PEI sorbents. With the addition of potassium carbonate into PEI/Silica sorbent, Wang and Song reported that not only the gravimetric and volumetric capacities increased greatly, the amine efficiency and sorbent cyclic stability were also improved, which was attributed to the change of amine-CO_2 reaction chemistry with the presence of K_2CO_3 [125].

2.3 CO_2 Sorption Mechanism Over Supported PEI Sorbents

In general, under dry conditions, CO_2 capture over the supported amine sorbents through the formation of carbamate type zwitterions via the interaction between CO_2 and N-groups in the primary and/or secondary amines, followed by deprotonation of zwitterions by another amine to produce a ammonium carbamate [16, 50, 126–130], as shown in Scheme 2.2. Thus two amine sites are required to react with one CO_2 molecule, giving a theoretical maximum amine efficiency of 0.5, which is defined as the ratio of moles of CO_2 captured to the moles of nitrogen in the material and is a useful metric to quantify the effectiveness of adsorbents for CO_2 capture.

Within PEI, amine groups are not only packed at a high dense form but also close to each other, which makes it unique in CO_2 capture with high amine efficiency. However, the theoretical maximum value of 0.5 has hardly been achieved, suggesting that a proper distance between amine sites is also critical. Under humid conditions, however, water can change the reaction pathway to produce ammonium bicarbonate [128], which can lead to the theoretical amine efficiency of 1. It should also be pointed out that the amine efficiency can be influenced by many factors including reaction conditions such as temperature, CO_2 concentration, amine structure, sorbent properties, diffusion limitations, etc. For instance, tertiary amines react with CO_2 only under humid conditions. At low CO_2 partial pressures such as atmospheric CO_2, the amine efficiency is always lower, far away from the

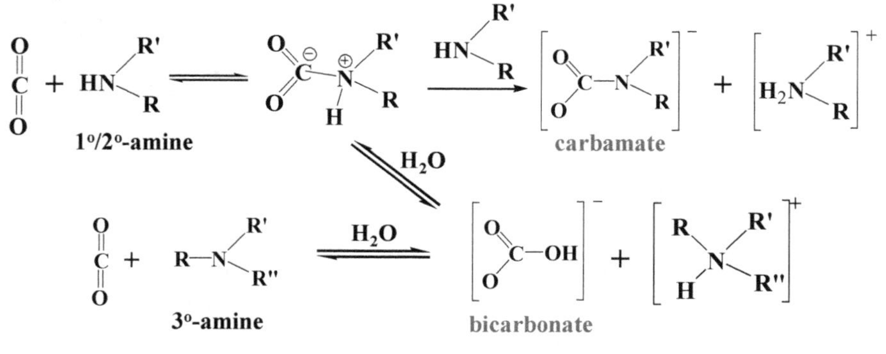

Scheme 2.2 General reaction mechanism of supported amines including primary, secondary and tertiary amine groups reacting with CO_2 in the absence and presence of water

theoretical maxima. Thus, besides the reaction of CO_2 and amine sites, the understanding of the sorption mechanism for CO_2 sorption over the supported PEI sorbents is also crucial, which can improve the fundamental understanding of CO_2 sorption kinetics and thus benefit the further sorbent development, the CO_2 sorption process design and the practical application.

Figure 2.4 demonstrates the effect of PEI loading and sorption temperature on CO_2 sorption capacity of PEI/SBA-15 sorbents [60, 115]. The CO_2 sorption capacity increased with the increase in PEI loading and sorption temperature and the maximum CO_2 capacity was obtained at PEI loading of 50–60 wt% and the temperature of 60–80 °C, which has been confirmed by many researcher [5, 25, 50, 59, 131]. However, when the temperature was decreased under the same CO_2 flow, the CO_2 capacity actually increased and highest CO_2 capacity was obtained at lower

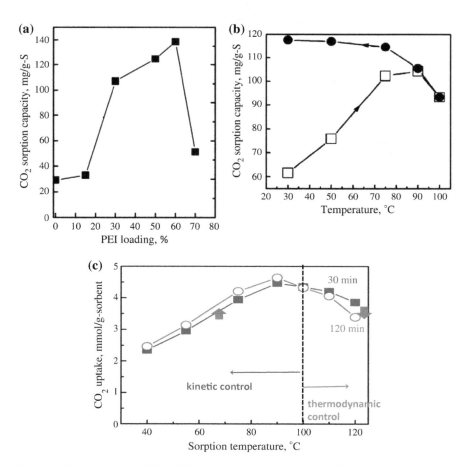

Fig. 2.4 CO_2 capacity of PEI-50/SBA-15 as a function of **a** PEI loading at 75 °C and **b** temperature with temperature increase (open square) and decrease (filled circle) under CO_2 flow [60] and **c** at different sorption time of (filled square) 30 min and (open circle) 120 min [115]

temperature (Fig. 2.4b, solid circle curve). Furthermore, the sorption time also affects the CO_2 sorption capacity, which varies with temperature as shown in Fig. 2.3c. Below 90 °C, extending the sorption time resulted in the increase of the capacity. On the contrary, at the temperature over 90 °C, longer sorption time caused the decrease of the capacity. The higher the sorption temperature, the more the CO_2 capacity decreased at the prolonged sorption time. These results clearly show that both the thermodynamics and kinetics impact CO_2 sorption over PEI/SBA-15.

Ma et al. first reported that there are two types of sorption (type-I and type-II) for CO_2 sorption on PEI by using a semi-empirical quantum chemical calculation method, as shown in Fig. 2.5a [52]. The sorption of CO_2 is through an interaction of the C atom in CO_2 with one N atom (type-I) and two N atoms (type-II), corresponding to CO_2 sorption on the PEI surface and in the PEI bulk, respectively. The transfer of CO_2 from exposed surface of PEI into the bulk of PEI has to over the diffusion barrier, i.e., energy difference between type-I and type-II which is relatively high. Wang and Song applied the CO_2-TPD technique and confirmed the existence of the two sorption types [115]. However, the contributions of two sorption types on CO_2 capacity were not always the same. As shown in Fig. 2.5b, c, it changes with the change of PEI loading and sorption temperature. The type-II sorption contribution increased with the increase of PEI loading, suggesting more diffusion barrier for CO_2 sorption at higher PEI loading. Increasing temperature, however, decreased the contribution of type-II sorption, indicating the diffusion barrier is more alleviated with higher temperature. Since the PEI loading did not change, it can be attributed to the increase of type-I sorption and the decrease of type-II sorption. In other words, increasing temperature generates more exposed PEI phase and less bulk PEI phase. One reason is that with the increase of temperature, the viscosity of PEI polymer gets lower and PEI molecules become more flexible and expandable as proved by FTIR characterization [132, 133], allowing more amine sites exposed and accessible for CO_2, resulting in the increase in the capacity.

Holewinski et al. directly probed the spatial distribution of the supported PEI polymer with the small angle neutron scattering (SANS) technique [134, 135], which is shown in Fig. 2.6a. The SANS spectra were normalized. The differences in SANS intensity directly reflect the changes of the volumetric fill fraction of PEI polymer. With the increase of PEI addition, the total scattered intensity increased and several unique changes to the shape of the scattering pattern were observed, which provide a signature of the PEI deposition morphology. Based on the data, they proposed the pore filling model as illustrated in Fig. 2.6b. In this model, PEI first forms a thin conformal coating on the pore walls, then all additional polymer aggregates into plug that grow along the pore axis. This model is consistent with the observed trends in pore size distributions and CO_2 sorption capacity with PEI loadings. It also points out that CO_2 sorption can be hampered by diffusion through PEI plug within pore channels.

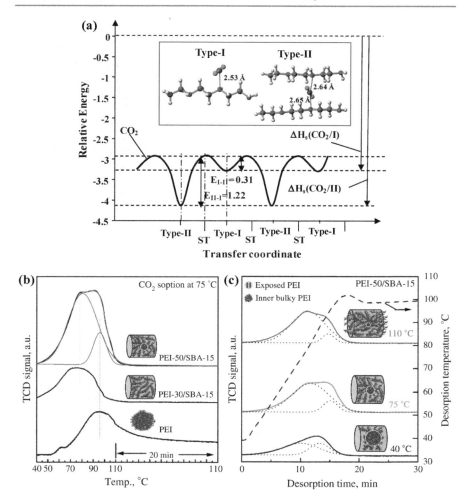

Fig. 2.5 a Potential energy surface for sorption and transfer of CO$_2$ on site-I and Site-II in PEI (inset: CO$_2$ sorption conformation on site-I and site-II) [52]. **b** CO$_2$-TPD profiles over PEI, PEI-30/SBA-15 and PEI-50/SBA-15 at 75 °C and **c** CO$_2$-TPD profiles over PEI-50/SBA-15 at sorption temperature of 40, 75 and 110 °C [115]

On the basis of those studies, a two-layer model was proposed to describe the CO$_2$ sorption mechanism over the supported PEI sorbents, which is illustrated in Scheme 2.3 [115]. In this model, the inner bulky PEI plug layer is enclosed by the exposed PEI layer forming a layer-plug pattern. During the sorption, CO$_2$ molecules are sorbed first in the exposed PEI layer, then diffuse into the bulky PEI-plug layer. Within the exposed PEI layer, the amine sites are widely open and mostly accessible to CO$_2$, while the amine sites are highly packed inside the bulky PEI-plug layer. Consequently, the bulky PEI-plug layer has a much higher barrier

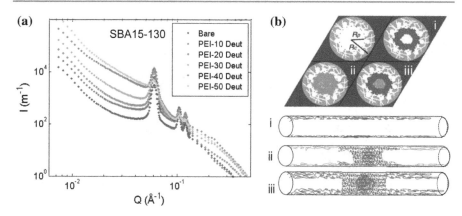

Fig. 2.6 **a** SANS patterns for the 0–50 wt% d-PEI/SBA-15 samples and **b** Schematic of the proposed SBA-15 morphology and various PEI-filling motifs. Adapted from literature [134].

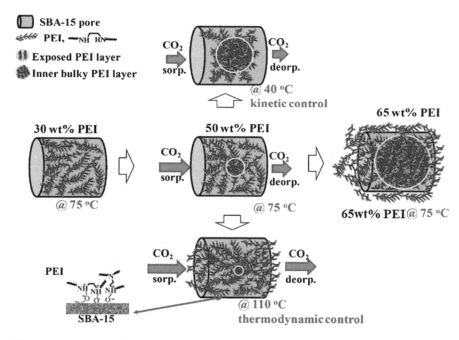

Scheme 2.3 Proposed CO_2 sorption mechanism over the supported PEI sorbents using two-layer model [115]

for CO_2 diffusion than the exposed PEI layer. Additionally, both layers vary with the change of PEI loading and sorption temperature. When PEI loading is lower, only the exposed PEI layer exists, forming a thin coating on the pore surface. If the

PEI content is more than 50 wt%, the higher the PEI loadings, the larger the bulky PEI-plug layer. When the sorption temperature changes, the loaded PEI polymer changes accordingly. Take 50 wt% PEI loading as an example. At low temperature, PEI packs densely within pores because of the molecular interactions, resulting in high content of the bulky PEI-plug layer. With increasing temperature, the PEI polymer becomes more flexible and mobile. As a result, the bulky PEI-plug layer shrinks, while the exposed PEI layer expands. With taking the CO$_2$ sorption kinetics and thermodynamics into the consideration, the proposed two-layer model could be able to reasonable interpret the effects of the PEI loading and sorption temperature on the CO$_2$ capture of the supported PEI sorbents.

2.4 CO$_2$ Sorption Kinetics Over Supported PEI Sorbents

Besides the sorption capacity, the sorption kinetics is another crucial parameter for practical applications, as it is directly related to the process design. Kinetic analysis allows determination of the residence time required for completion of the sorption process and thus the cyclic time of the CO$_2$ sorption process. This is a prerequisite to determine the performance of fixed-bed or any other flow-through process [136].

Early studies evaluated the CO$_2$ sorption kinetics of MCM-41 supported PEI sorbents by calculating the gradients of the sorption curves [51, 78, 137]. Monazam et al. [138] proposed a shrinking core model (SCM) to elucidate the importance of pore diffusion and surface chemical reaction in controlling the rate of reaction, as shown in Fig. 2.7. It consists of chemical reaction at the interface and the diffusion of gaseous reactant and products through the solid-product layer and through the boundary layer at the external surface of the solid. The data analysis suggested a surface reaction-controlled mechanism for the temperature below 40 °C, and a diffusion mechanism at higher temperature. While at the intermediate temperatures, the CO$_2$ sorption rate is a mixed contribution of both diffusion and chemical reaction rates.

Many different kinetic models have been studied by different researchers [86, 136–138, 145–149]. The most popular kinetic models used for CO$_2$ sorption over the supported PEI sorbents are summarized in Table 2.2. Serna-Guerrero and Sayari

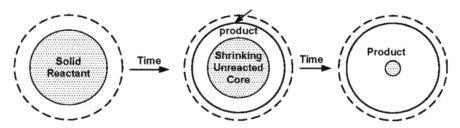

Fig. 2.7 Illustration of the shrinking core model (SCM): gas-solid proceeding with the formation of a shrinking unreacted core [138]

Table 2.2 The typical kinetic model used for CO_2 sorption on supported PEI sorbents[a]

Kinetic model	Diff. form	Integr. form	Rate equation	References
Pseudo-1st order	$\frac{\partial q_t}{\partial t} = k_1(q_e - q_t)$	$q_t = q_e(1 - e^{-k_1 t})$	$r_1 = k_1 q_e$	[139, 140]
Pseudo-2nd order	$\frac{\partial q_t}{\partial t} = k_2(q_e - q_t)^2$	$q_t = \frac{q_e V_0 t}{q_e + V_0 t}$	$r_2 = V_0 = k_2 q_e^2$	[141, 142]
Avrami's equation	$\frac{\partial q_t}{\partial t} = k_A^n t^{n-1}(q_e - q_t)$	$q_t = q_e(1 - e^{-(k_A t)^n})$	$r_A = k_A q_e$	[143]
Elovich's equation	$\frac{\partial q_t}{\partial t} = a \exp(-b q_t)$	$q_t = \frac{1}{\beta} \ln(\alpha \beta t + 1)$	$r_E = \alpha$	[144]

[a]q_e and q_t are the sorption capacity at equilibrium and at time, respectively and k is the rate constant; α and β are the model constant

[145] showed that the pseudo-first and pseudo-second order kinetic models had some limitations in describing the CO_2 adsorption and suggested that the Toth isotherm and the Avrami's fraction order model fitted better to adsorption data. Later, Sayari's group then developed a modified Avrami's equation to describe the CO_2 adsorption over the PEI-impregnated pore-expanded MCM-41, where the adsorption rate is directly proportional to the nth power of the driving force and mth power of the adsorption time as follows [136]:

$$\frac{\partial q_t}{\partial t} = k_n t^m (q_e - q_t)^n \tag{2.1}$$

where, k_n, n and m are the model constants and the value of n reflects the pseudo-order of the reaction with respect to driving force. The extended model was found to be in good agreement with experimental data under a wide range of conditions including different PEI loadings, CO_2 partial pressure and adsorption temperatures [136]. However, it deviated at the initial adsorption stage, which the authors considered as those data not reliable. Bollini et al. [137, 146] conducted similar studies to SBA-15 and showed that the kinetics analyzed from thermogravimetric analysis (TGA) curves were not able to be extended to other applications, but could be used to compare the adsorption kinetics of different adsorbents.

Andreoli et al. [149] analyzed the sorption kinetics of dry and wet CO_2 on PEI-impregnated single walled carbon nanotube (PEI-SWNT), graphite oxide (PEI-GO) and fullerene C_{60} (PEI-C_{60}) using six different kinetic models: Elovich, pseudo-1st order, pseudo-2nd order, pseudo-nth order, modified Avrami and extended model. As illustrated in Fig. 2.8, they found that PEI-SWNT followed a pseudo-2nd order kinetics in dry and wet CO_2, whereas PEI-GO followed a modified Avrami model. The kinetics of PEI-C_{60} was more complicated and slower due to gas diffusion limitations.

Jung et al. [147] investigated the adsorption/desorption kinetics of silica supported 1,2-epoxybutane modified PEI adsorbent by using experimental data from a miniature isothermal fixed-bed reactor system called Autochem. They found that the sorption data were fit well by the Langmuir isotherm and the Langmuir kinetic model multiplied by an exponential term that decreases the overall rate as CO_2

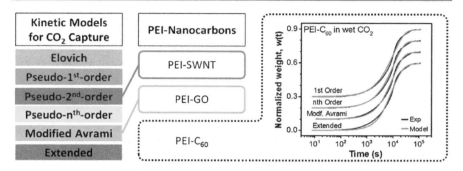

Fig. 2.8 Kinetic models fit best for PEI-SWNT, PEI-GO and PEI-C$_{60}$ sorbents [149]

sorption increases, which outperforms the other reported kinetic models including the pseudo-1st order and Avrami models.

In our laboratory, we have also conducted CO_2 sorption kinetics study over the PEI-based molecular basket sorbents (MBS) with different nanoporous supports in a fixed-bed flow system. Figure 2.9a presents a typical breakthrough curve for CO_2 sorption over MBS at 75 °C (red curve). It captured CO_2 100% before breakthrough. After breakthrough, the outlet CO_2 concentration quickly reached the concentration in the feed.

When it converts to the plot of the capacity versus time (Fig. 2.9a, blue curve), it clearly displays two stages of sorption process: fast and slow sorption stages, respectively. The CO_2 sorption capacity increased quickly and linearly with the sorption time in the first fast sorption stage (Fig. 2.9a, the blue curve in red region), normally within the initial 5 min of sorption, which accounted for more 85% of the total sorption capacity, although the percentage may vary with PEI loading and sorption temperature over different MBS. CO_2 sorption in this stage follows pseudo-1st order. After the fast sorption stage, the CO_2 uptake kept increasing but at a much slower rate, as shown in Fig. 2.9a, the blue curve in blue region, which is

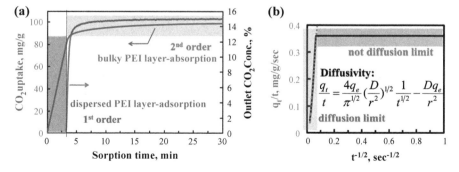

Fig. 2.9 Proposed two stages—pseudo 1st–2nd order mix model for CO_2 capture over PEI based molecular basket sorbents

identified as the second slow sorption stage. CO_2 sorption in the second stage follows pseudo-2nd order. It should be noted that the CO_2 uptake during the second slow stage may not be ignored and could be even higher than that in the first fast sorption stage, particularly for those sorbents with high PEI loading and/or high PEI molecular weight PEI at lower temperature [148].

Several studies applied a double-exponential model to describe the adsorption curves with fast and slow steps, in which they attributed the two steps to different adsorption sites in the amine-functionalized adsorbents as a result of different CO_2 diffusion conditions [92, 150, 151]. Here, considering the CO_2 sorption mechanism over the supported PEI sorbents as discussed in detail in the Sect. 2.3 that CO_2 sorption involves CO_2 sorption firstly in the exposed PEI layer (i.e., the first fast sorption stage), followed by the diffusion and sorption in the bulky PEI-plug layer (i.e., the second slow sorption stage), we reasonably propose a two-stages pseudo 1st–2nd order model to describe the CO_2 sorption kinetics over MBS materials. The proposed model fits the sorption data much better than the other kinetic models reported including the pseudo-1st order and modified Avrami's fraction models.

On the basis of the proposed model, we are able to calculate the apparent diffusion constant, which is estimated by Parabolic diffusion equation [152, 153]. The Parabolic diffusion equation can be expressed as [154]:

$$\frac{q_t}{q_e} = \frac{4}{\pi^{1/2}} \left(\frac{Dt}{r^2}\right)^{1/2} - \frac{Dt}{r^2} - \frac{1}{3\pi^{1/2}} \left(\frac{Dt}{r^2}\right)^{3/2} + \cdots \qquad (2.2)$$

where, q_e and q_t (mg/g) are the adsorption capacities at equilibrium and time t (sec), respectively. D (cm^2/s) is an "apparent" diffusion coefficient and r (cm) is the particle radius. For relative short time in most experiments, the third and subsequent terms may be ignored, and thus:

$$\frac{q_t}{q_e} = \frac{4}{\pi^{1/2}} \left(\frac{Dt}{r^2}\right)^{1/2} - \frac{Dt}{r^2} \quad \text{or} \quad \frac{q_t}{t} = \frac{4q_e}{\pi^{1/2}} \left(\frac{D}{r^2}\right)^{1/2} \frac{1}{t^{1/2}} - \frac{Dq_e}{r^2} \qquad (2.3)$$

Then, the apparent diffusion constant, D/r^2 (s^{-1}) can be obtained. At 75 °C, the apparent diffusion constant for CO_2 sorption over PEI(50)/SBA-15 and PEI(50)/TUD-1 sorbents was 1.38×10^{-3} s^{-1} and 1.45×10^{-3} s^{-1} [79], respectively. In literature, the diffusion constant of some adsorbents for CO_2 sorption was reported as 1.3×10^{-3} s^{-1} for MOF-5 [155], 0.43×10^{-3} s^{-1} for amine-grafted SBA-15 at 25 °C [156] and 2.74×10^{-3} s^{-1} for TRI-PE-MCM-41 adsorbent at 75 °C [145] estimated by linear driving force (LDF) method. As seen, the calculated diffusion constant based on our proposed sorption kinetic model is comparable to those values reported in the literature.

2.5 Regenerability and Stability of Supported PEI Sorbents

In addition to CO_2 capacity, selectivity, and sorption kinetics, the regenerability and stability of a sorbent in sorption-desorption cycles is also crucial for its practical application, as it can save the energy and cost in the system operation.

Table 2.3 lists the regeneration and stability of some typical supported PEI sorbents in cyclic CO_2 sorption-desorption operations at different conditions. As seen, the regeneration was normal performed by a combined temperature and pressure swings, under a temperature of 50–150 °C in an inert gas atmosphere such as N_2, He or Ar. In most cases, the sorbents suffered from a loss of CO_2 capacity, especially when the regeneration temperature. The 100% recovery in the CO_2 capacity up to 180 cycles was only reported at the temperature of 55 °C for regeneration [95]. The higher the regeneration temperature, the more significant the loss of the CO_2 capacity. One reason is the leaching of PEI polymer during the cyclic sorption-desorption process as there is no strong chemical interaction between the support and PEI polymer loaded by impregnation method [157]. Amine leaching is closely related to the molecular weight of PEI and the operation temperature. More leaching will happen when the PEI molecular weight is low and the regeneration temperature is high.

Another reason is the degradation of PEI with the cyclic CO_2 sorption-desorption operation. Drage et al. [157] conducted temperature programmed adsorption of CO_2 on PEI-loaded silica and observed a weight increase at the temperature over 135 °C, which was due to the chemical reaction between CO_2 and amine groups leading to the formation of urea. Sayari and Belmabkhout studied the regeneration and stability of amine supported CO_2 adsorbent under dry and wet conditions [158]. They observed that the under dry conditions with mild operation temperature (70 °C for adsorption and 70 °C for desorption), the adsorbent was stable over dozens cycles, but deactivated gradually over hundreds of cycle to reach

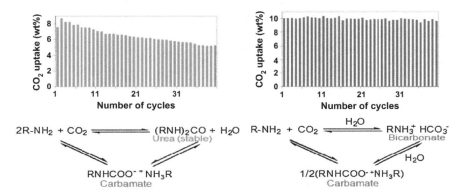

Fig. 2.10 CO_2 adsorption-desorption cycles and the corresponding mechanisms over amine supported CO_2 adsorbent under dry conditions and in the presence of moisture [158]

15% loss after about 750 cycles. As shown in Fig. 2.10, in couple with ^{13}C CP-MAS NMR and in situ FTIR techniques, they identified the formation and accumulation of stable urea species under dry condition is responsible for the deactivation of the amine-containing CO_2 adsorbent. In addition, the formation of urea species are favored at higher operation temperatures. While the presence of moisture could strongly inhibit the formation of urea, by completely reversing the reaction via hydrolysis. As a result, the same adsorbent was still stable even after 700 cycles under the humid condition (20 °C as dew point, 7.5% RH). In a follow-up study, the authors further claimed that primary amines are more likely prone to urea deactivation than secondary and tertiary amines [159]. The difference

Table 2.3 The regeneration and stability performance of supported PEI sorbents reported in literature

Support	PEI type[a]	Regeneration condition				References
		T (°C)	Purge gas	Cycle #	Capacity remained (%)	
Fumed silica	LPEI (Mw $\sim$ 25,000)	55	N_2	180	100	[95]
MCM-41	BPEI (Mw $\sim$ 600)	75	He	10	98	[61]
HMS	LPEI (Mw $\sim$ 600)	75	N_2	4	98	[58]
KIT-6	LPEI (Mw $\sim$ 600)	75	N_2	3	100	[25]
Hierarchical silica monolith	LPEI (Mw $\sim$ 600)	75	N_2	5	>99	[82]
MCF	LPEI (Mn $\sim$ 423)	75	Ar	8	>99	[102]
SBA-15	LPEI (Mn $\sim$ 423)	100	Ar	12	86	[101]
KIT-6	BPEI (Mw $\sim$ 800)	100	Vac.	10	93	[77]
Silica foam bead	LPEI (Mw $\sim$ 800)	100	N_2	4	89	[83]
Porous silica	BPEI (Mw $\sim$ 600)	100	N_2	20	95	[103]
Carbon black	LPEI (Mn $\sim$ 423)	100	N_2	10	92	[93]
MCF	BPEI (Mw $\sim$ 800)	105	N_2	50	93	[107]
SBA-15	LPEI (Mn $\sim$ 423)	110	N_2	20	>99	[52]
SBA-15	BPEI (Mw $\sim$ 800)	110	Vac.	4	93	[76]
Grade Q-10 silica	BPEI (Mw $\sim$ 800)	110	steam	20	98	[109]
Mesopor. silica nanotube	BPEI (Mw $\sim$ 800)	110	N_2	10	97	[97]
Mesopor. multilamellar silica	BPEI (Mw $\sim$ 600)	110	N_2	10	96	[90]
MCG	BPEI (Mw $\sim$ 1300)	115	Ar	10	68	[99]
Precipitated silica	BPEI (Mw $\sim$ 600)	120	N_2	30	89	[87]
Precipitated silica	BPEI (Mw $\sim$ 800)	120	N_2	30	81	[88]
PE-MCM-41	BPEI (Mn $\sim$ 600)	130	N_2	30	28	[159]
PE-MCM-41	LPEI (Mw $\sim$ 2500)	130	N_2	30	50	[159]
PE-MCM-41	LPEI (Mw $\sim$ 2500)	150	N_2	30	22	[159]

[a]Linear and branched PEI are denoted as LPEI and BPEI, respectively

in the stability of primary versus secondary and tertiary amines was associated with the intermediate isocyanate species toward the formation of urea groups as only primary amines can be precursors to isocyanate in the presence of CO_2 [160].

Li et al. [87] studied regenerability and stability of the precipitated silica supported PEI sorbents under two conditions, i.e., 120 °C in N_2 and 135 °C in CO_2. They found that the capacity loss was due mainly to PEI loss when regeneration in N_2. Whereas regeneration at 135 °C in CO_2, both the urea formation and PEI loss were the main contributors to the loss of CO_2 capacity. Subagyono et al. [99] assessed the regenerability of MCF-supported branched and linear PEI sorbents and found linear PEI was more stable than branched PEI under dry conditions. Adding water significantly reduced the capacity loss of branched PEI in regeneration. With steam stripping, a highly CO_2-enriched stream can be achieved at the outlet after steam condensation. However, PEI may leach out or reorganized on the support as PEI is soluble to water [161]. Sandhu et al. [109] applied steam to regenerate the PEI-impregnated commercial silica at 110 °C. After 20 cycles, a 9% of the CO_2 capacity loss was found. They attributed the capacity to the thermal degradation of the sorbent during drying under N_2 after steam stripping as no evidence of PEI leaching or changes in surface morphology and amine functionalities was observed. Chaikittisilp et al. [110] reported that mesoporous alumina supported PEI was more stable than PEI/SBA-15. Under 24 h accelerated steam testing, it retained more than 80% of the initial CO_2 capture for 10% CO_2 in N_2.

The presence of oxygen in the gas streams including flue gas and the air can cause the gradual deactivation of supported PEI sorbents via the oxidation of amine groups. Bollini et al. [162] assessed the oxidative degradation of different supported amine adsorbents under accelerated oxidizing conditions and found that both amine type and proximity had a significant effect on oxidative degradation rates. The supported primary and tertiary amines proved to be more stable that the secondary amine at elevated temperatures (105–135 °C). Similar results were also reported by Calleja et al. [163] in aging of amine-functionalized silica in air. Ahmadalinezhad and Sayari [164] examined the oxidative degradation of SBA-15 supported linear and branched PEI sorbents by exposing to air flow at different temperatures. Although both materials deactivated, it was found that linear PEI is less prone to oxidation than branched PEI. Analysis of the products of oxidative degradation of linear and branched PEI showed a number of predominantly structural fragments of imine and carbonyl species containing C=O and –CH=N– groups as demonstrated in Scheme 2.4 [164]. Zhang et al. [95] examined the stability of the fumed silica supported linear PEI sorbent in N_2, CO_2 and air. The sorbent was stable at 70 °C for 20 h in N_2 and CO_2. Whereas in air, the sorbent showed a noticeable decrease in CO_2 sorption capacity for 20 h at 70 °C, which became even more pronounced at 100 °C due to the oxidative degradation.

Our group also investigated the long-term stability of PEI-I(50)/SBA-15, PEI-II (50)/SBA-15 and PAA(50)/SBA-15 by heat treatment at 120 in a vacuum oven for a prolonged period of time to 260 h [118], which is presented in Fig. 2.11. The structure of PEI-I, PEI-II and PAA are described in Fig. 2.1. The total weight loss after heat treatment was 31 and 21% for PEI-I(50)/SBA-15 and PEI-II(50)/SBA-15,

Scheme 2.4 Oxidative degradation of linear and branched PEI to imine and carbonyl groups [164]

respectively. Whereas it was only 16% for PAA(50)/SBA-15, indicating PAA is more stable. In addition, after treatment, the CO_2 capacity of PEI-I(50)/SBA-15 and PEI-II(50)/SBA-15 only 2.2 and 2.5 mg/g, respectively, suggesting the complete deactivation of PEI. While PAA(50)/SBA-15 remained a good capacity of 50 mg/g. The FTIR analysis disclosed the oxidative degradation of PEI to carbonyl species while the evidence of oxidation on PAA was much insignificant (Fig. 2.11b, c), which is consistent with Sayari et al.'s findings [164].

More recently, Min et al. discovered that the polymeric amines contain p.p.m.-level metal impurities including Fe and Cu, which catalyze amine oxidation [124]. As illustrated in Fig. 2.12, they thus developed the oxidation-stable CO_2 adsorbent

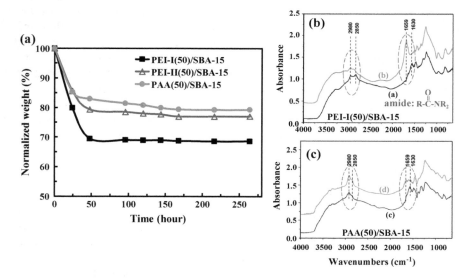

Fig. 2.11 **a** The normalized weight of PEI-I(50)/SBA-15, PEI-II(50)/SBA-15 and PAA(50)/SBA-15 as a function of time during the heat treatment at 120 °C in a vacuum oven, and DRIFT spectra of **a** PEI-I(50)/SBA-15 and **b** PAA(50)/SBA-15 before and after 260 h of heat treatment at 120 °C in vacuum oven. **a** PEI-I(50)/SBA-15, **b** PEI-I(50)/SBA-15-treated, **c** PAA(50)/SBA-15, and **d** PAA(50)/SBA-15-treated. Adapted from literature [118]

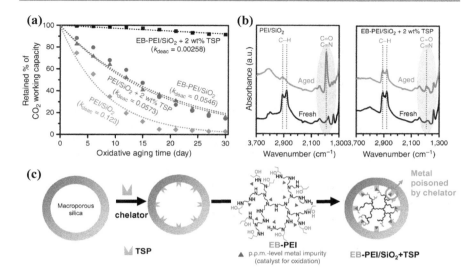

Fig. 2.12 **a** CO₂ capacity of sorbents after aging under 3% O₂, 15% CO₂, 10% H₂O in N₂ at 110 C. **b** FTIR spectra of sorbents before and after 30-days aging. **c** Synthesis of oxidation-stable CO₂ adsorbent. Adapted from literature [124]

which showed unprecedentedly high oxidative stability, by a combination of PEI functionalization with 1,2-epoxybutane and adding small amounts of chelators (<2 wt%) into a silica support before the impregnation of functionalized PEI as a catalyst poison to significantly suppress the rate of amine oxidation. The resultant sorbent showed a minor loss of CO₂ working capacity (8.5%) even after 30 days aging in O₂-containing flue gas at 110 °C, while a conventional PEI/silica showed a complete loss of CO₂ uptake after the same treatment [124].

Besides CO₂, O₂ and water vapor, post-combustion flue gas emitted from fossil fuel-fired power plants normally contains small percentages of nitrogen oxides and sulfur oxides as well. For example, flue gas from a coal plant typically contains 10–15% CO₂, 3–10% O₂, 4–5% water vapor, 2000 ppm SO₂ (before flue gas desulfurization), 1500 ppm NO (before selective catalytic reduction) balanced with N₂ [165, 166]. Xu et al. [167] applied the PEI/MCM-41 sorbent in the separation of CO₂ from the flue gas of a natural gas-fired boiler, which contains 7.4–7.7% CO₂, 14.6% H₂O, ~4.45% O₂, 200–300 ppm CO, 60–70 ppm NOₓ, and 73–74% N₂. They found that the sorbent simultaneously removed CO₂ and NOₓ from flue gas. Although the captured CO₂ amount was about 3000 times larger than that of NOₓ, very little NOₓ desorbed after sorption. Khatri et al. [166] studied the effect of SO₂ on CO₂ sorption over solid amine adsorbent and found that SO₂ reacted irreversibly with the sorbent and thus degraded the sorbent and inhibited CO₂ sorption, although the rate of SO₂ sorption is slower that of CO₂. Fan et al. [168] studied the cyclic stability of silica supported PEI hollow fiber sorbents in the presence of SO₂ and NO. Exposure to NO at 200 ppm did not lead to a significant degradation of the sorbent, but 200 ppm SO_2 caused 60% loss in CO_2 breakthrough capacity after 120 cycles. In further studies, Rezaei and Jones [169] examined the CO_2 sorption performance of PEI-impregnated sorbent materials in the presence of SO_2, NO and NO_2 impurities. They reported that the presence of NO_x impurities at low concentrations do not significantly influence the CO_2 sorption capacity, with NO having no significant impact even at high concentration of 200 ppm and NO_2 having a minimal impact at more typical concentrations of 20 ppm. Although high concentrations of SO_2 would result in CO_2 capacity loss over many cycles, the sorbent maintained the CO_2 capacity at low SO_2 concentration, indicating that the supported PEI sorbents are reasonably stable under dry flue gas with realistic concentrations of impurity gases. In addition, they claimed that the secondary amine is more stable to SO_x and NO_x impurities in CO_2 capture process than primary amine. According to those studies, that nitrogen oxides need to be treated by catalytic reactions and sulfur oxides need to be removed by desulfurization before CO_2 capture is recommended to prevent the irreversible reactions of nitrogen oxides and sulfur oxides with PEI.

2.6 Large-Scale Process Development Using Supported PEI Sorbents

Although adsorption process has been considered as an alternative to the commercial amine-scrubbing process, it is still in an early stage of development. Currently, most research work focuses on developing high performance, stable solid sorbent. There have been only a few large-scale process studies with amine-functionalized sorbents.

Sjostrom et al. [170] tested two amine-functionalized solid sorbents in a circulating fluidized bed unite that can process 1 kW-equivalent flue gas. One supported amine sorbent achieved 90% CO_2 removal repeatedly. Zhao et al. [171] examined supported amine sorbents in a 200 kWth scale pilot plant with moving bed process. They reported the process behaviors and the performance under various operating conditions and different regeneration strategies including thermal regeneration in pure CO_2 stream, vacuum regeneration and steam-stripping regeneration.

Recently our laboratory has teamed up with RTI International to develop molecular basket sorbents (MBS) based bench-scale CO_2 capture process [172–175]. The MBS materials were successfully scaled up to produce large quantities (over 100 kg) for testing in the bench-scale reactors. The engineering analysis suggested the dual fluidized, moving-bed reactors (FMBRs) design, in which the sorbent is continuously circulated between a CO_2 Sorber (for CO_2 capture) and a sorbent Regenerator (for CO_2 release and concentration) as shown in Fig. 2.13a. The system was able to process flue gas between 300–900 SLPM, solid circulation of 75–450 kg/h with filling capacity of 75 kg sorbent, and process roughly 150 kg-CO_2/day. Figure 2.13b showed CO_2 capture result during the 100-h test. The rate of CO_2 capture was maintained at 90% $\pm2\%$ for the majority of the testing

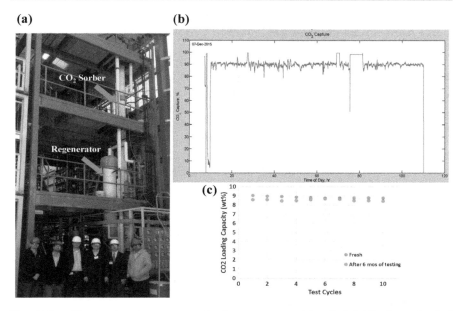

Fig. 2.13 **a** The bench-scale contract evaluation system containing dual fluidized, moving-bed reactors (FMBRs) with main team-members standing in the front. **b** Measured rate of CO$_2$ capture during the 100-h test in the bench-scale system. **c** MBS sorbent stability [172]

period, even as the operating conditions were changed multiple times. Meanwhile, the sorbent stability was checked routinely using a lab-scale fixe-bed reactor. As shown in Fig. 2.13c, the MBS sorbent maintained desired CO$_2$ capture performance stability even after 5 months of testing. In addition, the MBS sorbent showed no signs of significantly attrition during the test and no sorbent make-up had been required even after 5 months of testing.

In the subsequent study, the technology developed by RTI and Penn State team was demonstrated in an operating cement plant in Brevik, Norway to assess the long-term stability, CO$_2$ capture and sorbent regeneration performance under cement plant flue gas exposure [173, 175]. The system has been operated for 150 + h of CO$_2$ capture experimentation and data analysis is ongoing. One suggestion from that study is the addition of 100 ppm SO$_2$ in the flue gas caused the CO$_2$ capacity drop by roughly 30% over 100 sorption-desorption cycles, thus deep scrubbing of SO$_2$ will be required. This is consistent with the earlier study at Penn State with coal-fired and natural gas-fired pilot plant flue gases [176, 177].

Based on the achieved sorbent and process performance and assumptions including sorbent working capacity, sorbent makeup rate, operating temperatures, gas and solid residence time and regeneration CO$_2$ partial pressure, the techno-economic analysis of the process has been performed [173]. The cost was determined to be \$43.3/t-CO$_2$ for CO$_2$ capture from a coal-fired power plant, showing the potential to substantially reduce the total cost compared to conventional aqueous amine CO_2 scrubbing technology and to achieve the U.S. DOE's Carbon Capture Program's goal of >90% CO_2 capture rate at a cost of <$40/t-$CO_2$ by 2025.

As for CO_2 capture from the atmosphere, it was commercialized in the 1950s as pre-treatment for cryogenic air separation. Recently, some different prototypes have been developed for direct CO_2 capture from ambient air by a number of companies using the supported amine sorbents [8, 178]. In 2012, Kulkarni and Sholl [179] applied aminosilica sorbents on structured cordierite monolithic contactors, which can recover CO_2 at 1–3 t/unit/year for a steam-based process. Global Thermostat is developing technology based on porous amine sorbents supported on a monolithic contactors [180–183]. The sorbent is embedded in the walls of the contactor and is very effective at capturing CO_2 from the atmosphere at 400 ppm. The commercial modules of Global Thermostat have the capacity to capture up to 50,000 t-CO_2/year and can be further scaled up if needed. Climeworks built the world's first commercial plant to capture CO_2 from air in Switzerland in 2017 [178]. The plant is currently capturing 1000 t-CO_2/year and will reach 10 Mt-CO_2/year by 2025.

2.7 Summary and Prospects

The mitigation of CO_2 emissions to remedy the global climate change is both a difficult challenge and an urgent task for the sustainable development in energy and environment. To overcome the drawbacks of aqueous amine scrubbing process, scientists and engineers are seeking the alternatives for CO_2 capture, both from the concentrated sources and the atmosphere. The amine based solid sorbents exhibit great promise in the near-future application for CO_2 capture owing to their high CO_2 capacity even at extremely low CO_2 concentration (e.g., 400 ppm), excellent CO_2 sorption selectivity, no need for moisture pre-removal (moisture even showed promotion effect), lower energy consumption, less corrosion and easy handling compared to liquid amine. Among them, PEI-based sorbents have been considered as one of the most promising candidates and have been intensively studied.

Various types of porous materials with different pore properties have been studied as the supports to prepare PEI-base sorbents for CO_2 capture. Based on the studies, it is agreed that the pore properties including large surface area (to improve the dispersion), large pore volume (to accommodate more PEI), large pore size and pore structure like 3D interconnected pores) (to facilitate CO_2 diffusion inside pore channels) are important to achieve an outstanding PEI sorbent. In addition, the particle size also plays an important role. Smaller particle size and larger external surface area are preferred. On the other hand, the PEI structure and type should also be considered. As for CO_2 sorption capacity, linear PEI with low molecular weight is better than the branched PEI with high molecular weight. In terms of thermal stability, higher molecular weight PEI with linear structure would be preferred. Furthermore, adding the additives including PEG, surfactants, 1,2-epoxybutane, and even potassium carbonate can enhance CO_2 sorption on the supported PEI sorbents, but also promote the CO_2 sorption kinetics.

With aid of modern characterization techniques, the spatial distribution and deposition morphology of PEI in the supported PEI sorbent could be identified, which involves the exposed PEI-player and the bulky PEI-plug layer. Accordingly, the two-stage CO_2 sorption mechanism and two-stage pseudo 1st–2nd order kinetic model were proposed. The diffusion constant can thus be estimated.

The long-term regenerability and stability studies showed that the presence of CO_2 at elevated temperature could deactivate the sorbent due to the formation of urea groups. Adding moisture can inhibit the urea formation, thus significantly improve the stability of PEI sorbents. The supported PEI sorbents will gradually deactivate in the presence of oxygen due to the oxidation of amine groups to imine and carbonyl species, especially at high temperatures. It was found that primary amines are more likely prone to urea deactivation than secondary and tertiary amines, while primary and tertiary amines are proved to be more oxidation-resistant than the secondary amine at elevated temperatures. Moreover, the presence of SO_x and NO_x can irreversibly deactivate PEI sorbents. Pre-removal of SO_x and NO_x is thus recommended for CO_2 capture from flue gas over supported PEI sorbents.

The supported PEI sorbents have also been tested in pilot plant for CO_2 capture from simulated and real flue gas. The results showed the potential to substantially reduce the total cost compared to conventional aqueous amine CO_2 scrubbing technology and to achieve the U.S. DOE's Carbon Capture Program's goal of >90% CO_2 capture rate at a cost of <\$40/t-CO_2 by 2025.

Although great progress had been made in the past two decades, there are still some challenges to be addressed for the development and deployment of the technology using supported-amine sorbents for CO_2 capture from both the flue gas and the atmosphere:

(1) the sorbent cost for materials and preparation. Excellent CO_2 capture performance is usually achieved on novel materials which requires costly synthesis procedure. Using inexpensive raw materials with cheap preparation procedure would be highly desired.

(2) CO_2 sorption kinetics on supported PEI-based sorbents is not well addressed. Currently, there are very little data showing how fast the sorbent can deal with CO_2 capture at flue gas emission rate or direct air capture (huge flow due to extremely low CO_2 concentration). In addition, the effect of moisture on the CO_2 sorption kinetics and the CO_2 desorption kinetics are seldom studied. When steam-stripping is applied for sorbent regeneration and CO_2 and heat recovery, it would be particularly important.

(3) Long-term stability is still under study. The best case so far is 8.5% capacity loss over 30 days. More assessment, especially operation in the bench-scale size would be valuable for the deployment of the technology.

Acknowledgements The authors gratefully acknowledge the financial supports by the US Department of Energy, National Energy Technology Laboratory and the Pennsylvania State University on various portions of the CO_2 research. We also acknowledge the RTI International for the joint DOE project on pilot plant demonstration of the CO_2 MBS.

References

1. Climate change 2014 synthesis report summary for policymakers. IPCC. https://www.ipcc.ch/pdf/assessment-report/ar5/syr/AR5_SYR_FINAL_SPM.pdf
2. Seneviratne SI, Donat MG, Pitman AJ, Knutti R, Wilby RL (2016) Allowable CO_2 emissions based on regional and impact-related climate targets. Nature 529:477
3. Smith MR, Myers SS (2018) Impact of anthropogenic CO_2 emissions on global human nutrition. Nat Clim Change 8:834–839
4. Keith DW (2009) Why capture CO_2 from the atmosphere? Science 325:1654–1655
5. Song CS (2006) Global challenges and strategies for control, conversion and utilization of CO_2 for sustainable development involving energy, catalysis, adsorption and chemical processing. Catal Today 115:2–32
6. Lackner KS (2003) A guide to CO_2 sequestration. Science 300:1677–1678
7. Lackner KS, Brennan S, Matter JM, Park A-HA, Wright A, van der Zwaan B (2012) The urgency of the development of CO_2 capture from ambient air. Proc Natl Acad Sci 109:13156–13162
8. Sanz-Pérez ES, Murdock CR, Didas SA, Jones CW (2016) Direct capture of CO_2 from ambient air. Chem Rev 116:11840–11876
9. National Academies of Sciences, Engineering, and Medicine (2018) Negative emissions technologies and reliable sequestration: a research agenda. The National Academies Press, Washington, DC
10. IEA (2010) Energy technology perspectives: scenarios & strategies to 2050. In: I.E.A. OECD/IEA, Paris
11. OECD (2012) OECD environmental outlook to 2050
12. USEPA (ed) (2016) Inventory of U.S. greenhouse gas emissions and sinks: 1990–2014
13. Rochelle GT (2009) Amine scrubbing for CO_2 capture. Science 325:1652–1654
14. Rochelle GT (2016) Conventional amine scrubbing for CO_2 capture. In: Feron PHM (ed) Absorption-based post-combustion capture of carbon dioxide. Woodhead Publishing, pp 35–67
15. Tontiwachwuthikul P, Idem R (2013) Recent progress and new developments in post-combustion carbon-capture technology with reactive solvents, pp 2–8
16. Darunte LA, Walton KS, Sholl DS, Jones CW (2016) CO_2 capture via adsorption in amine-functionalized sorbents. Curr Opin Chem Eng 12:82–90
17. Haszeldine RS (2009) Carbon capture and storage: how green can black be? Science 325:1647–1652
18. Chu S (2009) Carbon capture and sequestration. Science 325:1599
19. Plaza MG, Pevida C, Arenillas A, Rubiera F, Pis JJ (2007) CO_2 capture by adsorption with nitrogen enriched carbons. Fuel 86:2204–2212
20. Aaron D, Tsouris C (2005) Separation of CO_2 from flue gas: a review. Sep Sci Technol 40:321–348
21. Ding Y, Alpay E (2000) Equilibria and kinetics of CO_2 adsorption on hydrotalcite adsorbent. Chem Eng Sci 55:3461–3474
22. Yong Z, Mata V, Rodriguez AE (2001) Adsorption of carbon dioxide onto hydrotalcite-like compounds (HTlcs) at high temperatures. Ind Eng Chem Res 40:204–209
23. Iyer MV, Gupta H, Sakadjian BB, Fan LS (2004) Multicyclic study on the simultaneous carbonation and sulfation of high-reactivity CaO. Ind Eng Chem Res 43:3939–3947

24. Zelenak V, Badanicova M, Halamova D, Cejka J, Zukal A, Murafa N, Goerigk G (2008) Amine-modified ordered mesoporous silica: effect of pore size on carbon dioxide capture. Chem Eng J 144:336–342
25. Son WJ, Choi JS, Ahn WS (2008) Adsorptive removal of carbon dioxide using polyethyleneimine-loaded mesoporous silica materials. Microporous Mesoporous Mater 113:31–40
26. Siriwardane RV, Shen MS, Fisher EP (2003) Adsorption of CO_2, N_2, and O_2 on natural zeolites. Energy Fuels 17:571–576
27. Takamura Y, Narita S, Aoki J, Hironaka S, Uchida S (2001) Evaluation of dual-bed pressure swing adsorption for CO_2 recovery from boiler exhaust gas. Sep Purif Technol 24:519–528
28. Hiyoshi N, Yogo K, Yashima T (2005) Adsorption characteristics of carbon dioxide on organically functionalized SBA-15. Microporous Mesoporous Mater 84:357–365
29. Gray ML, Soong Y, Champagne KJ, Pennline H, Baltrus JP, Stevens RW, Khatri R, Chuang SSC, Filburn T (2005) Improved immobilized carbon dioxide capture sorbents. Fuel Process Technol 86:1449–1455
30. Huang LL, Zhang LZ, Shao Q, Lu LH, Lu XH, Jiang SY, Shen WF (2007) Simulations of binary mixture adsorption of carbon dioxide and methane in carbon nanotubes: temperature, pressure, and pore size effects. J Phys Chem C 111:11912–11920
31. Razavi SS, Hashemianzadeh SM, Karimi H (2011) Modeling the adsorptive selectivity of carbon nanotubes for effective separation of CO_2/N_2 mixtures. J Mol Model 17:1163–1172
32. Zhang ZJ, Zhao YG, Gong QH, Li Z, Li J (2013) MOFs for CO_2 capture and separation from flue gas mixtures: the effect of multifunctional sites on their adsorption capacity and selectivity. Chem Commun 49:653–661
33. Torrisi A, Bell RG, Mellot-Draznieks C (2010) Functionalized MOFs for enhanced CO_2 capture. Cryst Growth Des 10:2839–2841
34. Gonzalez-Zamora E, Ibrra IA (2017) CO_2 capture under humid conditions in metal–organic frameworks. Mater Chem Front 1:1471–1484
35. Yu CH, Huang CH, Tan CS (2012) A review of CO_2 capture by absorption and adsorption. Aerosol Air Qual Res 12:745–769
36. Hicks JC, Drese JH, Fauth DJ, Gray ML, Qi GG, Jones CW (2008) Designing adsorbents for CO_2 capture from flue gas-hyperbranched aminosilicas capable, of capturing CO_2 reversibly. J Am Chem Soc 130:2902–2903
37. Rosenholm JM, Linden M (2007) Wet-chemical analysis of surface concentration of accessible groups on different amino-functionalized mesoporous SBA-15 silicas. Chem Mat 19:5023–5034
38. Rosenholm JM, Penninkangas A, Linden M (2006) Amino-functionalization of large-pore mesoscopically ordered silica by a one-step hyperbranching polymerization of a surface-grown polyethyleneimine. Chem Commun 37:3909–3911
39. Tsuda T, Fujiwara T (1992) Polyethyleneimine and macrocyclic polyamine silica-gels acting as carbon-dioxide absorbents. J Chem Soc Chem Commun 22:1659–1661
40. Tsuda T, Fujiwara T, Taketani Y, Saegusa T (1992) Amino silica-gels acting as a carbon-dioxide absorbent. Chem Lett 21:2161–2164
41. Kumar P, Guliants VV (2010) Periodic mesoporous organic-inorganic hybrid materials: applications in membrane separations and adsorption. Microporous Mesoporous Mater 132:1–14
42. Belmabkhout Y, Sayari A (2009) Effect of pore expansion and amine functionalization of mesoporous silica on CO_2 adsorption over a wide range of conditions. Adsorpt J Int Adsorpt Soc 15:318–328
43. Zelenak V, Halamova D, Gaberova L, Bloch E, Llewellyn P (2008) Amine-modified SBA-12 mesoporous silica for carbon dioxide capture: effect of amine basicity on sorption properties. Microporous Mesoporous Mater 116:358–364

44. Huang HY, Yang RT, Chinn D, Munson CL (2003) Amine-grafted MCM-48 and silica xerogel as superior sorbents for acidic gas removal from natural gas. Ind Eng Chem Res 42:2427–2433
45. Hiyoshi N, Yogo K, Yashima T (2004) Adsorption of carbon dioxide on amine modified SBA-15 in the presence of water vapor. Chem Lett 33:510–511
46. Wang YM, Wu ZY, Shi LY, Zhu JH (2005) Rapid functionalization of mesoporous materials: directly dispersing metal oxides into as-prepared SBA-15 occluded with template. Adv Mater 17:323–327
47. Yue MB, Chun Y, Cao Y, Dong X, Zhu JH (2006) CO_2 capture by as-prepared SBA-15 with an occluded organic template. Adv Funct Mater 16:1717–1722
48. Qi GG, Fu LL, Choi BH, Giannelis EP (2012) Efficient CO_2 sorbents based on silica foam with ultra-large mesopores. Energy Environ Sci 5:7368–7375
49. Liang Z, Fadhel B, Schneider CJ, Chaffee AL (2008) Stepwise growth of melamine-based dendrimers into mesopores and their CO_2 adsorption properties. Microporous Mesoporous Mater 111:536–543
50. Xu XC, Song CS, Andresen JM, Miller BG, Scaroni AW (2003) Preparation and characterization of novel CO_2 "molecular basket" adsorbents based on polymer-modified mesoporous molecular sieve MCM-41. Microporous Mesoporous Mater 62:29–45
51. Xu XC, Song CS, Miller BG, Scaroni AW (2005) Influence of moisture on CO_2 separation from gas mixture by a nanoporous adsorbent based on polyethylenimine-modified molecular sieve MCM-41. Ind Eng Chem Res 44:8113–8119
52. Ma XL, Wang XX, Song CS (2009) "Molecular basket" sorbents for separation of CO_2 and H_2S from various gas streams. J Am Chem Soc 131:5777–5783
53. Zhang ZH, Ma XL, Wang DX, Song CS, Wang YG (2012) Development of silica-gel-supported polyethylenimine sorbents for CO_2 capture from flue gas. AIChE J 58:2495–2502
54. Yang SB, Zhan L, Xu XY, Wang YL, Ling LC, Feng XL (2013) Graphene-based porous silica sheets impregnated with polyethyleneimine for superior CO_2 capture. Adv Mater 25:2130–2134
55. Tanthana J, Chuang SSC (2010) In situ infrared study of the role of PEG in stabilizing silica-supported amines for CO_2 capture. ChemSusChem 3:957–964
56. Liu YM, Shi JJ, Chen J, Ye Q, Pan H, Shao ZH, Shi Y (2010) Dynamic performance of CO_2 adsorption with tetraethylenepentamine-loaded KIT-6. Microporous Mesoporous Mater 134:16–21
57. Wang DX, Sentorun-Shalaby C, Ma XL, Song CS (2011) High-capacity and low-cost carbon-based "molecular basket" sorbent for CO_2 capture from flue gas. Energy Fuels 25:456–458
58. Chen C, Son WJ, You KS, Ahn JW, Ahn WS (2010) Carbon dioxide capture using amine-impregnated HMS having textural mesoporosity. Chem Eng J 161:46–52
59. Xu XC, Song CS, Andresen JM, Miller BG, Scaroni AW (2002) Novel polyethylenimine-modified mesoporous molecular sieve of MCM-41 type as high-capacity adsorbent for CO_2 capture. Energy Fuels 16:1463–1469
60. Wang XX, Ma XL, Song CS, Locke DR, Siefert S, Winans RE, Mollmer J, Lange M, Moller A, Glaser R (2013) Molecular basket sorbents polyethylenimine-SBA-15 for CO_2 capture from flue gas: characterization and sorption properties. Microporous Mesoporous Mater 169:103–111
61. Choi S, Drese JH, Jones CW (2009) Adsorbent materials for carbon dioxide capture from large anthropogenic point sources. ChemSusChem 2:796–854
62. D'Alessandro DM, Smit B, Long JR (2010) Carbon dioxide capture: prospects for new materials. Angew Chem Int Ed 49:6058–6082
63. Lin YC, Kong CL, Chen L (2016) Amine-functionalized metal-organic frameworks: structure, synthesis and applications. RSC Adv 6:32598–32614

64. Didas SA, Choi S, Chaikittisilp W, Jones CW (2015) Amine-oxide hybrid materials for CO_2 capture from ambient air. Acc Chem Res 48:2680–2687
65. Dutcher B, Fan MH, Russell AG (2015) Amine-based CO_2 capture technology development from the beginning of 2013—a review. ACS Appl Mater Interfaces 7:2137–2148
66. Chen C, Kim J, Ahn WS (2014) CO_2 capture by amine-functionalized nanoporous materials: a review. Korean J Chem Eng 31:1919–1934
67. Gargiulo N, Pepe F, Caputo D (2014) CO_2 adsorption by functionalized nanoporous materials: a review. J Nanosci Nanotechnol 14:1811–1822
68. Olajire AA (2017) Synthesis of bare and functionalized porous adsorbent materials for CO_2 capture. Greenh Gas 7:399–459
69. Yue MB, Sun LB, Cao Y, Wang Y, Wang ZJ, Zhu JH (2008) Efficient CO_2 capturer derived from As-synthesized MCM-41 modified with amine. Chem Eur J 14:3442–3451
70. Goeppert A, Czaun M, May RB, Prakash GKS, Olah GA, Narayanan SR (2011) Carbon dioxide capture from the air using a polyamine based regenerable solid adsorbent. J Am Chem Soc 133:20164–20167
71. Heydari-Gorji A, Belmabkhout Y, Sayari A (2011) Polyethylenimine-impregnated meso-porous silica: effect of amine loading and surface alkyl chains on CO_2 adsorption. Langmuir 27:12411–12416
72. Cogswell CF, Jiang H, Ramberger J, Accetta D, Willey RJ, Choi S (2015) Effect of pore structure on CO_2 adsorption characteristics of aminopolymer impregnated MCM-36. Langmuir 31:4534–4541
73. Gargiulo N, Peluso A, Aprea P, Pepe F, Caputo D (2014) CO_2 adsorption on polyethylenimine-functionalized SBA-15 mesoporous silica: isotherms and modeling. J Chem Eng Data 59:896–902
74. Heydari-Gorji A, Yang Y, Sayari A (2011) Effect of the pore length on CO_2 adsorption over amine-modified mesoporous silicas. Energy Fuels 25:4206–4210
75. Vilarrasa-García E, Cecilia J, Moya E, Cavalcante C, Azevedo D, Rodríguez-Castellón E (2015) "Low cost" pore expanded SBA-15 functionalized with amine groups applied to CO_2 adsorption. Materials 8:2495
76. Sanz R, Calleja G, Arencibia A, Sanz-Pérez ES (2010) CO_2 adsorption on branched polyethyleneimine-impregnated mesoporous silica SBA-15. Appl Surf Sci 256:5323–5328
77. Kishor R, Ghoshal AK (2016) High molecular weight polyethyleneimine functionalized three dimensional mesoporous silica for regenerable CO_2 separation. Chem Eng J 300:236–244
78. Wang DX, Wang XX, Ma XL, Fillerup E, Song CS (2014) Three-dimensional molecular basket sorbents for CO_2 capture: Effects of pore structure of supports and loading level of polyethylenimine. Catal Today 233:100–107
79. Wang XX, Song CS, Gaffney AM, Song RZ (2014) New molecular basket sorbents for CO_2 capture based on mesoporous sponge-like TUD-1. Catal Today 238:95–102
80. Vilarrasa-Garcia E, Moya EMO, Cecilia JA, Cavalcante CL, Jiménez-Jiménez J, Azevedo DCS, Rodríguez-Castellón E (2015) CO_2 adsorption on amine modified mesoporous silicas: effect of the progressive disorder of the honeycomb arrangement. Microporous Mesoporous Mater 209:172–183
81. Khader MM, Al-Marri MJ, Ali S, Qi G, Giannelis EP (2015) Adsorption of CO_2 on polyethyleneimine 10 k-mesoporous silica sorbent: XPS and TGA studies. Am J Anal Chem 6:11
82. Chen C, Yang S-T, Ahn W-S, Ryoo R (2009) Amine-impregnated silica monolith with a hierarchical pore structure: enhancement of CO_2 capture capacity. Chem. Commun. 24:3627–3629
83. Han Y, Hwang G, Kim H, Haznedaroglu BZ, Lee B (2015) Amine-impregnated millimeter-sized spherical silica foams with hierarchical mesoporous–macroporous structure for CO_2 capture. Chem Eng J 259:653–662

84. Goeppert A, Meth S, Prakash GKS, Olah GA (2010) Nanostructured silica as a support for regenerable high-capacity organoamine-based CO_2 sorbents. Energy Environ Sci 3:1949–1960
85. Ebner AD, Gray ML, Chisholm NG, Black QT, Mumford DD, Nicholson MA, Ritter JA (2011) Suitability of a solid amine sorbent for CO_2 capture by pressure swing adsorption. Ind Eng Chem Res 50:5634–5641
86. Monazam ER, Shadle LJ, Miller DC, Pennline HW, Fauth DJ, Hoffman JS, Gray ML (2013) Equilibrium and kinetics analysis of carbon dioxide capture using immobilized amine on a mesoporous silica. AIChE J 59:923–935
87. Li K, Jiang J, Tian S, Yan F, Chen X (2015) Polyethyleneimine–nano silica composites: a low-cost and promising adsorbent for CO_2 capture. J Mater Chem A 3:2166–2175
88. Li K, Jiang J, Yan F, Tian S, Chen X (2014) The influence of polyethyleneimine type and molecular weight on the CO_2 capture performance of PEI-nano silica adsorbents. Appl Energy 136:750–755
89. Minju N, Abhilash P, Nair BN, Mohamed AP, Ananthakumar S (2015) Amine impregnated porous silica gel sorbents synthesized from water–glass precursors for CO_2 capturing. Chem Eng J 269:335–342
90. Zhang L, Zhan N, Jin Q, Liu H, Hu J (2016) Impregnation of polyethylenimine in mesoporous multilamellar silica vesicles for CO_2 capture: a kinetic study. Ind Eng Chem Res 55:5885–5891
91. Chen C, Bhattacharjee S (2017) Trimodal nanoporous silica as a support for amine-based CO_2 adsorbents: improvement in adsorption capacity and kinetics. Appl Surf Sci 396:1515–1519
92. Chen Z, Deng S, Wei H, Wang B, Huang J, Yu G (2013) Polyethylenimine-impregnated resin for high CO_2 adsorption: an efficient adsorbent for CO_2 capture from simulated flue gas and ambient air. ACS Appl Mater Interfaces 5:6937–6945
93. Wang D, Ma X, Sentorun-Shalaby C, Song C (2012) Development of carbon-based "molecular basket" sorbent for CO_2 capture. Ind Eng Chem Res 51:3048–3057
94. Wang J, Wang M, Zhao B, Qiao W, Long D, Ling L (2013) Mesoporous carbon-supported solid amine sorbents for low-temperature carbon dioxide capture. Ind Eng Chem Res 52:5437–5444
95. Zhang H, Goeppert A, Prakash GKS, Olah G (2015) Applicability of linear polyethylenimine supported on nano-silica for the adsorption of CO_2 from various sources including dry air. RSC Adv 5:52550–52562
96. Yan W, Tang J, Bian Z, Hu J, Liu H (2012) Carbon dioxide capture by amine-impregnated mesocellular-foam-containing template. Ind Eng Chem Res 51:3653–3662
97. Niu M, Yang H, Zhang X, Wang Y, Tang A (2016) Amine-impregnated mesoporous silica nanotube as an emerging nanocomposite for CO_2 capture. ACS Appl Mater Interfaces 8:17312–17320
98. Subagyono DJN, Liang Z, Knowles GP, Chaffee AL (2011) Amine modified mesocellular siliceous foam (MCF) as a sorbent for CO_2. Chem Eng Res Des 89:1647–1657
99. Subagyono DJN, Marshall M, Knowles GP, Chaffee AL (2014) CO_2 adsorption by amine modified siliceous mesostructured cellular foam (MCF) in humidified gas. Microporous Mesoporous Mater 186:84–93
100. Le Thi MU, Lee S-Y, Park S-J (2014) Preparation and characterization of PEI-loaded MCM-41 for CO_2 capture. Int J Hydrogen Energy 39:12340–12346
101. Yan X, Zhang L, Zhang Y, Yang G, Yan Z (2011) Amine-modified SBA-15: effect of pore structure on the performance for CO_2 capture. Ind Eng Chem Res 50:3220–3226
102. Yan X, Zhang L, Zhang Y, Qiao K, Yan Z, Komarneni S (2011) Amine-modified mesocellular silica foams for CO_2 capture. Chem Eng J 168:918–924
103. Zeng W, Bai H (2016) High-performance CO_2 capture on amine-functionalized hierarchically porous silica nanoparticles prepared by a simple template-free method. Adsorption 22:117–127

104. Liu X, Gao F, Xu J, Zhou L, Liu H, Hu J (2016) Zeolite@Mesoporous silica-supported-amine hybrids for the capture of CO_2 in the presence of water. Microporous Mesoporous Mater 222:113–119
105. Ma J, Liu Q, Chen D, Wen S, Wang T (2014) CO_2 adsorption on amine-modified mesoporous silicas. J Porous Mater 21:859–867
106. Li W, Bollini P, Didas SA, Choi S, Drese JH, Jones CW (2010) Structural changes of silica mesocellular foam supported amine-functionalized CO_2 adsorbents upon exposure to steam. ACS Appl Mater Interfaces 2:3363–3372
107. Liu Z, Pudasainee D, Liu Q, Gupta R (2015) Post-combustion CO_2 capture using polyethyleneimine impregnated mesoporous cellular foams. Sep Purif Technol 156:259–268
108. Qi G, Wang Y, Estevez L, Duan X, Anako N, Park A-HA, Li W, Jones CW, Giannelis EP (2011) High efficiency nanocomposite sorbents for CO_2 capture based on amine-functionalized mesoporous capsules. Energy Environ Sci 4:444–452
109. Sandhu NK, Pudasainee D, Sarkar P, Gupta R (2016) Steam regeneration of polyethylenimine-impregnated silica sorbent for postcombustion CO_2 capture: a multicyclic study. Ind Eng Chem Res 55:2210–2220
110. Chaikittisilp W, Kim H-J, Jones CW (2011) Mesoporous alumina-supported amines as potential steam-stable adsorbents for capturing CO_2 from simulated flue gas and ambient air. Energy Fuels 25:5528–5537
111. Chaikittisilp W, Khunsupat R, Chen TT, Jones CW (2011) Poly(allylamine)–mesoporous silica composite materials for CO_2 capture from simulated flue gas or ambient air. Ind Eng Chem Res 50:14203–14210
112. Zhang W, Liu H, Sun C, Drage TC, Snape CE (2014) Capturing CO_2 from ambient air using a polyethyleneimine–silica adsorbent in fluidized beds. Chem Eng Sci 116:306–316
113. Cai H, Bao F, Gao J, Chen T, Wang S, Ma R (2015) Preparation and characterization of novel carbon dioxide adsorbents based on polyethylenimine-modified Halloysite nanotubes. Environ Technol 36:1273–1280
114. Zhang L, Wang X, Fujii M, Yang L, Song C (2017) CO_2 capture over molecular basket sorbents: effects of SiO_2 supports and PEG additive. J Energy Chem 26:1030–1038
115. Wang XX, Song CS (2012) Temperature-programmed desorption of CO_2 from polyethylenimine-loaded SBA-15 as molecular basket sorbents. Catal Today 194:44–52
116. Alkhabbaz MA, Khunsupat R, Jones CW (2014) Guanidinylated poly(allylamine) supported on mesoporous silica for CO_2 capture from flue gas. Fuel 121:79–85
117. Bali S, Chen TT, Chaikittisilp W, Jones CW (2013) Oxidative stability of amino polymer-alumina hybrid adsorbents for carbon dioxide capture. Energy Fuels 27:1547–1554
118. Wang D, Wang X, Song C (2017) Comparative study of molecular basket sorbents consisting of polyallylamine and polyethylenimine functionalized SBA-15 for CO_2 capture from flue gas. ChemPhysChem 18:3163–3173
119. Arenillas A, Smith KM, Drage TC, Snape CE (2005) CO_2 capture using some fly ash-derived carbon materials. Fuel 84:2204–2210
120. Meth S, Goeppert A, Prakash GKS, Olah GA (2012) Silica nanoparticles as supports for regenerable CO_2 sorbents. Energy Fuels 26:3082–3090
121. Sakwa-Novak MA, Tan S, Jones CW (2015) Role of additives in composite PEI/oxide CO_2 adsorbents: enhancement in the amine efficiency of supported PEI by PEG in CO_2 capture from simulated ambient air. ACS Appl Mater Interfaces 7:24748–24759
122. Wang J, Long D, Zhou H, Chen Q, Liu X, Ling L (2012) Surfactant promoted solid amine sorbents for CO_2 capture. Energy Environ Sci 5:5742–5749
123. Choi W, Min K, Kim C, Ko YS, Jeon JW, Seo H, Park Y-K, Choi M (2016) Epoxide-functionalization of polyethyleneimine for synthesis of stable carbon dioxide adsorbent in temperature swing adsorption. Nat Commun 7:12640
124. Min K, Choi W, Kim C, Choi M (2018) Oxidation-stable amine-containing adsorbents for carbon dioxide capture. Nat Commun 9:726

125. Wang XX, Song CS (2014) New strategy to enhance CO_2 capture over a nanoporous polyethylenimine sorbent. Energy Fuels 28:7742–7745
126. Pinto ML, Mafra L, Guil JM, Pires J, Rocha J (2011) Adsorption and activation of CO_2 by amine-modified nanoporous materials studied by solid-state NMR and $^{13}CO_2$ adsorption. Chem Mat 23:1387–1395
127. Mebane DS, Kress JD, Storlie CB, Fauth DJ, Gray ML, Li K (2013) Transport, zwitterions, and the role of water for CO_2 adsorption in mesoporous silica-supported amine sorbents. J Phys Chem C 117:26617–26627
128. Didas SA, Sakwa-Novak MA, Foo GS, Sievers C, Jones CW (2014) Effect of amine surface coverage on the Co-adsorption of CO_2 and water: spectral deconvolution of adsorbed species. J Phys Chem Lett 5:4194–4200
129. Yu J, Chuang SSC (2016) The structure of adsorbed species on immobilized amines in CO_2 capture: an in situ IR study. Energy Fuels 30:7579–7587
130. Shen XH, Du HB, Mullins RH, Kommalapati RR (2017) Polyethylenimine applications in carbon dioxide capture and separation: from theoretical study to experimental work. Energy Technol 5:822–833
131. Li KM, Jiang JG, Tian SC, Chen XJ, Yan F (2014) Influence of silica types on synthesis and performance of amine-silica hybrid materials used for CO_2 capture. J Phys Chem C 118:2454–2462
132. Wang XX, Schwartz V, Clark JC, Ma XL, Overbury SH, Xu XC, Song CS (2009) Infrared study of CO_2 sorption over "molecular basket" Sorbent Consisting of Polyethylenimine-Modified Mesoporous Molecular Sieve. J Phys Chem C 113:7260–7268
133. Wang XX, Ma XL, Schwartz V, Clark JC, Overbury SH, Zhao SQ, Xu XC, Song CS (2012) A solid molecular basket sorbent for CO_2 capture from gas streams with low CO_2 concentration under ambient conditions. Phys Chem Chem Phys 14:1485–1492
134. Holewinski A, Sakwa-Novak MA, Jones CW (2015) Linking CO_2 sorption performance to polymer morphology in aminopolymer/silica composites through neutron scattering. J Am Chem Soc 137:11749–11759
135. Holewinski A, Sakwa-Novak MA, Carrillo J-MY, Potter ME, Ellebracht N, Rother G, Sumpter BG, Jones CW (2017) Aminopolymer mobility and support interactions in silica-PEI composites for CO_2 capture applications: a quasielastic neutron scattering study. J Phys Chem B 121:6721–6731
136. Heydari-Gorji A, Sayari A (2011) CO_2 capture on polyethylenimine-impregnated hydrophobic mesoporous silica: experimental and kinetic modeling. Chem Eng J 173:72–79
137. Bollini P, Didas SA, Jones CW (2011) Amine-oxide hybrid materials for acid gas separations. J Mater Chem 21:15100–15120
138. Monazam ER, Shadle LJ, Siriwardane R (2011) Equilibrium and absorption kinetics of carbon dioxide by solid supported amine sorbent. AIChE J 57:3153–3159
139. Lagergren S (1898) About the theory of so-called adsorption of soluble substances. K Sven Vetenskapsakad Handl 24:1–39
140. Yuh-Shan H (2004) Citation review of Lagergren kinetic rate equation on adsorption reactions. Scientometrics 59:171–177
141. Ho Y-S (2006) Review of second-order models for adsorption systems. J Hazard Mater 136:681–689
142. Ho YS, McKay G (1999) Pseudo-second order model for sorption processes. Process Biochem 34:451–465
143. Avrami M (1940) Kinetics of phase change. II transformation-time relations for random distribution of nuclei. J Chem Phys 8:212–224
144. Low MJD (1960) Kinetics of chemisorption of gases on solids. Chem Rev 60:267–312
145. Serna-Guerrero R, Sayari A (2010) Modeling adsorption of CO_2 on amine-functionalized mesoporous silica. 2: kinetics and breakthrough curves. Chem Eng J 161:182–190
146. Bollini P, Brunelli NA, Didas SA, Jones CW (2012) Dynamics of CO_2 adsorption on amine adsorbents. 2. Insights into adsorbent design. Ind Eng Chem Res 51:15153–15162

147. Jung W, Park J, Lee KS (2018) Kinetic modeling of CO_2 adsorption on an amine-functionalized solid sorbent. Chem Eng Sci 177:122–131
148. Meng Y, Jiang J, Gao Y, Yan F, Liu N, Aihemaiti A (2018) Comprehensive study of CO_2 capture performance under a wide temperature range using polyethyleneimine-modified adsorbents. J CO_2 Utiliz 27:89–98
149. Andreoli E, Cullum L, Barron AR (2015) Carbon dioxide absorption by polyethylenimine-functionalized nanocarbons: a kinetic study. Ind Eng Chem Res 54:878–889
150. Al-Marri MJ, Kuti YO, Khraisheh M, Kumar A, Khader MM (2017) Kinetics of CO_2 adsorption/desorption of polyethyleneimine-mesoporous silica. Chem Eng Technol 40:1802–1809
151. Loganathan S, Tikmani M, Mishra A, Ghoshal AK (2016) Amine tethered pore-expanded MCM-41 for CO_2 capture: experimental, isotherm and kinetic modeling studies. Chem Eng J 303:89–99
152. Sparks DL (1989) Kinetics of soil chemical process. Academic Press, New York
153. Sparks DL (1998) Kinetics and mechanisms of chemical reactions at the soil mineral/water interface. In: Sparks DL (ed) Soil physical chemistry. CRC Press, pp 135–191
154. Crank J (1976) The mathematics of diffusion. Oxford University Press, London
155. Zhao ZX, Li Z, Lin YS (2009) Adsorption and diffusion of carbon dioxide on metal-organic framework (MOF-5). Ind Eng Chem Res 48:10015–10020
156. Stuckert NR, Yang RT (2011) CO_2 capture from the atmosphere and simultaneous concentration using zeolites and amine-grafted SBA-15. Environ Sci Technol 45:10257–10264
157. Drage TC, Arenillas A, Smith KM, Snape CE (2008) Thermal stability of polyethylenimine based carbon dioxide adsorbents and its influence on selection of regeneration strategies. Microporous Mesoporous Mater 116:504–512
158. Sayari A, Belmabkhout Y (2010) Stabilization of amine-containing CO_2 adsorbents: dramatic effect of water vapor. J Am Chem Soc 132:6312–6314
159. Sayari A, Heydari-Gorji A, Yang Y (2012) CO_2-induced degradation of amine-containing adsorbents: reaction products and pathways. J Am Chem Soc 134:13834–13842
160. Sayari A, Belmabkhout Y, Da'na E (2012) CO_2 deactivation of supported amines: does the nature of amine matter? Langmuir 28:4241–4247
161. Hammache S, Hoffman JS, Gray ML, Fauth DJ, Howard BH, Pennline HW (2013) Comprehensive study of the impact of steam on polyethyleneimine on silica for CO_2 capture. Energy Fuels 27:6899–6905
162. Bollini P, Choi S, Drese JH, Jones CW (2011) Oxidative degradation of aminosilica adsorbents relevant to postcombustion CO_2 capture. Energy Fuels 25:2416–2425
163. Calleja G, Sanz R, Arencibia A, Sanz-Pérez ES (2011) Influence of drying conditions on amine-functionalized SBA-15 as adsorbent of CO_2. Top Catal 54:135–145
164. Ahmadalinezhad A, Sayari A (2014) Oxidative degradation of silica-supported polyethylenimine for CO_2 adsorption: insights into the nature of deactivated species. Phys Chem Chem Phys 16:1529–1535
165. Chi S, Rochelle GT (2002) Oxidative degradation of monoethanolamine. Ind Eng Chem Res 41:4178–4186
166. Khatri RA, Chuang SSC, Soong Y, Gray M (2006) Thermal and chemical stability of regenerable solid amine sorbent for CO_2 capture. Energy Fuels 20:1514–1520
167. Xu X, Song C, Miller BG, Scaroni AW (2005) Adsorption separation of carbon dioxide from flue gas of natural gas-fired boiler by a novel nanoporous "molecular basket" adsorbent. Fuel Process Technol 86:1457–1472
168. Fan Y, Labreche Y, Lively RP, Jones CW, Koros WJ (2014) Dynamic CO_2 adsorption performance of internally cooled silica-supported poly(ethylenimine) hollow fiber sorbents. AIChE J 60:3878–3887

169. Rezaei F, Jones CW (2014) Stability of supported amine adsorbents to SO_2 and NO_x in postcombustion CO_2 capture. 2. Multicomponent adsorption. Ind Eng Chem Res 53:12103–12110

170. Sjostrom S, Krutka H, Starns T, Campbell T (2011) Pilot test results of post-combustion CO_2 capture using solid sorbents. Energy Proc 4:1584–1592

171. Zhao W, Veneman R, Chen D, Li Z, Cai N, Brilmana DWF (2014) Post-combustion CO_2 capture demonstration using supported amine sorbents: design and evaluation of 200 kWth pilot. Energy Proc 63:2374–2383

172. Nelson T, Kataria A, Soukri M, Farmer J, Mobley P, Tanthana J, Wang D, Wang X, Song C (2015) Bench-scale development of an advanced solid sorbent-based CO_2 capture process for coal-fired power plants. DOE report. https://www.osti.gov/servlets/purl/1301858

173. Nelson TO, Kataria A, Mobley P, Soukri M, Tanthana J (2017) RTI's solid sorbent-based CO_2 capture process: technical and economic lessons learned for application in coal-fired, NGCC, and cement plants. Energy Proc 114:2506–2524

174. Nelson TO, Coleman LJI, Kataria A, Lail M, Soukri M, Quang DV, Zahra MRMA (2014) Advanced solid sorbent-based CO_2 capture process. Energy Proc 63:2216–2229

175. Nelson TO, Coleman LJI, Mobley P, Kataria A, Tanthana J, Lesemann M, Bjerge L-M (2014) Solid sorbent CO_2 capture technology evaluation and demonstration at Norcem's cement plant in Brevik, Norway. Energy Proc 63:6504–6516

176. Song C, Xu X, Andresen JM, Miller BG, Scaroni AW (2004) Novel nanoporous "molecular basket" adsorbent for CO_2 capture. In: Park S-E, Chang J-S, Lee K-W (eds) Studies in surface science and catalysis. Elsevier, pp 411–416

177. Xu X, Song C, Andresen JM, Miller BG, Scaroni AW (2004) Adsorption separation of CO_2 from simulated flue gas mixtures by novel CO_2 "molecular basket" adsorbents. Int J Environ Technol Manage 4:32–52

178. National Academies of Sciences, Engineering, and Medicine (2018) Direct air capture and mineral carbonation approaches for carbon dioxide removal and reliable sequestration: proceedings of a workshop–in brief. The National Academies Press, Washington, DC

179. Kulkarni AR, Sholl DS (2012) Analysis of equilibrium-based TSA processes for direct capture of CO_2 from air. Ind Eng Chem Res 51:8631–8645

180. Choi S, Drese JH, Chance RR, Eisenberger PM, Jones CW (2013) Application of amine-tethered solid sorbents to CO_2 fixation from air. U.S. Patent 8491705 B2

181. Eisenberger PM, Chichilnisky G (2014) System and method for removing carbon dioxide from an atmosphere and global thermostat using the same. U.S. Patent 8894747 B2

182. Eisenberger PM (2012) Carbon dioxide capture/regeneration structures and techniques. U.S. Patent 8163066 B2

183. Eisenberger PM (2013) System and method for carbon dioxide capture and sequestration. U. S. Patent 8500855 B2

Technical and Industrial Applications of CO$_2$

Jan Vansant and Peter-Wilhem Koziel

Abstract
These last decades, carbon dioxide emissions have drawn a lot of attention because of the greenhouse effect. According to the Kyoto protocol, the 15 European countries committed themselves to reduce their carbon dioxide emissions by an average of 8% with respect to the 1990 level within the period 2008–2012. The more recent Paris Agreement [1] urges all parties to ratify and implement the second commitment period of this Kyoto Protocol up to 2020. In this context, the goal of this chapter is to analyze the technical and industrial applications of carbon dioxide as a possible contribution to CO$_2$ mitigation by supplanting less environmentally friendly technologies based on chemicals having a higher impact on soil, water and atmosphere. In this Chapter we make a distinction between different approaches on how and where carbon dioxide is being captured and used. The present text is an update of the previously published Chapter "Vansant J, Carbon Dioxide Emission and Merchant Market in the European Union." In: Aresta M. (eds) Carbon Dioxide Recovery and Utilization, 2003, 3–50, Springer, Dordrecht.

Chapter 3 is an update from the previously published chapter Vansant J. (2003) Carbon Dioxide Emission and Merchant Market in the European Union. In: Aresta M. (eds) Carbon Dioxide Recovery and Utilization. Springer, Dordrecht, pp. 3–50, https://doi.org/10.1007/978-94-017-0245-4_1.

J. Vansant (✉)
VFB Bvba, 3000 Leuven, Belgium
e-mail: vfb.nv@telenet.be

P.-W. Koziel
Alter Graben 8, D-33014 Bad Driburg, Germany

© Springer Nature Switzerland AG 2019
M. Aresta et al. (eds.), *An Economy Based on Carbon Dioxide and Water*,
https://doi.org/10.1007/978-3-030-15868-2_3

3.1 On-Site Industrial Large Scale Utilization of Gaseous CO_2

Several approaches to CO_2 conversion are at the industrial exploitation level or very close to it. While this is not part of this Chapter, we mention some of them to highlight the concept of "large scale" that refers to several Mt/y of used CO_2.

Mineralization [2] permanently binds CO_2 and produces valuable end-products: it is gaining more attention. This will be treated in more detail in Chap. 4 in this book.

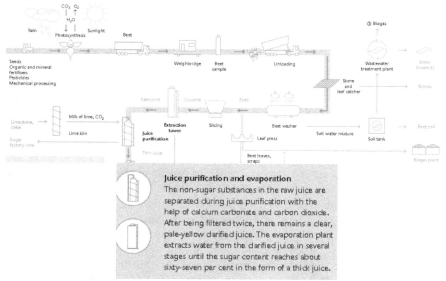

CO_2 is also used as an active component in the **sugar extraction from sugar beets**. A schematic summary of this process is shown below [3].

Methanol production: Mitsubishi Heavy Industries promotes the increased production rate from natural gas and steam by (re)injection of CO_2. More details are presented elsewhere in this Book.

Methionine production: The whole process has been running at Evonik-Degussa plants in Germany, in Antwerp and in Singapore (with Linde), for over 50 years and contributes with a capacity of 580,000 annual tons (Q4 2014) to 60% of the DL-methionine worldwide capacity of about 1 million annual tons.

However against the background of decreasing fossil resources and the stronger environmental constraints (hazardous intermediates and waste), more sustainable processes based on natural resources are gaining more and more interest [4].

PCC or **Precipitated Calcium Carbonate** is produced with CO_2 from flue-gases enriched with L-CO_2 up to proper concentration levels to yield $CaCO_3$ from CaO + CO_2. The total European PCC market exceeds 2 Mt/y, the majority of which is used in the paper industry [5]. There PCC acts as whitening agent thanks to the proper and uniform sizing of granules.

CO_2 and methanol can also be used to produce **dimethyl carbonate**, itself a precursor in the synthesis of polycarbonates.

3.2 CO_2 Liquefaction and Conditioning for On- or Off-site Use

In this subchapter we present the CO_2 production and product quality for the merchant market.

3.2.1 CO_2 Production for the Merchant Market

Carbon dioxide is a naturally occurring gas that exists in small amounts in the atmosphere (400 ppm). It also exists underground, either in a relatively pure state, or blended with water, natural gas or petroleum. With the exception of production from natural wells, manufacturing of carbon dioxide is a secondary operation integrated with the production of other products. Purification and liquefaction of carbon dioxide is capital and energy intensive. For the industrial merchant market, the major sources of by-product carbon dioxide are: (i) concentrated waste streams such as ammonia production, (ii) natural wells, (iii) fermentation during the production of alcohol, and (iv) chemical/petrochemical manufacturing (e.g. H_2 production). Together, these sources supply more than 85% of the carbon dioxide commercialised in the merchant market.

In ammonia and hydrogen plants, both hydrogen and carbon dioxide are generated from the steam cracking of hydrocarbons. In both cases, the CO_2 must be separated from the final gas stream. In many NH_3 plants, a large fraction of the CO_2 output is consumed captively in the production of urea, $(NH_2)_2CO$, which is used extensively as a fertilizer. Direct and production enhancement applications related to the urea production represent an estimated consumption of 50 Mt/y of CO_2.

A smaller quantity is purified and liquefied for sale on the merchant market as liquid CO_2. The growing uncertainty about the reliability of these traditionally very important sources for raw CO_2 gas has been well illustrated by the dramatic shortages of L-CO_2 during the summer of 2015 in the US and during late spring and the summer of 2018 in Western Europe [6] and in Mexico. During the last years also Japan has experienced severe shortages of L-CO_2 and dry ice in the early summer periods as up to 50% of all raw gas producers were out because of maintenance activities. Such shortages are mainly caused by the maintenance periods for ammonia plants planned during their low season, i.e. summertime, that, however, is the high season for L-CO_2 consumption. The US merchant market, estimated at over 10 Mt/y, suffers less from these outages as the sources for raw CO_2 gas are more diversified.

When manufactured by fermentation, carbon dioxide is a by-product of ethyl alcohol or bio-methane production. In contrast to the ammonia process, the fermentation process is carried out at relatively low temperatures so that cooling is not required. It is becoming clear that bio-ethanol and bio-methane sources are the future. New capacity is needed to support future demands. To reduce seasonality of supply, other waste streams are being considered such as the new TPI (Siad) plant in the Netherlands recovering CO_2 from the flue gases that are emitted by the waste-to-energy plant [7]. Other examples of this are linked to the Quad-Generation approach being promoted especially in regions with limited or unreliable CO_2 supply. Amongst others, Coca Cola has been implementing this in some of its bottling plants [8]. Recent developments have shown that it is feasible and affordable to capture CO_2 from the atmosphere for less than 100 US$/t. This is obviously not competitive in areas where sufficient CO_2-rich waste streams are available but attractive elsewhere [9]. Other initiatives such as www.NewCO2.eu try to fill the gap between available CO_2 production sites and ad hoc demands.

3.2.2 Product Quality

The CO_2 source determines the possible contaminants such as nitrogen- or sulphur-components and hydrocarbons that need to be removed for most applications. ISBT, CGA, EIGA define standards for purity in order to guarantee that for food-grade CO_2 all possible contaminants have been removed to a level where they will not influence taste or odour and where they will be harmless for the health of consumers. Other species such as water (humidity) and oxygen are also removed to a level, which is acceptable to the process requirements at the plants that subsequently process the L-CO_2. Quality control and quality assurance measures currently implemented by the producers of L-CO_2 for food and beverage applications are a lot more stringent than even 10 years ago. Detection levels for some of the possible contaminants now go as low as the ppb (parts per billion) threshold.

3.3 Carbon Dioxide Merchant Market

CO_2 finds many "technological" industrial applications. Being able to rely on guaranteed supply of good quality CO_2 is the key element for the successful development of industrial CO_2 applications. CO_2 can be used under different conditioning forms: liquid bulk, cylinders and dry ice. The applications can be divided in two categories: (i) those that temporarily use CO_2 but do not convert it into other chemicals (bubbles, cooling, …) and (ii) those that remove CO_2 from the atmosphere by converting it into other products (pH control of waste water and mineralisation as described here above…) which may have a quite different life-time: from months (urea), to decades (polymers) and centuries (inorganic carbonates). Today, the merchant market utilization of CO_2 (close to 220 Mt/y) represents less than 0.6% of the CO_2 released into the atmosphere (37 000 My/y). It is clear that, in order to make the utilisation option relevant for carbon dioxide mitigation, the amount of carbon dioxide used should be increased by at least one order of magnitude. Nevertheless, even if this limit is not reached in the short term, environmental benefits are obtained by the industrial implementation of existing and innovative technologies. This specific issue needs to be evaluated carefully as it may be very easy to underestimate its potential. Interesting to note is that for instance the German CO_2 industry started in 1885 with just 122 t of CO_2; in 1899 it was 15.000 t and in 1918 the CO_2 merchant market exceeded 33.000 t. Today we are around 1.0 Mt of CO_2 consumed just in Germany [10]. In 2012 the total consumption of CO_2 in Europe including the eastern part (Poland etc.) reached about 4.0 Mt/y. It is currently estimated to be above 6 Mt/y. Excluding Enhanced Oil Recovery and on-site uses of gaseous CO_2, the US merchant market is estimated to exceed 10 Mt/y (representing a value of 1.5 Billion US$) and the global market could well reach 30 Mt/y. The merchant market players (Industrial gases companies as well as international suppliers of CO_2 production and recovery equipment) are seeing lots of consolidations in the recent years. See for instance Air Liquide acquiring Airgas [11], and the 70 BUS$ Linde & Praxair merger. After the acquisition of ACP by Air Products, there will be only one single-product company left linked to CO_2, CARBO Kohlensäure in Germany. MHI (Mitsubishi Heavy Industries) now owns TNSC (Taiyo Nippon Sanso Corp) which itself owns Matheson, which acquired Continental Carbonic in 2014 and 4 spin-off plants from the Airgas-AL merger in the US. TNSC is also intending to acquire the European divestments linked to the Praxair-Linde merging. Together with CVC Capital Partners, the Messer Group is planning to acquire the US part to then become MG Industries.

CO$_2$ finds applications [12a–d] in the industry in nearly all sectors of activities. Indeed, CO_2 has a lot of properties (fizz, inert, chemical, acid, pressure, coolant, high density, electrically non-conducting, solvent, bacteriostatic, photosynthesis, pesticide leaving no residue, sublimation (not melting)) that makes it important for industrial application. We defined the following market segmentation.

3.3.1 Food Processing

Carbon Dioxide has use throughout the food processing industry from stunning livestock prior to slaughtering through to delivery to the store or to the household. CO_2 is suitable for use on nearly every foodstuff including red meat, poultry, sea food, fresh vegetables and bakery. Six main applications constitute this market segment, which in the USA accounts for more than 52% of the merchant CO_2 market (excluding captive on-site productions such as urea production and re-injection in oil-wells):

- Refrigeration: Chilling food products during transportation
- Modified Atmosphere Packaging (MAP/CAP)
- Freezing: Chilling food products during processing
- Supercritical Fluid Extraction
- Stunning: Slaughter of Swine, Stunning of Poultry.
- Decontamination of food stuff.

Food safety is an area of increasing importance for L-CO_2 customers. Food processors all over the world have implemented procedures for hazard analysis of critical control points, or HACCP, to identify and correct steps in the processing chain where foods can go off spec. Innovative CO_2-based cooling technologies are, literally, taking the heat off these trouble spots. In refrigeration, freezing and packaging applications, CO_2 competes with nitrogen. Both gases are non-toxic and leave no residue on/in the product, so the choice between the two is based upon cost and availability.

3.3.1.1   Refrigeration: Maintaining the "Cold Chain" for Food Products During Transportation

Liquid carbon dioxide is used to refrigerate food both prior to shipment and during the trip. It is sometimes used to pre-cool trailers equipped with conventional refrigeration systems. It is also used in complete carbon dioxide portable refrigeration systems. These can be fed with either dry ice or contain special boxes that will accumulate CO_2 snow, injected into them from a liquid supply tank. When L-CO_2 is directly injected, some of the liquid carbon dioxide also deposits as layers of snow directly on the food or in plastic containers at the top of the containers. Such snow gradually sublimates during shipment to provide additional refrigeration. The modest equipment costs associated with the use of carbon dioxide chilling and its rapid effect are the main factors behind carbon dioxide's popularity in this use. YARA-Thermoking [13] uses CO_2 evaporation instead of burning more fuel in order to refrigerate food transportation trucks and trailers. Although it is used in tandem with conventional refrigeration systems, fully carbon dioxide-based refrigeration systems are gaining market share at the expense of mechanical refrigeration systems due to both environmental and cost considerations. Carbon dioxide systems do not employ chlorofluorocarbon (CFC) refrigerants, which are being phased out due to environmental considerations. These systems also eliminate the emissions associated with the diesel fuel used to operate mechanical systems.

Additionally, carbon dioxide systems are silent and require little maintenance. Finally, carbon dioxide systems reach the target temperature more rapidly and hold it longer. Unfortunately the acceptance of this application seems to stagnate with virtually no growth since 2013; the main reason being the lack of availability of re-filling stations for L-CO_2. Although the general trend over the last decade has been toward the use of liquid carbon dioxide in food refrigeration, solid carbon dioxide (dry ice) has also retained some applications in the food industry (See Part 4 below).

3.3.1.2 Modified Atmosphere Packaging (MAP/CAP)

In packaging uses, gaseous carbon dioxide, generated from liquid, is used as an inert medium to prevent flavour loss, deterioration by oxidation, and bacterial growth in food. Carbon dioxide is used for packaging coffee and as an inert atmosphere in the processing and transporting of fruits, vegetables and cereals. Carbon dioxide is completely non-toxic and leaves no residue on the product. Carbon dioxide competes with nitrogen in this use, with the choice of gas determined by costs and regional availability. Modified or controlled atmosphere packaging (MAP/CAP), is focussed on the absolute control over processing and packaging conditions, and employs breathable flexible film in combination with gas flushing (carbon dioxide, nitrogen, or blends of the two) to significantly extend the shelf life of packaged foods. This reduces the need for preservatives and additives, as well as waste, extends the range of distribution for food growers and processors, and satisfies consumer demands for high quality, minimally processed, easy-to-prepare foods. This technology is still developing, but has already been applied to baked goods, meat and poultry, produce and prepared foods (e.g., lasagna and fresh pasta [14, 15]) as examples of optimized gas mixtures and their application of the shelf life of the packaged goods for food MAP [16]. Even if for the most common inert applications, nitrogen can be a substitute for CO_2, in the more advanced application CO_2 (or blends of CO_2 with other gases) provide specific properties which cannot be achieved by just substituting CO_2 with nitrogen. An interesting evolution is the addition of new sensations via this inert gas, such as e.g. the Freshline® Aroma MAP™ from Air Products.

Recommended gas mixtures for meat and meat products

Product	Gas mixture	Gas volume	Typical shelf-life		Storage temp.
		Product volume	Air	MAP	
Raw red meat	60–80% O_2 + 20–40% CO_2	100–200 ml 100 g meat	2–4 days	5–8 days	2–3 °C
Raw light poultry	40–100% CO_2 + 0–60% N_2	100–200 ml 100 g meat	4–7 days	16–21 days	2–3 °C
Raw dark poultry	70% O_2 + 30% CO_2	100–200 ml 100 g meat	3–5 days	7–14 days	2–3 °C
Sausages	20–30% CO_2 + 70–80% N_2	50–100 ml 100 g prod.	2–4 days	2–5 weeks	4–6 °C
Sliced cooked meat	30% CO_2 + 70% N_2	50–100 ml 100 g prod.	2–4 days	2–5 weeks	4–6 °C

Recommended gas mixtures for fish and seafood

Product	Gas mixture	Gas volume	Typical shelf-life		Storage temp.
		Product volume	Air	MAP	
Raw fish	40–90 % CO_2 + 10 % O_2 + 0–50 % N_2	200–300 ml 100 g fish	3–5 days	5–14 days	0–2 °C
Smoked Fish	40–60 % CO_2 + 40–60 % N_2	50–100 ml 100 g fish	15 days	30 days	0–3 °C
Cooked fish	30 % CO_2 + 70 % N_2	50–100 ml 100 g fish	7 days	30 days	0–3 °C
Prawns (peeled, cooked)	40 % CO_2 + 60 % N_2	50–100 ml 100 g prod.	7 days	21 days	4–6 °C

Recommended gas mixtures for dairy products

Product	Gas mixture	Gas volume	Typical shelf-life		Storage temp.
		Product volume	Air	MAP	
Hard cheese	80–100% CO_2 + 0–20% N_2	50–100 ml 100 g cheese	2–3 weeks	4–10 weeks	4–6 °C
Hard cheese, (sliced, grated)	40% CO_2 + 60% N_2	50–100 ml 100 g cheese	2–3 weeks	7 weeks	4–6 °C
Soft cheese	20–60% CO_2 + 40–80% N_2	50–100 ml 100 g cheese	8 days	21 days	4–6 °C
Yogurt	0–30% CO_2 + 70–100% N_2		10–14 days	22–25 days	4–6 °C

3.3.1.3 Freezing: Chilling Food Products During Processing

CO_2 can be used to control the temperature [17] of products during mixing by direct injection of liquid CO_2, or by use of snow horns introducing dry ice snow directly into the mixing bowl, as none of the CO_2 is retained in the mixture it can be much more convenient than using water or water ice to achieve the control. There are many uses for CO_2 in the freezing and chilling of foods, depending on the process requirement. The product can be chilled prior to packing, frozen for storage or crust frozen to allow better control of slicing operations. The principal cooling mechanism is the injection of pressurised liquid CO_2 that expands at atmospheric pressure to produce solid CO_2 at ~ -78 °C. As the latter sublimes, cools down the system. Forced movement of the cold gas provides additional cooling energy. Equipment using liquid CO_2 as cryogenic agent is available to suit batch or continuous processes. There are cabinets that accept product placed on trolleys, also many different types of tunnels available, these can be single or multi pass conveyors inside a freezing zone, or spiral freezers allowing a large belt length to be accommodated in a much smaller footprint. Cryogenic freezing systems take up less space inside a production facility as they do not require an associated fridge unit.

Dohmeyer – Tunnel for Hamburgers

Dohmeyer - Tumbler Detour 2018

IQF Freezing

Another area where CO_2 is ideal within the food freezing industry is IQF, Individual Quick Frozen: small pieces of product that are prone to sticking together such as Pizza toppings, are kept moving in rotating equipment as injected CO_2 is mixed with the product. Temperature is lowered as solid CO_2 sublimes allowing the cold gas to mix with the product, with freezing of the product whilst keeping it free flowing, This process also works well with coated products. Even more CO_2 technologies may move from field testing to food processing plants if governments mandate changes to avoid E. coli and other bacteria. For all these cooling and freezing applications, the use of CO_2 (liquid or solid) is in direct competition with either mechanical cooling or the use of liquid nitrogen (LIN). CO_2 can be also used to control the temperature, keeping the product in the ideal temperature zone for absorption by proteins, during massaging for brine addition used for flavour enhancement and preservation.

3.3.1.4   Supercritical Fluid Extractions (SFE)

A substance such as carbon dioxide is in a supercritical state when it exists at a temperature and pressure greater than its "critical point." In this state, carbon dioxide is not properly a gas or a liquid; rather, it has some qualities of both. This supercritical product has unique solvent properties, dissolving substances in a manner similar to a liquid solvent, although the process is more rapid due to its gas-like qualities. Unlike liquids (which are non-compressible), supercritical carbon dioxide becomes more dense as pressure is increased. Although other chemicals can be used for SFE applications, carbon dioxide is ideal because it is non-toxic, non-flammable, inexpensive and readily available. Additionally, carbon dioxide reaches a supercritical state at a temperature of only 31 °C. Supercritical fluid extraction processes, can be used in the decaffeination of coffee and the production of flavours, natural colours and essential oils. This method is popular in coffee decaffeination because it is non-toxic, does not employ solvents such as methylene chloride, and leaves no trace in the final product. Carbon dioxide SFE extraction is also gaining favour for other food extraction applications such as hop flavours [18]. Non-food applications of SC-CO_2 processing are described in § 3.3.2 below.

3.3.1.5   Stunning: Slaughter of Swine, Stunning of Poultry

Since many years, CO_2 has been used in slaughterhouses for stunning of pigs instead of using conventional electrical stunning. The use of CO_2 offers a wide variety of advantages, such as reduced animal stress, less injuries, etc., with a better meat quality and yield as result. For poultry, gas mixtures are used. CO_2 is also used to rapidly kill all birds (chicken or turkey) in contaminated farms in case of bird's deceases such as bird flu epidemics.

3.3.1.6   Decontamination of Foodstuff

In a process called PoroCrit^{SM}, liquid foods and liquid medicines are sterilized and preserved by contact with compressed CO_2 at room temperature. This process using membrane technology, which is the most effective way to bring CO_2 into contact with the liquids to be sterilized, was unfortunately abandoned after the death of the inventor. SC-CO_2 pasteurization is also gaining acceptance in the treatment of for instance carrot juice, apple juice or orange juice [19].

3.3.1.7   Inert Blanket or Propellant in Food Processing Applications

Neutral gases such as nitrogen and CO_2 are commonly used as a propellant to dispense food or beverage from its vessel into production lines and dosing units.

3.3.2 Carbonated Beverages

By far the dominant use of CO_2 is in carbonated soft drinks [20], although carbon dioxide is also used in the production of some types of sparkling wines and other carbonated beverages. Growth in carbon dioxide volumes basically follows soft drink consumption, although the introduction of new carbonated beverages, such as carbonated fruit juices and artificially carbonated sparkling wine, also contributes to market growth. Carbonated soft drinks continue to benefit from a general shift in consumer preference toward non-alcoholic beverages. Consumers currently favour beverages that are low in calories, cold, convenient and sweet. These trends explain the stagnation and/or decline in demand for certain traditional beverages, including tea, milk, coffee, beer and distilled spirits. Soft drink consumption is expected to moderate somewhat from strong growth over the last decade. Beverages are carbonated by use of a carbonator or saturator. Pressurized gas supplied to the carbonation machine is evaporated from L-CO_2. Water cooled to about 5 °C, pumped with gas to the top of the carbonator, flows over baffles under pressure where it is saturated with carbon dioxide. It is then ready for mixing with additives and bottling. Carbon dioxide requirements for soft drinks range from about 2.5 vol.% of gas per volume of beverage to 4.5 vol.% of gas per volume of liquid for highly carbonated beverages such as ginger beer and tonic waters. The total CO_2 volumes consumed by this industry is higher than this because of losses at the point of use by bottlers, use of carbon dioxide for non-carbonation operations such as cleaning and water treatment (see pH control), and the use of carbon dioxide for carbonating non-soft drink beverages such as wines and fruit juices not included in soft drink consumption figures. Furthermore, CO_2 is used in this industry as a propellant gas for emptying tanks and as a shielding gas for preserving the drinks quality (exclusion of oxygen).

Production of beers and ales does not require external carbon dioxide injection because, especially in large breweries, the brewing process generates adequate quantities of carbon dioxide and this is recovered and purified on-site for further use. Very large breweries in the Benelux and Germany for instance have drastically reduced their purchases of L-CO_2 on the merchant market over the last couple of decades and focussed on longer production runs and better utilisation of their own fermentation CO_2. Controlling dissolved oxygen concentration in beer and in other carbonated beverages is crucial for the shelf life. When dissolved oxygen concentration is too high, noticeable chemical and biochemical changes take place after packaging. Bottled beer becomes subject to oxidation, fermentation and microbiological spoilage. Tasting panels can easily recognize beers that have been oxidized. Clearly, the character of beer (clarity, color, odor and taste) is determined by the concentration of dissolved oxygen and by the quantity of oxygen in the headspace.

Therefore, beer should be protected from the atmospheric oxygen throughout processing and inside the closed package. During bottling and canning, this protection can be achieved by dosing a controlled quantity of dry ice into the empty bottle (before filling) or/and in the headspace (after filling). The dry ice sublimates and transfers into heavy CO_2 vapors that effectively purge the air out of the bottle and prevent ambient air to penetrate inside the bottle. The dosing of the dry ice is achieved by using liquid CO_2 dozers CARBO + designed by VBS Europe [21] that release a precise quantity of liquid CO_2 in each container, at the high speeds needed for industrial packaging plants. Because CO_2 is an efficient gas for modified atmosphere packaging, recently, this dosing of dry ice has also been applied to industrial packaging of other food products than beverages.

Liquid carbon dioxide used in beverages must be odourless and as pure as possible, as impurities may affect the taste of the beverage. Licensors of proprietary carbonated beverages attempt to maintain a uniform taste worldwide, and many licenses for carbonated beverages contain a clause stating that the "*license may be revoked without recourse by the licensee if taste satisfactory to the licensor is not maintained*". Supply of consistently high quality carbon dioxide to beverage producers is therefore critical. Many debates focus on the definition of "Natural CO_2" versus synthetic CO_2. Question then is if underground wells (consequence of fermentation processes which happened a very long time ago) generate more natural CO_2 than current fermentation processes such as ethanol or bio-methane production or should such wells be closed after all as they bring CO_2 into the atmosphere which was well stored underground for many, many years. A growing application in the beverage market is certainly the delivery of liquid CO_2 in so called mini-bulk systems. These fixed containers are placed inside the premises of a pub, restaurant or gas station shop and are filled directly from cryogenic tanks on distribution trucks. This CO_2 is then used to either pressurize beverage dispensing tools or to mix in situ the soft drinks starting from water and syrup. For dispensing beer from smaller kegs, Heineken has developed a system that does not require the presence of CO_2 (cylinders) at the dispensing location using a bladder-in-keg design [22].

A recent new tendency is represented by SODASTREAM [23]. Home-made sparkling water production represents already a market of 300 M€/y, mostly in Europe and especially in Germany and Northern Europe where over 2.2 million users consume over 11 million small cylinder each year. The importance of this is best illustrated by the recent acquisition of the SodaStream company by PepsiCo for 3.2 Bn $. It is also clear that the utilisation of CO$_2$ in this beverage markets is the main cause of the seasonality noticed in the L-CO$_2$ merchant market data; and there is no alternative.

3.3.3   Chemical Industry

3.3.3.1   CO$_2$ as Feedstock

As a feedstock, carbon dioxide is used for the production of basic lead carbonate (white lead), as well as sodium, potassium, ammonium and hydrogen carbonates and bicarbonates. Sodium salicylate (an intermediate in the manufacture of aspirin) is produced by the reaction of carbon dioxide with sodium phenate. Some sectors of the chemical industry are already trying to develop new chemical and biochemical processes, which use carbon dioxide in the production of high-value chemicals (vide infra). Such processes have the added benefit of avoiding potential environmental penalties for carbon dioxide emissions. If new or improved processes could be developed, carbon dioxide usage by the chemical industry could be increased by a factor of 10 or more. New approaches presently under development for using carbon dioxide include: production of algae-based biomass from power plant stack gas (see Chap. 12); pigment production with reduced toxic by-products; polycarbonate production from supercritical carbon dioxide; dimethylcarbonate synthesis from CO$_2$ and methanol; urethane production using carbon dioxide instead of phosgene; and the direct synthesis of gasoline from carbon dioxide." Many of these processes use the gaseous waste streams from industry directly.

European projects with a multitude of research institutes are looking even deeper into the possibility of using CO$_2$ as a feedstock to produce for instance: oxalic acid [24] or lactate and isobutene [25] as well as several routes to the reduction of CO$_2$ to CO, CH$_4$, CH$_3$OH [26]. More than 40 industrial and research stakeholders have recently joined forces to launch a new European association dedicated to CO$_2$ utilisation named "CO$_2$ Value Europe" (www.co2value.eu) About carbon dioxide utilization in the synthesis of chemicals, materials and fuels more can be found in Chaps. 4–12 below.

3.3.3.2   Supercritical Fluid Extractions

As in the food industry, carbon dioxide has an emerging use in the chemical industry for supercritical fluid distillation and the extraction of heat sensitive products. Carbon dioxide SFE is especially attractive for pharmaceutical production because of the lack of solvent residues in the product and the low temperature of the process. Cannabis extraction is a growing application in the US [27]. It can also be used to remove contaminants from chemical process streams. Supercritical carbon dioxide can be used to vaporize non-volatile substances at moderate temperatures, at which they are normally non-distillable. The ability of supercritical carbon dioxide to vaporize non-volatile compounds at moderate temperatures reduces the energy requirements relative to conventional extraction and distillation. Many other applications (such as cleaning of semiconductors or selective precipitation of "narrow-molecular weight distribution" fractions of polymer) have been developed using the unique properties of SC-CO_2.

3.3.3.3 Other Applications

Liquid and solid carbon dioxide are also used for direct injection into chemical reaction systems to control temperature. As an inert gas, carbon dioxide is used in the chemical industry to purge and fill reaction vessels, storage tanks and other equipment to prevent the formation of explosive gas mixtures and to protect easily oxidized chemicals from contact with air.

3.3.4 Metal Fabrication

3.3.4.1 Shielding Gas for Welding

The major use for carbon dioxide in the metals industry is for welding operations in private ships, on construction sites and in numerous types of manufacturing operations. Used either alone or with other gases such as Argon, carbon dioxide is used in the gas shielded arc as a shielding gas to protect the welding zone from the deleterious effects of oxygen, nitrogen and hydrogen. Carbon dioxide is one of only a few gases suitable for this purpose, and competes primarily with argon and helium. Helium provides the hottest arc and argon allows greater stability in AC (alternating current) arcs and provides a cleansing function on the welding surface. Although slightly less effective in this application than helium and argon, carbon dioxide provides major cost savings, and thus also finds wide use. When used as a shielding gas with semi-automatic micro-wire welding equipment, carbon dioxide provides welding speeds up to ten times faster than those obtainable with conventional equipment and eliminates the cleaning or wire brushing of the welds. Carbon dioxide used for welding must be in gaseous form; thus this is commercialized in cylinders that contain liquid CO_2 but deliver the gas in equilibrium with liquid at *ca.* 6 MPa. However, some large welding operations may purchase bulk liquid and convert it to gas at the point of use.

3.3.4.2 Foundries

For the production of sand moulds and cores for casting iron and other metals, CO_2 can be used in combination with other additives to form the binding agent between the sand grains. This application goes down rapidly: whereas foundries consumed 15% of the CO_2 market in Germany in 1970, this has gone down to below 1% today.

3.3.4.3 Separating Waste

Messer Group GmbH (Bad Soden, Germany) commercialises a CO_2 process for recycling sludge generated during metalworking. The waste consists of lubricant oil used for machining and metal fillings. When mixed with CO_2, the oil dissolves into the supercritical gas leaving behind clean metal particles. Both oil and metal are suitable for reuse, and the CO_2 itself is recaptured. Also contaminated soils can be cleaned this way.

3.3.5 Agriculture

3.3.5.1 Fumigate Grain Silos

Carbon dioxide is used as a non-toxic pesticide in grain silos. As a fumigant or insecticide, carbon dioxide kills insects by desiccation, and offers significant advantages over competitive chemical pesticides such as phosphine or methyl bromide. For instance, insects cannot develop immunity to it. Additionally, carbon dioxide leaves no residue, and thus calls for no special aeration procedures. Carbon dioxide is registered with the EPA as a non-restricted use pesticide. Within this application, carbon dioxide finds most of its use in the non-toxic fumigation of grain silos. In this use, liquid carbon dioxide is converted to a gas onsite, and then injected into the silo, where the gas displaces the existing atmosphere with one fatal to all stages of insect life. Because an atmosphere of 60% carbon dioxide is lethal to insects, the silo needs not be airtight for the method to be effective.

3.3.5.2 Additive to Irrigation Water

Carbon dioxide also may be added to irrigation water to enhance the absorption of nutrients by plants. In the latter use, carbon dioxide is effective because, although it has weak nutrient properties of its own, it acts as an adjunct to other nutrients by changing the pH of the soil, which can increase the nutrient absorption by plants.

3.3.5.3 Photosynthesis in Air: Additive to Greenhouse Atmosphere for Additional Plant Productivity and Consistent Quality [28]

Plants need water, light, warmth, nutrition and CO_2 to grow. By increasing the CO_2 level in the greenhouse atmosphere (typical to 600 ppm instead of normal 400 ppm value), the growth for some plants can be stimulated in an important way, with yield increases up to 20%, especially for tomato, cucumber, lettuce, strawberry, and also for potted plants and cut flowers. This is an important application in some parts of Europe, e.g. Benelux and Poland. A nice example is LINDE's OCAP [29] pipeline (85 km transport pipeline and 300 km distribution network) which distributes well over 450 kt/year of gaseous CO_2 to over 550 greenhouses linked to it. This gaseous CO_2 originates from a Shell refinery and the AlcoBioFuel bio-ethanol plant both located in the Rotterdam area [28, 29].

In the US this application has recently started to grow (>40% in the last 7–8 years). One important incentive has been the safe conditions required to grow legally allowed cannabis.

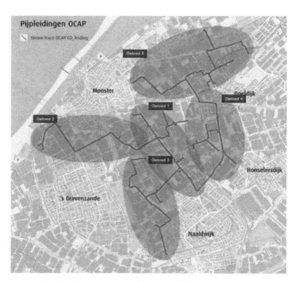

An interesting development seen in largely populated cities (e.g. in Japan) is vertical farming. This offers the opportunity to grow farmable plants on the outside (roof and walls) of buildings without requiring additional plots of very scarce land. Estimates go as far as predicting a CO_2 consumption of 600 t/y building [12a].

3.3.5.4 Photosynthesis in Water Environment: Production and Processing of Algae for Industrial Applications

The food, pharmaceutical and other sectors show an increasing demand for fine chemicals that have a renewable origin and natural character. At the same time, the agricultural sector in the EU faces small margins on production and tightened rules to reduce the environmental impact. Algae appear to be a very promising link between fine chemical demand and changes for the agricultural sector and are therefore covered in more detail in a separate chapter in this book. Reliable market figures for this rapidly growing application are hard to find as in many of the applications it is either a mixed application with traditional greenhouses or waste gas streams are used directly without going for either intermediate enrichment or liquefaction of the CO_2 used to enrich the water in which the algae are grown. In these applications sources of waste heat are used to accelerate the process (see Chap. 12 for more information about aquatic biomass growth and exploitation).

3.3.6   Rubber and Plastics Processing

3.3.6.1   Blowing Agent

The primary challenge for the 100 kt/y foam blowing industry has been to find suitable alternatives to chlorofluorocarbon (CFC) and hydrocarbon blowing agents, which are being phased out due to their allegedly detrimental effect on the Earth's ozone layer. Therefore, an emerging application for carbon dioxide is its use as a blowing agent. Carbon dioxide gas under pressure is introduced into rubber and plastic mixes, and foams the material upon pressure release. The foam blowing industry is divided into two markets, thermoplastics and thermosets. Thermoplastics, which are used in residential housing as sheeting and roofing, as well as in commercial roofing applications, are the smaller of the two. Although polystyrene is the biggest section of this market, there are sizable segments in polypropylene and polyvinyl chloride as well as several smaller segments. Thermosets are a much larger market, taking about four times as much business as thermoplastics. This industry closely follows the plastics market. Polyurethane agents go into residential sheeting, refrigerator doors, insulating foam for pipes, spray foam for electrical outlets, and sandwich panels–the thin layer of foam which is placed between layers of metal in walk-in refrigerators, storage trucks and other larger refrigerating devices. A new tendency is the use of chemicals produced from waste CO_2 streams. In its 2017 Annual report, Recticel for instance announces its intention to use a CO_2-based polyol in its flexible foam production process. CO_2 systems nowadays match the majority of existing blowing machines (Cannon, Hennecke, Beamech, etc.)

3.3.6.2   De-flashing of Molded Rubber and Plastic Articles

Flashing (i.e., thin protrusions of material at mould joints) can be quickly removed by tumbling these articles with liquid carbon dioxide or crushed dry ice.

3.3.6.3   Blow Molding

By injecting liquid CO_2 into the moulded product immediately after the blowing step, the cooling time can be reduced in an important way. Cycle times of existing machines can therefore be reduced and capacity increased. Due to increase in performance of today's blowing techniques, the use of CO_2 has reduced in popularity, but still shows significant advantages when used with plastic moulds having an important wall thickness.

3.3.7   Other Uses as Solvent

3.3.7.1   Spray Painting

In *spray painting*, supercritical carbon dioxide is emerging as a functional alternative to VOCs. This technology has been developed by Union Carbide and is marketed under the tradename UNICARB. Although other gases could be used for this application, carbon dioxide is used because it mixes well with many paint polymers.

3.3.7.2 Aerosols

CO_2 has since long been used as a propellant in many aerosol applications, as a replacement of conventional propellants, which are allegedly known to damage the earth's ozone layer.

3.3.7.3 Cleaning Parts Using Ultrapure CO_2

Supercritical carbon dioxide is being used for cleaning precision parts, electronic components, as well as for the removal of hydrocarbon machine coolants from metal parts. Specifically, supercritical fluid technology is currently being investigated for possible application in a vast range of cleaning, extraction and thin metal film deposition applications for semiconductors and electronics. In particular, the future of supercritical fluid technology to precision clean semiconductor substrates and electronic boards and parts is promising with potential commercial applications in the areas of information storage devices, semiconductor devices, electrical and electronic components, inertial guidance systems, precision optical devices, medical equipment and devices, and metal finishing. While competition from established cleaning methods may seem severe, there are a number of attractive features which singles out the supercritical carbon dioxide process as a unique method for precision cleaning. These include environmentally-compatible cleaning, low carbon dioxide cost and overall cost, rapid processing, ability to clean large areas and complex shapes, ability to form homogeneous cleaning supercritical fluids for general cleans as well as the ability to incorporate additives to tailor more specialised and specific cleaning. In all cleaning applications, the increasing demand is for greater efficiency and effectiveness in contaminant removal. There is a market drive for environmentally conscious semiconductor and electronic parts cleaning, in particular due to stringent deadlines and aggressive scheduling imposed by legislative bodies to phase out the production and use of ozone-depleting substances such as CFCs and a reduction in VOC emissions from of organic solvents use. In addition, reducing worker exposure to volatile organic compounds has also become increasingly important in the electronic industry. Existing cleaning processes available to the industry e.g. RCA based cleans for semiconductors, flux cleaning for oxides in electronics, solvent cleaning of flux residues in electronics, present severe limitations including adequate level of cleanliness, volumes of waste solvents generated, worker exposure to solvents, costly chemicals and expensive equipment. As such, there is a requirement for the development and integration of a closed loop, environmentally and substrate compatible, cost effective technology resulting in significant savings in solvent and associated energy costs. Recent work on supercritical fluids indicate that as well as successively removing organic contaminants, the technology may also be used to remove or extract metals by using metal complexing ligands. These discoveries open the potential for a multi-purpose (organic and metallic or ionic contaminant removal) clean compatible with environmentally conscious manufacturing and the process requirements of contaminant-sensitive materials.

3.3.7.4 Dry Cleaning

Dry cleaning of clothes is a large industrial activity in Europe conducted in 60,000 dry cleaning shops, employing 180,000 workers and cleaning 2.2 million tonnes of textiles each year. The dry cleaning sector currently has 75,000 cleaning machines using perchloro-ethylene (PERC) or CFC 113 as cleaning solvents. Some time ago, the hope was that a major portion of the PERC and CFC 113 might be replaced by SCDC with CO_2 if the right technology, apparatuses and detergents were developed. Important aspects for the cleaning effect are the removing of soil and stains on the one hand without re-deposition and garment damages on the other hand. Therefore parameters such as mechanical action, the fluid/garment ratio, the fluid flow, the type of detergent, the detergent concentration, application of additives, relative humidity, time, temperature and pressure have to be investigated, separately and in combination, in order to obtain an optimal process which meets the criteria for cleaning effect, greying, colourfastness, shrinkage, appearance and surface structure of the textiles/leathers. Also the behaviour of different textiles, leathers and other materials on the garments in relation to different types of soil and stains will be a part of the investigation. Current status and suggestions for improvement can be found in the Master thesis of S. Sutane @ TU Delft (7 April 2014) [30]. So far, despite the benefit associated with an eventual substitution of PERC with CO_2, CO_2 dry cleaning never really took off as an industrially viable process and much research is still necessary.

3.3.7.5 Production of Fine Particles and Polymer Fractionation Based on Pressure Differentiation

Several companies, such as Separex in France and Messer have also developed a number of technologies that use supercritical CO_2 in the production of extremely fine particles. One recently commercialised process dissolves compounds ranging from pharmaceuticals to paint pigments in supercritical CO_2. When the pressure is suddenly released, the material precipitates as particles smaller than 10 μm with a very homogeneous size distribution. The same process allows for controlled precipitation of polymers with narrow bands of molecular weight distribution.

3.3.8 Water Treatment

3.3.8.1 Re-Carbonation of Sweet Surface Waters to Be Used as Tap Water

Modern water works use CO_2 together with lime to raise the hardness of drinking water, especially with soft surface waters or desalted or highly soft waters from desalination plants. At the right pH-level, the mentioned treatment allows the formation of a protective surface layer in water-mains pipes, thus avoiding corrosion and improving water quality. Remineralisation is also needed to make these waters healthy and tasty.

3.3.8.2   Water Treatment Market: pH Control [31]

Some industries create a lot of highly alkaline waste water: iron and steel making, textile and dying, pulp and paper, glass bottle washing at beverage filling plants, and many more create effluents with a pH value above 11. Carbon dioxide can be used for treatment of these wastewater streams. Carbon dioxide is non-toxic and leaves no residues upon evaporation. Indeed, lowering the pH of alkaline water with CO_2 is on the rise. Thanks to the natural buffering capacity of bicarbonate, which, along with carbonic acid and carbonate, is produced when CO_2 and water mix, Neutralizing with the gas lowers corrosivity and makes overshooting the pH range below 5 difficult; overdosing CO_2 will produce sparkling water which is still not very harmful. The technology has been around for over 30 years, but in the last decade we've seen a significant increase in its use as people become more aware of the dangers of storing and handling sulphuric acid. This method offers thus many advantages towards the use of classical acids, such as less reagent, less equipment and monitoring devices, less downtime and increased safety, equipment life, etc. CO_2 used in H_2O system is known to also dissolve $CaCO_3$ deposit from the pipes. This is commercially not used yet. CO_2 addition as a way to maintaining the quality of tap water is also commonly used in Japan [32].

3.3.8.3   Water Well Rehabilitation

The water well rehabilitation technique consists in injecting gaseous carbon dioxide at the desired depth in the treatment area. This produces a highly abrasive carbonic acid solution that penetrates far into the surrounding formation. Liquefied carbon dioxide is then injected at various temperatures and pressures. When it comes in contact with water, it expands rapidly, producing tremendous agitation. The continued, controlled, injection of the liquefied carbon dioxide assures the freezing of water within the formation around the well, resulting in superior disinfection and dislodging of mineral encrustation. After treatment the well is mechanically developed using surge/airlift methods to remove the newly dislodged particulate matter from the well and formation. The well pump is then reinstalled and the well returned to service, providing and increased supply of water for its intended use.

3.3.8.4   Salt Water Desalination

An important growth of CO_2 utilisation is noticed in the desalination of seawater to produce drinking water. This is especially the case of Israel and the Middle East. Today more than 80% of the desalination plants are located in the Golf states; allegedly, Saudi Arabia will invest 2 billion \$ in the next 5 years in this application. When trying to value the CO_2 consumption for this application, a ratio of 33 gr of CO_2 per m^3 of desalinated water is used.

3.3.9   Nuclear

CO_2 is used in the nuclear energy sector as a cooling medium for power plant cooling.

3.3.10 Well Re-injection

3.3.10.1 Secondary Oil-Recovery and Re-injection [33]

In most of the oil fields already exploited, there is some oil left in place underground. As oil can be miscible with CO$_2$ under the correct pressure and temperature conditions in the reservoir, CO$_2$ is injected into suitable depleted oil reservoirs in order to enhance oil recovery by typically 10–15%. This technique is called Enhanced Oil Recovery (EOR). Part of the injected CO$_2$ will be recovered at the production well together with oil, while *ca.* 40+% of CO$_2$ will be trapped in the depleted reservoir. Enhanced oil recovery by CO$_2$ injection is already used at lots of sites in the USA, and well over 80 Mt/year of CO$_2$ is being delivered to these fields for injection.

3.3.10.2 Coal Mining Coupled with Methane Recovery [34]

Methane is present both in the pores of coal, but is mostly adsorbed on the coal surface. CO$_2$ injection in the coal seam decreases methane partial pressure; methane is desorbed and can be recovered at the production well. Unlike in enhanced oil recovery, CO$_2$ remains stored underground, adsorbed on the coal surface instead of methane. Moreover, coal can adsorb twice as much CO$_2$ by volume than methane, which means that the storage capacity exceeds the quantity of methane recovered. This technique is called Enhanced Recovery of Coal Bed Methane (ERCBM), and is proven feasible due to a demonstration plant in New Mexico, USA, where 100.000 ton CO$_2$ have been injected in 3 years, none of which has found its way to the production wells. A field test of ERCBM using CO$_2$ and N$_2$ mixtures is being carried out by the Alberta Research Council under an international project facilitated by the IEA Greenhouse Gas R&D Programme.

3.3.10.3 Shale Gas [35]

Hydraulic fracturing with pure L-CO$_2$ injection has been used since several decades; it is only recently that supercritical CO$_2$ is being tested as a fracturing technique. In addition to the mechanical action following the injection of the CO$_2$, the fact that it absorbs significantly better in shale than the CH$_4$ leads to secondary methane recovery and possibly yields permanent bonding of the CO$_2$ in these underground layers.

3.3.10.4 Re-injection of Waste CO$_2$ in Depleted Natural Gas Wells [36]

Allegedly, this safe storage method would also stop the land sinking in the sea [37]. Combined with the OCAP pipeline, the port of Rotterdam has this plan to also inject megatons of CO$_2$ in the depleted gas field in front of its coast line.

3.3.11 Cylinder Filling

CO_2 can be used in cylinders for several applications, such as:

(i) **Horeca**: As a dispensing gas for mainly beer taps in bars and restaurants, CO_2 cylinders have been in use since over a century.

(ii) **Fire fighting**: Consumer use and industrial applications. CO_2 smothers fires without damaging or contaminating materials and is used especially when water is ineffective or undesirable.

(iii) **Consumer use**: Cartridges for carbonating (see §3.2). Carbon dioxide also finds use in gas-operated firearms.

(iv) **Medical use**: Finally, the medical market is a fast growing consumer of industrial gases, with the primary gases used being oxygen (for respiratory therapy, CAT scans and numerous other uses), carbon dioxide (laser surgery), helium (magnetic resonance imaging), and small volume specialty gases (laser surgery).

3.3.12 Other Applications (Not Using Large Quantities of L-CO_2 Yet)

CO_2 continues to substitute for traditional halogenated-carbon refrigerants that eat up atmospheric ozone or have even more severe greenhouse effects than CO_2. Norsk Hydro ASA (Oslo) and a Norwegian university have developed a CO_2-based cooling system for automobiles. Daimler Chrysler AG (Stuttgart) tested the Mobile air Conditioning 2000, and Norsk claims the technology competes favourably in price, weight, space and energy efficiency. Likewise, Sanyo Electric Co. (Tokyo) developed a closed-type rotary compressor that uses CO_2 as the coolant. Linde promotes CO_2 as cooling medium for airco systems in e.g. cars and cooling systems in supermarkets (see Linde brochure on R744) [38]. In life science blends of 10% Ethylene oxide and 90% CO_2 are used for Cold Sterilisation applications [39]. Another interesting development is the research focused on Supercritical CO_2 Power Cycle technology, especially for applications in the greener power plants of the future. A large test plant is being started in San Antonio, Texas [40].

3.4 CO_2 Used as Dry Ice

3.4.1 Production

Dry ice, or carbon dioxide (CO_2) in its solid form, is produced commercially in dry ice production machines. Liquid CO_2, which is stored in bulk storage tanks, is transferred from the storage tank to the dry ice production machines. The production machines that are widely used are commonly referred to as pelletizers (see Fig. 3.1). Liquid CO_2 is injected into chambers within these machines and returns to atmospheric pressure. Once exposed to the atmospheric pressure, the liquid CO_2 expands and results in the formation of nearly equal portions of CO_2 snow and CO_2 gas. Sometimes the gas is recovered and converted back into liquid form. The snow is compressed in a cylinder and extruded or pressed at high pressure, resulting in the manufacture of various types of dry ice. Pellets and nuggets are formed when the snow is extruded through die plates, which can be interchanged to create various sizes. Dry ice is commonly manufactured in four forms: pellets (3–19 mm), nuggets (3–19 mm), blocks and slices/slabs (see Fig. 3.2).

Fig. 3.1 Pelletizers are common dry ice production equipment (Cold Jet)

Fig. 3.2 Dry ice comes in four forms. Pellets, nuggets, blocks and slices/slabs (Cold Jet)

Dry ice is commonly loaded directly into a specially designed, well-insulated container, which is essential in order to minimize product loss. These containers preserve dry ice quality, minimize sublimation and are ideal for storage and transportation of dry ice and for maintaining product quality at time of delivery. For dry ice manufactures and distributors, it is essential to maintain an efficient distribution system. Dry ice immediately begins to sublimate as soon as it is produced.

3.4.2 Properties

Dry ice is the solid phase of CO_2, which is a gas found naturally in the atmosphere. CO_2 is a stable, linear molecule that is non-polar. It does not possess a positive or negative charge; it is also an inert, non-flammable and non-toxic molecule. The temperature of dry ice is −78.9 °C (−109.3 °F). Dry ice undergoes the process of deposition or sublimation at the critical temperature of −78.5 °C (−109.3 °F). As CO_2 reaches temperatures below −78 °C, it transitions directly from a gaseous to a solid state (i.e. dry ice forms) in a process called deposition. In contrast, as CO_2 reaches temperatures above −78.5 °C, it transitions directly from a solid to gaseous state in a process called sublimation. During the processes of deposition and sublimation, dry ice never reaches a liquid phase; thus, there is no residue or aqueous waste associated with dry ice. Dry ice is a relatively soft media and is therefore non-abrasive to most surfaces. The hardness of dry ice was found to be 1.5–2.0 Mohs (Mohs Scale of Hardness) (see Fig. 3.3).

Fig. 3.3 Mohs hardness scale for minerals (Cold Jet)

Mohs Hardness Scale for Minerals

1 – Talc

2 – Gypsum, Dry Ice (Fingernail, Baking Soda ~ 2.5)

3 – Calcite (a penny)

4 – Fluorite (Corn Cob ~ 4.5)

5 – Apatite (Glass Beads & Nut Shells ~ 5.5)

6 – Orthoclase, Feldspar, Spectrolite (Steel File ~ 6.5-7.5)

7 – Quartz, Amethyst, Citrine, Agate (Garnet ~ 7.5)

8 – Topaz, Beryl, Emerald, Aquamarine

9 – Corundum, Ruby, Sapphire (Alum. Oxide ~ 8.5)

10- Diamond

The density of dry ice varies, but typically will measure around 1.56 kg/dm^3. The heat vaporisation of dry ice will measure around 645 kJ/kg (152 kcal/kg) at 0 ° C (twice the refrigerating capacity of water ice per unit weight). Dry ice is a food grade medium and is EPA, FDA and USDA approved for use around food manufacturing. It is colourless, tasteless, odourless and non-toxic.

3.4.3 Applications

Dry ice is used in practically all type of industries and for many types of applications, mainly because of its cooling properties. Dry ice is most commonly utilized commercially for:

- Cold chain management
- Dry ice blasting.

3.4.3.1 Dry Ice as a Cooling Medium

Due to its low temperature, dry ice is utilized in many industries as a cooling medium. It can be used or supplied as pellets, nuggets, slabs or blocks to suit industry and production needs. Due to the fact that dry ice sublimates and therefore does not create residue or water, as well as its higher cooling capacity and lower temperature, makes dry ice an excellent alternative to water ice for most cooling applications. Nearly 90–95% of all the dry ice that is produced is utilized in transport cooling of various products. Cooling products or parts during manufacturing, processing and shipping processes is a common application for dry ice. Examples are:

Fig. 3.4 Dry ice slabs or slices are commonly used for cooling purposes (Cold Jet)

- **(i) Airline catering**: Most airlines rely on dry ice for on-board cooling, especially on long-haul flights. Dry ice slabs (see Fig. 3.4) are placed in food trolleys to keep food and drinks chilled or frozen until it is time to prepare/serve.
- **(ii) Food Packaging and Distribution**: Dry ice is recommended for transporting frozen products (0 °C to −20 °C), such as ice cream or steaks, especially with long durations of two or more days [41]. Home Delivery Services (online ordered frozen or cooled products delivered at home in insulated boxes containing dry ice as coolant) are seeing a fast growing volume of dry ice being used.
- **(iii) Medical Samples and Pharmaceutical Distribution**: Dry ice is commonly used to transport biological samples (i.e. tissue, blood) and temperature-sensitive biomedical/ pharmacological products during the shipping process.
- **(iv) Food Processing**: Meat processors use dry ice to keep meat temperatures low and reduce spoilage. It has been proven to preserve freshness, colour and flavour.
- **(v) Winemaking (Cryomaceration)**: Dry ice is utilized in the maceration stage of winemaking (i.e. the time the juice from the grapes is left in contact with the skins and stems), to inhibit the beginning of fermentation by rapidly reducing the temperature below 10 °C. This technique is called 'pre-fermentative cryo-maceration', or using low temperatures to delay fermentation.
- **(vi) Baking**: The baking industry incorporates dry ice into mixing processes for a few reasons: to slow yeast growth, control dough temperature and impede bacteria.
- **(vii) Composite Material Manufacturing**: Dry ice is used in composite material production to keep components cool and malleable.

3.4.3.2 Dry Ice as a Blasting Media for Surface Cleaning, Surface Preparation and Parts Finishing

The dry ice blasting technique was originally developed to remove paint and coating from airplanes in the 1970's. The technology was developed commercially by Cold Jet in 1986. Dry ice blasting is similar in concept to sand, bead and soda blasting in that it cleans surfaces using a media accelerated in a pressurized air

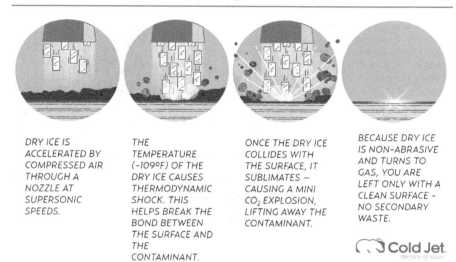

DRY ICE IS ACCELERATED BY COMPRESSED AIR THROUGH A NOZZLE AT SUPERSONIC SPEEDS.

THE TEMPERATURE (-109°F) OF THE DRY ICE CAUSES THERMODYNAMIC SHOCK. THIS HELPS BREAK THE BOND BETWEEN THE SURFACE AND THE CONTAMINANT.

ONCE THE DRY ICE COLLIDES WITH THE SURFACE, IT SUBLIMATES – CAUSING A MINI CO$_2$ EXPLOSION, LIFTING AWAY THE CONTAMINANT.

BECAUSE DRY ICE IS NON-ABRASIVE AND TURNS TO GAS, YOU ARE LEFT ONLY WITH A CLEAN SURFACE - NO SECONDARY WASTE.

Cold Jet.
the force of nature

Fig. 3.5 Dry ice blasting relies on the Pellet Kinetic Energy Effect, Thermal Effect and Gas Expansion Effect to clean surfaces (Cold Jet)

stream. It differs in that dry ice cleaning uses solid CO$_2$ pellets or MicroParticles, accelerated to high velocities to impinge on the surface and clean it. The particles sublimate upon impact, lifting dirt and contaminants off the underlying substrate without damage. Dry ice has unique thermal and sublimation effects. The cleaning process of dry ice blasting is based on three effects (see Fig. 3.5).

- **(i) Pellet Kinetic Energy Effect**: The energy delivered by the impact of the dry ice particle onto a substrate is known as the Pellet Kinetic Energy Effect. This effect has the largest contribution to the cleaning process when substrates are at ambient temperatures or below.
- **(ii) Thermal Effect**: The inherently low temperature of dry ice aids removal of the contaminant. Its' low temperature causes the contaminant to embrittle and shrink, creating rapid micro-cracking and causing the bond between the contaminant and the substrate to fail.
- **(iii) Gas Expansion Effect (sublimation)**: Upon impact, the CO$_2$ particles will sublimate or expand instantly and return to its natural gas state. During this phase transition from solid to gas, the volume of dry ice expands up to 800 times and lifts the contaminant off of the substrate.

Dry ice blasting is used in three primary applications:

- **(i) Surface cleaning**: Dry ice blasting is a non-abrasive, non-conductive, non-toxic cleaning method that does not create secondary waste. It is used to

Fig. 3.6 Dry ice blasting can be used to clean many types of surfaces while not damaging the underlying substrate or creating secondary waste (Cold Jet)

Fig. 3.7 The dry ice blasting process is utilized to clean parts before applying paint. It can also be integrated into existing production lines (Cold Jet)

clean many types of surfaces in a multitude of commercial and manufacturing settings. Surfaces of tooling, machinery, and finished parts can all be cleaned with dry ice (see Fig. 3.6).

- **(ii) Surface preparation**: Dry ice cleaning is a dry process and eliminates the need for aqueous or chemical solutions when preparing a surface for painting. Coatings/paint can be applied immediately after cleaning because the surface is left completely dry. Dry ice cleaning will also leave a slight profile on the surface as well, which improves the bonding of the paint or sealant that is being applied (see Fig. 3.7).

- **(iii) Parts finishing**: The process can safely remove flash and burrs from a variety of materials: PEEK, PBT, Acetal, Nylon, LCP, ABS, UHMWPE, Nitinol and more. Due to its non-abrasive nature, dry ice leaves the surface undamaged and free of residual media (see Fig. 3.8).

Fig. 3.8 Dry ice blasting is ideal for removing burrs or flash from finished parts (Cold Jet)

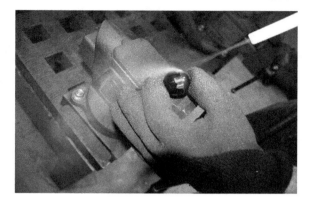

Dry ice blasting is used to clean production equipment in many different types of industries, including: Aerospace, Automotive, Contract Cleaning, Composite Tool Cleaning, Electric Motor, Fire Restoration, Food and Beverage, Foundry, Oil and Gas, General Maintenance and Facilities, Historical Restoration, Medical Device Manufacturing, Mold Remediation, Packaging, Plastics, Power Generation, Printing, Rubber, Textile and Engineered Wood. In most cases, substantial savings can be obtained by cleaning with dry ice blasting over alternative cleaning methods. The process commonly reduces cleaning time, labour required for cleaning and production downtime associated with the cleaning process. Dry ice blasting is also used in disaster remediation projects (fire and smoke damage removal, mold remediation, water damage and odour elimination) and historical restoration projects. The first patents regarding development and design of modern-day single-hose dry ice blasting technology were awarded to Cold Jet in 1986. Cold Jet has since developed and refined the technology. Cold Jet has two distinct lines of business centered around the use of dry ice. We provide environmental cleaning, surface preparation and parts finishing systems to global manufacturing industries. These systems utilize particles of dry ice as a blasting medium. Secondly, we produce systems for the production, metering and packaging of dry ice. These systems enable the consistent production of a controlled range of dry ice products for food transportation, cold chain management and dry ice cleaning. Customers are using our technology-based solutions to replace outdated processes that are inefficient and harmful to health and safety.

3.4.3.3 Smoke Effects in Theatres and Bars

When dry ice particles are thrown in a liquid at elevated temperature, the dry ice sublimes rapidly into CO_2 gas. This is heavier than air, and spreads itself slowly as a white misty cloud around the source. This is used in some bars to produce foaming cocktails or in theatres and disco's. As an example, the annual dry ice market in Germany has grown to 80 kT/yr (equivalent to 200.000 t/y of L-CO_2) and still keeps on growing at an annual rate of 5–8% [12a].

3.5   Volume of the CO_2 Merchant Market [42]

Determining the volume of each segment of the CO_2 merchant market is problematic for the simple reason that those CO_2 volumes are confidential. For the carbonated beverage sector we can evaluate the volume of the CO_2 merchant market by collecting information on the production of carbonated beverages as there is no alternative in that application to the use of CO_2. For the other segments, we could identify some studies where data on the CO_2 merchant market are gathered. They are mainly realized by American societies, and therefore the results are very often limited to the American continent. What is clear is that by far the largest merchant market segment for L-CO_2 in the US is food processing; this accounts for 53% of the merchant market versus only 22% in Europe. For carbonated beverages the tendency is completely opposite with 51% of the merchant market selling to that industry in Europe compared to only 15 in the US. Other industrial applications account for roughly 30% of the merchant market both in the US and Europe. However some of these studies also include the European CO_2 merchant market. The results of those studies are not available. Indeed, as this data collection was realized by private companies, the reports are not free of charge and are prohibitively expensive. Three studies where data on the CO_2 merchant market are gathered are presented hereunder:

Merchant market volumes:	2004-2005	2010-2011	2016-2017	Average price
	kT/yr	kT/yr	kT/yr	€/T
Germany	720	800	850	100 - 110
France	500	540	580	100 - 110
Italy	410	420	450	130
Poland	210	220	230	< 100
Belgium	155	160	165	100 - 110
Netherlands	150	150	155	100 - 110
Czech Republic	75	80	90	
Austria	48	50	70	150

   It is clear that the modest growth rate of this merchant market, even if attractive by itself, will not solve the dramatic increase of man-made CO_2 gas emissions into the atmosphere.

References

1. The Paris Agreement. www.climatefocus.com
2. www.green-minerals.nl
3. https://www.nordzucker.com/fileadmin/downloads/Verbraucher/NZ_Sugar_production_EN.pdf
4. Willke T (2014) Appl Microbiol Biotechnol 98(24):9893–9914

5. (a) http://www.co2reuse.eu/images/stories/pdf-files/5-Schyvinck_CO2reuse_and_Minerals_ Oct2012.pdf. (b) http://www.scielo.org.za/pdf/sajc/v70/01.pdf
6. Sampson J (2018) CO_2 supply crisis. Gasworld, June 19
7. https://www.siad.com/
8. Owen-Jones J (2018) CO_2 from quad-generation. Gasworld, Aug 24
9. https://www.gasworld.com/large-scale-capture-of-co2-is-feasible-and-affordable-/2014903. article?utm_source=dlvr.it&utm_medium=linkedincompanies
10. Koziel P. Private communication
11. Cockerill R (2018) Mergers in CO_2 business. Gasworld, June 5
12. (a) Koziel P (2018) CO_2 market development till 2025, Linde Report 2018. (b) Koziel P (2018) Varbrauch und anwendungen, Linde Report 2018. (c) Koziel P (2018) Saisonal Schwankungen, LINDE Report 2018. (d) Koziel P (2018) Market study of CO_2, 2018
13. www.bodan.be
14. https://www.messergroup.com/documents/20182/701246/Modified_Atmosphere_Packaging. pdf/2b9398a2-e63a-4018-a6f2-42dd35b95e87
15. MAP gas composition vs Product. https://dansensor.com/. https://www.modifiedatmos pherepackaging.com/
16. https://www.linde-gas.com/en/images/MAPAX%20brochure_tcm17-4683.pdf
17. (a) van Damme F (2018) Freezing food with CO_2. Dohmeyer Report 2018, August 20. (b) http://www.globalfoodtechnology.com/machines-by-manufacturer/dohmeyer/
18. http://www.supercriticalfluids.com/company-information/about-sft-inc/
19. (a) CRA (2010)Fruit juice pasteurization. Innov Food Sci Emerg Technol 2010 11:477–484. (b) CRA (2016) Carrot juice pasteurization. Int J New Technol Res (IJNTR) 2016 2(2):71–77. ISSN: 2454-4116
20. Rushing SA (2017) CO_2 for beverages. Gasworld, November 3
21. Kerckx P. VBS Europe. https://www.vbseurope.com/en/food-beverage
22. http://www.brewlocksystem.com/
23. https://www.sodastream.com.au/how-it-works/
24. http://avantium.com/
25. De Wever H (2018) VITO, H2020 „BioReCO2Ver" Project. see: https://vito.be/en/news/ bioreco2ver-turning-co2-chemicals-using-bioconversion
26. Essencia BG (2016) "Technology watch" Project, "Carbon Dioxide Conversion». See: http:// www.essenscia.be/en/Document/Download/16288
27. http://www.supercriticalfluids.com/products/supercritical-fluid-extraction-products/ cannabissfe/
28. https://www.the-lindegroup.com/en/clean_technology/clean_technology_portfolio/co2_ applications/greenhouse_supply/index.html
29. (a) https://www.ocap.nl/nl/co2-smart-grid/index.html. (b) https://www.the-linde-group.com/ en/news_and_media/press_releases/news_130311_2.html
30. (a) Sutanto S (2014) Textile dry cleaning using carbon dioxide: process, apparatus and mechanical action. Master Sci Chem Eng, TU Delft, Delft, April 7. (b) https://www.google. be/search?ei=aIqGXNDaG8PPwALwjazgAQ&q=stevia+sutanto+Master+of+Science+in +Chemical+Engineering%2C+TU+Delft+7+April+2014/
31. PRAXAIR. Total capabilities in CO_2 water. https://www.praxair.com/-/media/corporate/ praxairus/documents/specification-sheets-and-brochures/industries/water-and-wastewater-treatment/p-10419-carbon-dioxide-to-reduce-ph.pdf?la=en&rev= c6e886a58dc54615afa864b999a7b2eb
32. (a) Owen-Jones J (2017) Maintaining tap water quality in Japan. Gas Rev 438. (b) https:// www.gasworld.com/co2-maintaining-tap-water-quality/2013300.article
33. Mitsubishi Heavy Ind. Ltd., EOR, and ECBM Technology: see https://www.mhi.com/ products/energy/co2_enhanced_oil_recovery.html
34. Mitsubishi Heavy Ind. Ltd., ECBM Technology, see: https://repository.lib.ncsu.edu/bitstream/ handle/1840.4/8998/Lempert%2C%20Michelle%20final.pdf?sequence=1&isAllowed=y

35. Gandossi L (2013) European Commission, L-CO$_2$ Hydraulic Fracturing for shale gas production, Report EUR 26347 EN
36. (a) Report #C2051/MD-MV20090186. www.dhv.com. (b) TNO-Netherlands (2005) CO$_2$ storage in natural gas reservoirs. Oil Gas Sci Technol Rev IFP 60(3):527–536
37. http://www.clingendaelenergy.com/files.cfm?event=files.download&ui=6BE1ABB2-5254-00CF-FD03BD1B4B16E664
38. LINDE Report. CO$_2$ as refrigerant: R744 Refrigerant grade CO$_2$. http://www.r744.com/files/252_Linde_R744.pdf
39. Rushing SA (2017) CO$_2$ and the life sciences. Gasworld, June 1
40. https://www.gastopowerjournal.comtechnologyainnovation/item/9020-construction-starts-for-119m-co2-pilot-power-plant-in-texas/
41. (a) LINDE Patent EP3,173,715 A1. (b) Private contribution (with permission to publish) coordinated by Christian Rogiers and his team at Cold Jet LLC. See https://www.coldjet.com/en/index.php#. (c) Cachon R, Girardon P, Voilley A (eds). Gases in agro-food processes. ISBN: 9780128124659. Elsevier, Academic Press. Section 9: CO$_2$ Blasting for Cleaning. Under publication
42. (a) Cockerill R (2017) The US merchant CO$_2$ market—a shifting landscape. Gasworld, April 5. (b) Barr J (2017) US CO$_2$ market report 2016. Gasworld, May 1

Mineral Carbonation for Carbon Capture and Utilization

4

Tze Yuen Yeo and Jie Bu

Abstract

The appeal of mineral carbonation (MC) as a process technology for scalable and long-term CO_2 reduction, is that it is a solution that has the sequestration capacity to match the amount of CO_2 emitted from energy generation and industrial activities [1–3]. Many inorganic materials such as minerals [4, 5], incineration ash [6, 7], concrete [8, 9] and industrial residues [10, 11] are potentially huge sinks for anthropogenic CO_2 emissions. These materials are typically abundant sources of alkaline and alkaline-earth metal oxides, which can react naturally with CO_2 to form inorganic carbonates and bicarbonates. In addition, their products are thermodynamically stable and relatively inert at ambient conditions. On paper, MC should be able to fully sequester all anthropogenic CO_2 emissions, since the abundance of magnesium and calcium atoms on Earth far exceeds the total amount of carbon atoms [12, 13]. However, despite the apparently favorable pre-conditions, we still observe a net accumulation of CO_2 in the atmosphere because the rates of reaction to form (bi)carbonates in nature are too slow compared to the current rate at which CO_2 is being emitted [14, 15]. If left to their own devices, thousands of years are needed to achieve any substantial sequestration of CO_2 [16]. This is clearly not rapid enough to solve the pressing problem of climate change that is already affecting us now. Therefore there is a need to employ mineral carbonation as an artificial method to accelerate the rates of CO_2 sequestration. In this chapter, we will take a look into the chemistry and thermodynamics of mineral carbonation and discuss some of the main obstacles to large scale MC implementation. Additionally, we highlight the types of starting materials from which basic alkaline-earth metal oxides can be obtained and discuss how their abundance and

T. Y. Yeo · J. Bu (✉)
Institute of Chemical and Engineering Sciences, A*STAR, 1 Pesek Road, Jurong Island, Singapore 627833, Singapore
e-mail: bu_jie@ices.a-star.edu.sg

© Springer Nature Switzerland AG 2019
M. Aresta et al. (eds.), *An Economy Based on Carbon Dioxide and Water*,
https://doi.org/10.1007/978-3-030-15868-2_4

properties affect MC performance. We will also give a short review of current research in the area to develop MC into viable and economic processes, with some focus on the main categories of process designs and their working principles. We will then look at MC from a techno-economic standpoint and assess the opportunities to integrate MC into the existing industrial and environmental landscape. Lastly, we conclude the chapter with a hypothetical scenario of MC deployment in Singapore, an economically developed but land-scarce country under threat by rising sea levels.

4.1 Background and Key Concepts in Mineral Carbonation

In this section, we introduce and discuss several of the key concepts behind mineral carbonation. This is to provide the reader with sufficient background information to understand and appreciate the science and basic guiding principles of MC. Hopefully this will inspire the reader to utilize this knowledge to develop new ideas and exploit new opportunities that may emerge from it. Here, we will briefly explain the chemistry, reaction characteristics, thermodynamics, and material sources for mineral carbonation.

4.1.1 The Chemistry of Mineral Carbonation

There is often a common misconception that CO_2 is an unreactive molecule that is difficult to functionalize and convert into other industrially relevant chemicals. This is demonstrably untrue, as CO_2 is in fact capable of participating in a wide variety of reactions [17]. For example, we can easily observe its effects on the gradual chemical erosion of calcium and magnesium silicate materials at ambient conditions [18]. At its very core, mineral carbonation is a simple reaction between an acid (CO_2) and a base (usually alkaline or alkaline-earth metal oxides/hydroxides):

$$M(OH)_2 + CO_2 \rightarrow MCO_3 + H_2O$$

This reaction occurs spontaneously in nature, and can be considered as a subset of natural weathering processes [18]. This phenomenon is responsible for the formation of limestone and Mg/Ca salinity in the environment. In a natural setting, rain and flowing water can dissolve CO_2 and other acidic gases from the air. When CO_2 is dissolved in water, it is hydrated and carbonic acid is formed:

$$CO_{2_{(g)}} + H_2O_{(l)} \rightarrow H_2CO_{3_{(aq)}}$$

When this acidic water comes into contact with minerals with high Mg or Ca content, the dissolved CO_2 reacts with the alkaline-earth metals and forms soluble aqueous bicarbonate species. For example, when wollastonite (a calcium-rich silicate mineral) reacts with carbonic acid, aqueous calcium bicarbonate is formed:

$$CaSiO_{3(s)} + 2H_2CO_{3(aq)} \rightarrow Ca(HCO_3)_{2(aq)} + SiO_{2(s)} + H_2O_{(l)}$$

These dissolved bicarbonates are then carried off downstream and are eventually re-precipitated as solid carbonates when the local conditions change (i.e. warmed, pH raised, CO_2 partial pressure lowered, evaporated etc.):

$$Ca(HCO_3)_{2(aq)} \rightarrow CaCO_3 + CO_{2(g)} + H_2O_{(l)}$$

The net result is that the original mineral is altered and magnesium or calcium carbonate sediments are formed elsewhere:

$$CaSiO_{3(s)} + CO_{2(g)} \rightarrow CaCO_{3(s)} + SiO_{2(s)}$$

The dissolved aqueous bicarbonates may also be carried off by rivers into the ocean as well, where the calcium can be taken up by marine organisms to form shell structures or coral which eventually settle on the ocean floor or wash up on seashores as beach sand.

The chemistry and speciation of CO_2 in aqueous solutions are ultimately governed by the pH, which in turn depends on many factors such as the temperature, total dissolved solids (TDS) content, and the partial pressure of CO_2 in equilibrium with the solution etcetera [19]. As carbonic acid is a diprotic acid, it can be deprotonated once and twice into the bicarbonate and carbonate anions respectively. Also, since carbonic acid is a weak acid, the individual deprotonations occur gradually over a range of pH values (as opposed to a clear, sharp and complete deprotonation within a narrow pH range for strong acids). This leads to the observation that different inorganic carbon species can potentially coexist in the solution, especially at moderate pH values (between 4 and 12). The speciation can be visualized in a Bjerrum plot, showing the relative amount of each species at equilibrium within the solution at any given pH value [20]. As shown in Fig. 4.1, in a closed system, the bicarbonate anion is the predominant inorganic carbon species at neutral pH. Dissolved CO_2 (simplified as carbonic acid here) makes up most of the remaining minor fraction, and carbonate anions are nearly non-existent.

At higher pH values, the second deprotonation occurs more extensively, and carbonate anions begin to predominate under these conditions. At lower pH values, the bicarbonate anion is re-protonated and forms carbonic acid, which has a strong tendency to revert to water and gaseous or dissolved CO_2.

The pKa values for the two deprotonation reactions are:

$$pK_{a1} = \frac{[HCO_3^-][H^+]}{[H_2CO_3]} \approx 6.3 \quad pK_{a2} = \frac{[CO_3^{2-}][H^+]}{[HCO_3^-]} \approx 10.3$$

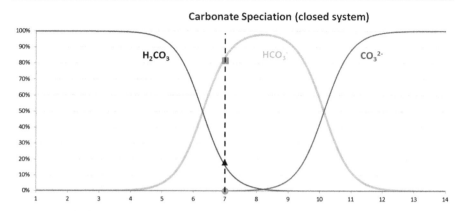

Fig. 4.1 Bjerrum plot showing the speciation of inorganic carbon species in aqueous solutions versus pH, within a closed system. Plot obtained and modified from GWCarb v1.0 Carbonate Speciation Tool [20]

As can be seen from the equations above, the distribution of carbon species in solution is highly dependent on the pH value. An increase in $[H^+]$ shifts the speciation towards more protonated species such as bicarbonates and carbonic acid. Conversely, increasing the pH encourages more deprotonated species to form (in other words increasing $[HCO_3^-]$ or $[CO_3^{2-}]$) in order to maintain the equilibrium within the system.

A fundamental understanding of the factors affecting this equilibrium and how they determine the behavior of inorganic carbon species in solution is important for the effective design of mineral carbonation processes. Firstly, it is known that the carbonate/bicarbonate system acts as a buffer to resist drastic changes in ocean chemistry [21]. The oceans are a huge carbon sink that absorbs significant quantities of CO_2 emitted into the atmosphere every year [22]. The amount of CO_2 absorbed into the oceans is estimated to be around 7.5 billion tonnes of CO_2 annually. Yet despite this, the average oceanic pH has "merely" dropped 0.1 pH units from pre-industrial values thanks to the exceptional buffering capacity of the carbonate/bicarbonate system [23]. However, this apparently small change has already begun to affect the efficiency of shell and coral formation in many biological systems. It has also been predicted that the ocean pH will drop by another 0.3–0.4 units by the year 2100 under business-as-usual scenarios, with irreversible and devastating consequences on marine biota [24].

Secondly, many bicarbonate compounds are found only in aqueous solutions. Notable exceptions are sodium, potassium and ammonium bicarbonate, which can exist as solids under ambient conditions. Even so, these salts are notoriously hygroscopic and thermally unstable. For example, NH_4HCO_3 decomposes to release ammonia, water and CO_2 when mildly heated, thus making it an unsuitable reservoir to reliably store captured CO_2 [25]. Carbonates in general are often more stable than their bicarbonate counterparts, meaning that they are less inclined to re-release the CO_2 after sequestration. Therefore it is desirable that the captured CO_2 is sequestered as a carbonate species. However, in order to facilitate the rapid formation of carbonate products within an aqueous phase, a high concentration of carbonate anions must be present, implying the need to adjust the solution pH to high values to shift the speciation and increase its concentration.

Thirdly, since we know that many bicarbonate compounds are relatively temperature sensitive, it is possible to dissociate and precipitate them as carbonates from aqueous solutions simply by heating them [26, 27]. As the solution is heated, some bicarbonate anions dissociate to give water and CO_2, which is less soluble at high temperatures and escapes the solution as a gas. As CO_2 is removed from the aqueous system, the equilibrium is perturbed and this results in the formation of carbonates which are also usually much less soluble. Thus, as the carbonate product precipitates, it is removed from the aqueous system and a new equilibrium in the aqueous phase is re-established. This is in fact what leads to the formation of lime scale in kettles and boilers that come into contact with hard water.

The behaviors listed above are crucial to our understanding of the working principles of mineral carbonation, and they are the basis of the chemistry that supports nearly all reasoning that goes into the design, analysis and development of MC processes.

4.1.2   The Rates and Mechanisms of Mineral Carbonation

Although the conversion of CO_2 into carbonates and bicarbonates occurs naturally in the environment, the rates of reaction are still too slow to match current rates of CO_2 emissions from industrial activities. The imbalance in the rates of CO_2 emission and sequestration leads to an accumulation of the greenhouse gas in the atmosphere, at an average rate of approximately +2 ppm per year in the 21st century. Thus, the acceleration of mineral carbonation reactions will enable CO_2 to be removed from the atmosphere in a quicker manner, and hopefully tip the balance back to a more manageable equilibrium. As with any set of chemical reactions, the rates of mineral carbonation can be analyzed through the lens of conventional reaction engineering principles.

In nature, mineral carbonation is technically a three-phase reaction, which involves the dissolution of CO_2 (a gas) and alkaline-earth metal cations from minerals (a solid) into water (a liquid) where they react to form the solid carbonate product [28, 29]. The rate limiting step is usually the CO_2 dissolution step, since CO_2 is only very slightly soluble in pure water (around 1.5 grams per liter at 1 bar CO_2 pressure) [30]. The very low concentration of CO_2 in the atmosphere (which currently stands at around 410 ppm or 0.041%) further retards the dissolution kinetics.

The dissolution of CO_2 into a liquid phase also depends on several other factors such as pH and temperature. As with many other gases, higher temperatures decrease the solubility of CO_2 in water, and vice versa. However, the effect of temperature on CO_2 solubility under normal conditions is usually negligible, since the solubility value does not change much within the range of ambient temperatures commonly encountered in nature [31]. A higher pH in the aqueous solution helps to stabilize dissolved CO_2 as (bi)carbonates, and prevents degassing from the liquid phase. In addition, higher pH values reduce the CO_2 chemical potential at the liquid side of the gas-liquid interface, thus eliminating mass transfer resistances and speeds up the uptake of CO_2 into solution [32–35].

On the other hand, the natural dissolution of the alkaline-earth cations from minerals is comparatively more favored due to the common presence of organic acids in groundwater, which promotes the dissolution reaction [36, 37]. The organic acids in groundwater originate from the metabolism and decomposition of terrestrial organisms, and are occasionally supplemented by inorganic species such as nitric and sulfuric acid from acid rain. However, as we will see later, as a result of the mineral structures and dissolution mechanisms, these rates gradually taper off and slow down significantly as time goes by.

A simplified reaction network chart showing the relationships and interactions between various carbon and alkaline-earth metal species is given in Fig. 4.2.

As mentioned previously, when CO_2 gas comes into contact with liquid water, it can dissolve and speciate into various compounds in solution. These interactions can be roughly described as the CO_2 dissolution (*1a*), CO_2 hydration (*1b*) and sequential deprotonation (*1c* and *1d*) reactions. All five species are often found to coexist, and their abundance and equilibrium positions are highly dependent on environmental conditions under which they are present.

On the other hand, the alkaline-earth metal oxides in silicate minerals have to be liberated from the mineral matrix for the carbonation reaction to occur. In nature, this dissolution is usually promoted by the presence of weak acids in the water in contact with the minerals. For example, the hydration of dissolved CO_2 yields carbonic acid, which can react with and leach out calcium or magnesium cations

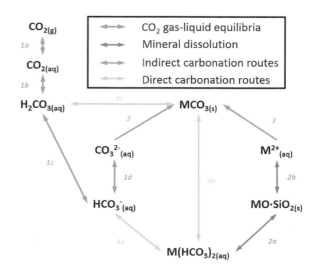

Fig. 4.2 Reaction network chart showing relationships and interactions between various carbon species and alkaline-earth metal oxides in mineral carbonation. More in-depth explanations are given in the text, with reference to items shown in this figure

from the minerals (*2a*). As shown in Fig. 4.2, this results in a dilute solution of metal bicarbonates. Under weakly acidic conditions, the leaching reactions are often quite selective and only the more alkaline components in the mineral are dissolved. Stronger acids can also be employed to accelerate and intensify the metal dissolution reactions (*2b*), though these come at a cost of loss in selectivity and more difficult reagent regeneration steps, as will be shown later.

The dissolution of alkaline-earth metal oxides from silicate minerals also suffers from a secondary problem, which is the gradual passivation of the mineral surface by unreactive silica residues left behind after acid leaching [38, 39]. Many studies have been done on the actual mechanisms of acid leaching of minerals, and the common consensus is that the dissolution reaction follows a shrinking core model (SCM). In the SCM, metal silicates react with acid to give the soluble metal salt and unreacted silica. The selective removal of the metal atoms from the mineral matrix results in the formation of vacancies in the material, effectively forming a layer of porous silica on the surface of the particle. At first, the layer is thin enough to allow acid species to rapidly penetrate and access the metal oxides situated deeper in the particle. However, as the reaction proceeds, the silica layer becomes increasingly thick, and the increased tortuosity and mass transfer resistance slows down the rate of acid diffusion into the particle. As a result, the dissolution rates drop off sharply and the reaction conversions plateau as a result. Figure 4.3 depicts the changes in spatial composition of the mineral particle according to SCM as the leaching reaction proceeds.

Dissolved alkaline-earth metal cations can react directly with carbonate anions in solution to precipitate solid carbonates (*3*). The precipitation reaction is usually near instantaneous, and yields can be close to 100% under the right reaction conditions. Any soluble counter-ions that are present in the solution (Na^+, K^+, NH_4^+, Cl^-,

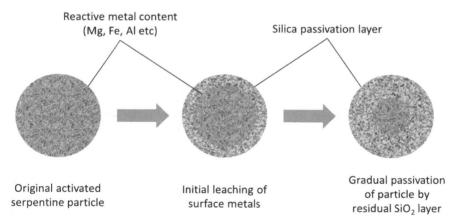

Fig. 4.3 Shrinking core mechanism showing the general dissolution behaviour of metal oxides from metal silicates into solution. A passivating layer of silica residue accumulates on the surface of the leached particle, which inhibits acid penetration and reaction with the core metal oxides

SO_4^{2-}, NO_3^- etc.) during the reaction will combine to give a salt product, which is discarded or regenerated into the respective acids and bases for further reuse. This technique for mineral carbonation, first through a dissolution step followed by a precipitation step in separate reactors (*2b + 3*), is traditionally termed as an indirect method [40]. The acids used in indirect carbonation processes are usually strong mineral acids, and the metal salt product is typically separated from the silica residue (and occasionally purified) prior to carbonation in a second reactor [41–44].

In contrast, direct carbonation methods supply pressurized CO_2 into a single reactor containing the minerals, where the dissolution/precipitation reactions take place simultaneously [45, 46]. Here, a high CO_2 pressure forces an increase in the absolute amount of bicarbonate and carbonate species, which can react to form aqueous metal bicarbonate salts (*4a*), which in turn forms solid carbonates when the conditions are changed, for example when the system is heated or the pH is raised. Raising the pH disproportionates the bicarbonate compounds into two carbonate molecules (*4b*), while heating degasses the bicarbonate solution and shifts the equilibrium towards more carbonate formation (*4c*). In effect, both changes in the pH and temperature drive the precipitation reaction by removing one mole equivalent of carbonate anions from the system.

Based on these observations, it should be apparent by now that the conditions for dissolution and precipitation in mineral carbonation are clearly contradictory. The various chemical species that react in mineral carbonation processes are involved in an intricate interplay that necessarily leads to sub-optimal performances, especially if the process is conducted in a single steady-state reactor.

4.1.3 The Thermodynamics of Mineral Carbonation

The sign (positive or negative) of the change in Gibbs free energy (ΔG_{rxn}) is an indicator of the spontaneity of a reaction. A negative ΔG_{rxn} value points to a spontaneous reaction, and vice versa. ΔG_{rxn} is a function of the change in enthalpy (ΔH_{rxn}) of a reaction, and also of the temperature (T) and change in entropy (ΔS) of the system:

$$\Delta G_{rxn} = \Delta H_{rxn} - T\Delta S$$

Based on this equation, we can infer that reactions with certain characteristics lend themselves more easily to spontaneous occurrence. Reactions that are highly exothermic (negative ΔH_{rxn}) tend to occur naturally, as are those that result in increased disorder in the system (positive ΔS, usually by increasing the number of chemical species or molecules) [47]. By applying our understanding of this basic interpretation of the second law of thermodynamics, we can make a few observations with regards to mineral carbonation. Firstly, since the chemical weathering of magnesium and calcium silicate minerals to form carbonates is known to occur spontaneously in nature, it logically follows that the ΔG_{rxn} of the overall reactions is negative and therefore the products are less energetic than their starting materials.

In other words the carbonates are obviously less energetic and more stable than CO_2 and the starting mineral silicates [48].

Secondly, since the carbonation reaction is one that produces solid products from (partly) gaseous reactants, this suggests that the net entropy (disordering) of the reaction system is one that is decreased at the end. In other words, the ΔS value of carbonation reactions is almost always negative. Bearing in mind that the pre-requisite of a spontaneous reaction is that ΔG_{rxn} is negative, it thus follows that carbonation reactions should have a highly negative value for ΔH_{rxn} to compensate for the always positive $(-T\Delta S)$ term [49]. Put differently, for ΔG_{rxn} to be negative the following needs to be true when ΔS values are negative:

$$|\Delta H_{rxn}| \gg |-T\Delta S|$$

It is thus logical to infer that carbonation reactions should give out a relatively large amount of heat, based on what has been mentioned above. This is in fact what is observed in accelerated carbonation experiments [50]. On the other hand, it is also useful to consider the thermodynamic aspects of mineral carbonation from perspective of the carbon atom, i.e. what actually happens to it as it is converted from a component in an organic molecule to CO_2, and finally to a component within a carbonate solid. An analysis of this energy level pathway will quickly illustrate one of the key strengths of mineral carbonation compared to other CO_2 utilization technologies.

CO_2 is formed when materials containing carbon are oxidized. The most prominent examples of these oxidation reactions include fuel combustion and the in vivo metabolism of glucose to produce energy. For instance, when an energetic molecule like methane is burned, the oxidation state of the carbon atom is increased from $4-$ to $4+$ and a huge amount of heat energy is released:

$$CH_{4(g)} + 2O_{2(g)} \rightarrow CO_{2(g)} + 2H_2O_{(g)}$$

$$\Delta H = -802.5 \frac{kJ}{mol}$$

This change in enthalpy is the dominant driving force that pushes the reaction forward, since the ΔS value of this reaction is miniscule (on the order of a few Joules per Kelvin, as the net number of molecules does not increase, nor is any phase change apparent). Also, since methane can be oxidized into CO_2 through numerous combinations of reactions, it is often neither practical nor necessary to calculate and tabulate the thermodynamic components for all the possible combinations. What really matters in our analysis is how the "embodied" Gibbs free energy of formation in the carbon-containing species changes, and by how much. This concept can be illustrated by mapping the Gibbs free energy of formation of various carbon-based molecules, in relation to CO_2. We can use the following equation to calculate the relative Gibbs free energy of formation of the various carbon-based compounds:

$$\Delta G_r = \left(\frac{\Delta G_m}{N_m}\right) - \Delta GCO_2$$

In the equation, ΔG_r is the relative Gibbs free energy of formation of the product molecule, ΔG_m is the Gibbs free energy of formation of one mole of the product molecule, N_m is the number of carbon atoms in the product molecule, and ΔG_{CO_2} is the Gibbs free energy of formation of one mole of CO_2. Implicit in the equation is the assumption that all the carbon atoms in the product molecule are obtained from the reformation of CO_2. This is expressed in our calculations through the $\Delta G_m/N_m$ term in the equation (i.e. one mole of CO_2 can be converted into one mole of calcium carbonate, or half a mole of ethylene, or a quarter of a mole of butanol, and so forth). While this is clearly not realistic, this simplification is helpful to illustrate the relation between CO_2 and its potential reformed products. A positive ΔG_r indicates that the formation of that particular product molecule requires a net energy input, and vice versa.

As can be seen in Fig. 4.4, all of the organic molecules are situated above CO_2 in the figure, meaning that if these molecules were to be synthesized from CO_2, a net amount of energy input is always required, no matter how efficient the process is. In other words, the reformation of CO_2 into organic compounds is always thermo-dynamically unfavorable. For example, once we account for the heats of formation for water and the effects of temperature, we find the difference in Gibbs free energy between methane and CO_2 to be around 330 kJ/mol at STP. This value is the relative Gibbs free energy of formation between these two chemical species, and can be considered as the absolute minimum amount of energy input required to convert one mole of CO_2 into methane.

On the other hand, the conversion of CO_2 into the carbonate anion has a negative ΔG_r, meaning that the conversion actually releases energy into the system. This value decreases further when it is stabilized by a counter-cation such as Mg, Ca or Na. As mentioned earlier, this large and negative ΔG_r value entails the release of significant amounts of heat. If properly and effectively harnessed, this energy can supplement part of the overall energy consumption in mineral carbonation pro-cesses [51].

Obviously, a thermodynamic analysis of mineral carbonation processes should not be limited to the carbon and alkaline-earth metal species only. The utilization of additives and manipulation of reaction conditions necessarily implies an energy input, either in the form of chemical potentials (regeneration of acid and base catalysts), thermal energy (reactor heating, mineral thermal activation etc.) or physical work (agitation, grinding, CO_2 compression etc.). Proper analysis and accounting of these embodied energies is crucial to ensure that any proposed mineral carbonation process is truly effective at sequestering more CO_2 than it emits.

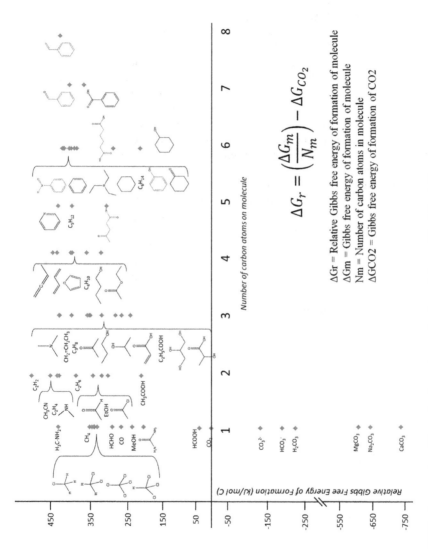

$$\Delta G_r = \left(\frac{\Delta G_m}{N_m}\right) - \Delta G_{CO_2}$$

ΔGr = Relative Gibbs free energy of formation of molecule
ΔGm = Gibbs free energy of formation of molecule
Nm = Number of carbon atoms in molecule
$\Delta GCO2$ = Gibbs free energy of formation of CO2

Fig. 4.4 Map of Gibbs free energies of formation for various molecules, relative to CO_2 (kJ/mol C)

4.1.4   Raw Materials for Mineral Carbonation

As implied by the name, the most widely available raw materials for mineral carbonation are alkaline-earth silicate minerals that are highly abundant in the earth's crust. The preferred silicate raw materials are those termed as ultramafic minerals, whose name refers to their high magnesium and ferric content (hence the term *mafic*) [52]. As a very general rule of thumb, ultramafic minerals typically contain around 40% MgO, 40% SiO_2, 10% Fe_2O_3 and 10% other metal oxides in dry weight, though obviously this is subject to major fluctuations depending on the quality, locality and age of the mineral deposits. The amount of these minerals far exceed the total quantity of combustible carbon on the planet, meaning that theoretically it is possible to store all of the anthropogenic CO_2 into these minerals even if all of the fossil fuels were burned for industrial uses or energy production [53]. These minerals are also rather widespread, with known deposits being found in easily accessible locations that are near to urban centers and major point sources of CO_2 emissions (Fig. 4.5).

In addition to the chemical composition of the minerals, their physical structure and crystalline habits also play determinative roles in their reactivity and suitability for mineral carbonation reactions. Olivine (Mg_2SiO_4) and serpentine $(Mg_3Si_2O_5(OH)_4)$ are two groups of ultramafic minerals that are commonly studied to explore their potential as sources of magnesium for mineral carbonation. Olivine is an orthosilicate mineral; its structure contains individual isolated tetrahedral $[SiO_4]^{4-}$ units that are coupled to interstitial cations $(Mg^{2+}, Fe^{2+}, Ni^{2+}$ etc.). On the other hand, serpentine is a phyllosilicate mineral, and has parallel and interconnected sheets of silica that contain metal cations in between the layers (Fig. 4.6).

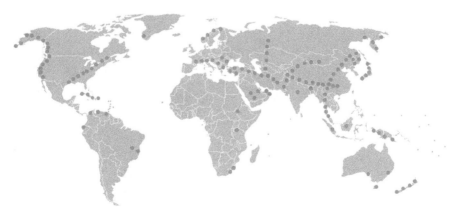

Fig. 4.5 World map with red dots showing regions where ultramafic minerals are known to be commonly found. The amount of accessible ultramafic minerals is estimated at approximately 90 trillion tonnes, enough to store around 22 trillion tonnes of CO_2. For comparison, the total amount of potential anthropogenic CO_2 from proven reserves is calculated to be around 10 trillion tonnes [53]

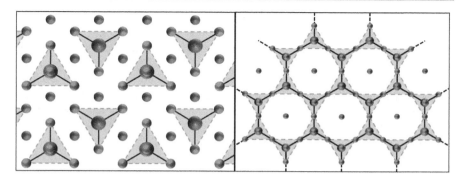

Fig. 4.6 Examples of crystal structures of orthosilicate minerals (left) and phyllosilicate minerals (right). Orthosilicates have individual silica tetrahedrals coupled with interstitial divalent cations, while phyllosilicates have lamellar (sheeted layers) networks of interconnected silica, with the metal cations sitting in the centre

Olivine and serpentine are closely related, in the sense (loosely speaking) that one is the parent of the other. Olivine is an igneous rock, meaning that it originates from tectonic activity in the earth's mantle or crust. It is one of the most common minerals on the planet, with most of it being found in sub-surface deposits that have little contact with the atmosphere. When olivine is exposed to the elements, the mineral quickly weathers to serpentine where it changes its structure and composition, and also undergoes hydration [54]. As such, serpentine is termed as a metamorphic rock, where it is changed from a previous different mineral to its current form.

Serpentine has three notable and distinct morphologies; these are antigorite, lizardite and chrysotile [51, 55, 56]. Antigorite and lizardite differ only slightly in terms of their silicate arrangements in the crystal habit, and their bulk structures remain fairly similar to each other. Both are green-grey minerals with a scaly texture, and have a Moh's hardness index of around 2–4. Chrysotile on the other hand, is notably different from the other two, being a soft fibrous mineral that can exist in long white strands. In fact, chrysotile is better known as asbestos, a cheap and highly effective thermal insulator that was widely used prior to its identification as a carcinogenic material. Some reports have suggested that chrysotile is much more reactive compared to its other two counterparts [57, 58].

The metamorphosis of olivine into serpentine also incorporates water and hydroxyl groups into the altered crystal structure. These hydrates and hydroxyl groups in serpentine can be removed via heating, which increases the porosity of the material [59, 60]. The act of dehydration also often destroys the crystallinity of the mineral in the process and makes it more reactive. As a result, the utilization of serpentine in modern mineral carbonation processes usually starts with a thermal activation step, where the mineral is heated to around 600 to 700 °C for the dehydration reaction to occur. The dehydration step must be done in a controlled manner, as excessive temperatures or long heating times may result in the recrystallization of the mineral into forsterite, which is a member of the olivine family. Forsterite is much less reactive than activated serpentine, hence it is important to optimize the thermal activation step to maximize the reactivity of the minerals [61, 62].

In addition to natural minerals, many other artificial sources of alkalinity can also be used for mineral carbonation. Notable examples of these include red mud (the waste product from bauxite processing), steel slag, waste concrete, incineration bottom ash and many others [6, 46, 63–72].

In the Bayer process, sodium hydroxide is used to dissolve aluminum from bauxite, the primary ore from which it is obtained. The resulting slurry is filtered to obtain a sodium aluminate solution and the unreacted residue. The sodium aluminate is sent downstream to be processed into aluminum metal while the residue is discarded and accumulates in disposal ponds. This residue, known as red mud, is highly alkaline in nature as it still contains an excess of unrecovered sodium hydroxide in the filter cake. The red coloration arises from the presence of iron hydroxides in the residue. More than 150 million tonnes of red mud is generated every year as a byproduct of aluminum production. The CO_2 uptake capacity of red mud, depending on its quality, is reported to be in the range of 0.03–0.08 t CO_2/t red mud. This translates to around 4–12 million tonnes of CO_2 sequestered, in addition to the benefits of treatment and neutralization of the toxic material. Although relatively small, this amount is still considerable and should not be overlooked [73].

Slag from iron and steel production is an even larger source of industrial waste materials for mineral carbonation. During the smelting of iron ore, lime is added to the melt to remove impurities in the mixture. The added lime reacts with the silica and other contaminants to form a solid calcium silicate phase, which floats on top of the molten iron in the furnace and is skimmed off. This slag is alkaline in nature and can be used as a source of calcium to bind with captured CO_2. There are also other utilization routes for the material, especially in construction applications. With an average CaO content of around 45–50 wt%, iron and steel slags are excellent sinks for CO_2 emissions. Annual production of steel and iron slag amounts to around 470–610 million tonnes, which corresponds to a CO_2 sequestration capacity of approximately 143–186 million tonnes/a if fully utilized [74].

Waste concrete is generated when buildings are demolished, and a large portion of the waste is recycled as aggregates in new construction projects. However, there is still a fraction of waste concrete that is disposed of every year, and these can be utilized as carbon sinks for CO_2 as well. Concrete contains significant amounts of calcium, and is capable of capturing CO_2 on its own as it settles and hardens. The natural carbonation of concrete occurs on the surface, which gradually retards the penetration of CO_2 into the deeper layers and maintains a highly basic environment in the interior. The alkaline pH in the interior sections of concrete helps to prevent corrosion of the steel reinforcing bars in the structure, ensuring that the construction remains structurally sound for a long time. When buildings are demolished, the concrete is broken down into smaller pieces, exposing the interior parts to air and binding CO_2 to the calcium content in the concrete waste [75].

Incineration bottom ash (IBA) is also a potential source of alkalinity for the sequestration of CO_2 through carbonation. Combustion and gasification of the organic content in municipal solid waste (MSW) leave behind a solid ash residue. This ash consists of non-volatile materials such as ceramics, glass, metals and metal oxides. A rudimentary magnetic separation step is usually employed to remove most of the ferrous metals in the ash for recycling. The remains typically contain around 10–25% of alkaline-earth oxide content, though this is highly variable and will change depending on fluctuations in municipal waste compositions. In order to meet environmental standards, the flue gas from incineration is usually scrubbed with lime to remove NO_x and SO_x prior to release into the atmosphere. The lime is usually added in excess, to prevent the pollutants from escaping into the environment. The spent lime is not recycled, and is typically mixed with the incineration bottom ash. This further increases the calcium content in the ash residue, and may be made available for the capture and sequestration of CO_2. The calcium oxide content in the mixed ash can be as high as 50% by weight, though a sizeable portion of this may already be bound as nitrates, carbonates and sulfates. Joseph et al. estimates that globally, a total of around 1.3 billion tonnes of MSW is generated every year. Of this, 10% (130 million tonnes) are incinerated and the rest are landfilled. The incinerated MSW is reduced in mass and volume, and generally the ash residue is around 10% of the original weight of the disposed waste (equating to around 13 million tonnes of IBA). This also means that IBA in effect acts as a renewable source of alkalinity for CO_2 capture and sequestration via carbonation. A very rough calculation would suggest that IBA from MSW incineration would then have the potential to absorb around 8–10 million tonnes of CO_2 per year at current rates of MSW disposal and incineration. The carbonation of IBA would also stabilize the material and inhibit the leaching of heavy metals from the ash to a certain degree, thus reducing the environmental impacts of its disposal [68, 76–78].

A comparison of the indicative compositions of the abovementioned materials is given in Table 4.1.

Table 4.1 Indicative compositions of common materials that can be used to supply alkalinity for mineral carbonation reactions

	Serpentine	Wollastonite	Steel slag	Cement waste	Incineration bottom ash
MgO	**40%**	Minor	4.5%	2%	2.5%
Al_2O_3	5%	0.5%	2.5%	5%	5%
SiO_2	40%	50%	18%	20%	12%
CaO	5%	**45%**	**50%**	**65%**	**50%**
Fe_2O_3	10%	Minor	2%	2.5%	10%
Other notable components	Cr, Mn, Ni and others	K, Mn, Ti, P and others	Mn, Ti, S and others	K, P, Ti, Mn and others	Na, P, S, Cl, K and others

The major alkaline components in these materials are highlighted in bold
Weight percentages may not add up entirely to 100%, and these should be taken as approximate values only

4.2 Overview of Basic Process Designs for Mineral Carbonation

In the second part of the chapter, we will discuss the practical application of the theoretical understanding and concepts as described in Part One. Based on these ideas, many chemical processes have been developed to exploit and accelerate the reactions, in order to enable mineral carbonation to become a feasible pathway for large scale CO_2 capture and sequestration. This section will act both as a review of recent developments in mineral carbonation process designs, and also as a study in how the previously introduced concepts are exploited and applied in practical examples for mineral carbonation.

Traditionally speaking, there are two schools of mineral carbonation technologies that were developed in parallel; these are the *direct* (or sometimes called "pressure") carbonation and *indirect* (occasionally described as "chemically enhanced") carbonation routes [40, 79]. These process designs are not strictly incompatible or inherently better than the other, and sometimes a process can even contain a little bit of both routes such that they may defy categorization into the conventional ways that we usually classify mineral carbonation processes.

4.2.1 Pressure Carbonation

Pressure carbonation, as the name implies, involves mineral carbonation reactions that are conducted under pressure. In one of the earliest iterations of pressure carbonation processes, pure CO_2 and steam were supplied at high pressure (115–150 bars) into a heated autoclave to react with activated serpentine (i.e. milled and heat-treated). This process was developed by NETL in the mid-2000s [45, 80–82]. The reaction was designed as a one-pot process, where the CO_2 and activated serpentine were converted to silica and magnesium carbonate:

$$3MgO \cdot 2SiO_2 + 3CO_2 \rightarrow 3MgCO_3 + 2SiO_2$$

Other types of minerals such as olivine and wollastonite were also used as well, and the products were obtained as a mixture. The reaction conversions were in the range of 36–92%, depending on the type of raw materials used. The authors reported that lizardite was the least reactive with conversions around 40%. Olivine, wollastonite and antigorite were much more effective for mineral carbonation, attaining conversions in excess of 80%. In addition to CO_2 and the minerals, 1 mol/L NaCl and 0.64 mol/L $NaHCO_3$ were also added to the reaction system promote the mineral carbonation reactions. The authors also concluded that the best reaction conditions for various mineral types were as follows: olivine (185 °C, 150 bar, with additives); wollastonite (100 °C, 40 bar without additives); and for activated serpentine (155 °C, 115 bar, with additives). A simplified flow diagram for this process is given in Fig. 4.7.

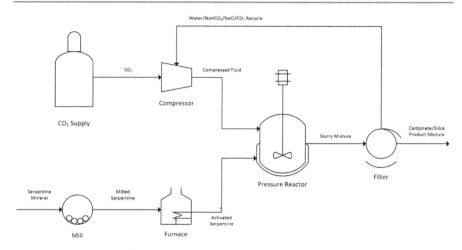

Fig. 4.7 Simplified diagram of pressure carbonation process developed by NETL in the mid-2000s

In the NETL process, high pressures and temperatures were necessary to push CO_2 into the supercritical phase to facilitate mass transfer between the reactants. The high pressure also shifts the CO_2 speciation equilibrium in the system, by forcing more CO_2 to convert into carbonic acid and thus lowering its pH (think of it as a supercharged club soda). This acidified environment, combined with the high temperatures, helped to leach out the alkaline contents in the mineral and made them available for reaction. The shifted equilibrium also resulted in more carbonate and bicarbonate anions being present in the system, which rapidly combines with the leached alkaline-earth cations to form solid carbonates. The mixture exits the reactor after 1 h of reaction time to be filtered. The water and unreacted CO_2 are recycled and recompressed, and the carbonate and silica product mixture is separated.

We now know that in pressure carbonation, there are two distinct reactions that occur that ultimately result in the precipitation of inorganic carbonates from the reaction of CO_2 with minerals. These two reactions are contradictory, and their optimum reaction conditions are opposites of each other [83]. As described earlier in the first part of the chapter, the first reaction is a dissolution reaction, where an alkaline-earth metal oxide is extracted (preferably) into an aqueous phase. Acidic conditions and lower temperatures (dissolution is typically an exothermic reaction) are favorable for this. However, for practical purposes, the dissolution reaction is often conducted under moderate heating, since the gains in reaction rates far outweigh the benefits from the small driving force due to heat removal. Once dissolved, the alkaline-earth metals can be precipitated from solution as well. Precipitation is favored under high pH and temperature conditions, where carbonate anions predominate and easily react to form insoluble solids.

With a better understanding of the basic mechanisms of mineral carbonation, it is tempting to draw a conclusion that the one-pot reactor method for direct carbonation is inefficient. The conditions applied in the process are not conducive for either reaction, and both compete against each other. Thus, a more rational design is needed for a direct mineral carbonation process. From this analysis, researchers at Royal Dutch Shell came up with an improved and more refined version of the process developed by NETL. In their version of the direct carbonation process, the fundamental reaction mechanisms remained more or less the same: pressurized CO_2 is supplied into a heated reactor to extract magnesium from activated serpentine. The extracted magnesium is then precipitated to give magnesium carbonates, and the fluid phase is recycled [84–86].

However, the main difference was that the individual reactions were conducted in separate reactors, and this allowed them to isolate the reactants from each other and optimize the conditions in order to promote the desired reactions in each step. By compartmentalizing the reactions, the reactions could be conducted under less harsh conditions. In their design, 5 bars of CO_2 pressure was sufficient to allow

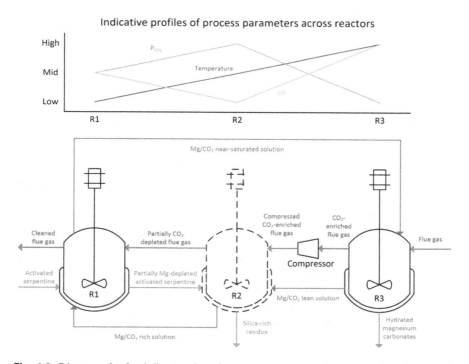

Fig. 4.8 Diagram of refined direct carbonation process as proposed by researchers from Shell: note the overall counter-current design embedded in the system to maximise the driving forces (chemical potential gradients) between reactions. The liquid phase is recycled instead of being discarded. **R1**: Dissolution reactor; **R2**: optional dissolution enhancement reactor; **R3**: precipitation reactor

reasonably high degrees of Mg extraction as bicarbonate species (compared to operating conditions of >100 bars for the NETL process) in the first step. In the second step, near-total recoveries of magnesium (as hydrated carbonates) were achievable at sub-boiling temperatures up to 90 °C. One version of their process also had the dissolution reaction occur in two stages (two separate reactors) to maximize the conversion of Mg into the solution [86]. The entire process was designed as a counter-current setup, where activated serpentine minerals enter from one end and is gradually depleted in magnesium content, while flue gas enters from the other end and is enriched with CO_2 and then consumed before it exits the system. The counter-current design in the process ensures that the driving force in each reactor is maximized to enhance reaction efficiencies. A simplified diagram of their refined process is given in Fig. 4.8.

As can be seen, the refined process by Shell is perhaps more elegant, and clearly has an advantage in terms of operating conditions.

4.2.2   Chemically-Assisted Carbonation

Chemically-assisted carbonation processes are another sub-category of mineral carbonation technologies. The basic sequence of dissolution-precipitation still applies in chemically-assisted carbonation, although there are some notable differences between the two process categories in several aspects. Firstly, the reactions are accelerated primarily via the addition of chemical energy instead of mechanical or thermal energies. Secondly, chemically-assisted processes typically proceed at much lower temperatures and pressures (mild heating and ambient pressure). Thirdly, raw minerals can be used directly without the need for intensive comminution and heat treatment. Fourthly, a regeneration step is usually needed to convert the spent salt by-products into reactants again for recycling back into the process [87–90].

A common feature of chemically-assisted carbonation processes are the use of highly energetic chemical species such as strong acids and bases [67, 91]. These are added to the system to react with CO_2 or the alkaline-earth containing species to form a reactive intermediate. Many chemically-assisted carbonation processes follow a general template which involves dissolution steps to bring the (bi)carbonate and alkaline-earth species into the same aqueous system followed by rapid precipitation of the solid carbonate product. For example, a common way to facilitate the dissolution reaction is to use aqueous HCl to dissolve Mg from serpentine into the aqueous phase:

$$Mg_3Si_2O_5(OH)_{4(s)} + 6HCl_{(aq)} \rightarrow 3MgCl_{2(aq)} + 2SiO_{2(s)} + 5H_2O_{(l)}$$

The reaction is technically a neutralization reaction, where HCl is consumed and converted into a dissolved chloride salt ($MgCl_2$). Also, since silica is practically inert against acid attacks, it remains as a solid residue in the mixture. A filtration step usually follows, where the liquid and solid phases are separated. On the other hand, CO_2 can be contacted with an alkali such as sodium hydroxide to form carbonate or bicarbonate species:

$$CO_{2(g)} + 2NaOH_{(aq)} \rightarrow Na_2CO_{3(aq)} + H_2O_{(l)}$$

This reaction is clearly also a neutralization reaction, where the overall effect is to form a salt (Na_2CO_3) from stronger acid/base parents. Following these steps, the aqueous salt products are mixed and the precipitation reaction takes place almost immediately:

$$MgCl_{2(aq)} + Na_2CO_{3(aq)} \rightarrow MgCO_{3(s)} + 2NaCl_{(aq)}$$

In the case shown here, sodium chloride is obtained as a spent by-product, which obviously may not be reused directly in the process. The spent by-product has to be regenerated in an additional step, where the salt is converted into its acid and base parents. The regeneration step is often the most energy intensive step in the entire chemically-assisted carbonation process. As such, recent trends have tended towards the use of reagents that can be thermally regenerated at lower temperatures. On the other hand, electrochemical regeneration methods are uncommon, but not unheard of [92–94]. A systematic analysis of process thermodynamics can help chemists and engineers to determine the most suitable combination of materials and reactants for a proper design of chemically-assisted carbonation processes.

Since there are near-infinite combinations of acids and bases that can be used to accelerate the dissolution reactions, there is a plethora of various chemically-assisted carbonation processes that have been reported in the literature [19, 40, 79, 95, 96]. The wide array of reagent choices and combinations available for a chemist to use in a process can seem bewildering at first, and it is impossible to properly study each and every possible combination. Fortunately, there are some general rules of thumb and process design heuristics for this:

- Begin with the identification of a salt compound that is easily dissociated or hydrolyzed at relatively low temperatures, and then determine if the acid and base parents are sufficiently reactive and soluble in an aqueous phase.
- The intermediate salt compounds (i.e. $MgCl_2$, Na_2CO_3 etc.) should be highly soluble to minimize the amount of water needed in the system.
- Heating usually helps in the alkaline-earth dissolution steps, but is not mandatory. On the other hand, ambient or low temperatures generally facilitate the CO_2 absorption step.
- Inorganic acids such as HCl, H_2SO_4 and HNO_3 are commonly used to dissolve alkaline-earth metal oxides from minerals. The reason for this is that their conjugate bases are often more resistant to deterioration during the thermal regeneration of their spent salts compared with organic acids.
- Monoprotic bases are commonly used to absorb CO_2, since their (bi)carbonate salts are usually highly soluble.

- Impurities in the feed materials will have significant effects on the quality of carbonate products, and have to be accounted for when estimating the mass and energy balance for the process.
- Multiple separation/purification steps are often included in the process design for economic reasons. These steps allow the recovery of high purity products that can be sold to recover the (usually) higher costs of chemically-assisted carbonation processes.

As mentioned earlier, the bulk of chemically-assisted carbonation processes described in literature follow a similar acid leaching/CO_2 absorption/carbonate precipitation sequence. The large number of interesting processes based on this template are too numerous to list and describe in detail. In lieu of that, we would like to bring the reader's attention to some of the more recent and unique processes that deviate from this basic structure. These will serve as examples to highlight and explain how the rules of thumb come into play in chemically-assisted carbonation processes, and also how others have sought to bypass the major limitations of the standard processes through innovative process design.

Researchers in Abo Akademi, Finland have developed a process (the "AA route") utilizing ammonium sulfate as a leaching reagent for mineral carbonation [43, 97–99]. Ammonium sulfate is easily dissociable at moderately low temperatures. When mixed and heated with milled serpentine, the sulfate salt melts and decomposes to form ammonium bisulfate and gaseous ammonia:

$$(NH_4)_2SO_{4(s)} \rightarrow NH_{3(g)} + NH_4HSO_{4(l)}$$

The molten and acidic bisulfate salt reacts with the serpentine mineral to form magnesium sulfate and release more ammonia gas. The resultant mixture comprising $MgSO_4$, SiO_2, unreacted bisulfates and impurities is then washed with water. The wash water forms a sulfate salt solution which is filtered off, and the prior degassed ammonia is recombined with the solution to precipitate magnesium hydroxides:

$$MgSO_{4(aq)} + 2NH_{3(g)} + 2H_2O_{(l)} \rightarrow Mg(OH)_{2(s)} + (NH_4)_2SO_{4(aq)}$$

As a result, ammonium sulfate is regenerated as an aqueous solution, and separated from magnesium hydroxide for reuse. The regenerated solution is evaporated to recover solid $(NH_4)_2SO_4$, and magnesium hydroxide is used as the sorbent for CO_2 capture and carbonation. The carbonation reaction in this case is conducted at high temperature and pressures to increase the reaction rates. A simplified illustration of the AA route is given in Fig. 4.9.

The biggest benefit of the AA route is that it allows the amount of water used in the process to be properly controlled and minimized to avoid excessive energy use. The steps requiring the highest temperatures (dissolution, gas-solid carbonation) are substantially free of water, which helps to minimize energy use associated with vaporization.

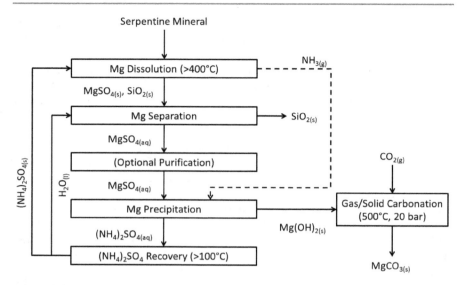

Fig. 4.9 Simplified flow diagram of the AA route as developed by Zevenhoven et al.

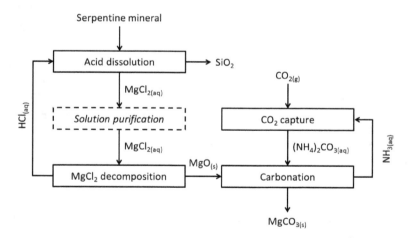

Fig. 4.10 Simplified block diagram of ICES carbonation process. The chloride species is regenerated into hydrochloric acid before the carbonation step to allow for easier separation and recovery

The regeneration step can also be improved by preventing the formation a salt product that has high thermodynamic stability such as NaCl, or has acid/base parents that are hard to separate ($NH_4Cl \rightarrow NH_3$ and HCl, for example). Researchers at the Institute of Chemical and Engineering Sciences (ICES), Singapore have developed a chloride-based process based on this concept [89]. The process is depicted in Fig. 4.10.

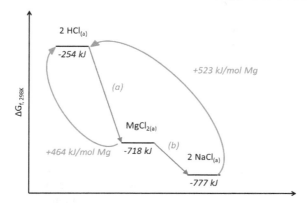

Fig. 4.11 Energy changes from the perspective of the chloride species in the chemically-assisted mineral carbonation system. Blue arrows depict: **a** Formation of MgCl$_2$ from HCl in the dissolution step; and **b** conversion of MgCl$_2$ into NaCl in the carbonation step. Regeneration of the hydrochloric acid from MgCl$_2$ is energetically less demanding than from NaCl

In the ICES process, serpentine mineral is first reacted with aqueous HCl to dissolve magnesium out from the material. The chloride solution is then purified by removing residual silica and any co-extracted metals from the aqueous phase. After that, the resultant magnesium chloride solution is concentrated and pyrolyzed to obtain MgO, and HCl gas is regenerated as well. Heat can be recovered from the steam generated in the pyrolysis step, and the condensed water is used to absorb HCl to regenerate an acid solution for recycling back into the process. MgO from the pyrolysis step is brought forward for reaction in a carbonate or bicarbonate solution, which regenerates a base (typically Na$^+$- or NH^{4+}-based) for CO$_2$ capture from flue gas.

In this process, regeneration step is done "halfway" into the process, before the two intermediates interact to form the carbonate product and stable salt by-product. This effectively isolates the acid and base components from each other, and avoids complicated regeneration and separation regimes usually associated with the lysis of highly stable chemical compounds. An illustration of the concept in thermodynamic terms is given in Fig. 4.11.

The processes mentioned above adhere to and cleverly exploit many of the rules of thumb mentioned earlier in this section, yet do not resemble any of the "standard" chemically-assisted process templates. This goes to show that designing a process for mineral carbonation is an art as much as it is science.

4.2.3   Other Carbonation Processes

Many other less conventional mineral carbonation processes have been reported. These do not conform strictly to the categories as described previously, yet they do adhere to the general principles and mechanisms of mineral carbonation reactions.

The concept of bio-catalyzed mineral carbonation has been studied by several research groups. These processes revolve around the use of an enzyme called carbonic anhydrase to increase the concentrations of carbonate anions in aqueous systems, in order to facilitate the actual carbonation reaction [100–102]. Carbonic anhydrase is an enzyme that is commonly found in many organisms, and its primary function is to quickly convert dissolved carbonic acid or bicarbonate species into hydrogen cations and carbonate anions:

$$H_2CO_{3(aq)} \rightarrow H_{(aq)}^{+} + HCO_{3(aq)}^{-} \rightarrow 2H_{(aq)}^{+} + CO_{3(aq)}^{2-}$$

As mentioned previously in the first section in the chapter, the rates of the actual carbonation reaction are dependent on the concentrations of the carbonate anions and alkaline-earth cations in the system. This reaction is often limited by the rate of CO_2 dissolution, hydration and deprotonation in the aqueous phase, thus the use of carbonic anhydrase can help to overcome one of the biggest rate-limiting obstacles to mineral carbonation. A simple and effective way to exploit the benefits of this biomimetic carbonation mechanism is to introduce carbonic anhydrase into hard water, which already contains relatively high levels of magnesium or calcium cations in solution. The added enzyme will quickly convert the bicarbonate counter-ions into carbonate anions, which then interact and bind with the dissolved alkaline-earth metals and precipitate from solution. Power et al. have reported an increase in carbonation rates of up to 240% using carbonic anhydrase, and this could be improved even more by increasing the supply of CO_2 into the system [103].

However, since enzymes are proteins, they have a very narrow range of ideal operating conditions and are also very sensitive to changes in local conditions. Fortunately, the optimum operating conditions for enzyme usage are often under ambient conditions, and generally do not require additional energy or material inputs to the system. On the other hand, there are also several major drawbacks of using a biomimetic reaction system for mineral carbonation. Enzymes are susceptible to significant changes in pH and temperature, and will undergo irreversible denaturation and permanently lose their function once this occurs. Secondly, an enzyme is a catalyst, which does not alter the equilibrium of the system. This means that the reactor must be optimized in such a way that the systemic equilibrium is favored towards product formation, and not allowing the products to revert into the reactants. This usually means a continuous removal of the solid carbonate products, or a constant supply of excess reactants into the reactor to ensure that the equilibrium is pushed towards the right. Other major drawbacks of using enzymes are that they are costly to replace, and not really volume efficient compared with other mineral carbonation processes. Therefore, while enzyme mediated mineral carbonation is an attractive prospect from an energy use standpoint, it remains to be seen if the process can be implemented in a practical and economic manner.

Apart from bio-catalysis, the hydration and conversion of CO_2 into reactive aqueous species can also be accomplished using metal-based catalysts. Nickel and zinc complexes have been reported to be reasonably efficient for CO_2 hydration reactions, though some have questioned their effectiveness in real life situations [104–106]. They can also be applied for the reverse reaction, where dissolved bicarbonates are decomposed (ultimately) into gaseous CO_2 to precipitate solid carbonates from solution (see Fig. 2).

Another interesting way to conduct mineral carbonation is to couple it with other existing reaction systems, either as a complement or to improve the main process. One notable possibility is the integration and utilization of mineral carbonation for sorption-enhanced water-gas shift reactions [107]. The water-gas shift reaction involves the conversion of carbon monoxide and steam into carbon dioxide and hydrogen gas.

$$CO_{(g)} + H_2O_{(g)} \leftrightarrow CO_{2(g)} + H_{2(g)}$$

While the water-gas shift reaction can achieve reasonably high conversions under specific conditions, it is still an equilibrium-controlled reaction where a high concentration of products can retard the forward reaction. By continuously removing the CO_2 byproduct from the product stream, two benefits are realized simultaneously: Firstly, the equilibrium in the system is shifted and higher conversions are achievable. Secondly, the hydrogen gas is purified as well, from a maximum of 50 v/v% in a non-sorption enhanced system, to more than 90 v/v% with proper removal of CO_2. Depending on the reaction conditions, the removal of CO_2 is typically effected through the use of in situ chemisorbents such as calcium oxide or magnesium oxide [108].

4.3 Techno-economic Aspects of Mineral Carbonation

In this section of the chapter, we will look at the general technical and economic aspects of mineral carbonation and see how these can influence the practicality and implementation of MC technologies on an industrial scale. More specifically, there are three major techno-economic aspects that are of interest to quickly evaluate and screen for viable process alternatives for mineral carbonation. These are the CO_2 penalty, capital costs (CapEx) and operating costs (OpEx). These helpful indicators allow researchers to conduct a reality check on their ideas, and rapidly pinpoint major issues with their proposed processes. In addition, this section will also briefly discuss how mineral carbonation can be integrated with material upgrading processes that can generate multiple revenue streams to offset the costs associated with the sequestration process itself.

4.3.1 The CO_2 Penalty

As with all other green technologies, a mineral carbonation process is clearly bound by some material and energy use limits to ensure that there is a net sequestration of CO_2 when the process is implemented. In other words, an MC process must be designed in such a way that it does not emit more greenhouse gases than it sequesters throughout its life cycle. However, it is inevitable that a certain amount of energy has to be supplied to enable the reactions to take place. These energies can take the form of electrical, mechanical or thermal energies [109].

The use of fossil fuel-derived energy to drive these reactions will result in CO_2 emissions that will offset the gains in an MC process. These emissions reduce the efficiency of mineral carbonation, via the imposition of a CO_2 penalty on the process. One simple way of defining the CO_2 penalty is to state it as the amount of CO_2 emitted for every unit of CO_2 sequestered.

$$CO_2\ Penalty = \frac{CO_2\ emitted\ from\ process}{CO_2\ sequestered\ by\ process}$$

The units can be in terms of mass per mass (i.e. t/t) CO_2, though it can also be defined as a percentage value for a more straightforward indication. In this form, the equation can be used as a gauge of the efficiency of an MC process, but its usefulness is rather limited. The equation can be expanded to take into account the amount of energy use in a mineral carbonation process, and also the type of fuel the energy is derived from.

$$CO_2\ Penalty = \frac{\sum (EU \times EF \times Eff)}{CO_2\ sequestered\ by\ process}$$

In this form, the amount of CO_2 emitted from the process is expressed as the total sum of the products of the energy use (EU) in each process step, their respective emission factors (EF) and the efficiency (Eff) of that particular unit operation involved. The units of energy use are typically in joules (or its multiples, i.e. kJ, MJ, GJ etc.) or watt-hours (also its multiples, i.e. kWh, MWh etc.). It is often more convenient to express thermal energy use in units of joules, and electrical or mechanical energy use in watt-hours, since these are the units that are most commonly encountered when discussing the respective energy outputs. An emission factor is a measure of the amount of CO_2 emitted when a unit of energy is produced and consumed [110]. As such, the units associated with an emission factor are in the form of mass of CO_2 per energy unit (i.e. t CO_2 per GJ, kg CO_2/kWh etc.). Multiplication of the amount of energy use with an emission factor would thus give the amount of CO_2 emitted from a particular process step, and their sum is then compared against the amount of CO_2 sequestered as a result of implementing the MC process.

This ratio, the CO_2 penalty, is therefore an indication of the efficiency of an MC process. Logically, any combination of process steps that result in 100% or more

Table 4.2 Approximate or representative values of common unit operations or phenomena encountered in mineral carbonation processes

Process step/unit operation	Approximate/representative value	References
Mineral crushing (to ~500 µm)	2 kWh$_{el}$/tonne mineral	Gerdemann et al. [45]
Mineral grinding (to <75 µm)	13 kWh$_{el}$/tonne mineral	
Mineral grinding (to <38 µm)	83 kWh$_{el}$/tonne mineral	
Mineral grinding (to <10 µm)	233 kWh$_{el}$/tonne mineral	
Serpentine activation (700 °C)	0.55 GJ$_{th}$/tonne mineral	Balucan et al. [111]
CO$_2$ compression (5 bar)	38 kWh$_{el}$/tonne CO$_2$	ASPEN Plus simulations
CO$_2$ compression (30 bar)	83 kWh$_{el}$/tonne CO$_2$	
CO$_2$ compression (73 bar)	108 kWh$_{el}$/tonne CO$_2$	
Specific heat of water	4.2 MJ/tonne H$_2$O	Perry's Chemical Engineer's Handbook [112]
Latent heat (vap.) of water	2257 MJ/tonne H$_2$O	

The actual energy use for CO$_2$ compression depends a lot on the design and specifications of the CO$_2$ stream and compression train. For discussion purposes, it is assumed that the compression train involves 4 stages of isentropic compressors, with intercooling to 35 °C and the CO$_2$ enters at 30 °C and 1 bar pressure. The notations kWh$_{el}$ and GJ$_{th}$ refers to electrical and thermal units of energy respectively

will mean that the process will have no net sequestration of CO$_2$, and in fact will be better off without its implementation. It thus follows that this number should be as low as possible. With the near-infinite combinations of process steps and reaction conditions that can be implemented in an MC process, it is up to the researcher to determine the energy use in their own process designs. However, there are some approximate rules of thumb that are readily available for reference, and these usually relate to process steps that are common to nearly all mineral carbonation processes. A table listing the approximate energy consumption for these steps is shown below (Table 4.2).

These numbers serve as a rough guideline to quickly estimate the energy use in each process step in mineral carbonation. They are not meant to be precise, as slight changes in the system variables will yield a different value for each item listed above. For example, CO$_2$ entering a compression train at a relatively high pressure (let's say 50 bars) and compressed to supercritical state will obviously consume less energy than when it's coming in at 1 bar, and depending on the impurities present, it may also vary by an order of magnitude or more. Another example is that similar classes of minerals from different sources (i.e. serpentine from China vs from Finland) will have slight differences in grindability and thermal dehydration properties, and also mean that they have grinding and activation energies that are particular to that sampling of materials. If a more thorough and precise study of the energy use of a mineral carbonation process is needed, researchers are advised to conduct experiments to measure the relevant mass and energy balances. In addition, some values such as the energy requirements for CO_2 compression and salt solution heating can also be referenced from published standards or data curves available in the literature.

On the other hand, the emission factors for energy use are much more straightforward and easier to estimate. If renewable energy is used to drive any of the process steps, then the related emission factors can be safely taken to be zero [110]. However, if fossil fuels are used to supply energy into the process, then some care must be taken to calculate their impact on the CO_2 penalty of the process. In general, energy is supplied as heat or mechanical energy, which is in turn derived from electricity generated from power plants. As a general rule, the use of electricity in a mineral carbonation process should be kept to a minimum for a couple of reasons [19]. Firstly, by diverting its electricity output into MC processes, the power plant has to generate more electricity to compensate for the reduced amount that can be sold to household or other industrial consumers. This lowers the profitability of the plant and indirectly leads to increased CO_2 output at the source [113]. Secondly, fossil fuel-derived electrical energy has much higher emission factors compared with thermal energy use. The most efficient commercially available turbines level out at around 50–60% efficiency, meaning that every joule of energy supplied as heat from fossil fuel combustion yields at most about half a joule of electrical energy and the rest is lost as waste heat. Additionally, in most countries, electricity is generated and supplied through a grid that links various power plants. These plants have different designs, efficiencies and may even run on different fuel types. As such, the emission factor for electricity use in a particular country is highly dependent on its fuel mix, and the value is usually available from environmental reports prepared by non-governmental sources.

In addition to the type of energy used, the kind of fossil fuel used to generate that energy also plays a major role in determining the value of the emission factors. Common fuels include natural gas, oil, coal, municipal solid waste and biomass, and each has a characteristic and representative emission factor related to the fuel type. Natural gas and biomass have the lowest emission factors; crude oil is somewhere in the middle, and coal and municipal solid waste have the highest CO_2 emissions per unit energy supplied. Approximate or indicative values for the emission factors of various fuel types can be found in Table 4.3.

For illustration purposes, let us apply the concepts that we have just discussed to evaluate a process for obtaining activated serpentine for use in mineral carbonation. This typically involves grinding the mineral to 100 microns and heat activation at 700 °C. For convenience we assume that all the energy requirements are supplied entirely by coal. It is also assumed that four tonnes of minerals thusly prepared are needed to sequester one ton of CO_2. Finally, we ignore all efficiency factors in the unit operations and assume that these are 100% efficient. The CO_2 penalty of this process is therefore:

Table 4.3 Approximate CO_2 emission factors from energy use based on fossil fuel type

Energy and fuel type	Indicative emission factors	Reference
Electricity (average efficiency)	*(kg CO_2/MWh$_{el}$)*	U.S. Environmental Protection Agency (EPA) [114]
Coal (33%)	1000	
Crude oil (31%)	800	
Natural gas (42%)	430	
Heat	*(kg CO_2/GJ$_{th}$)*	
Coal	90	
Oil	70	
Natural gas	50	

$$\frac{\left(4\,t\,\text{minerals} \times 10\,\frac{\text{kWh}_{el}}{t\,\text{minerals}} \times 1\,\frac{\text{kg CO}_2}{\text{kWh}_{el}}\right) + \left(4\,t\,\text{minerals} \times 0.55\,\frac{\text{GJ}_{th}}{t\,\text{minerals}} \times 90\,\frac{\text{kg CO}_2}{\text{GJ}_{th}}\right)}{1\,\text{tonne CO}_2\,\text{sequestered}}$$

$$= \frac{238\,\text{kg CO}_2\,\text{emitted}}{\text{tonne CO}_2\,\text{sequestered}} = 23.8\%$$

Based on these estimates, we find that a coal-powered process for mineral grinding and heat activation will incur a CO_2 penalty of around 24%. Further downstream unit operations to actually react CO_2 with the activated minerals will also add to the CO_2 penalty. It can also be seen that the use of cleaner sources of energy will greatly help with lowering the CO_2 penalty, making the mineral carbonation process much more efficient and capable of sequestering larger amounts of CO_2.

4.3.2 The Economics of Mineral Carbonation

CO_2 emissions are a negative externality of industrial activity; this means that their economic impacts are often not properly accounted for in conventional economic analyses [115]. In addition, since carbon capture and storage (CCS) technologies generally do not produce tangible products that can be sold to generate revenue, the financial motivations for capturing and sequestering CO_2 are often weak and not influenced by market forces [113]. As such, without laws or taxation schemes that put a price on carbon emissions, it is difficult to determine what an appropriate or acceptable target cost is for CO_2 sequestration.

Therefore the economic justifications for implementing CCS technologies are usually context-dependent, and are highly reliant on local government policies for carbon emissions pricing [116, 117]. The development and implementation of CCS technologies will be hampered by negligible carbon taxes, since this may mean that it is cheaper to simply pay the tax than to install and run expensive processes to prevent CO_2 emissions. The reverse is also true; a high carbon tax incentivizes industry players to seek out cheaper alternatives to paying a hefty tax on excessive CO_2 emissions. It is thus worth bearing in mind that any CCS process must have a CO_2 sequestration cost that is less than the local carbon tax/price. Even so, we will see that this is not strictly true in some very special cases. Certain processes that capture and utilize CO_2, such as enhanced oil recovery and CO_2-to-chemicals will generate a saleable product and can bring economic benefits [118–121]. As a result, the drive to develop, adopt and implement them is of greater interest for the private sector. Similar to these, mineral carbonation processes can also be relatively independent from government policy on carbon pricing, and we will discuss this later in the chapter.

Since the targeted readers of this chapter are researchers, it is likely that any ideas the audience has in mind are in the relatively early stages of process development. At this stage, it is impossible to obtain precise or reliable costing data, seeing as the information available on hand is likely to be scant anyway. However, this does not mean that the proposed MC process cannot be evaluated properly. To this end, we will discuss some shortcut techniques to quickly estimate the capital expenditure (CapEx) and operating expenditure (OpEx) of MC processes in order to reach a ballpark figure. These techniques can also be generalized and applied to other carbon capture and storage/utilization processes as well.

The unit operations typically encountered in mineral carbonation processes are usually quite traditional; these often include milling, calcination, filtration, gas compression, conventional reactors etc. As such, the optimization, sizing and costing of MC processes are usually quite straightforward. Sizing equations and costing curves that correlate size and equipment costs are widely available, and these are easily found in textbooks or the internet. This aspect will not be discussed in-depth here.

While it is not that difficult to calculate the CapEx of a mineral carbonation process for a fixed scale, repeated calculations for scenarios with different plant capacities can be quickly become tedious. Fortunately, there are shortcut methods to extrapolate the CapEx from one set of data easily, using a lumped parameter approach. One particular method of note is called the *six-tenths rule* [122–124]. The six-tenths rule relates the sizes and costs of two plants at different scales through the following equation:

$$\frac{C_1}{C_2} = \left(\frac{S_1}{S_2}\right)^{0.6}$$

In the equation, S_1 and S_2 are the capacities of two similarly configured chemical plants. One of these is the capacity for which a chemical plant has been adequately designed and costed for (S_1), while the other is the desired scale for which an estimated cost is to be obtained (S_2). On the other side of the equation, C_1 and C_2 are the costs of the respective plants with the specified capacities. Since S_1 is the evaluated and well characterized scenario, C_1 is the known variable in the equation.

As such, C_2 can be obtained through a fairly straightforward calculation. However, there are several things that should be noted while using this equation:

- The two plants being compared should differ only in scale. Their general designs should not vary too much; otherwise the results are very likely to be invalid.
- This equation is applicable to within reasonable capacity limits only; it is usually not used to estimate the costs of two plants that are more than two orders of magnitude apart in size.
- The equation can be used to size and cost individual pieces of equipment, though it is more useful to apply it to entire plants to minimize systematic errors in the results.
- The value of the exponential factor can be variable. The reason a value of 0.6 is used is that it is a good approximation of the relationship between surface areas to volumes of equipment. When used to size individual pieces of equipment, the exponential can be anything from 0.2 to 0.8; recommended values for these can be found elsewhere [125].
- There are limits to the maximum sizes of individual unit operations. The process scales cannot be magnified indefinitely, and once the maximum size of a unit operation is reached, an additional unit has to be added in parallel in order to enable further scaling. This clearly will change the process configuration and have a non-negligible impact on the results.

The six-tenths rule can be used to quickly conduct sensitivity analyses on how the CapEx will affect the economics of mineral carbonation, and also to determine the minimum scale of a viable MC process. Researchers familiar with this equation will also note that economies of scale are a major consideration for the implementation of MC technologies, since CapEx increases at a slower rate than the plant scale. This means that the relative contribution of CapEx to the costs of MC is much lower at larger scales, and therefore the unit costs of sequestered CO_2 are also reduced.

On the other hand, the operating expenses (OpEx) for a mineral carbonation process can be calculated in a manner similar to the CO_2 penalty as described earlier. The OpEx can be calculated by summing up the amounts of energy and materials used, multiplied by their unit costs:

$$\text{OpEx} = \frac{\sum (\text{Amount consumed} \times \text{unit cost})}{\text{Amount of } CO_2 \text{ sequestered}}$$

Since the amounts of energy or materials consumed are usually quite static relative to the unit amounts of CO_2 processed, OpEx tends to be independent of process scale and can be quite significant compare to CapEx. This is especially true for bulk chemical processing or at larger scales [126]. Furthermore, the unit costs of energy and materials for a MC process are obviously dependent on local conditions, and these will have to be determined via market research.

By applying these techniques, we can thus estimate the approximate unit costs of mineral carbonation. A very rough way to put all these together to estimate the costs of processing one unit of CO_2 can be expressed as follows:

$$\text{Unit cost} = \left(\frac{\text{CapEx} \times 1.1}{\text{Plant Capacity} \times 20} \right) + \left(\frac{\text{Operating Labor} \times 1.7}{\text{Plant Capacity}} \right) + \text{OpEx}$$

The units for each term in the equation above should be noted to be as follows: CapEx, \$; Plant Capacity, tonnes/year; Operating Labor, \$/year; OpEx, \$/tonne CO_2 sequestered. The coefficient 1.1 linked to the CapEx term takes into account maintenance, utilities, and insurance etc., while the coefficient 1.7 attached to the operating labor term considers admin, tech support and others. A plant life of 20 years is assumed here, which is similar to or typical for many large scale petrochemical plants. We emphasize again that these numbers are a very approximate guide and may vary depending on local situations.

A cursory analysis of the equation can provide a few insights into the costs of mineral carbonation. Firstly, MC processes can clearly benefit from economies of scale, since we have seen previously from the six-tenths rule that CapEx does not increase linearly with process scale. This is not surprising, as the principle holds true for most bulk chemical manufacturing processes. Secondly, operating labor costs tend to contribute to only a small fraction of the unit costs in modern bulk chemical plants. This is especially true as automation reduces the need for manual labor in many aspects of chemical manufacturing. Under these circumstances, the contribution of operating labor to the unit costs typically does not exceed 5–10% of the total costs. Thirdly, OpEx is usually the main contributor to the costs of bulk chemical manufacturing, since these are usually pegged to each unit of CO_2 processed and less related to gains in efficiency through scaling. Therefore, we can conclude that MC process development should focus more on reducing operating costs more than anything else, since these costs are easy targets to quickly achieve major improvements.

4.3.3 Integrated Mineral Carbonation Processes

Even the most well-designed mineral carbonation process is ultimately still dependent on an external and often non-technical factor such as a carbon price. This puts the development of MC processes (and most green technologies in general) at the mercy of policymakers and funding agencies under pressure to deliver tangible outcomes from public spending. Furthermore, in the absence of a profit motive, there is a practically zero chance of attracting the attention of private sector entities and get them to take up the process for commercialization.

With this obstacle in mind, some startup companies have recently begun a trend that seeks to integrate mineral carbonation processes with a mineral upgrading side operation to upgrade residual wastes or carbonation by-products such as silica and transition metal oxides into value-added chemicals for sale [127–130]. The sale of

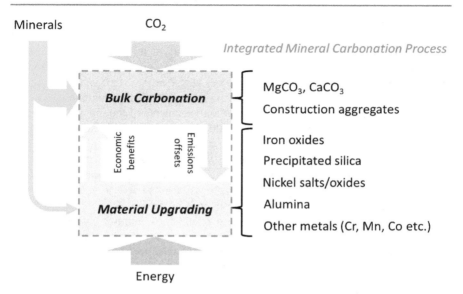

Fig. 4.12 Concept of integrated mineral carbonation, linking together a bulk carbonation process and a mineral upgrading process to bring both environmental and economic benefits from implementation of MC technologies

these products can then finance the larger and more costly MC process, and may even turn a profit to attract investments into the technology. Generally speaking, the side processes to upgrade mineral carbonation residues tend to be rather energy consuming and emit non-negligible amounts of CO_2 on their own. However, the coupled bulk carbonation process can help to offset the CO_2 emissions from the upgrading process, and result in a net negative amount of emissions. With a proper techno-economic analysis to determine the appropriate scales for both processes, it may be possible to develop a symbiotic integrated process that brings both environmental and financial benefits to society. An illustration of the concept of an integrated mineral carbonation process is given in Fig. 4.12.

The relationship between the bulk carbonation and material upgrading processes has to take into account a few considerations. Firstly, the source of materials subject to upgrading has to be defined clearly. These materials can be the by-products of the bulk carbonation process, or a small quantity of raw or processed minerals diverted from the input at the start. Examples of such materials include magnetic fractions, Mg- or Ca-depleted residue after carbonation, or even crushed minerals straight from the mine.

Secondly, the nature of the material to be upgraded will play a large determining role in the design of the upgrading process, and whether it can generate sufficient revenue to turn a profit for the overall integrated process. Non-homogeneous and complex materials such as natural minerals or incineration ash will usually require complicated designs for the upgrading process, but they are also rich and diverse sources of highly valuable metals such as nickel and cobalt [131–133]. Simpler materials such as concrete waste or red mud have fewer components and thus require fewer separation steps, but they often lack the flexibility to adapt to changes in market demands (especially when the primary product sees a huge drop in prices). Generally speaking, a modular process where certain steps can be turned on or off in response to market trends would be most suitable for this task.

Thirdly, the techno-economic aspects of both segments of the integrated process must be congruent and balanced. This would mean that the bulk carbonation process must be sufficiently large in order to be able to sequester enough CO_2 to offset the emissions from the upgrading process, yet not so massive that the costs are exorbitant and are difficult to be recovered. Similarly, the upgrading process must be relatively small (or as energy efficient as possible), yet able to generate enough revenue to bring economic benefits to the overall process.

The ubiquitous presence of some oxide species (silica, alumina and ferric oxide) in these materials means that they are common to many material upgrading processes. Depending on the quality and grades, refined versions of these products can be sold for relatively high prices. These are briefly discussed below.

4.3.3.1 Precipitated Silica

Precipitated silica is used in a variety of applications; among the largest ones is its use as a specialty additive into rubber products. Precipitated silica with different characteristics can be added during the manufacturing process to impart various desirable properties to the rubber; these include improvements in mechanical strength, heat resistance, friction control and visual appearance. The market demand for precipitated silica is estimated to be around 1.5–2 million tonnes a year. Although relatively small compared to the global need for CO_2 reductions, this is nevertheless a large enough market to accommodate a portion of the value-added products from an integrated process (remember that the material upgrading segment in an integrated process is intended to be a small part in the overall scheme). Depending on the market and location, the price of precipitated silica usually falls in the range of $500–$1000 per ton [134, 135].

4.3.3.2 Pigment Grade Ferric Oxide

Metal oxides can also be recovered from the by-products of the bulk carbonation process. These include the oxides or salts of many transition metals, which are useful as catalysts, pigments and raw materials for metal and alloy production. Iron oxides are likely the most abundant and common species found in the residues of mineral carbonation. Low grade ferric oxide intended for iron or steel production will not command a high price, since there is a lot of competition from mined iron ores. However, pigment grade ferric oxide can be sold for $1300–$1500 per ton, although its market demand is much lower compared to the raw material intended for steel refining. The annual demand for pigment grade Fe_2O_3 is estimated to be in excess of 2 million tonnes [136].

4.3.3.3 Transition Metal Oxides and Alumina

Other transition metals (and their oxides or salts) that can be refined through a material upgrading process include nickel, chromium, manganese, cobalt, and copper. The relative abundance of these transition metals obviously varies according to the source material, but they are usually commonly found in slags, incineration ash and natural minerals. Needless to say, these are much more valuable products that have much higher selling prices compared to ferric oxide pigments or precipitated silica.

The bulk of alumina produced every year (>90%) goes into the production of aluminum metal; apart from that there are not many applications that involve its direct use in manufacturing processes. As with iron oxide, alumina obtained from mineral carbonation residues may face stiff competition from commercial Bayer process operations and thus may not be cost competitive in the traditional sense. However, there may be opportunities for process innovations that allow for the co-recovery of alumina with embedded or impregnated metal oxides. These composite materials may find use as catalysts in petrochemical plants, and is also possibly easier to produce compared to traditional catalyst syntheses [137].

4.3.3.4 Magnesium and Calcium Carbonates

Depending on the process choice and designs, highly pure magnesium or calcium carbonates can also be obtained from a bulk carbonation process. The sales of these carbonates can help to recoup the costs of mineral carbonation; however it must be noted that the amounts of pure carbonates produced as a result of mineral carbonation will greatly exceed their demand and therefore only a small part of these can be sold. These carbonates have a wide variety of industrial applications, though some of them (for example their use in soil pH adjustments or as refractory materials) will ultimately result in the release of the bound CO_2 back into the atmosphere.

The differences between magnesium and calcium carbonates are generally minimal, especially when it comes to their use as fillers or additives in various manufactured goods. Despite this, these carbonates are not strictly interchangeable but dependent on the particular formulation to achieve the desired properties in the final product. As additives, both are used as whiteners in paper manufacturing, as well as fire-retardant fillers in plastics and rubbers. Food grade carbonates can be used in personal care products and cosmetics, as antacids, as well as excipients for pharmaceuticals. Obviously this would be much more challenging to produce (due to likely heavy metal contamination from minerals) to meet the stringent safety requirements in these applications.

However, it should be noted that unlike calcium carbonates, magnesium carbonates can come in a variety of hydrated species such as hydromagnesite $(4MgCO_3 \cdot Mg(OH)_2 \cdot 4H_2O)$, nesquehonite $(MgCO_3 \cdot 3H_2O)$, artinite $(MgCO_3 \cdot Mg(OH)_2 \cdot 3H_2O)$ and many other forms. This should be factored into the techno-economic analyses for MC processes, since the mass balances and economic factors can be rather different when the carbonate product is not exactly not $MgCO_3$.

Again, depending on the quality and location, the prices of these carbonates can vary significantly. Food or pharma grade magnesium carbonates can be expected to sell at $2000–$2500 per tonne, while lower grades can fetch prices of around $500–$1000 per tonne [138]. Calcium carbonate is unfortunately much cheaper (due to competition from abundant and easily-mined resources) and is typically sold at less than $300 per tonne [139].

4.4 Synergy Between Mitigation and Adaptation Actions Against Climate Change, via Mineral Carbonation

In this penultimate section, we will discuss some of the big-picture aspects of mineral carbonation. By envisioning a scenario where the technology is implemented on a moderately large scale, we will see why mineral carbonation has the potential to play a massive role in the fight against climate change. To this end, we will study a couple of hypothetical case studies involving the implementation of mineral carbonation technologies in Singapore, an affluent but land scarce island nation in Southeast Asia.

4.4.1 A Brief Overview of Singapore's Circumstances

Singapore is a prosperous and highly developed country situated at the southern tip of the Malay Peninsula. It is one of the smallest sovereign nations in the world, consisting of one main island and 62 other much smaller islets that fall under its jurisdiction. The country is densely populated, with (at the time of writing) around 5.6 million residents living on approximately 722 km^2 of land. Most of the population lives on the main island, though there are small communities that still inhabit the outlying islands. As a result, the main island is highly urbanized and has one of the highest population densities (~ 7800 p/km^2) in the world. Most of the country is less than 5 m above sea level, and a large proportion of the key infrastructure (i.e. the city core, financial district, airport and seaport) lies in areas that are very vulnerable to rising sea levels.

Being extremely land-scarce, the country has continuously sought to increase its habitable land area by reclaiming land from the sea (see Fig. 4.13 for a map of modern Singapore). Since its separation from Malaysia in 1965, the country has reclaimed more than 130 square kilometers of land, first by using its own soil (from excavated hills and tunnel construction projects), and subsequently through imported sand from neighboring countries. More recently, the government has unveiled plans to further expand the total land area to 766 km^2 to accommodate a projected increase of an additional 1 million residents by 2030 [140–142].

The importation of these materials have led to rising tensions, and as of 2018 export bans on these aggregates are in effect in several countries. However, discrepancies between reported sand imports and the amounts consumed locally are

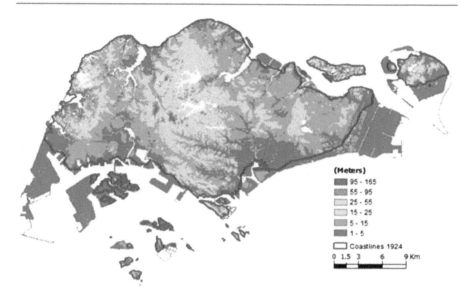

Fig. 4.13 Elevation map of modern Singapore (2012), with blue outlines marking the original coastline as mapped in 1924. Figure is adapted from Wang et al. [142]

also rather significant, which suggests that sand smuggling is a serious issue. This has also occasionally led to criticism from environmental groups [143–146].

Being situated on one of the busiest shipping routes that connect China and Europe, Singapore has traditionally been a busy shipping hub as well. It is also one of the most affluent countries, with a GDP per capita of around $98,000. Almost all (95%) of the electricity in Singapore is generated from natural gas-fired power plants; as such the country's CO_2 emissions from this sector are relatively low [147]. Despite having no natural resources of its own, Singapore is nevertheless the world's third largest crude oil refining center. Most of the refining capacity is located on Jurong Island and Bukom Island, where the country's chemical manu-facturing industries are based. The government has pushed hard for Singapore to become a smart and clean city, and has invested heavily in high tech manufacturing industries. As a result, apart from oil refining and some chemical industries, there are practically no other heavy polluting industries (other than a handful of steel recycling and waste incineration plants) in the country. All these factors mean that Singapore contributes less than 0.1% of the world's total CO_2 emissions. Even so, the country's per capita CO_2 footprint is rather significant (slightly more than 10 tonnes per person per year), which is typical of residents of highly developed and affluent countries [148].

As a land scarce country, Singapore also does not have the luxury of allocating large amounts of precious land area for landfilling purposes. As a result, municipal and industrial waste is extensively recycled here, with overall recycling rates in excess of 60%. The remaining unrecycled waste is sent to four waste-to-energy incineration plants in the country, where it is burnt to generate some electricity and physically converted to incineration bottom ash (IBA) in order to reduce the disposal volumes. Despite the high recycling rates, up to half a million tonnes of IBA is still generated annually from the incineration of the remainder unrecycled solid waste. This IBA is then disposed of in Semakau Island, an artificial offshore island designed for this specific purpose. Semakau Island was built in 1999, and is expected to last until 2035 where it will become fully utilized and a new site will have to be found to handle additional IBA in the future [149].

Singapore is also a signatory of the Paris Agreement, and has pledged to reduce their CO_2 emissions intensity (amount of CO_2 emitted per unit GDP) by 36% from 2005 levels by the year 2030. To this end, the government has introduced a carbon tax beginning at \$5/t CO_2, which takes effect in 2019. The tax is planned to increase to around \$15 - \$20/t CO_2 in the near future. The government's advisory panel on climate change (NCCS) has also proposed a multi-pronged approach that involves the use of renewables such as solar power and improvements in energy efficiency in the housing, commercial and industrial sectors [150–152].

4.4.2 A Pilot Scale Mineral Carbonation Plant in Singapore

In the case of refining, power and chemical plants in Singapore, there is likely little room for improvement to reduce their CO_2 emissions any further. This is because these plants are relatively modern and already tap on some of the more efficient technologies available during their construction and operation. In addition, these are large scale facilities that are relatively inflexible in changing their configuration or adapting retrofits onto their systems [153].

Mineral carbonation can help to reduce their CO_2 emissions by capturing and converting the greenhouse gas into solid inorganic carbonates. The quality and quantity of CO_2 emissions are quite diverse across the industries, and the choice of mineral carbonation technology would have to be tailored to the particular scenario in question. However, for discussion purposes we assume for the time being that there is a clear and mature MC process technology that is available and applicable to these varying flue gas streams. With this in mind, we can begin to look at the implementation of MC in Singapore from a high level perspective. A back of the envelope calculation can be very helpful to demonstrate the utility of mineral carbonation under these circumstances.

Let's suppose that a decision has been made to construct and operate a pilot scale mineral carbonation plant in Singapore that captures and converts one million tonnes of CO_2 annually. This amount is equivalent to roughly 1.3% of the total CO_2 emissions in Singapore. The logistics of the project would depend a lot on the actual design and efficiency of the MC process, but we can roughly estimate the amount of material flows for this project. As mentioned previously, a good estimation of the overall mass balance would be to assume that four tonnes of minerals are needed to sequester one tonne of CO_2 in mineral carbonation. This produces five tonnes of material in total. This also means that four million tonnes of minerals have to be imported every year from overseas, which is a relatively small amount in mining terms (for comparison, the port of Newcastle in Australia alone handles approximately 160 million tonnes of coal every year).

This amount of mixed carbonate products (five million tonnes) has an equivalent volume of around 3 million cubic meters, assuming that their packed density is around 1.6 tonnes per m^3. With an average depth of 30 m in the surrounding waters of Singapore, these materials would be sufficient to reclaim 10 hectares of land every year, which is equivalent to the area of 14 football (soccer) fields. Of course, larger scales of mineral carbonation will produce more materials for land reclamation purposes, and using these materials for land reclamation in Singapore would reduce the country's reliance on contentious foreign imports of sand.

On the other hand, incineration bottom ash can serve as a partial substitute for minerals for MC in Singapore. Although smaller in capacity, incineration bottom ash nevertheless represents a constant and readily available source of alkalinity to sequester CO_2 from industrial emitters in Singapore. Based on our previous estimates, we can expect that the 500 kilotonnes of IBA generated every year in Singapore to be sufficient to capture up to around 350 kt CO_2 and produce 850 kt of treated solids every year. The toxic metal leaching characteristics of these solids would ideally be inhibited, and this would allow them to be used in various construction industries with minimal risk to the population. This latter criterion is especially crucial in a small island country like Singapore, since any contamination of the soil or water by heavy metals would seriously impact its inhabitants due to the high population density of the country.

In addition to reclaiming land, these solid materials can also be used to construct levees and seawalls, and to raise floodwater-threatened infrastructure to adapt against storm surges and rising sea levels. The implementation of MC in this manner would thus not only mitigate climate change by reducing CO_2 emissions, but also adapt against it by using it to defend against its worst effects.

Let's suppose again that 1% of the total mineral imports were diverted into a material upgrading process to generate some revenue for the overall integrated MC process. This is equivalent to a small plant that processes 40,000 tonnes of minerals every year, which is miniscule compared with typical mining and refining operations. Assuming that the mixed raw materials can be refined into their pure components, we can expect that the upgrading process would produce around 16,000 tonnes of precipitated silica (assuming a price of $1000 per tonne, therefore worth a total of $16 million), 4000 tonnes of pigment grade iron oxide ($1500 per tonne, worth $6 million), and another 16,000 tonnes of MgO ($500 per tonne, worth $8 million). The sales of these products can thus generate up to $30 million dollars in revenue; in other words this will help to reduce the net costs of CCU by $30/t CO_2 sequestered in this case. Of course, if the net costs of the integrated process are less than $30/t CO_2, then the entire operation will actually be earning a gross profit. An overview of a hypothetical pilot scale mineral carbonation operation and its potential benefits is shown in Fig. 4.14.

It is also possible to upgrade incineration bottom ash in a manner similar to the mineral fraction in an integrated mineral carbonation process. Since IBA is

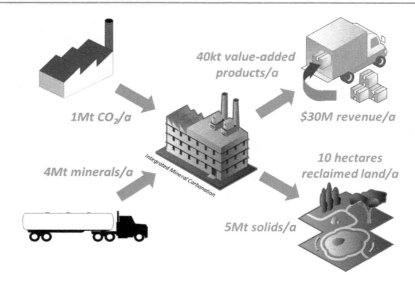

Fig. 4.14 Overview of a hypothetical 1Mt CO_2/year integrated mineral carbonation process in Singapore. Red symbols indicate inputs, and green symbols indicate expected benefits

relatively richer in valuable transition metal oxides, the volumes that need to be processed to obtain equivalent amounts of value-added products are often much smaller. However, IBA is also much more complex compared with natural minerals, and it may not be practical to separate and recover each and every element in the ash (at least not without incurring excessive costs).

It is worth bearing in mind that this scenario ignores the higher value transition metal products such as nickel, chromium, cobalt etc. that can be refined from the minerals. Their recovery and sales can generate even more revenue for the overall operations (whether their quantities and recovery costs are worth the effort will remain a debate for another time). It is also worth noting that the amounts of value-added products generated in this scenario are far smaller than the world's annual market demand for them, meaning that there is almost no risk of market flooding and price destabilization for these products.

It should also be noted that this scenario is not particular to or designed specifically for Singapore's specific circumstances. Since a majority of the world's cities are densely populated, energy intensive and situated along major rivers or coasts, they face many of the same problems that affect Singapore. Thus, they will also benefit from the implementation of MC technologies as outlined previously in this section. To take this one step further, collaborative partnerships between large emitters and vulnerable regions threatened by rising water levels can mutually benefit both parties from the dual action of mitigation and adaptation afforded by MC technologies. Of course, these partnerships are not limited to the national level but can also involve international cooperation treaties.

4.5 Conclusions and Outlook

Mineral carbonation technologies can play an important role in our collective actions against climate change. Comparatively speaking, MC has numerous unique strengths that are not often found in other carbon capture and utilization technologies. These include:

- **Scalability**: Mineral carbonation can be scaled to match the amounts of CO_2 emissions from all anthropogenic activities on the planet. There is, in theory, sufficient alkaline-earth species that can be made available to react with all the CO_2 from fossil fuel combustion.
- **Permanence**: The solid carbonates formed from mineral carbonation are highly stable and not susceptible to decomposition under ambient conditions. This ensures that the products are permanent sinks to sequester CO_2 without fear of re-emission.
- **Favorable thermodynamics**: This is related to the second point above. On paper, the carbonation reaction has favorable thermodynamics, as the products are lower in energy than the starting reactants. This raises the possibility of harnessing this energy difference to drive the reactions forward, though this is much more difficult in practice.
- **Economic potential**: The raw materials used for mineral carbonation often contain many valuable elements, and they can be refined into saleable products to enhance the overall economics of MC.
- **Synergetic with adaptation actions**: The solid carbonate products can contribute to land reclamation and fortify against rising sea levels and storm surges. Proper utilization of these materials enables small island nations or city waterfronts to adapt against the worst effects of climate change.

However, there are still some obstacles that prevent MC from being ready for deployment on a wide scale. The natural reaction rates of mineral carbonation are very slow, and they have to be accelerated to ensure that the technologies are feasible and practical to suit the task of large scale CO_2 removal and sequestration. To this end, an energy input is often needed, and these typically involve the use of chemicals, pressure or higher temperatures to enhance the reaction rates. This energy use results in a decrease in process efficiencies, and currently most MC processes still do not appear to meet the energy-use criteria needed to justify their widespread implementation.

While economies of scale can likely help to bring the costs of MC down, there are also financial-related obstacles to its large scale implementation. The recoverable amounts of value-added products from an upgrading process are limited by market forces, meaning that this aspect of the integrated process is not as scalable as bulk carbonation processes in general. Therefore, under these conditions, there are limited avenues for revenue generation from value-added byproducts to supplement the economics of MC. There will be a point where the value-added products will begin to flood the market and affect their selling prices, thus negating the entire point of generating revenue from these byproducts.

Since the idea of mineral carbonation was conceived more than twenty years ago, the field has gradually grown and broadened to include the use of other less conventional materials such as industrial waste or incineration ash as sources of alkalinity to sequester CO_2. These materials represent the low hanging fruit from which alkaline calcium or magnesium species can be obtained easily, and thus possibly reduce the energy use associated with their extraction (compared to natural minerals). An added benefit is that carbonation can help to neutralize the effects of toxic heavy metal leaching from these materials as well. Although these schemes are much smaller in scope than conventional mineral carbonation processes (due to the limited amounts of waste available), their success in the short-to-medium term can help to pave the way towards adapting the learnt experience for large scale carbonation technologies using more abundant minerals.

In addition to that, by finding outlets for the utilization of the products from mineral carbonation, the prospects of implementing these technologies on a larger scale can be enhanced. One way to do this would be to engineer the properties of the carbonate and silica products to allow them to meet the requirements for direct utilization in construction applications, thus providing a minor (but nevertheless significant) economic incentive to help offset the costs. This approach enables mineral carbonation to support climate change adaptation efforts as well.

References

1. Lackner KS (2003) A guide to CO_2 sequestration. Science 300(5626):1677–1678
2. Huijgen WJJ, Comans RNJ (2003) Carbon dioxide sequestration by mineral carbonation. Literature review. Energy research Centre of the Netherlands ECN2003
3. Zevenhoven R, Eloneva S, Teir S (2006) Chemical fixation of CO_2 in carbonates: routes to valuable products and long-term storage. Catal Today 115(1–4):73–79
4. Geerlings H, Zevenhoven R (2013) CO_2 mineralization—bridge between storage and utilization of CO_2. Ann Rev Chem Biomol Eng 4(1):103–117
5. Sanna A, Maroto-Valer MM (2017) CO_2 sequestration by ex-situ mineral carbonation
6. Rendek E, Ducom G, Germain P (2006) Carbon dioxide sequestration in municipal solid waste incinerator (MSWI) bottom ash. J Hazard Mater 128(1):73–79
7. Wee J-H (2013) A review on carbon dioxide capture and storage technology using coal fly ash. Appl Energy 106:143–151
8. Mo L, Panesar DK (2013) Accelerated carbonation—a potential approach to sequester CO_2 in cement paste containing slag and reactive MgO. Cem Concr Compos 43:69–77
9. Xi F et al (2016) Substantial global carbon uptake by cement carbonation. Nat Geosci 9:880
10. Si C, Ma Y, Lin C (2013) Red mud as a carbon sink: variability, affecting factors and environmental significance. J Hazard Mater 244–245:54–59
11. Kirchofer A, Becker A, Brandt A, Wilcox J (2013) CO_2 mitigation potential of mineral carbonation with industrial alkalinity sources in the United States. Environ Sci Technol 47 (13):7548–7554
12. Mcdonough WF, Teisseyre ER, Majewski E (2000) Earthquake thermodynamics and phase transformations in the Earth's interior

13. McDonough WF, Sun S-S (1995) The composition of the Earth. Chem Geol 120(3–4):223–253
14. Aresta M (2010) Carbon dioxide as chemical feedstock. Wiley
15. Berner RA (2003) The long-term carbon cycle, fossil fuels and atmospheric composition. Nature 426(6964):323
16. White AF (2003) 5.05—natural weathering rates of silicate minerals A2—Holland, Heinrich D. In: Turekian KK (ed) Treatise on geochemistry, Pergamon, Oxford, pp 133–168
17. Aresta M, Dibenedetto A, Angelini A (2013) The changing paradigm in CO_2 utilization. J CO_2 Utilization, 3:65–73
18. White AF, Brantley SL (2003) The effect of time on the weathering of silicate minerals: why do weathering rates differ in the laboratory and field? Chem Geol 202(3–4):479–506
19. Yuen YT, Sharratt PN, Jie B (2016) Carbon dioxide mineralization process design and evaluation: concepts, case studies, and considerations. Environ Sci Pollut Res 23(22):22309–22330
20. Taulis M (2012) GWCarb v1. 0: carbonate speciation tool
21. Stumm W, Morgan, JJ (2012) Aquatic chemistry: chemical equilibria and rates in natural waters. Wiley
22. Wanninkhof R et al (2013) Global ocean carbon uptake: magnitude, variability and trends. Biogeosciences 10(3):1983–2000
23. Zeebe RE, Zachos JC, Caldeira K, Tyrrell T (2008) Carbon emissions and acidification. Science 321(5885):51–52
24. Orr JC et al (2005) Anthropogenic ocean acidification over the twenty-first century and its impact on calcifying organisms. Nature 437:681
25. Wen N, Brooker MH (1995) Ammonium carbonate, ammonium bicarbonate, and ammonium carbamate equilibria: a Raman study. The J Phys Chem 99(1):359–368
26. Zhao Y, Zhu G (2007) Thermal decomposition kinetics and mechanism of magnesium bicarbonate aqueous solution. Hydrometallurgy 89(3–4):217–223
27. Keener TC, Frazier GC, Davis WT (1985) Thermal decomposition of sodium bicarbonate. Chem Eng Commun 33(1–4):93–105
28. Casey WH, Banfield JF, Westrich HR, McLaughlin L (1993) What do dissolution experiments tell us about natural weathering? Chem Geol 105(1–3):1–15
29. Oelkers EH, Gislason SR, Matter J (2008) Mineral carbonation of CO_2. Elements 4(5):333–337
30. Dreybrodt W, Lauckner J, Zaihua L, Svensson U, Buhmann D (1996) The kinetics of the reaction $CO_2 + H_2O \rightarrow H+ + HCO_3-$ as one of the rate limiting steps for the dissolution of calcite in the system H_2O–CO_2–$CaCO_3$. Geochim Cosmochim Acta 60(18):3375–3381
31. Duan Z, Sun R (2003) An improved model calculating CO_2 solubility in pure water and aqueous NaCl solutions from 273 to 533 K and from 0 to 2000 bar. Chem Geol 193(3–4):257–271
32. Cullinane JT, Rochelle GT (2004) Carbon dioxide absorption with aqueous potassium carbonate promoted by piperazine. Chem Eng Sci 59(17):3619–3630
33. Bishnoi S, Rochelle GT (2000) Absorption of carbon dioxide into aqueous piperazine: reaction kinetics, mass transfer and solubility. Chem Eng Sci 55(22):5531–5543
34. Ma'mun S, Svendsen HF, Hoff KA, Juliussen O (2007) Selection of new absorbents for carbon dioxide capture. Energy Convers Manag 48(1):251–258
35. Astarita G, Savage DW, Longo JM (1981) Promotion of CO_2 mass transfer in carbonate solutions. Chem Eng Sci 36(3):581–588
36. Xie Z, Walther JV (1994) Dissolution stoichiometry and adsorption of alkali and alkaline earth elements to the acid-reacted wollastonite surface at 25 C. Geochim Cosmochim Acta 58(12):2587–2598
37. Brantley, SL (2008) Kinetics of mineral dissolution. Kinetics of water-rock interaction. Springer, pp 151–210

38. Teir S, Revitzer H, Eloneva S, Fogelholm C-J, Zevenhoven R (2007) Dissolution of natural serpentinite in mineral and organic acids. Int J Miner Process 83(1–2):36–46
39. Apostolidis C, Distin P (1978) The kinetics of the sulphuric acid leaching of nickel and magnesium from reduction roasted serpentine. Hydrometallurgy 3(2):181–196
40. Sanna A, Uibu M, Caramanna G, Kuusik R, Maroto-Valer M (2014) A review of mineral carbonation technologies to sequester CO_2. Chem Soc Rev 43(23):8049–8080
41. Hemmati A, Shayegan J, Bu J, Yeo TY, Sharratt P (2014) Process optimization for mineral carbonation in aqueous phase. Int J Miner Process 130:20–27
42. Hemmati A, Shayegan J, Sharratt P, Yeo TY, Bu J (2014) Solid products characterization in a multi-step mineralization process. Chem Eng J 252:210–219
43. Zevenhoven R, Slotte M, Koivisto E, Erlund R (2017) Serpentinite carbonation process routes using ammonium sulfate and integration in industry. Energy Technol 5(6):945–954
44. Sanna A, Steel L, Maroto-Valer MM (2017) Carbon dioxide sequestration using $NaHSO_4$ and NaOH: a dissolution and carbonation optimisation study. J Environ Manage 189:84–97
45. Gerdemann SJ, O'Connor WK, Dahlin DC, Penner LR, Rush H (2007) Ex situ aqueous mineral carbonation. Environ Sci Technol 41(7):2587–2593
46. Ghacham AB, Cecchi E, Pasquier L-C, Blais J-F, Mercier G (2015) CO_2 sequestration using waste concrete and anorthosite tailings by direct mineral carbonation in gas–solid–liquid and gas–solid routes. J Environ Manage 163:70–77
47. Smith JM (1950) Introduction to chemical engineering thermodynamics. ACS Publications
48. Lackner KS, Wendt CH, Butt DP, Joyce EL Jr, Sharp DH (1995) Carbon dioxide disposal in carbonate minerals. Energy 20(11):1153–1170
49. Zevenhoven R, Kavaliauskaite I (2010) Mineral carbonation for long-term CO_2 storage: an exergy analysis. Int J Thermodyn 7(1):23–31
50. Huijgen WJ, Ruijg GJ, Comans RN, Witkamp G-J (2006) Energy consumption and net CO_2 sequestration of aqueous mineral carbonation. Ind Eng Chem Res 45(26):9184–9194
51. Goff F, Lackner K (1998) Carbon dioxide sequestering using ultramafic rocks. Environ Geosci 5(3):89–101
52. Alexander E, Wildman W, Lynn W (1985) Ultramafic (serpentinitic) mineralogy class 1. Mineral classification of soils, no. mineralclassifi, pp 135–146
53. Matter JM, Kelemen PB (2009) Permanent storage of carbon dioxide in geological reservoirs by mineral carbonation. Nat Geosci 2(12):837
54. Moody JB (1976) Serpentinization: a review. Lithos 9(2):125–138
55. Rinaudo C, Gastaldi D, Belluso E (2003) Characterization of chrysotile, antigorite and lizardite by FT-Raman spectroscopy. The Can Mineral 41(4):883–890
56. Groppo C, Rinaudo C, Cairo S, Gastaldi D, Compagnoni R (2006) Micro-Raman spectroscopy for a quick and reliable identification of serpentine minerals from ultramafics. Eur J Mineral 18(3):319–329
57. Evans BW (2004) The serpentinite multisystem revisited: chrysotile is metastable. Int Geol Rev 46(6):479–506
58. Lacinska AM et al (2016) Acid-dissolution of antigorite, chrysotile and lizardite for ex situ carbon capture and storage by mineralisation. Chem Geol 437:153–169
59. Farhang F, Rayson M, Brent G, Hodgins T, Stockenhuber M, Kennedy E (2017) Insights into the dissolution kinetics of thermally activated serpentine for CO_2 sequestration. Chem Eng J 330:1174–1186
60. Benhelal E et al (2018) Study on mineral carbonation of heat activated lizardite at pilot and laboratory scale. J CO_2 Utilization 26:230–238
61. Rashid MI et al (2018) ACEME: direct aqueous mineral carbonation of dunite rock. Environ Prog Sustain Energy
62. Dlugogorski BZ, Balucan RD (2014) Dehydroxylation of serpentine minerals: Implications for mineral carbonation. Renew Sustain Energy Rev 31:353–367
63. Yadav VS, Prasad M, Khan J, Amritphale S, Singh M, Raju C (2010) Sequestration of carbon dioxide (CO_2) using red mud. J Hazard Mater 176(1–3):1044–1050

64. Bonenfant D et al (2008) CO_2 sequestration by aqueous red mud carbonation at ambient pressure and temperature. Ind Eng Chem Res 47(20):7617–7622
65. Huijgen WJ, Witkamp G-J, Comans RN (2005) Mineral CO_2 sequestration by steel slag carbonation. Environ Sci Technol 39(24):9676–9682
66. Santos RM, Van Bouwel J, Vandevelde E, Mertens G, Elsen J, Van Gerven T (2013) Accelerated mineral carbonation of stainless steel slags for CO_2 storage and waste valorization: effect of process parameters on geochemical properties. Int J Greenhouse Gas Control 17:32–45
67. Dri M, Sanna A, Maroto-Valer MM (2013) Dissolution of steel slag and recycled concrete aggregate in ammonium bisulphate for CO_2 mineral carbonation. Fuel Process Technol 113:114–122
68. Meima JA, van der Weijden RD, Eighmy TT, Comans RN (2002) Carbonation processes in municipal solid waste incinerator bottom ash and their effect on the leaching of copper and molybdenum. Appl Geochem 17(12):1503–1513
69. Lin WY, Heng KS, Sun X, Wang J-Y (2015) Accelerated carbonation of different size fractions of MSW IBA and the effect on leaching. Waste Manag 41:75–84
70. Lin WY, Heng KS, Sun X, Wang J-Y (2015) Influence of moisture content and temperature on degree of carbonation and the effect on Cu and Cr leaching from incineration bottom ash. Waste Manag 43:264–272
71. Pan S-Y, Chang E, Chiang P-C (2012) CO_2 capture by accelerated carbonation of alkaline wastes: a review on its principles and applications. Aerosol Air Qual Res 12(5):770–791
72. Bobicki ER, Liu Q, Xu Z, Zeng H (2012) Carbon capture and storage using alkaline industrial wastes. Prog Energy Combust Sci 38(2):302–320
73. Paramguru R, Rath P, Misra V (2004) Trends in red mud utilization–a review. Mineral Process Extr Metall Rev 26(1):1–29
74. Shi C (2004) Steel slag—its production, processing, characteristics, and cementitious properties. J Mater Civ Eng 16(3):230–236
75. Papadakis VG, Vayenas CG, Fardis MN (1991) Fundamental modeling and experimental investigation of concrete carbonation. Mater J 88(4):363–373
76. Meima JA, Comans RN (1999) The leaching of trace elements from municipal solid waste incinerator bottom ash at different stages of weathering. Appl Geochem 14(2):159–171
77. Baciocchi R et al (2010) Accelerated carbonation of different size fractions of bottom ash from RDF incineration. Waste Manag 30(7):1310–1317
78. Li X, Bertos MF, Hills CD, Carey PJ, Simon S (2007) Accelerated carbonation of municipal solid waste incineration fly ashes. Waste Manag 27(9):1200–1206
79. Olajire AA (2013) A review of mineral carbonation technology in sequestration of CO_2. J Petrol Sci Eng 109:364–392
80. O'Connor W, Dahlin D, Rush, G, Gerdemann, S, Penner, L, Nilsen D (2005) Aqueous mineral carbonation. Albany Research Center: Albany, OR
81. O'Connor WK, Dahlin DC, Rush G, Gerdemann SJ, Penner L (2004) Energy and economic considerations for ex-situ and aqueous mineral carbonation. Albany Research Center (ARC), Albany, OR
82. O'Connor WK, Dahlin DC, Rush G, Dahlin CL, Collins WK (2001) Carbon dioxide sequestration by direct mineral carbonation: process mineralogy of feed and products. Albany Research Center (ARC), Albany, OR
83. O'Connor WK, Dahlin DC, Nilsen DN, Walters RP, Turner PC (2000) Carbon dioxide sequestration by direct aqueous mineral carbonation. Albany Research Center (ARC), Albany, OR
84. Geerlings JJC, Wesker E (2010) Process for sequestration of carbon dioxide by mineral carbonation. Google Patents
85. Geerlings JJC, Van Mossel GAF, Veen BCMIT (2010) Process for sequestration of carbon dioxide. Google Patents

86. Werner M, Hariharan S, Mazzotti M (2014) Flue gas CO_2 mineralization using thermally activated serpentine: from single-to double-step carbonation. Phys Chem Chemcal Phys 16 (45):24978–24993
87. Sun Y, Yao M-S, Zhang J-P, Yang G (2011) Indirect CO_2 mineral sequestration by steelmaking slag with NH4Cl as leaching solution. Chem Eng J 173(2):437–445
88. Bai P, Sharratt P, Yeo TY, Bu J (2011) Production of nanostructured magnesium carbonates from serpentine: implication for flame retardant application. J Nanoeng Nanomanuf 1 (3):272–279
89. Bu J, Yeo TY, Sharratt P (2018) Method of producing metal carbonate from an ultramafic rock material. Google Patents
90. Alexander G, Maroto-Valer MM, Gafarova-Aksoy P (2007) Evaluation of reaction variables in the dissolution of serpentine for mineral carbonation. Fuel 86(1–2):273–281
91. Wang X, Maroto-Valer MM (2011) Dissolution of serpentine using recyclable ammonium salts for CO_2 mineral carbonation. Fuel 90(3):1229–1237
92. Littau KA, Torres FE (2013) System and method for recovery of CO_2 by aqueous carbonate flue gas capture and high efficiency bipolar membrane electrodialysis. Google Patents
93. Shuto D et al (2015) CO_2 fixation process with waste cement powder via regeneration of alkali and acid by electrodialysis: effect of operation conditions. Ind Eng Chem Res 54 (25):6569–6577
94. Van der Zee S, Zeman F (2016) Production of carbon negative precipitated calcium carbonate from waste concrete. The Can J Chem Eng 94(11):2153–2159
95. Naraharisetti PK, Yeo TY, Bu J (2019) New classification of CO_2 mineralization processes and economic evaluation. Renew Sustain Energy Rev 99:220–233
96. Liu Q, Maroto-Valer MM, Sanna A (2017) Mineral carbonation technology overview. CO_2 sequestration by ex-situ mineral carbonation: World Scientific, pp 1–15
97. Fagerlund J, Nduagu E, Romão I, Zevenhoven R (2012) CO_2 fixation using magnesium silicate minerals part 1: process description and performance. Energy 41(1):184–191
98. Romão I, Nduagu E, Fagerlund J, Gando-Ferreira LM, Zevenhoven R (2012) CO_2 fixation using magnesium silicate minerals. Part 2: energy efficiency and integration with iron-and steelmaking. Energy 41(1):203–211
99. Highfield J, Lim H, Fagerlund J, Zevenhoven R (2012) Activation of serpentine for CO_2 mineralization by flux extraction of soluble magnesium salts using ammonium sulfate. RSC Adv 2(16):6535–6541
100. Mirjafari P, Asghari K, Mahinpey N (2007) Investigating the application of enzyme carbonic anhydrase for CO_2 sequestration purposes. Ind Eng Chem Res 46(3):921–926
101. Power IM, Harrison AL, Dipple GM, Southam G (2013) Carbon sequestration via carbonic anhydrase facilitated magnesium carbonate precipitation. Int J Greenhouse Gas Control 16:145–155
102. Jo BH, Kim IG, Seo JH, Kang DG, Cha HJ (2013) Engineered *Escherichia coli* with periplasmic carbonic anhydrase as a biocatalyst for CO_2 sequestration. Appl Environ Microbiol pp AEM. 02400-13
103. Power IM, Harrison AL, Dipple GM (2016) Accelerating mineral carbonation using carbonic anhydrase. Environ Sci Technol 50(5):2610–2618
104. Seo S, Perez GA, Tewari K, Comas X, Kim M (2018) Catalytic activity of nickel nanoparticles stabilized by adsorbing polymers for enhanced carbon sequestration. Sci Rep 8 (1):11786
105. Ramsden JJ, Sokolov IJ, Malik DJ (2018) Questioning the catalytic effect of Ni nanoparticles on CO_2 hydration and the very need of such catalysis for CO_2 capture by mineralization from aqueous solution. Chem Eng Sci 175:162–167
106. Hu G, Xiao Z, Smith K, Kentish S, Stevens G, Connal LA (2018) A carbonic anhydrase inspired temperature responsive polymer based catalyst for accelerating carbon capture. Chem Eng J 332:556–562

107. Zevenhoven R, Virtanen M (2017) CO_2 mineral sequestration integrated with water-gas shift reaction. Energy 141:2484–2489
108. Chein R-Y, Yu C-T (2017) Thermodynamic equilibrium analysis of water-gas shift reaction using syngases-effect of CO_2 and H_2S contents. Energy 141:1004–1018
109. Naraharisetti PK, Yeo TY, Bu J (2017) Factors influencing CO_2 and energy penalties of CO_2 mineralization processes. ChemPhysChem 18(22):3189–3202
110. UEPA (2014) Emission factors for greenhouse gas inventories. Stationary combustion emission factors. US Environmental Protection Agency
111. Balucan RD, Dlugogorski BZ, Kennedy EM, Belova IV, Murch GE (2013) Energy cost of heat activating serpentinites for CO_2 storage by mineralisation. Int J Greenhouse Gas Control 17:225–239
112. Perry JH (1950) Chemical engineers' handbook. ACS Publications
113. David J, Herzog H The cost of carbon capture. In: Fifth international conference on greenhouse gas control technologies, Cairns, Australia, 2000, pp 13–16
114. USEPA (1 November 2018) GHG emissions factors hub. Available: https://www.epa.gov/sites/production/files/2018-03/documents/emission-factors_mar_2018_0.pdf
115. Rezai A, Foley DK, Taylor L (2012) Global warming and economic externalities. Econ Theor 49(2):329–351
116. Wang Q, Chen X (2015) Energy policies for managing China's carbon emission. Renew Sustain Energy Rev 50:470–479
117. Pezzey JC, Jotzo F, Quiggin J (2008) Fiddling while carbon burns: why climate policy needs pervasive emission pricing as well as technology promotion. Aust J Agric Resour Econ 52 (1):97–110
118. Damen K, Faaij A, van Bergen F, Gale J, Lysen E (2005) Identification of early opportunities for CO_2 sequestration—worldwide screening for CO_2-EOR and CO_2-ECBM projects. Energy 30(10):1931–1952
119. Mendelevitch R (2014) The role of CO_2-EOR for the development of a CCTS infrastructure in the North Sea Region: a techno-economic model and applications. Int J Greenhouse Gas Control 20:132–159
120. Aresta M, Dibenedetto A, Angelini A (2013) Catalysis for the valorization of exhaust carbon: from CO_2 to chemicals, materials, and fuels. Technological use of CO_2. Chem Rev 114(3):1709–1742
121. Song C (2006) Global challenges and strategies for control, conversion and utilization of CO_2 for sustainable development involving energy, catalysis, adsorption and chemical processing. Catal Today 115(1–4):2–32
122. Sinnott R (1999) Coulson & Richardson's chemical enginering: volume 6/chemical engineering design. Elsevier Butterworth Heinemann
123. Lieberman MB (1984) The learning curve and pricing in the chemical processing industries. Rand J Econ 15(2):213–228
124. Tribe M, Alpine R (1986) Scale economies and the "0.6 rule". Eng Costs Prod Econ 10 (1):271–278
125. Whitesides RW (2005) Process equipment cost estimating by ratio and proportion. Course notes, PDH Course G, vol 127
126. Anderson J (2009) Determining manufacturing costs. CEP, pp 27–31
127. Liu W et al (2018) Optimising the recovery of high-value-added ammonium alum during mineral carbonation of blast furnace slag. J Alloys Compd
128. Pasquier L-C, Kemache N, Mocellin J, Blais J-F, Mercier G (2018) Waste concrete valorization; aggregates and mineral carbonation feedstock production. Geosciences 8 (9):342

129. Chiang P-C, Pan S-Y (2017) Aggregates and high value products. Carbon dioxide mineralization and utilization. Springer, pp 327–334
130. Yeo TY, Bu J (2017) MC process scale and product applications. CO_2 sequestration by ex-situ mineral carbonation: World Scientific, pp 133–165
131. Di Maria F, Micale C, Sordi A, Cirulli G, Marionni M (2013) Urban mining: quality and quantity of recyclable and recoverable material mechanically and physically extractable from residual waste. Waste Manag 33(12):2594–2599
132. Tunsu C, Petranikova M, Gergorić M, Ekberg C, Retegan T (2015) Reclaiming rare earth elements from end-of-life products: a review of the perspectives for urban mining using hydrometallurgical unit operations. Hydrometallurgy 156:239–258
133. Cossu R, Salieri V, Bisinella, V (2012) Urban mining: a global cycle approach to resource recovery from solid waste. CISA Publication
134. USGS (1 November 2018) 2015 Minerals Yearbook (Silica). Available: https://minerals. usgs.gov/minerals/pubs/commodity/silica/myb1-2015-silic.pdf
135. Bai P, Sharratt P, Yeo TY, Bu J (2014) A facile route to preparation of high purity nanoporous silica from acid-leached residue of serpentine. J Nanosci Nanotechnol 14 (9):6915–6922
136. USGS 2015 Minerals Yearbook (Iron Oxide Pigments). Available: https://minerals.usgs. gov/minerals/pubs/commodity//iron_oxide/myb1-2015-feoxi.pdf
137. Ashok J, Das S, Yeo T, Dewangan N, Kawi S (2018) Incinerator bottom ash derived from municipal solid waste as a potential catalytic support for biomass tar reforming. Waste Manag 82:249–257
138. USGS (1 November 2018) 2015 Minerals yearbook (magnesium compounds). Available: https://minerals.usgs.gov/minerals/pubs/commodity/magnesium/myb1-2015-mgcom.pdf
139. USGS (1 November 2018) 2015 Minerals yearbook (lime). Available: https://minerals.usgs. gov/minerals/pubs/commodity/lime/myb1-2015-lime.pdf
140. Khoo HH et al (2011) Carbon capture and mineralization in Singapore: preliminary environmental impacts and costs via LCA. Ind Eng Chem Res 50(19):11350–11357
141. Lai S, Loke LH, Hilton MJ, Bouma TJ, Todd PA (2015) The effects of urbanisation on coastal habitats and the potential for ecological engineering: a Singapore case study. Ocean Coast Manag 103:78–85
142. Wang T, Belle I, Hassler U (2015) Modelling of Singapore's topographic transformation based on DEMs. Geomorphology 231:367–375
143. Kog Y-C (2006) Environmental management and conflict in Southeast Asia–Land reclamation and its political impact
144. Franke M (2014) When one country's land gain is another country's land loss…: the social, ecological and economic dimensions of sand extraction in the context of world-systems analysis exemplified by Singapore's sand imports. Working Paper, Institute for International Political Economy Berlin
145. Torres A, Brandt J, Lear K, Liu J (2017) A looming tragedy of the sand commons. Science 357(6355):970–971
146. Gavriletea M (2017) Environmental impacts of sand exploitation. Analysis of sand market. Sustainability 9(7):1118
147. Finenko A, Cheah L (2016) Temporal CO_2 emissions associated with electricity generation: Case study of Singapore. Energy Policy 93:70–79
148. Lean HH, Smyth R (2010) CO_2 emissions, electricity consumption and output in ASEAN. Appl Energy 87(6):1858–1864
149. Chan JKH (2016) The ethics of working with wicked urban waste problems: the case of Singapore's Semakau Landfill. Lands Urban Plan 154:123–131

150. Deng Y, Li Z, Quigley JM (2012) Economic returns to energy-efficient investments in the housing market: evidence from Singapore. Reg Sci Urban Econ 42(3):506–515
151. Chung W, Hui Y, Lam YM (2006) Benchmarking the energy efficiency of commercial buildings. Appl Energy 83(1):1–14
152. Kannan R, Leong K, Osman R, Ho H (2007) Life cycle energy, emissions and cost inventory of power generation technologies in Singapore. Renew Sustain Energy Rev 11(4):702–715
153. Allcott H, Greenstone M (2012) Is there an energy efficiency gap? J Econ Perspect 26(1): 3–28

Catalytic CO_2 Conversion to Added-Value Energy Rich C_1 Products

5

Jangam Ashok, Leonardo Falbo, Sonali Das, Nikita Dewangan, Carlo Giorgio Visconti and Sibudjing Kawi

Abstract

Carbon-dioxide emission from various sources is the primary cause of rapid climate change. Its utilization and storage are becoming a pivotal issue to reduce the risk of future devastating effect. The conversion of carbon-dioxide as an abundant and inexpensive feedstock to valuable chemicals is a challenging contemporary issue having multi-facets. There is a need to elucidate the process of utilizing CO_2 to gain a fundamental understanding to overcome the challenges. This chapter focuses on converting CO_2 to C_1 valuable chemicals via hydrogenation (methane, methanol, syngas, formic acid) and reforming reactions (syngas). The first four parts of this chapter cover the production of methane, methanol and formic acid via hydrogenation reaction and syngas via reverse water gas shift reaction. Moreover, the last part of the chapters consists of reforming whereby CO_2 acts as a mild oxidant for the production of syngas $(CO + H_2)$. The chapter covers different aspects, including the current challenges in the process, the state of the art and design of catalysts, and mechanistic consideration, all of which are critically evaluated to give the insight into each reaction.

J. Ashok · S. Das · N. Dewangan · S. Kawi (✉)
Department of Chemical and Biomolecular Engineering, National University of Singapore, Singapore 119260, Republic of Singapore
e-mail: chekawis@nus.edu.sg

L. Falbo · C. G. Visconti (✉)
Laboratory of Catalysis and Catalytic Processes, Department of Energy, Politecnico di Milano, Milan, Italy
e-mail: carlo.visconti@polimi.it

© Springer Nature Switzerland AG 2019
M. Aresta et al. (eds.), *An Economy Based on Carbon Dioxide and Water*,
https://doi.org/10.1007/978-3-030-15868-2_5

5.1 Introduction

Anthropogenic carbon dioxide production is widely accepted as a major reason for accelerated climate change and global warming. In recent years, there has been a wide amount of global interest in finding sustainable ways to reverse the increasing CO_2 levels in the atmosphere. Globally, treaties such as the Kyoto Protocol and the Paris Agreement identify reduction in carbon emissions as vital in preventing the potentially disastrous effects of further global warming. Carbon Capture and Utilization (CCU) is one of the key areas that can achieve CO_2 emission targets at the same time contributing to the production of energy, fuels and chemicals to support the increasing demand. In CCU, carbon dioxide is captured and separated from emission gases and then converted into valuable products. CO_2 may be used as a raw material for conversion into syngas and energy products such as methane, methanol, dimethyl ether etc., or as chemical feedstock for production of inorganic or organic carbonates, carboxylates, urea etc. [1–3]. Many of the technologies to convert CO_2 into value-added products are still immature, and the focus of active research. In this chapter, we will cover the technologies for the reduction of CO_2 into C_1 molecules that are the fundamental building blocks of the fuel and chemical industry. Technologies covered are CO_2 hydrogenation to methane, syngas, methanol, and formic acid, and methane reforming to syngas production.

5.2 CO_2 Hydrogenation to Methane

A noteworthy product of CO_2 hydrogenation is Synthetic (or Substitute) Natural Gas (SNG). The SNG, which is mainly constituted of methane, has attracted much attention in the last few years because of its wide potential utilization market and because it is the only CO_2 hydrogenation product with an existing distribution network available in many industrialized countries. In principle, SNG can be injected in the existing low, medium or high-pressure natural gas grid, so as to be promptly distributed or stored.

The production of SNG from CO_2 and renewable H_2 is commonly referred to as "Power to Gas" (PtG) process [4–6]. Low-cost hydrogen can be produced by water electrolysis, exploiting the excess (renewable or nuclear) electricity when available in the power grid. In this regard, CO_2 can be considered as a low-cost carbon vector, that is converted into methane to allow more effective storage of H_2 and, at the same time, assist the stabilization of the electric grid. For this reason, CO_2 methanation is considered as an enabling technology for the long term (chemical) storage of electricity [5].

5.2.1 Thermodynamic Consideration

CO_2 methanation [Sabatier reaction, (5.1)] is strongly exothermic and brings about the molar contraction of the reacting mixture. Furthermore, the equimolar production of CO through the mildly endothermic reverse water gas shift [RWGS, (5.2)] reaction occurs during the synthesis.

$$CO_2 + 4H_2 \rightarrow CH_4 + 2H_2O \quad \Delta H^{\circ}_{298\,K} = -164\,\text{kJ mol}^{-1} \tag{5.1}$$

$$CO_2 + H_2 \rightarrow CO + H_2O \quad \Delta H^{\circ}_{298K} = 41.2\,\text{kJ mol}^{-1} \tag{5.2}$$

Due to the exothermic nature and the decrease in number of moles, thermodynamics indicates that low temperature and high pressure boost the CH_4 yield (Fig. 5.1).

As shown in Fig. 5.1, CO is the only important by-product of the CO_2 methanation. C_{2+} hydrocarbons can be also found in the products pool, but their content is usually negligible, especially when operating at low to moderate pressures.

The development of a CO_2 methanation technology is historically associated to the production of H_2 from syngas. Indeed, when pure hydrogen has to be produced from fossil fuels, usually natural gas, syngas is first produced by one or two reforming steps (usually steam reforming + autothermal reforming), then CO is converted into CO_2 in high and low-temperature water gas shift units, most of CO_2 is captured through adsorption processes and finally the remaining traces of CO_x are hydrogenated to methane. In this latter process, methanation reaction is used as a purification step, and the process conditions are very different from those of interest for the methanation of concentrated CO_2 streams: the inlet concentration of CO_x is very low during H_2 purification steps, while H_2 is in large excess. Under these highly diluted conditions, thermodynamics allows CO_x conversion close to 100%, and regardless of the high exothermicity of the CO_x hydrogenation reactions, low duties have to be managed in the reactor.

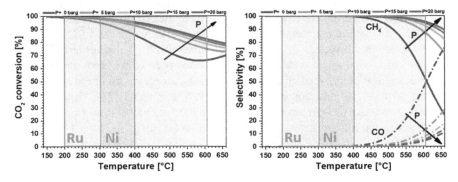

Fig. 5.1 CO_2 conversion, CH_4 and CO selectivity as predicted by thermodynamic equilibria of reactions (5.1) and (5.2) as a function of temperature

Massive production of SNG through CO_x methanation is more demanding: very active catalysts are needed, that are able to work at lower temperature to prevent strong thermodynamic limitations. Also, reactors able to effectively remove high reaction duties must be designed. Nevertheless, CO_x hydrogenation processes aimed at the production of SNG gained importance during the oil crisis in the late 1970s, when methane was produced by using syngas obtained from coal gasification as feedstock [7]. Several plants were built worldwide in that period and some of them are still in operation [8, 9]. Furthermore, during the last years, the increase in biomass utilization led to the construction of new plants for the conversion of biomass to methane [7].

Massive production of SNG through CO_2 hydrogenation is much more recent, and is related, as previously discussed, both to the need of effectively storing excess renewable power for long periods, and to the increased sensibility on the need of limiting CO_2 emissions in the atmosphere. The typical feedstock of these methanation plants is the carbon dioxide separated from biogas or from the flue gases of power plants [10–12]. However, more interest in this process has been also demonstrated by energivorous industries, such as steel manufactory, which are important CO_2 emitters. The first pilot plant for CO_2 methanation was built in the 1990s, and nowadays the first commercial applications are available [10].

5.2.2 Catalysts

5.2.2.1 Nature of Active Center

At the process conditions where CO_2 methanation is thermodynamically favored, significant kinetic limitations exist. Thus, catalyst design plays a pivotal-role to achieve acceptable process performance.

At the beginning of the 20th century, Sabatier and Senderens [13] discovered that metallic nickel is able to catalyze the hydrogenation of carbon oxides producing methane and water. In the following decades, CO_2 methanation was investigated over several suitable catalytic systems based on VIII B group metals [14]. Cobalt, iron and almost all the noble metals were also found to be able to activate CO_2. Different selectivity to methane were measured [7]. Nickel [15, 16], cobalt [17, 18] and noble metals [19, 20] produce mainly CH_4 and CO, while Fe also catalyzes the formation of significant amounts of C_{2+} hydrocarbons [21, 22].

Although several suitable catalytic systems have been considered, the Ni-based catalysts are those more studied and the only ones currently used at industrial scale because of their low cost and wide availability of this metal [23]. Usually, commercially used catalysts consist of high concentrations of nickel dispersed on a high surface area support. Unsupported Ni nanoparticles [18, 24], Ni-Raney [14, 23, 25] and hydrotalcite-like $NiAl(O)_x$ samples [26] have been also proposed as effective catalysts, but their effective performances (activity/selectivity/stability/cost) cannot outperform that of conventional catalyst when used in real conditions. Typically, Ni-based catalysts are active at temperatures where CO_2 conversion is extensively limited by the thermodynamics (green areas in Fig. 5.1). Moreover, due to the rather high process temperature, significant amount of CO is produced during CO_2 methanation, in addition to traces of C_{2+}. Also, at the typical process conditions, Ni-based catalysts are rather prone to deactivation [27]. Sintering phenomena are observed due to the high process temperature [28]. Furthermore, these catalysts suffer from carbonyls and carbides formation, resulting in deactivation by active phase volatilization [7, 27–30]. Fouling phenomena have been observed, and carbon deposition leading to whisker carbon formation has been reported [29, 31]. In the case of contaminated feed stream, severe sulfur poisoning has been also shown [29].

Efforts are currently ongoing in many laboratories worldwide to design more active nickel catalysts, that can be operated at process conditions where thermodynamics is more favorable, and temperature driven deactivation mechanism are slowed-down. Important progress has been reported in the very last years, but optimization work is still needed before introducing a new generation of Ni-based methanation catalysts.

Noble metals are much more active than nickel [14, 23], therefore low-temperature operations are possible, even using metal loading lower than 1 wt%. Solymosi and Erdöhelyi [19], working noble metal alumina-supported catalysts prepared using metal chlorides, have shown that the specific CO_2 methanation rate decreases in the order Ru > Rh $\gg$ Pt $\sim$ Ir $\sim$ Pd, with turnover numbers for Ru and Rh two orders of magnitude higher than for the other noble metals. Also, Ru- and Rh-based catalysts have been found to be more selective to methane than Pt- and Pd-based ones, which lead instead to high CO selectivity [7, 14]. Regarding the resistance to deactivation, noble metal-based catalysts are more tolerant than nickel to sulfur poisoning, carbon deposition or carbides formation [14, 32–37]. Also, high temperature treatments in hydrogen have been shown to be able to restore the initial activity of a Ru-based catalyst when carbonaceous species are accumulated on the surface [38–40].

The longer catalytic life and the higher activity and CH_4 selectivity are the essential characteristics that allow some catalysts based on precious metals to favorably compete with the cheaper Ni-based materials. On the basis of these premises, to date, Ru-based catalysts seem to be the most promising alternative to Ni-based catalysts for the intensification of SNG production from CO_2. Indeed, Ru-catalysts are able to operate with CO_2 yield to CH_4 over 99% already at atmospheric pressure, exploiting temperature range where the thermodynamics is particularly favorable (red areas in Fig. 5.1). This paves the way for once-through reactor configurations and for the production of a SNG that does not need complex and expensive purification steps to comply with the feed-in regulations of several countries' gas grid [6].

5.2.2.2 Role of Support

The stability, activity and selectivity of CO_2 methanation catalysts have been shown to depend on the crystal size [41, 42], on the shape of the metal particles dispersed on the support [43, 44], as well as on the interaction between the active metals and the oxide supports [45, 46]. In the case of Ni-based catalysts, the support has been shown to have a significant influence on the morphology of the active phase, as well as on the reactants adsorption and on the catalytic properties [47]. Several preparation methods have been developed to control the dispersion of the active phase [44, 48, 49]. On the contrary, the production of highly dispersed noble metals supported catalyst is more straightforward. Due to the low-metal loading and the strong interaction with the support, atomic dispersion is often approached [50]. Here, the problem may be that of the non-homogeneous distribution of the active phase in the support pellet, which may result in egg-shell configurations. If this represents an issue, solutions can be easily found based on the use of competitors during the impregnation phase, or by changing the pH of the impregnating solution.

In addition to those more "conventional" characteristics, that are common for many supports used in heterogeneous catalysis, it has been shown that during CO_2 methanation the support has a crucial role in granting CO_2 adsorption and activation. Szanyi and coworkers [50] reported that no CO_2 is hydrogenated by using a catalyst made of 1 wt% of Pd supported on carbon nanotubes, while methanation activity was observed on 1 wt% Pd supported on alumina. Similar results were reported with different active phases and supports, such as Ni on ceria-zirconia [51] or Ru on titania [52].

Although it has been shown that, at least on some metals, CO_2 methanation can occur also on unsupported catalysts, intensified catalyst performances can be achieved by exploiting the additional active centers that are created at the metal-support interface. These centers "at the interface" offer additional catalytic sites for CO_2 methanation [53, 54], where CO_2 molecules adsorbed on the support can react with hydrogen dissociatively adsorbed on the metal sites. As a result, the catalyst performances are boosted.

High surface area supports, usually oxides, are extensively used for the preparation of methanation catalysts. Among them, γ-Al_2O_3 is that most studied [55, 56] mostly due to its high surface area, well known properties, effective interaction with active metals, stability and low price. Other inorganic oxides, like TiO_2 [43, 52, 57, 58], SiO_2 [15, 59] and CeO_2 [49, 60–62], have been investigated, both as single or mixed oxide structures [63–65]. Furthermore, TiO_2, which is a good semiconductor support material, is widely studied especially for the CO_2 photocatalytic methanation [66]. Eventually, zeolites [67], metal-organic frameworks [68] have been investigated [44]. Detailed information on the performances of these supports can be found in recent reviews [41, 42, 44].

5.2.2.3 Role of Promoters

As previously discussed, conventional Ni-based catalysts suffer from remarkable catalyst deactivation during the CO_2 methanation reaction. In order to overcome such an issue, promoters may be added in the catalyst formulation which provide auxiliary functions, such as sulfur-, sintering- and carbon-resistance properties [29].

It has been also proposed that the addition of a second metal can enhance the stability and activity of Ni-based catalysts [44]. The addition of Fe increases the amount of adsorbed CO_2 and reduces the CO dissociation energy, thus favoring the methanation reaction [69]. The doping with La_2O_3 modifies the catalyst electronic properties, thus boosting the CO_2 activation [70]. Eventually, the addition of MgO increases the fouling resistance, enhances the thermal stability and minimize the sintering of Ni-based catalysts [7]. Both La_2O_3 [50] and MgO [71, 72] can be also effectively added to some noble-metal catalysts supported on materials that are not able to activate carbon dioxide. In this case, a bi-functional catalyst for CO_2 methanation is produced, where carbon dioxide is activated on the promoter, while hydrogen is dissociated on the noble metal [23]. Ru-based catalysts are usually unpromoted: their intrinsic activity and stability is already appropriate to catalyze CO_2 methanation for a long period of time.

5.2.3 Reaction Pathway and Kinetics

For what concerns the reaction pathway, CO_2 methanation mechanism is still debated and there is evidence that the nature of the metal, the typology of the support and the process conditions can strongly affect the reaction mechanism [14, 23, 29, 41]. Through in situ infrared spectroscopy it has been shown that, during CO_2 methanation, adsorbed CO is the key reaction intermediate [52, 54]. The CO formation mechanism, however, is debated as well. Most of the authors believe that it is produced via RWGS, where the CO_2 is adsorbed as bicarbonate on the support surface and then transformed into formate at the metal-support interface [53, 54]. Other authors suggest that CO_2 is adsorbed dissociatively, with the consequent formation of CO and some formates, that act as spectators [73, 74].

Regarding the mechanism of hydrogenation of adsorbed CO, both H-assisted [54] and unassisted [31, 75] dissociation pathways have been reported.

Efforts have been devoted to the development of empirical and mechanistic kinetic models, especially in the case of Ni-based catalysts [7, 76, 77]. One of the most comprehensive kinetic studies on the Sabatier process was proposed by Weatherbee and Bartholomew [78], who studied a 3% Ni/SiO_2 catalyst in a single-pass differential fixed-bed reactor working at low pressure. A second kinetic study worthy of note is the one reported by Xu and Froment [79], who modeled the process through 3 reactions happening at the same time: the methanation of carbon dioxide, the methane steam reforming (i.e. the reverse reaction with respect to the CO-methanation), as well as the water gas shift reaction. Although the model was developed and validated specifically for the steam reforming process on Ni-catalysts, it has been used also to describe experiments related to the CO_2 methanation over similar catalysts [80]. Mechanistic kinetic expression, based on the Langmuir-Hinshelwood-Hougen-Watson approach, have been also proposed [77]. Some of these expressions, which are frequently used in the literature for process or reactor modeling studies, are listed in Table 5.1.

Looking to CO_2 methanation on Ru-based catalysts, literature reports mainly power-law empirical equations [32, 81], even though some mechanistic equations are available [54]. Most of the empirical models, developed by exploiting experimental data collected in laboratory reactors operating under differential conditions [81], show that the reaction rate has a dependence on H_2 partial pressure (reaction orders in the range 0.3–2.5) stronger than on CO_2 (reaction order 0–1) at low CO_2

Table 5.1 Kinetic equations reported in the literature for CO_2 methanation

Kinetic equation	Catalyst	References
$r_{CO_2} = \dfrac{k P_{CO}^{0.5} P_{H_2}^{0.5}}{\left(1 + K_1 \frac{P_{CO_2}^{0.5}}{P_{H_2}^{0.5}} + K_2 P_{CO_2}^{0.5} P_{H_2}^{0.5} + K_3 P_{CO}\right)^2}$	3 wt% Ni/SiO_2	[78]
$r_1 = \dfrac{k_1}{P_{H_2}^{2.5}} \dfrac{\left(P_{CH_4} P_{H_2O} - \frac{P_{H_2}^3 P_{CO}}{K_{eq,1}}\right)}{(DEN)^2}$ $r_2 = \dfrac{k_2}{P_{H_2}} \dfrac{\left(P_{CO} P_{H_2O} - \frac{P_{H_2} P_{CO_2}}{K_{eq,2}}\right)}{(DEN)^2}$ $r_3 = \dfrac{k_3}{P_{H_2}^{3.5}} \dfrac{\left(P_{CH_4} P_{H_2O}^2 - \frac{P_{H_2}^4 P_{CO}}{K_{eq,3}}\right)}{(DEN)^2}$ $DEN = 1 + K_{CO} P_{CO} + K_{H_2} P_{H_2} + K_{CH_4} P_{CH_4} + K_{H_2O} \frac{P_{H_2O}}{P_{H_2}}$	15.2 wt% $Ni/MgAl_2O_4$	[79]
$r_{CO_2} = \dfrac{k P_{CO_2}^{0.5} P_{H_2}^{0.5} \left(1 - \frac{P_{CH_4} P_{H_2O}^2}{P_{CO_2} P_{H_2}^4 K_{eq}}\right)}{\left(1 + K_{OH} \frac{P_{H_2O}}{P_{H_2}^{0.5}} + K_{H_2} P_{H_2}^{0.5} + K_{mix} P_{CO_2}^{0.5}\right)^2}$	$NiAl(O)_x$	[26]
$r_{CO_2} = k \left\{ [P_{CO_2}]^n [P_{H_2}]^{4n} - \dfrac{[P_{CH_4}]^n [P_{H_2O}]^{2n}}{[K_{eq}]^n} \right\}$	0.5 wt. Ru/Al_2O_3	[32, 81]

conversion values. Unfortunately, no valuable information can be derived from these models at the conditions of industrial interest for PtG applications, i.e. at high conversion and concentrated reactant streams.

On the contrary, the kinetic model proposed by Lunde and Kester [32], which also accounts for the approach to thermodynamic equilibrium, is widely used for the description of CO_2 methanation over Ru-based catalysts at the process conditions of industrial interest. Recently, Falbo et al. [81] extended the validity of this kinetic model by proposing a new set of kinetic parameter able to describe the catalyst performances also under pressure (Table 5.1). This is particularly relevant for the design of modern optimized PtG processes aiming at producing pressurized SNG to be injected in the natural gas grid.

5.2.4 Engineering Challenges

A key-challenge in the engineering of the highly exothermic CO_2 methanation process is the temperature control in the reactor: assuming a space velocity of $5000\ h^{-1}$ and a complete CO_2 conversion, around $2\ MW/m^3_{cat}$ of heat need to be removed (methanol synthesis requires $0.6\ MW/m^3_{cat}$) [4]. Several reactor configurations have been proposed to grant the strong reaction exothermicity: adiabatic packed-bed with interstage cooling, multitubular packed-bed with external cooling, multitubular structured reactors with external cooling, fluidized-bed reactors.

For Ni-based catalysts, which are limited by thermodynamic equilibrium, adiabatic multistage fixed-bed reactors (from 2 to 5 catalyst stages) with interstage heat exchanger and unconverted reactants recirculation is the most viable and less expensive option [7].

For Ru-based catalysts, non-adiabatic reactors are most appropriate. Among those, fluidized-bed reactors with internal heat exchanger would grant the best isothermicity [82, 83]. Nevertheless, due to high mechanical stress resulting from fluidization, attrition processes may strongly affect the catalyst and the reactor performances. Multitubular packed-bed represent a viable option to this configuration, even though these reactors are less effective in preventing the formation of hot-spots, risk of pore diffusion limitation exists if big catalyst pellets are selected and inacceptable pressure drops are encountered when small pellets are employed. In order to further enhance the heat transfer and to prevent hot spots, microchannel reactors [84], or structured reactors based on highly conductive metallic honeycomb monoliths or open-cell foams (sponges) [85], coated with the catalyst, have been proposed. Due to their internal metallic structure, monolith reactors grant enhanced heat transport due to heat conduction within the continuous substrate. Also, the thin catalyst layer prevents pore diffusion limitations and the high void fraction in the reactor limits the pressure drops. Drawbacks of structured reactors are the demanding catalyst deposition on the structured substrate, as well as the difficulty of catalyst loading, unloading and replacement [4, 7].

A second key issue of PtG technology, which is common for all the PtX technologies, regards the fluctuating availability of excess renewable energy. In this regard, PtX processes can be designed so to operate either under steady-state or under dynamic conditions. For steady-state operation, a H$_2$-storage with high capacity is required to grant a constant H$_2$ flow to the methanator even when excess power is not available and water electrolysis must be stopped. However, this increases the PtG facility costs [4] and poses some operational limits. If dynamic operations are selected, catalysts must be designed and optimized so to withstand H$_2$-poor feed-streams and/or low inlet flow rates and/or low temperatures. Also, the reactor has to be designed to grant flexible operations, i.e. fast response to changes in the process variables and effective performances at low loadings. These requirements seem to be more applicable to Ru-based catalysts loaded in multitubular reactors loaded with highly conductive structured catalysts, that exploit a flow-independent heat transfer mechanism like conduction. Fluidized-bed reactors are indeed limited by superficial gas velocity within the reactor, which cannot be too low in order to assure minimum fluidization conditions and cannot be too high in order to avoid catalyst elutriation. Adiabatic reactors with Ni-catalysts, if fed with streams with flow-rates and/or composition far from the optimal, would operate far from the maximum reaction rate profile, with uncontrolled temperature in the adiabatic bed. Finally, multitubular packed-bed reactors would suffer in terms of convective heat transfer, that is worsened at low feed-loadings (low Reynold number would decrease both the effective radial conductivity and the wall hear transfer coefficient) and low H$_2$-contents (which would decrease the gas thermal conductivity).

5.3 CO_2 Hydrogenation to CO (Reverse Water Gas Shift Reaction)

Reverse water gas shift reaction is a crucial process for the production of CO which is a building block for the production of various useful chemicals such as methanol or other long-chain hydrocarbons [86]. CO that is considered to be the initial step of CO_2 hydrogenation on metal catalysts is the primary product from RWGS reaction. This reaction exists in conjunction with FT synthesis for the production of hydrocarbons from syngas [87]. Indeed, this process is more technically feasible as compared to the alternative technologies converting CO_2 to CO and gives an added versatility in the products obtained from CO transformation. The RWGS is also of great interest to be used in space exploration due to high (95%) atmospheric CO_2 concentration on Mars and availability of H_2 as a by-product of oxygen generation [88].

5.3.1 Thermodynamic Considerations

RWGS reaction is thermodynamically favourable at high temperature since it is an endothermic process. However, at low temperature region the WGS reaction is prominent as the reaction is exothermic. This reaction is also accompanied by other side reaction such as methanation that reduces the selectivity towards CO formation [89–93].

The RWGS reaction occurs according to (5.2).

$$CO_2 + H_2 \leftrightarrow CO + H_2O \quad \Delta H^\circ_{298\,K} = 41.2 \ kJ\,mol^{-1} \tag{5.2}$$

However, several undesired parallel and side reactions (5.1), (5.3) and (5.4) tend to occur as well

$$2CO \leftrightarrow C + CO_2 \quad \Delta H^\circ_{298\,K} = -172.6 \ kJ\,mol^{-1} \tag{5.3}$$

$$CO + 3H_2 \leftrightarrow CH_4 + H_2O \quad \Delta H^\circ_{298\,K} = -206\,kJ\,mol^{-1} \tag{5.4}$$

All reactions above occur simultaneously producing H_2O, CO, CO_2, H_2, and C in the reaction medium. For any other reaction generating CO as a main product, the reaction generally requires H_2:CO_2 ratio of about 2 [94]. This additional amount of H_2 imposes around 50% cost increase, thus not substantiating the overall process economy and feasibility. There are several other studies which compared the RWGS reaction with other processes producing CO, and among the investigated methods, RWGS reaction showed greater potential and higher efficiency when flue gas is the source of CO_2. Therefore, an additional reason to improve catalytic activity for the RWGS reaction at lower range temperatures is to reduce heat requirements necessary for the FT process which follows RWGS. In the lower temperature region of 600 °C, methanation reaction becomes prominent and only at a temperature higher than 700 °C RWGS can produce CO as a significant product. Therefore, a novel catalyst design is needed to overcome this problem and achieve higher activity at an even lower temperature.

5.3.2 Catalyst Types

Design of catalyst is crucial to obtain a high activity and selectivity. Generally, the catalyst is designed to promote the dual functionality of active metal/metal oxide sites and sites on the surface of the support. The correct choice for metal/metal oxide and supports promotes the better adsorption of reactants, followed by reaction and finally the desorption of products. There are mainly three categories of catalysts for RWGS, namely: supported metal catalyst, mixed metal oxide catalyst, and transition metal carbide catalyst.

5.3.2.1 Supported Metal Catalyst

There are several combinations of active metal and support which correspond to the production of a wide range of chemicals from hydrogenation of CO$_2$. For examples, Wambach et al. reported metals (Ni, Cu, Ag, Rh, Ru, Pt, Pd and Au) supported on ZrO$_2$ samples for CO$_2$ hydrogenation reaction [95]. In this study, it was found that Ag and Cu are favorable for methanol, whereas Ni and Ru lead to methane as the major product, and the rest produced mainly CO, methanol and methane. Dai et al. studied mesoporous M (Ni, Cu, Co, Fe, Mn)–CeO$_2$. Although the conversion for Ni–CeO$_2$ was higher among all other metal-CeO$_2$, however the selectivity was lower for Ni and Co. Several modifications such as alloying with Cu [91, 96–98], doping with alkali metals [93, 99], and enhanced metal-support interaction [100, 101] for Ni based catalysts have been reported to improve the selectivity towards desired product by methane suppression [102, 103]. Metals such as Cu, Fe, Mn showed almost 100% selectivity for CO during RWGS reaction [104].

From theoretical insights and experimental results, one of the pivotal criteria to consider while designing a catalyst for RWGS reaction is based on the electron properties of d-orbital holes of metals and the difference between desorption energy and dissociation barrier of metal carbonyls determined by the adsorption configuration [105]. The presence of incompletely filled d-orbitals leads to ease in the adsorption of reactants to form an intermediate, thereby improving the catalytic activity. Thus, noble metal based catalysts are considered as a highly important class of catalyst for RWGS reaction. Chen et al. studied the Pt/TiO$_2$ supported catalyst and reported that the presence of oxygen defects in the reducible TiO$_2$ support increase the number of interfacial sites and showed varying reactant adsorption at low and high temperature region as shown in Fig. 5.2 [106]. Additionally, the presence of other metals such as Ni, Co enhances the electronic property of Pt and forming bimetallic catalysts enhances the CO selectivity as compared to Pt alone which favors the formation of methane [107]. In contrast to reducible oxide supports, non-reducible supports such as Al$_2$O$_3$, SiO$_2$, and

Fig. 5.2 Pt particle size and
reaction temperature on the
selectivity of CO and CH_4
[106]

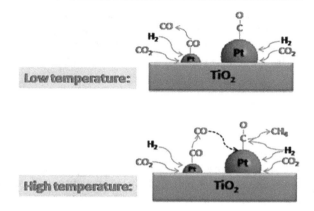

zeolites show lower activity [108, 109]. Besides, when reducible phase is added to
the parent metal oxide like MoO_x, it decreases the activation energy barrier, thus
enhancing the RWGS activity. Reducible oxide support for noble metals such as Pt,
improves the activity due to high oxygen mobility in the presence of oxidant CO_2
[110]. Addition of alkali metals such as potassium also enhances the stability of Pt
thereby preventing metal sintering [111], provides active site for formate decomposition, reduces the adsorption strength of CO and shows higher TOF than
un-doped Pt. Highly dispersed supported Au has been studied in depth for RWGS
reaction and showed promising performance in terms of stability of metal sites, and
decreased the required reaction temperature to below 400 °C.

Apart from noble metal supported catalysts, transition metals such as Cu and Fe
are considered favorable for RWGS reaction. One of the comparative studies
between Cu–ZnO and Cu–ZnO supported on alumina [112], showed the presence
of alumina increases the dispersion of CuO and ZnO species. Nickel catalysts are
rarely considered as effective RWGS catalysts because of their excellent hydrogenation to methane behavior [48, 49, 102, 113]. Catalysts with highly dispersed
Ni nanoparticles on supports with large oxygen exchanging capacity are still found
to be effective for RWGS reaction.

Single atom based catalyst showed a recent boom in terms of research interest to
understand its property and wide application. Metal particle size plays a unique role
in maintaining the stability of catalyst during the CO_2 hydrogenation reaction.
Matsubu et al., studied Rh/TiO_2 supported catalysts. During the reaction, it was
observed that Rh nanoparticles disintegrated to form isolated Rh sites. This change
in size, controls the changing reactivity with time on stream. A strong correlation
was observed between the reaction mechanism, TOF and number of Rh-isolated
sites and methanation reaction [114]. Another investigation revealed the importance
of Ru loading on Al_2O_3. With the loading percentage of less than 0.5%, the active
metal was mostly atomically dispersed and high selectivity for CO was obtained
[115]. Therefore, from all the studies done so far on supported metal catalyst for
RWGS reaction, it can be concluded that the metal particle size, type of support

(reducible and non-reducible), morphology of support and bimetallic catalyst showed distinctive behavior to enhance the activity and stability of catalyst under different reaction conditions.

5.3.2.2 Mixed Metal Oxide Catalyst

Transition metal oxides such as ZnO, Fe_2O_3, Cr_2O_3 and mixed oxide solid solutions have been reported to be the promising catalyst for the RWGS reaction. Due to high temperature reaction condition, ZnO based catalyst loses activity with time; however this metal oxide, when mixed with a proper ratio of Al_2O_3, forms a spinel phase, $ZnAl_2O_4$ at higher temperature and it was found to be stable at 600 °C for 100 h of operation [116]. Indeed, ZnO/Cr_2O_3 showed excellent performance and stability with no coke formation. On the other hand, Fe_2O_3/Cr_2O_3 showed slight deactivation. Catalyst deactivation due to coke deposition is one the main issues associated with RWGS reaction. Thus, utilizing high oxygen storage elements such as CeO_2 is another possible way to improve the performance [117]. Ce doped ZnO showed enhanced performance and stability even at the high temperature of 800 °C [118]. Co-CeO_2 mixed oxides prepared by co-precipitation method showed excellent performance and low coke deposition. The presence of well dispersed Co on CeO_2 support suppressed methane formation whereas high loading of cobalt led to an increase in the possibility of methanation reaction. Another group of mixed metal oxides extensively applied for RWGS are $Zn_xZr_{1-x}O_{2-x}$ [119], $Ni_xCe_{0.75}Zr_{0.25-x}O_2$ [120]. These catalysts offer promising properties such as high oxygen storage and reducibility during high temperature reaction condition. Under severe reaction conditions, the catalyst undergoes deactivation after several cycles. Zn replaces the Zr ions in the lattice to form a surface solid solution. This solid solution increases the reducibility by improving oxygen vacancy which improves the oxygen mobility, thereby suppressing carbon formation.

Another type of mixed oxides are in the form of perovskite. Perovskite type oxides are widely used for various catalytic applications including reverse water gas shift reaction. Mixed oxide perovskites consist of A- and B-site ions. This, combined with their structural stability, allows for the variation of composition, oxygen vacancies and oxidation state of metal ions. For these reasons, mixed oxide perovskites are suitable models to study the relation between the solid state chemistry and the catalytic activity of the mixed metal oxides. In order to break thermodynamic equilibrium limitation and suppress methane formation, another technique reported is coupling RWGS reaction with chemical looping (CL). Perovskite based catalyst possess several possibilities to be modified to achieve desired oxygen vacancy in the structure and thus show huge potential for the process of RWGS coupled with chemical looping applications. In this process, carbon dioxide is first captured from its emissions source or separated from air and purified. The RWGS-CL operation converts CO_2 and H_2 to produce separate streams of CO and water. The produced CO can then be combined with additional H_2 for liquid fuel production via FTS or methanol synthesis [121]. These separate product streams eliminate the possibility of methanation as a side reaction, because there is no direct interaction between CO_2 and H_2, and aid in avoiding thermodynamic limitations as

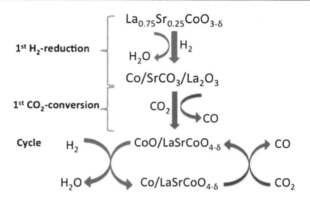

Fig. 5.3 Schematic representation of CO_2 conversion to CO on the oxygen deficient oxide system. A H_2 treatment reduces the perovskite-type oxides to metallic cobalt and base oxides while producing water. With CO_2 present, the reduced phases re-oxidize producing CO [122]

shown in Fig. 5.3. Daza et al. studied $La_{0.75}Sr_{0.25}CoO_{3-\delta}$ and $LaFeO_3$ for RWGS coupled with CL and found that, with the oxidation-reduction process, a higher CO production was achieved, with good recyclability of catalysts [122]. Ba–Zr-based perovskite catalysts with Zn, Y and Ce doped in the structure were also studied. Among the three doping elements, Zn- and Y-doped BZYZ catalyst showed an outstanding activity for the RWGS reaction at 600 °C, whereas Ce doped perovskite showed no positive effect on RWGS reaction [123, 124].

5.3.2.3 Transition Metal Carbide (TMC) Catalyst: An inexpensive and Emerging Class

Transition metal carbides are considered as an alternative to the noble metal based catalysts, with a low cost of materials. Pioneering work by Levy and Boudart showed addition of carbon to metal like tungsten modified the electronic properties, making it similar to the noble metal Pt [125]. Presence of different transition metal changes the surface properties. From previous studies, TMCs are considered as a promising substrate for metal dispersion, thereby enhancing the hydrogen dissociation and C=O scissoring [126]. There are several studies done on TMC which showed CO_2 activation over β-Mo_2C, where CO_2 binds to Mo_2C in bent configuration breaking the C=O bond [125, 127]. The dissociated CO desorbs while O interacts with Mo_2C to form Mo_2C–O that is then removed by H_2 to complete the cycle [128]. This oxy-carbide formation is crucial as it is a descriptor for determining the activity of WGS reaction [129]. TMC can also be used as a catalyst support for the dispersion of active metal sites to form small sized and stable nanoclusters. Zhang et al. studied Cu/Mo_2C prepared using Cu–MoO_3 as a precursor [130]. This catalyst showed excellent performance at higher temperature range of 600 °C and maintained 85% activity for 40 h of operation. The stability was attributed to the strong metal support interaction, enhancing Cu dispersion and preventing Cu agglomeration as shown in Fig. 5.4.

Fig. 5.4 Schematic showing the metal support interaction with reaction rate [130]

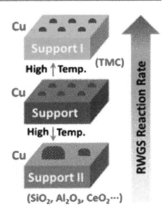

5.3.3 Mechanistic Considerations

Based on current development in understanding the kinetics and mechanism of RWGS reaction, there are several advanced techniques used, such as isotopic tracer method, operando DRIFTS and DFT calculations. For RWGS reaction, the reaction mechanism can be classified into two main reaction mechanisms as shown below in Table 5.2.

5.3.3.1 Surface Redox Mechanism

During redox mechanism, a rapid change on the active sites was observed in the presence of CO$_2$ and H$_2$ due to oxidation and reduction steps involved. The redox mechanism envisions absorbed CO over reduced metals that reacts with an oxygen atom contributed from the support to form CO$_2$. Reduced support re-oxidizes by H$_2$O, releasing hydrogen. Wang et al. investigated the redox mechanism using gold metal supported on CeO$_2$ as a catalyst for RWGS. The authors used TAP analysis to elucidate the mechanism and to show the interaction of CO$_2$ with the catalyst [110].

In this study it was found that the surface oxygen can be removed by reaction with H$_2$ shown in (5.5).

$$H_2 + O_{CeO_2} \rightarrow H_2O + \square_{CeO_2} \tag{5.5}$$

where O$_{CeO2}$ is the oxygen atom and $\square_{CeO_2}$ oxygen vacancies at the surface of the CeO$_2$ support.

In the second step, CO$_2$ acts as an oxidant for the partly reduced Au/CeO$_2$ catalyst surface, as shown in (5.6).

$$CO_2 + \square_{CeO_2} \rightarrow CO + O_{CeO_2} \tag{5.6}$$

A few studies reported the role of CO$_2$ as an oxidant even at low or mild temperature condition. Sharma et al. found that Pd/CeO$_2$, Pd/ZrO$_2$ and Pt/CeO$_2$ can partially be reduced by CO$_2$ at 350 and 200 °C, respectively (Fig. 5.5) [131]. A vast range of

Table 5.2 Reaction mechanism for RWGS reaction

Redox mechanism	Dissociative mechanism
$H_2 + 2^* \rightarrow H^* + H^*$	$H_2 + 2^* \rightarrow H^* + H^*$
$CO_2 + {}^* \rightarrow CO_2^*$	$CO_2 + {}^* \rightarrow CO_2^*$
$CO_2^* + {}^* \rightarrow CO^* + O^*$	$CO_2^* + H^* \rightarrow HCOO^*(COOH^*) + {}^*$
$H^* + O^* \rightarrow OH^* + {}^*$	$HCOO^*(COOH^*) + {}^* \rightarrow HCO^*(COH^*) + O^*HCO^*(COH^*) + {}^* \rightarrow CO^* + H^*$
$H^* + OH^* \rightarrow H_2O^*$	$COOH^* + {}^* \rightarrow CO^* + OH^*$
$OH^* + OH^* \rightarrow H_2O^* + O^*$	$H^* + OH^* \rightarrow H_2O^*$
$H_2O^* \rightarrow H_2O + {}^*$	$H_2O^* \rightarrow H_2O + {}^*$
$CO^* \rightarrow CO + {}^*$	$CO^* \rightarrow CO + {}^*$

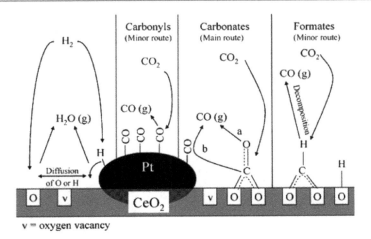

v = oxygen vacancy

Fig. 5.5 Redox reaction of the RWGS on Pt/Al$_2$O$_3$ [118]

studies were also focused on investigating Cu-based catalysts. In this system, CO_2 was dissociated on metallic Cu atoms as active sites and the reduction of the oxidized Cu catalyst was shown to be faster than the oxidation process [132], as shown in (5.7) and (5.8).

$$CO_{2(g)} + 2Cu^0{}_{(s)} \rightarrow CO_{(g)} + Cu_2O_{(s)} \tag{5.7}$$

$$H_{2(g)} + Cu_2O_{(s)} \rightarrow H_2O_{(g)} + 2Cu^0{}_{(s)} \tag{5.8}$$

Gines et al. studied the role of H_2/CO_2 ratio on CuO/ZnO/Al$_2$O$_3$ based catalysts. The change in ratio shifts the reaction from first order in H_2 to first order in CO_2 by varying the pressure of the two gases [133]. Fujita et al. concluded that surface redox mechanism pathway is caused by the presence of subsurface hydrogen trapped on the reconstructed copper surface caused by oxygen overlayer. This serves in making the surface more reactive to CO_2 adsorption [134].

5.3.3.2 Associative Mechanism

In this mechanism, the formation of intermediates such as formate, carbonates, and carbonyl, is crucial for RWGS reaction. An in-situ ATR-IRS study was used over Pt/Al$_2$O$_3$ based catalyst to show that RWGS reaction occurs at the sites of oxygen defects present on the thin surface of Al$_2$O$_3$ and a carbonate like species was formed on Pt sites, followed by CO_2 adsorption on the oxygen vacancy of Al$_2$O$_3$ near the interface of Pt which further reacts with H_2 to form CO [135]. The formate mechanism postulates a bidentate formate reaction intermediate produced through the reaction of CO with terminal hydroxyl groups over the oxide support. This intermediate decomposes to form H_2 and a mono-dentate carbonate. With Cu as the

Fig. 5.6 The proposed
reaction mechanism of
RWGS reaction over
Pd/Al$_2$O$_3$ catalyst

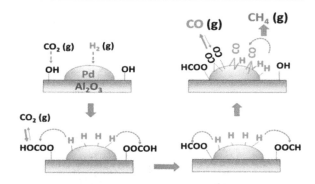

catalyst active site, the formate species derived from the association of H$_2$ and CO$_2$ is mainly proposed to be the key intermediate for CO production. Another technique involves mass spectroscopy and diffuse reflectance Fourier transform infrared spectroscopy (DRIFTS) during steady-state isotopic transient kinetic analysis (SSITKA) experiments to dynamically detect the surface species over Pt/CeO$_2$ catalyst [136]. By performing pulse-response TAP experiments, 1% CO$_2$ + 4% H$_2$ reaction mixtures containing isotopic CO$_2$ were introduced alternatively in a single reactor. Similarly, some other catalysts, Pd, Ni and Ru on Al$_2$O$_3$ also showed the formation of carbonates on alumina support [137].

In another study, the RWGS mechanism on Pd/Al$_2$O$_3$ was studied and it was shown from DRIFTS analysis that surface species and their evolution patterns are comparable during transient and steady-state experiments, during the switch of feed gases among CO$_2$, H$_2$, and CO$_2$ + H$_2$. Indeed, there was no direct dissociation of CO$_2$; instead, the CO$_2$ first reacted with surface hydroxyls on the oxide support. The formed bicarbonates react with adsorbed hydrogen dissociated on Pd particles to produce adsorbed formate species. Formates near the Pd particles rapidly reacts with adsorbed H to produce CO, which then adsorbs on the metallic Pd particles. In this analysis it was found that there are two types of Pd sites available-one with a weak interaction with CO and the other having stronger interaction with CO. The latter sites are reactive toward adsorbed H atoms on Pd, leading eventually to CH$_4$ formation, as shown in Fig. 5.6.

5.4   CO$_2$ Hydrogenation to Methanol

Global methanol production is in the region of 80 Mt/y. Methanol, a convenient and safe liquid (b.p.: 64.78 °C) at ambient conditions, can be used as:

- fuel, although it has half the volumetric energy density relative to gasoline or diesel,
- blending component for gasoline,

- feed for fuel cells, where it is oxidized with air to carbon dioxide and water to produce electricity,
- reactant for the MTO (methanol-to-olefins) process to produce ethylene or propylene,
- reactant for the MTG (methanol-to-gasoline),
- building block for the chemical industry,
- solvent,
- energy storage material.

Nowadays, most of the quantity of methanol is produced from a mixture of carbon monoxide, carbon dioxide, and hydrogen (syngas) at elevated pressures and moderate temperatures. In this process, CO$_2$ is the carbon source for methanol at the molecular scale, while CO reacts with the water produced in the process to form CO$_2$ and H$_2$ via the water-gas shift (WGS) reaction. These reactions are represented by (5.9) and (5.10):

$$CO_2 + 3H_2 \rightarrow CH_3OH + H_2O \quad \Delta H_{298K} = -49\,kJ\,mol^{-1} \tag{5.9}$$

$$CO + H_2O \rightarrow CO_2 + H_2 \quad \Delta H_{298K} = -41\ kJ\,mol^{-1} \tag{5.10}$$

The hydrogenation of CO$_2$ to produce methanol for use as both a fuel and a chemical building block is one possible route towards the development of an economy where CO$_2$ (from industrial effluents or the atmosphere) is regarded as an abundant carbon source (i.e. as alternative energy source), instead of being considered as a waste molecule. This "methanol economy" concept, proposed at the end of the '90s, is generally attributed to the Noble Laureate Olah [138], and is often referred to as the anthropogenic carbon cycle.

The hydrogenation of carbon dioxide can be carried out either by catalytic conversion with H$_2$ or by electrochemical reduction. This chapter will only focus on the first route, which appears to be closer to commercial application. For more details on the electrochemical reduction, the reader is referred to a recent review by Albo et al. and references therein [139].

Even though scientists discovered during the early 1960s that small amounts of CO$_2$ added to the feed enhance the yield of methanol [140], the first experiments on pure catalytic CO$_2$ hydrogenation to methanol were performed only in 1975 [141]. The interest at that time was the comprehension of the role of CO and CO$_2$ in the syngas hydrogenation to methanol. The utilisation of pure CO$_2$ as a carbon source for methanol synthesis is a more recent concept, that has been studied extensively, though it is yet to be commercialised on a large scale. In the following, an overview will be given on this process, with a special attention to gap analysis and needs assessment.

Key barriers to CO_2 hydrogenation to methanol include [142]:

- *Restriction by thermodynamic equilibria*

CO_2 hydrogenation to methanol is an exergonic process, which is favoured at low reaction temperatures or high reaction pressure. Even at 240–260 °C, which is typical of the low-T and low-P methanol synthesis process from syngas, the equilibrium constant of reaction (5.1) lies between 10^{-5} and 10^{-6}, allowing for a single-pass methanol yield of 15–25% and thus necessitating the implementation of costly recycling loops to achieve similar yields as in the syngas conversion in the presence of CO [143].

- *The need of highly active and robust catalysts against product inhibition*

In the light of the equilibrium limitations, the need for operation at lower temperature becomes apparent, and thus CO_2 hydrogenation to methanol requires highly active catalysts. Also, the absence of CO in the feed (i) prevents the role of CO as a scavenger of surface H_2O molecules (through the WGS reaction), which strongly stick on the catalyst surface, and thus resulting in a kinetic inhibition of the process [144]; (ii) activates the RWGS reaction, instead of the WGS. Thus, while the WGS equilibrium helps in the syngas conversion to keep the surface of the catalyst clean of adsorbed water, the RWGS suppresses the methanol formation rate because RWGS produces more water and consumes valuable hydrogen that cannot be used for methanol formation.

- *The need of highly selective catalysts*

Methanol production from CO_2 is indeed in competition with CO formation via the reverse water gas shift reaction, and with C–O bond dissociation and hydrogenation reactions. Side reactions yield CO, CH_4 and C_2–C_4 hydrocarbons, which incur significant downstream separation costs. Also, as pointed out in the previous point, RWGS is an undesired competitive reaction that consumes H_2 and forms H_2O, both of which kinetically and thermodynamically inhibit methanol synthesis from CO_2. H_2O also poses a criticism for the catalyst stability (see next point). Accordingly, the RWGS activity of catalysts for CO_2 hydrogenation to methanol should be suppressed.

- *The need of stable catalysts*

Research to date has been focused on finding selective catalysts that can activate CO_2 under mild reaction conditions with high selectivity towards methanol.

5.4.1 Catalyst Considerations

Since catalyst design is vital, a good catalyst must show [142]:

- strong adsorption and transportation of CO$_2$
- high concentration of hydrogenation sites
- ability to stabilise intermediates
- resistance towards water induced deactivation.

Wang et al. [46] recently reviewed the various catalytic systems that have been studied for the synthesis of methanol by CO$_2$ hydrogenation. The reader is referred to this paper for more details. In the following, the main aspects of most studied catalysts will be considered.

5.4.1.1 Cu-Based Catalysts

The catalysts used for industrial methanol synthesis (from syngas) are composed of copper, zinc, and alumina, and are based on the catalysts originally designed by ICI during the 1960s [145]. Even though the reaction mechanism during CO$_2$ hydrogenation to methanol may be rather different than those during syngas hydrogenation, due to its pivotal role in the methanol synthesis from syngas, Cu has been extensively studied also for CO$_2$ hydrogenation, often in the presence of secondary oxides such as Zn, Mn, Mg or Ce. Al$_2$O$_3$ is usually employed as the catalyst support, but TiO$_2$ and SiO$_2$ have been also considered. Cu/ZnO/Al$_2$O$_3$ catalyst proved to offer satisfactory performances at a pressure of 60 bar, space velocity of 22,000 h^{-1}, at temperatures between 260 and 270 °C. The CO$_2$ per-pass conversions were in the range of 35–45% and showed a slight decrease over time-on-stream.

It has been reported that three major strategies for optimization of Cu/ZnO-based for CO$_2$ hydrogenation to methanol are [143]:

- *Making smaller Cu particles*

There is a consensus on the fact that the methanol productivity is strongly correlated with the specific copper surface area of a catalyst [146]. This calls for new synthesis strategies, which utilize new precursor phases, advanced impregnation and thermal treatment methods or new synthesis approaches.

- *Improving the beneficial interaction between ZnO$_x$ and Cu*

Zn is the most studied secondary oxide for Cu-based catalysts. It has been shown that ZnO acts as structural support that helps to keep Cu in a highly dispersed state. Also, ZnO is an electronic promoter for Cu, due to its partial reducibility and to the resulting strong metal support interaction [147]. Accordingly, both the oxidation state of the copper and the copper–zinc interaction have an effect on the catalyst activity [145]. Doping of ZnO with small amounts of trivalent ions like Al^{3+}, Ga^{3+}

or Cr^{3+} has been shown to lead to promising results likely due to the enhanced reducibility of doped catalyst [148]. Also zirconia has been successfully used as a promoter for CO_2 hydrogenation over Cu/ZnO catalysts [149].

- *Increasing the defect concentration in metallic nanoparticles and amount of surface steps and edges of the Cu phase*

It has been proposed that this is possible by playing with the solid state chemistry of CuO reduction to Cu metal [150]. One of the major limitations of Cu/ZnO-based catalysts for CO_2 hydrogenation to methanol is their deactivation on-stream. This arises as catalytic activity is highly dependent on copper surface area whilst water, which is formed through the RWGS, induces sintering of copper nanoparticles [151], as a consequence of the modification of ZnO crystallization [152]. So, although the initial copper surface area is important, the ability to retain this surface area under reaction conditions appear to be the key consideration [153]. In conclusion, a lot of work remains to be done to develop a successful CO_2 hydrogenation catalyst based on Cu/ZnO [143]. The major challenges are related to synthetic inorganic chemistry, defect generation in metallic nanoparticles and surface/interface design between the two major catalyst components Cu and ZnO [143].

5.4.1.2 Pd-Based Catalysts

Pd has been also found to be an active metal for CO_2 hydrogenation [154], and its higher stability than Cu towards water-induced sintering has been also shown during methanol steam reforming [155]. Notably, Pd has attracted interest due to the property malleability derived from the presence of a nearly full d-band [156]. As for Cu, it has been shown that both the preparation method and Pd precursor may affect the selectivity of the products [154]. The catalytic activity of Pd is also strongly associated with the type of metal oxide support including ZnO, CeO_2, Ga_2O_3, TiO_2 and Al_2O_3 [156]. In fact, the proximity between metal and substrate stabilizes the intermediate formate species, which is often reported as the reaction intermediate. Notably, Pd on ZnO, TiO_2 and Al_2O_3 has low methanol productivity. On the contrary, catalysts containing alloyed PdZn nanoparticles with mean diameters in the region of 3–6 nm are characterized by an increased productivity. Those active centres are formed by pre-reduction of the catalyst at high temperature [156]. Also the presence of a $ZnTiO_3$ phase has been reported to boost the methanol yield. This phase is obtained when ZnO is reduced at a temperature of 650 °C, in the presence of TiO_2 as support [142]. As in many heterogeneous catalytic systems, the dispersion of Pd nanoparticles is important to grant activity to the catalyst, but this increases the risk of active area loss by sintering during the process.

5.4.1.3 In-Based Catalysts

Indium oxide (In_2O_3), as such or supported on ZrO_2, has been recently proposed as a very active, selective and stable catalyst for methanol synthesis by CO_2 hydrogenation [157]. The measured performances have been shown to be much better than a benchmark $Cu-ZnO-Al_2O_3$ catalyst, which has been found to be unselective rapidly deactivating at the same conditions where 100% methanol selectivity and stable performances were measured for the indium based catalyst. This behaviour has been attributed to rapid creation and annihilation of oxygen vacancies as active sites during CO_2 hydrogenation, which has been also simulated by DFT on pure In_2O_3 catalyst [158]. Notably, steady-state experiments carried out at 50 bar, at variable temperature around 300 °C, indicated a lower apparent activation energy for CO_2 hydrogenation than for the RWGS, which was used to explain the high methanol selectivity observed [159]. Water was found to strongly inhibit the reaction rate, and to deactivate the catalyst by sintering for partial pressures exceeding 1.25 bar [159]. Pd has been shown to be able to enhance the activity of indium oxide by facilitating hydrogen splitting [160].

5.4.1.4 Ag-Based Catalysts

Ag/ZrO_2 catalysts have been also proposed, showing interesting performances when Ag^+ cations are present in the vicinity of oxygen vacancies. In this case, the introduction of ZnO in the catalyst formulation was shown not to have significant effect on the methanol yield which differed from the behavior of Cu-based catalyst [161]. More complex catalyst formulation containing both Ag/ZrO_2 and Cu have been also proposed: it has been shown that the presence of Ag in the $Ag/CuO-ZrO_2$ catalyst changes the surface condition of metallic Cu via the formation of a Ag-Cu alloy, which imparts high activity and selectivity on CO_2-to-methanol hydrogenation [162].

5.4.2 Engineering Challenges

One of the main challenge associated with CO_2-to-methanol is that CO_2 is less reactive than CO, so that the yield of methanol is much lower than that obtained from syngas conversion under the same temperature and pressure [163]. This leads to larger and more expensive reactors. This is why, as discussed, many attentions are currently devoted to the development of more active catalysts.

Also, as already mentioned, due to the RWGS, more water is also produced when pure CO_2 is used instead of syngas to manufacture methanol. Crude methanol from the CO_2-based process contains approximately 30–40% water. Apart from the discussed issues related with catalyst deactivation, high water concentrations inhibit the process kinetics, further requiring for large reactors. Notably, the synthesis of methanol from CO_2 is less exothermic than the synthesis of methanol from CO. This simplifies the reactor design, which in the case of methanol from syngas is so critical to require multi-tubular reactors with external cooling (usually boiling water reactors, BWRs) or stage conversion reactors with intermediate quenching or cold-shot gas injection. In the case of methanol from CO_2, tube-cooled reactors are appropriate [164]. In those reactors the feed gas to the reactor controls the temperatures in the catalyst bed. Fresh feed gas enters typically at the bottom of the reactor and is preheated as it flows upwards through tubes in the catalyst bed.

The heated feed gas leaves the top of the tubes and flows down through the catalyst bed where the reaction takes place. The use of a tube-cooled reactor is advantageous over externally cooled reactor (like BWRs) in terms of the lower cost, higher efficiency, and relative simplicity of operation [164]. Additionally, tube-cooled reactors are preferred as being more efficient than adiabatic or cold-shot reactors which may require multiple reactors in series to achieve desired conversion rates. The flexibility with respect to load changes may be however a critical issue related to tube-cooled reactor.

The use of CO_2 instead of CO/CO_2 mixtures as C-source also limits the formation of by-products such as higher alcohols (mostly ethanol), esters, ethers (mostly dimethyl ether), and ketones (such as acetone and methyl ethyl ketone) [164]. It has been reported that CO_2-based process yields methanol in higher purity with five times lower by-product contents [152]. This can be partly explained by the high temperature sensitivity of the by-product formation reactions and by the better temperature control in the reactor during the less exothermic catalytic hydrogenation of CO_2. This simplifies the purification steps needed, limiting the number of distillation operations and consequently the energy consumption of the process. It has been reported that, when starting from pure CO_2, effective purification of methanol may be achieved by a single column separation (lean methanol leaves the column as bottom liquid) followed by a stripping operation on the condensed distillates to separate unconverted CO_2 from pure methanol. When dealing with separation, it is worth mentioning that there is typically more CO_2 in the crude methanol when pure CO_2 is hydrogenated [164].

5.4.3   Status of Industrial Development

Worldwide first demonstration of converting the greenhouse gas CO_2 to methanol as a useful chemical was reported by Lurgi in 1994 [165]. Another pilot plant was built in 1996 in Japan [166]. At that time, this new technology was attractive for producers with access to pure CO_2 and excess H_2, such as methyl tertiary butyl ether makers with dehydrogenation units, thus making the process as cost-effective as conventional methods [140]. Other technologies have been proposed for "CO_2 hydrogenation to methanol", but many of them were in reality technologies for the hydrogenation of CO/CO_2 mixtures with high CO_2 contents. A review of these technologies can be found in [140].

Since 2012, Carbon Recycling International (CRI) is operating the "George Olah" plant (Iceland, formerly known as Svartsengi plant), that has a capacity of 4000 tons per annum of methanol produced from carbon dioxide and renewable hydrogen. The CO_2 is extracted and purified from the flue gases of the nearby geothermal power plant, while the hydrogen required for the production is generated by alkaline water electrolysis using Iceland's entirely renewable grid electricity [164].

5.5 CO$_2$ Hydrogenation to Formic Acid

Another approach for utilizing CO$_2$ as C1 source for fuels/chemicals is to convert CO$_2$ into formic acid (FA) via CO$_2$ hydrogenation technique [1, 167]. FA is a valuable chemical commonly used as preservative and antibacterial agent [168, 169]. Additionally, it is an established hydrogen storage component. Upon decomposition, it generates CO$_2$ and H$_2$ with a possible reversible transformation back to regenerate formic acid. FA contains 53 gL^{-1} H$_2$ at room temperature and atmospheric pressure, and by weight, it contains 4.3 wt% of H$_2$ in pure formic acid. Being liquid at ambient conditions, its transportation and storage is more straightforward than that of molecular hydrogen. Additionally, formic acid can also be used in a formic acid based fuel cell, and/or can be further reduced to a carbon-based fuel. Generally, FA can be produced using photocatalytic reduction, electrochemical processes, enzymatic conversion, and hydrogenation methods [46, 170, 171]. Moreover, current industrial methods for producing formic acid include carbonylation of methanol to methyl formate followed by hydrolysis of methyl formate to generate formic acid [1, 172]. In comparison with conventional synthesis methods, the direct hydrogenation of carbon dioxide into formic acid serves two important purposes-CO$_2$ utilization as C$_1$ source and hydrogen storage in a liquid form.

5.5.1 Thermodynamic Considerations

Direct conversion of gaseous carbon dioxide and hydrogen into liquid formic acid is mildly exothermic, but it is an entropically disfavored reaction (5.11) due to the involvement of phase change from gaseous reactants into a liquid product [1].

$$H_{2(g)} + CO_{2(g)} \leftrightarrow HCO_2H_{(l)} \quad \Delta G°_{298\,K} = 32.9\,kJ\,mol^{-1};$$
$$\Delta H_{298\,K} = -31.2\,kJ\,mol^{-1} \quad (5.11)$$

$$H_{2(aq)} + CO_{2(aq)} \leftrightarrow HCO_2H_{(aq)} \quad \Delta G°_{298\,K} = -4\,kJ\,mol^{-1} \quad (5.12)$$

$$H_{2(aq)} + CO_{2(aq)} + NH_4 \leftrightarrow HCO_2^-{}_{(aq)} + NH_4^+{}_{(aq)} \quad \Delta G°_{298\,K} = -9.5\,kJ\,mol^{-1};$$
$$\Delta H_{298\,K} = -84.3\,kJ\,mol^{-1} \quad (5.13)$$

Alternatively, this reaction becomes slightly exothermic and favorable when operated in an aqueous phase as shown in (5.12). Moreover, if this reaction is carried out in a suitable solvent, it can be more favorable towards formation of formate/formic acid (5.13). The presence of suitable solvent alters the thermodynamics of the reaction. Thus, the most commonly adopted practice for transforming CO$_2$ into formic acid is to carry out the reaction with strong base additives such as amino or alkali/alkaline earth bicarbonates in water and/or alcohols (methanol,

ethanol) solvent. In liquid phase reaction, though the CO_2 hydrogenation to FA is thermodynamically favored, significant kinetic limitations exist. Thus, catalyst design and suitable solvent plays a pivotal-role to achieve acceptable process performance and selectivity towards desired product.

5.5.2 Catalytic Systems

The transformation of CO_2 into formic acid was first reported by Inoue et al. using homogenous Ru-based catalyst [173]. Since then, CO_2 hydrogenation into FA using homogeneous catalysts has been extensively studied [174]. For this reaction, Ru and Ir-based homogenous catalytic systems were much explored and reviewed by many researchers. The homogenous route offers the advantage of operating at milder reaction conditions. However, it also shows some disadvantages such as inefficient capture and recycling of the precious-metal catalysts, and the limited liquid-phase solubility of hydrogen. Comparatively, the research on heterogeneous catalysts is less explored than homogenous catalysts. However, in recent times, interest towards development of heterogeneous catalysts for CO_2 hydrogenation to formic acid is drawing interest due to the advantages of heterogeneous catalysts such as easy recovery after the reaction and application in a continuous reaction system in a large industrial process. Recently, Alvarez et al. [1] categorically reviewed the heterogeneous catalysts for CO_2 hydrogenation to formic acid/formate and methanol/DME synthesis. In this section, we will be covering the recent developments of heterogeneous catalysts for CO_2 hydrogenation to FA reaction. The unsupported/bulk metal particles and supported mono and bi-metallic catalysts for CO_2 hydrogenation reactions with respect to the catalytic performance and the acceptable reaction pathways over supported and bulk metal catalysts will be highlighted.

5.5.2.1 Unsupported/Bulk Metal Particles

In order to perform direct conversion of CO_2 and H_2 into formic acid in aqueous media, activation of both CO_2 and H_2 is required. As discussed before, CO_2 can be activated by additives such as amino or alkali/alkaline earth bicarbonates, and hydrogen molecule can be activated using metal species. Thus, during initial periods, pure metals such as Pd, Raney Ni, Au, and Ru were employed as heterogeneous catalysts for this reaction. Bredig and Carter et al. [175] synthesized formic acid using Pd black as a catalyst material using alkali/alkaline earth (bi)-carbonates as the CO_2 source in the presence of H_2 (reaction condition: 70−95 °C, 30−60 bar of H_2, 0−30 bar of CO_2). Similarly, Raney Ni was employed as a catalyst for synthesizing formaldehyde at 400 bars pressure in the presence of amines and alcohol as solvent [176]. Likewise, Takahashi et al. [177] reported selective formation of formic acid in a hydrothermal reactor using Ni powder as hydrogenating catalyst and K_2CO_3 as CO_2 source at 300 °C. Upon mixing with Fe powder, the selectivity towards methanol formation was observed. Jin et al. [178] reported a novel strategy to generate formic acid via CO_2 reduction using zero valent metal/metal oxide redox cycle under hydrothermal conditions. They have showed that metals such as Mg, Zn, Al in reduced form catalyze the formation of formic acid by CO$_2$ reduction in the presence of water at 300 °C. The yield of FA formation can be further improved in the presence of Ni and Cu metal particles. The oxidized metals can be regenerated back to reduced form using biochemical reductants such as glycerol. Unsupported Au nanoparticles were employed for CO$_2$ hydrogenation at 20 bars pressure in the presence of ternary amine and ethanol [179]. In most of the above reported catalytic systems, the turnover number (TON) was reported to be quite low, which is possible due to the lower availability of active metal centers to perform hydrogenation reaction. Recently, Srivastava et al. [180] reported the prominent catalytic activity of Ru nanoparticles, wherein the catalytically active Ru nanoparticles were synthesized in an ionic liquid. In-situ generated Ru NPs in [DAMI][NTf$_2$] ionic liquid was found to be highly active in terms of formic acid formation during the CO$_2$ hydrogenation reaction to other ionic liquid immobilized standing Ru NPs. They reported the TOF of 245 h^{-1} at 100 °C. Likewise, in another report, Ru nanoparticles prepared in a methyl alcohol solution under solvothermal condition were employed to perform the reaction with supercritical CO$_2$ in the presence of trimethylamine and water as promoter. They have achieved a high TON of 6351 after 3 h at 80 °C [181]. In both cases, the presence of water showed a positive effect in catalytic performance of Ru nanoparticles.

5.5.2.2 Supported Mono/Bimetallic Metal Catalysts

In supported metal catalysts, the role of support is mainly to disperse active metal species, which in turn helps to improve the number of available surface active metal species and prevents the agglomeration of metals during hydrogenation reaction. Several supports such as Al$_2$O$_3$, TiO$_2$, ZnO, CeO$_2$, Mg–Al$_2$O$_3$, activated carbon, CaCO$_3$ and BaSO$_4$ were employed for this reaction to disperse metals such as Pd, Au, Ru and Ni. For instance, Stalder et al. [182] reported Al$_2$O$_3$ supported Ru, Rh, Pd and Pt catalysts for conversion of aqueous sodium bicarbonate to sodium formate. Among them, Ru/Al$_2$O$_3$ showed better catalytic performance. In another work, Pd supported over activated carbon gave better catalytic performance than Pd supported over Al$_2$O$_3$, CaCO$_3$ and BaSO$_4$ [183]. Generally, positive effect on the catalytic performance for supported catalysts was reported as compared to bulk metal catalysts. Furthermore, Bi et al. [184] studied the effect of Pd loading for Pd/r-GO (reduced graphene oxide) catalysts by preparing several loadings of Pd (1– 5 wt%) supported catalysts. Among all, the lowest Pd (1 wt%) loaded r-GO showed better catalytic performance; according to the authors, it is caused by the large lattice strains. Similarly, Song et al. [185] observed that 0.25%Pd/chitin gave impressive catalytic performance with a TOF of 257 h^{-1}. They have reported that the dispersion of Pd particles was promoted by the acetamide group present in chitin support. In another report, Hao et al. [186] reported Ru supported on activated carbon, MgO and Al$_2$O$_3$ for this reaction. According to them, Ru/Al$_2$O$_3$ showed better catalytic performance due to the presence of higher surface hydroxyl group. The availability of surface hydroxyl groups on the support showed positive effect on catalytic performance. Other than Pd and Ru as the active metal component, Au supported catalysts were also explored for this reaction. Filonenko et al. [179] observed that the supported Au particles showed superior catalytic performance than unsupported Au particles. They have also reported that among all the supported (TiO_2, Al_2O_3, ZnO, CeO_2, MgAl−hydrotalcite, MgCr−hydrotalcite, and $CuCr_2O_4$) catalysts, Au/Al_2O_3 gave highest catalytic performance. According to them, the basic sites of the Al_2O_3 support play an important role, acting cooperatively with Au^0 nanoparticles in improving catalytic performance.

Besides monometallic supported catalysts, bimetallic supported catalysts were also utilized in this reaction. Takahashi et al. [177] examined the mixture of Ni and Fe oxide powder for CO_2 hydrogenation to formic acid using K_2CO_3 as CO_2 source at 300 °C. They have reported that the product selectivity can be controlled by varying the metal oxide compositions. In such a type of metal oxide mixture, the interaction between the active metal species is minimal, thus they can act as independent catalytic centers. In another work, Nguyen et al. [187] reported the synthesis of formic acid via CO_2 hydrogenation using PdNi alloys supported on carbon nanotube-graphene (PdNi/CNT-GR) catalyst in the absence of a base and water as solvent. Alloying Pd with Ni brought a significant enhancement in catalytic activity compared to the monometallic Pd catalyst. The composite support improved the dispersion and intimate contact between both metal species.

According to literature, among noble metals, Pd and Au are the most verified active metals for the synthesis of formic acid/formates and their catalytic activity can be enhanced by a proper choice of support material. Hydrophobic carbon-based materials are preferred choices as support for Pd catalysts, whereas more hydrophilic support materials such as Al_2O_3 and TiO_2 are preferably employed for Au catalysts (the same is also indicated for Ru catalysts). Among non-noble metals, Ni and Fe are the mostly explored active metals. Comparatively, noble metals showed more promising catalytic performance than non-noble metal based catalysts with respect to TON and longevity. Still, the number of studies on bare metal particles and supported metal particles as catalysts for CO_2 reduction reaction are limited and there are discrepancies on the fundamental aspects. Further investigations on this catalytic system for formic acid and formate synthesis are absolutely required to establish clearer catalyst structure-activity relationships.

5.5.3 Reaction Mechanism

As discussed before, formation of formic acid via direct CO_2 hydrogenation reaction is entropically disfavored when both reactants are in gaseous state. The reaction is feasible if both reactants are in aqueous state. Therefore, the role of suitable solvent medium and active metal species are critical. The presence of additives in reaction media mainly fixes the gaseous CO_2 in the form of an aqueous carbonates/bicarbonates and helps the dissolution of H_2 to reach metal species to carryout hydrogenation reaction. Actually, in formic acid synthesis, CO_2 is unlikely to be the reactant, but in most cases, carbonates/bicarbonates act as source for CO_2. The carbonates/bicarbonates can be generated by dissolving CO_2 in the base

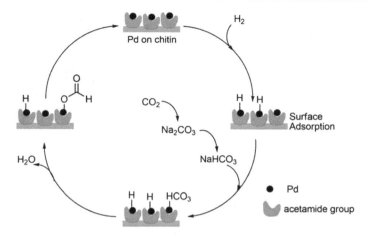

Fig. 5.7 Proposed CO_2 hydrogenation to formic acid reaction mechanism for Pd supported chitin catalyst [185]

solvents. Dissolved gaseous H_2 can be the source for hydrogen, in order to dissolve in the solvent, and the reaction should be operated at high pressure conditions. In general, the reaction mechanism of a typical heterogeneous catalysis involves four main steps. Firstly, the reactants diffuse and contact with the catalyst. Secondly, the reactants are adsorbed at an active site. Thirdly, the surface reaction occurs and finally, the formed product is desorbed from the active site. In literature, the reaction pathways for transformation of CO_2 into formic acid in the presence of unsupported, supported mono and bi-metallic catalysts are reported. For monometallic supported catalysts, Song et al. [185] proposed a plausible CO_2 hydrogenation to formic acid reaction pathway using Pd supported chitin catalysts (Fig. 5.7). According to them, the diffused hydrogen molecule is homolytically cleaved and attached to the vacant sites on Pd. The cleavage of hydrogen molecule can be promoted by the amino groups present on the catalyst surface. Similarly, CO_2 in bicarbonate form is then adsorbed to vacant sites on palladium. A hydrogen molecule is then inserted into the adsorbed carbonate anion, followed by a breakage of the C–O bond, the desorption of formate, and formation of formic acid.

Furthermore, the advantage of having bimetallic surfaces during CO_2 hydrogenation to formic acid is clearly highlighted by Nguyen et al. (Fig. 5.8) [187]. The reaction mechanism of CO_2 hydrogenation to formic acid over bimetallic PdNi supported over carbon nanotube-graphene was investigated and the proposed mechanism is illustrated in Fig. 5.8. Unlike the mono metal supported catalysts, in bi-metallic catalysts both metals can be involved in activating CO_2 and H_2 molecules. According to them, in the first step, an electron transfer from Ni to Pd atoms occurs. Therefore, Pd and Ni are in the electron-rich and—deficient state, respectively. It is followed by H_2 dissociative adsorption on Pd surface and CO_2 adsorption through its O-atoms on the Ni surface. Reaction between H on Pd and adsorbed CO_2 leads to the formation of adsorbed HCOOH.

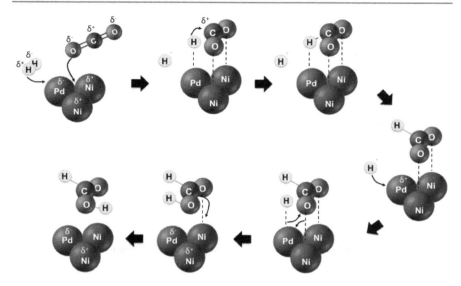

Fig. 5.8 Proposed scheme for CO_2 hydrogenation over PdNi bimetallic surface [187]

Thongnuam et al. [188] computationally investigated CO_2 hydrogenation to formic acid reaction on modified zeolites (ZSM-5, BEA, FAU) using density function theory (DFT) with the M06-L functional. According to them, the reaction proceeds in two steps on the surface of zeolite material. At first, abstraction of hydrogen atom by CO_2 forms a formate intermediate. After that, the intermediate takes another hydrogen atom to form formic acid. The formation of formate intermediate is observed to be the rate-determining step of the reaction for both the perfect and defect Sn-ZSM-5 zeolites. Similar result was also reported by Eseafili and Dinparast [189] over Ti-doped graphene nanoflakes (Ti-GNF) for CO_2 hydrogenation to formic acid reaction. According to them, the presence of large positive charge on the Ti atom can greatly regulate the surface reactivity of GNF. The hydrogenation of CO_2 over Ti-GNF also occurred as two-step reaction, i.e. (a) $H_2 + CO_2 \rightarrow HCOO + H$, and (b) $HCOO + H \rightarrow HCOOH$. Step (a) for the formation of formate intermediate is found to be the rate determining step. The activation energy for the first step was calculated to be 0.85 eV, while the second step could occur quickly due to the small reaction barrier (0.08 eV).

5.5.4 Technological Challenges

When comparing the technology readiness level (TRL) for heterogeneous CO_2 hydrogenation to formic acid with CO_2 hydrogenation to methane, methanol and CO, the TRL for this process is far lower (TRL 1–2). With the available state of art on catalytic systems, which have been proven to generate formic acid with decent reaction rates, the economic feasibility for this reaction needs to be demonstrated.

The technology of homogenously catalyzed synthesis of formic acid was first introduced by BO Chemicals in the 1980s and has been developed by BASF [190]. In this technology, a soluble Ru-complex is employed to catalyze CO$_2$ to formic acid using a mixture of ternary amine and alcohol at 50–70 °C and 10–12 MPa. The process must keep the expensive transition metal complex catalyst active yet avoid even traces of it being present in active form in formic acid distillation, because it can catalyze decomposition of the acid. Possible catalyst residues can be reversibly deactivated with CO. Such issues can be addressed comfortably by adopting heterogeneously catalyzed technology. For this, a better catalyst plays a crucial role; in many cases the nature of the active metal species and the structure of the catalyst material have been explored. However, there is a need to understand the catalyst structural changes occurring under the reaction conditions and their consequences to catalyst longevity and reusability. Moreover, recent advances in developing catalysts for dehydrogenation of formic acid to generate CO$_2$ and H$_2$ make FA an efficient chemical H$_2$ carrier [191]. The main advantages of formic acid over other H$_2$ carriers include easy handling, refueling, and transportation. For this, research for development of efficient catalyst system is required to catalyze both hydrogenating CO$_2$ to FA and decomposition to generate H$_2$ and CO$_2$ with minimal operational changes.

5.6 CO$_2$-Methane Reforming to Syngas

Dry reforming of methane (DRM) using CO$_2$ as an oxidant is one of the primary areas being researched for the transformation of CO$_2$ into higher chemicals through syngas (CO + H$_2$) as an intermediate. By principle, it is similar to the steam reforming of methane, which is an established and widely used technology for H$_2$ or syngas production, but DRM uses CO$_2$ as an oxidizing agent instead of steam, which makes it more attractive for CO$_2$ mitigation and a carbon-neutral economy. DRM consumes two major greenhouse gases, CO$_2$ and CH$_4$, and converts them to syngas which can then be converted into higher chemicals through Fischer-Tropsch synthesis or used as a source of H$_2$. The H$_2$/CO ratio in the product syngas is 1, which is lower than that achieved by steam reforming, and is suitable for subsequent usage in Fischer-Tropsch synthesis of long-chain hydrocarbons.

The major technological challenges for the industrial application of DRM are its high endothermicity, requiring high temperatures for appreciable conversion and rapid deactivation of catalysts under reaction conditions, caused by coking or sintering. The high energy input required by the process causes a serious penalty on the efficiency of CO$_2$ utilization, since the heat input is mainly provided through the combustion of fossil fuels. The option of using solar heating to supplement the energy requirement of DRM is a highly attractive green alternative and is one area of active research. For industrial application of DRM, there is a need to develop cost-effective catalysts that can maintain stable performance for extended durations in DRM. In the past decade, there has been a substantial increase in research focus on catalyst development for DRM and tremendous progress has been made in increasing the activity and stability of catalysts.

5.6.1 Reaction Thermodynamics

Dry Reforming of methane (5.14) is a highly endothermic reaction, since both CH_4 and CO_2 are very stable molecules with high bond dissociation energy (435 kJ/mol for CH_3–H and 526 kJ mol^{-1} for CO–O). High temperatures (>800 °C) are hence required to achieve good conversion of methane and carbon dioxide to syngas. DRM is more endothermic than steam reforming of methane or partial oxidation of methane or autothermal reforming of methane reactions.

$$CH_4 + CO_2 \rightarrow 2CO + 2H_2 \quad \Delta H_{298} = 248\,kJ\,mol^{-1} \qquad (5.14)$$

The ideal H_2/CO ratio in the DRM product is 1 but this is influenced by the simultaneous occurrence of Reverse Water Gas Shift reaction (5.2) which lowers the H_2/CO ratio to <1 by producing water. Based on the relative endothermicity of the DRM and RWGS reaction, the effect of RWGS on product selectivity is more significant in the temperature range of 400–800 °C [192]. Apart from RWGS, other significant side-reactions in DRM are methane decomposition (5.15) and the CO disproportionation (Boudouard reaction) (5.3). Both of these reactions result in the formation of solid carbon that can cover the catalytically active sites and cause rapid deactivation.

$$CH_4 \rightarrow C + 2H_2 \quad \Delta H_{298} = 75\,kJ\,mol^{-1} \qquad (5.15)$$

Wang et al. reported that CH_4 decomposition occurs above 557 °C, while the Boudouard reaction occurs below 700 °C [193]. In a temperature range of 557–700 °C, carbon deposition in DRM happens from both methane decomposition and CO disproportionation, which ultimately leads to a suppression of catalytic activity. At higher temperatures (>800 °C), the carbon deposition is derived mostly from methane decomposition, where the formed carbon species is relatively more reactive and can be easily oxidized. Wang et al. suggested that the optimum temperature at the feed ratio of CO_2/CH_4=1:1 is between 870 and 1040 °C, considering the conversion and carbon formation. Several researchers have conducted thermodynamic simulations for DRM under various process conditions and it has commonly been concluded that the operation of DRM at high temperatures above 850 °C and low pressures is required to attain high conversion [192].

5.6.2 Catalysts for DRM

A variety of catalysts such as supported metal catalysts including both noble metals like Pt, Pd, Ru, Rh and transition metals like Ni, Cu, Co etc. are active for dry reforming of methane. The overall activity of a catalyst in DRM depends on a combination of factors such as the active metal, nature of support, metal-support interaction, particle size, support surface area etc. Along with high activity, a crucial parameter for DRM catalyst is that it should be resistant to deactivation due to sintering at the high reaction temperature conditions or due to deposition of coke from the reforming reaction.

5.6.2.1 Type of Active Metal

Noble metals such at Pt, Pd etc. show higher resistance to sintering and coke formation in DRM compared to transition metals, but their usage is limited due to high cost. Due to easier availability and low cost, transition metals like Ni, Cu are gaining more interest as DRM catalysts, and tremendous efforts are being made to stabilize these catalysts under DRM conditions.

In terms of catalytic activity, it has been shown that the metals follow an order of Ru > Rh > Ni > Pt > Pd on SiO$_2$ support, Ru > Rh > Ni > Pd > Pt on a MgO support and Rh > Ni > Pt > Ir > Ru > Co on an Al$_2$O$_3$ support [194]. The higher activity of Ru and Rh than Ni, Pt, Pd of the same dispersion has also been proven by first principle calculations [195]. Alloying or promoting transition metals like Ni with noble metals to form bimetallic systems have been shown to significantly improve the activity and resistance to deactivation by coke formation or sintering [196]. For instance, doping trace amounts of noble metals like Pt, Pd, Rh can increase the reducibility of Ni by a hydrogen spill-over effect, wherein H$_2$ molecules preferentially dissociate on noble metal atoms to hydrogen atoms, which diffuse to the non-noble metals and enhance their reduction to create more active sites [196–198]. It has also been shown that based on the synthesis conditions and the reduction potential/kinetics of the different metals in a bimetallic catalyst, the two metals may exist in a uniform alloy phase or segregate to form surface enrichment of one metal species, that can affect the catalytic activity [199, 200]. Bimetallic systems involving only transition metals such as Ni–Co, Ni–Fe, Ni–Co have also shown considerable benefits over monometallic catalysts due to a synergistic effect on activity and coke resistance [196].

5.6.2.2 Role of Support

Active metal components are usually dispersed on metal oxide supports that provide high surface area for metal dispersion, exposing higher amounts of active sites and prevent metal agglomeration. A number of supports for these active metals have also been investigated, including SiO$_2$, La$_2$O$_3$, ZrO$_2$, TiO$_2$, CeO$_2$, Al$_2$O$_3$, and MgO [201]. Depending on the chemical nature of the support, it may also participate in the reaction mechanism of DRM and affect the product selectivity and tendency for coke formation. Inert supports like SiO$_2$ serve mainly as a medium to disperse the active metal and mostly do not contribute to the reaction pathway. On the other hand, basic or redox supports like MgO, CeO_2 etc. can actively reduce the formation of coke species in DRM by easily activating the CO_2 and providing more oxygen species for the removal of coke.

The specific surface area and porosity of the support plays a key role in the effective dispersion of the active metal phase. High surface area supports with ordered mesopores can stabilize very small metal nanoparticles within the pores while providing good accessibility to the reactant gas molecules. Silica based ordered mesoporous supports like SBA-15, MCM-41, KIT-6 etc. have been widely used to support metal nanoparticles, and can provide high metal dispersion, stability and sintering resistance by confining the nanoparticles within the mesopores [202, 203].

The interaction between the metal and support is another very important factor in determining the activity and stability of the catalyst in DRM. Stronger metal-support interaction reduces the mobility of metal particles on the support under high reaction temperatures and can thus yield sinter-resistant catalysts with higher metal dispersion [201]. On the other hand, too strong metal support interaction can make it extremely hard to reduce the metals to active metallic state, which causes a decline in activity. In case of reducible oxides, it has been shown that for high metal-support interaction, the support may partially cover the metal particles, thus blocking active sites and reducing the overall activity [204]. The metal support interaction is dependent on the catalyst preparation method and can be tweaked by changing synthesis conditions or catalyst pre-treatment conditions.

Some supports may also form different inorganic phases with the active metal under reaction conditions, that may affect the catalyst performance. For example, under high temperatures, Ni/Al_2O_3 catalysts may form a perovskite phase $NiAlO_3$ that is inactive for DRM and causes rapid deactivation [205].

5.6.2.3   Effect of Promoters

Promoters are non-active materials that can help in improving the catalytic activity or selectivity by inducing structural or electronic changes in the catalyst. Promoters may affect the metal-support interaction and hence, the dispersion of active metals on the catalyst. For instance, Sigl et al. showed that using V_2O_5 as a promoter forms an overlayer of VO_x on Rh/SiO_2 which breaks down the larger ensembles into smaller Rh particles, thereby increasing number of sites for activation of CH_4 [206]. Pan et al. observed that introducing Ga_2O_3 onto SiO_2 helped in activating CO_2 in the form of surface carbonate and bicarbonate species, and consequently promoted the coke resistance of the catalyst [207]. Addition of alkali metal and alkaline earth metal oxides such as Na_2O, K_2O, MgO etc. can neutralize the surface acidity of the catalyst, reduce methane dehydrogenation activity, increase CO_2 adsorption, and improve the coke elimination in DRM [208, 209]. Promoters may also affect the nature of coke formed during DRM [210]; for instance, Ag-promoted catalysts alter the type of coke formed on the catalyst surface from whisker to amorphous species [211].

5.6.3 Reaction Mechanism

CH₄ & CO₂ Activation

Methane activation is usually considered to be the most kinetically significant and rate determining step in the DRM process. CH_4 activation requires the presence of metals like Ni, Pt, Ru etc. which can adsorb and dissociate methane either directly or through intermediates like CH_x or formates. Methane activation is believed to happen through intermediate formation at lower temperatures (<550 °C) while direct dissociation is favoured at higher temperatures. It is usually agreed in literature that on catalysts using inert supports like SiO_2, DRM follows a monofunctional pathway where both CH_4 and CO_2 get activated on the metal surface. On acidic/basic supports like Al_2O_3, MgO, CeO_2 etc., the mechanism is usually bi-functional wherein CH_4 is activated on the metal and CO_2 is activated on the acidic/basic support. CO_2 activation occurs through the formation of formates on acidic supports with the surface hydroxyls and through oxy-carbonates/carbonates on basic supports. In such catalysts, the catalytic activity becomes a function of the interfacial area between the metal and support instead of the metal surface area alone [212]. The mechanism and the rate determining step strongly depend on the catalyst system, and a wide variety of mechanisms and rate determining steps have been reported for different systems [213, 214]. Some of the commonly reported mechanisms are described in Table 5.3.

Table 5.3 Reaction mechanism for dry reforming of methane

CH_4 activation	CO_2 activation
Direct decomposition	Direct decomposition on metal
1. $CH_4(g) + {}^* \leftrightarrow CH_4{}^*$	14. $CO_2(g) + {}^* \leftrightarrow CO_2{}^*$
2. $CH_4{}^* + {}^* \leftrightarrow CH_3{}^* + H^*$	15. $CO_2{}^* + {}^* \leftrightarrow CO^* + O^*$
3. $CH_3{}^* + {}^* \leftrightarrow CH_2{}^* + H^*$	H assisted activation
4. $CH_2{}^* + {}^* \leftrightarrow CH^* + H^*$	16. $CO_2{}^* + H^* \leftrightarrow COOH^*$
5. $CH^* + {}^* \leftrightarrow C^* + H^*$	Redox mechanism
Activation by Surface Hydroxyl species	17. $CO_2{}^* + O_{x-1} \leftrightarrow CO^* + O_x$ (O_x, O_{x-1} lattice oxygen and oxygen vacancy in support)
6. $CH_3{}^* + OH^* \leftrightarrow CH_3OH^*$	CO and H_2 formation
7. $CH_3OH^* + {}^* \leftrightarrow CH_2OH^* + H^*$	18. $CHO^* + {}^* \leftrightarrow CO^* + H^*$
8. $CH_2OH^* + {}^* \leftrightarrow CHOH^* + H^*$	19. $C^* + O^* \leftrightarrow CO^*$
9. $CHOH^* + {}^* \leftrightarrow COH^* + H^*$	20. $COOH^* + {}^* \leftrightarrow CO^* + OH^*$
10. $C^* + OH^* \leftrightarrow COH^*$	21. $C^* + O_x \leftrightarrow CO^* + O_{x-1}$
Activation by surface O species	22. $CO^* \leftrightarrow CO(g) + {}^*$
11. $CH_3{}^* + O^* \leftrightarrow CH_3O^*$	23. $H^* + H^* \leftrightarrow H_2{}^* + {}^*$
12. $CH_3O^* + {}^* \leftrightarrow CH_2O^* + H^*$	24. $H_2{}^* \leftrightarrow H_2(g) + {}^*$
13. $CH_2O^* + {}^* \leftrightarrow CHO^* + H^*$	

Kinetically, DRM has been mostly observed to be first order in CH_4, with the CO_2 reaction order varying from 0 on supports like Al_2O_3 to low positive values (<0.4) on supports like La_2O_3, MgO etc. [192, 213, 214]. Isotopic studies performed on Rh/Al_2O_3 by Wang et al. showed that the conversion of CH_4 was greater than the conversion of CD_4, suggesting that the dissociation of the C–H bond is the rate determining step in the reforming reaction [215]. Similar conclusions have been reported by a number of other researchers such as Wei and Iglesia [216]. The dissociation of CH_4 has also been observed to be very sensitive to the catalyst surface structure, with the rate of CH_4 decomposition reducing in the order of Ni (110) > Ni(100) > Ni(111) on different faces of Ni crystallites. However, on various systems, other elementary reactions such as surface reaction between CH_x and O species, carbon oxidation etc. have been observed to be the slower and rate-controlling step [213].

5.6.4   Reactor Systems for DRM

Continuous fixed bed reactors (FBR) are the most widely studied reaction systems for DRM. Isothermal operation is necessary due to the high endothermicity of the reaction. Several other experimental setups such as Fluidized Bed reactors (FIBR), membrane reactors (MR) etc. have also been studied by several groups [217]. In a single and multi-mode switching study on a FBR and FIBR with Ni/Al_2O_3 catalyst, Chen et al. showed that a FIBR showed higher methane and CO_2 conversion while also more efficiently suppressing coke formation [218]. However, for lab-scale studies, it is expensive and labour intensive to operate fluidized bed systems.

Membrane reactors with H_2 permeable membranes provide the benefit of simultaneously separating produced H_2 from the syngas, driving the reaction in the forward direction beyond thermodynamic equilibrium constraints. Pd-based hollow fiber membranes provide extremely high selectivity in separating pure H_2 and high surface to volume ratio for separation and coating of catalysts. A hollow fiber membrane reactor with Pd/Al_2O_3 composite membrane has been shown to have higher methane conversion (34% higher than thermodynamic equilibrium) than a fixed bed reactor in DRM [219].

Recently, unconventional systems such as plasma reactors or reactors with electric field have also been applied in DRM. Plasma reactors provide the benefit of activating the reactant molecules at very low temperature through electron impact excitation, dissociation and ionization, yielding high conversion but suffer from low energy utilization efficiency and low selectivity to syngas due to the formation of substantial amount of higher hydrocarbon side-products [220]. Since the first use of plasma reactors for DRM around 40 years back, different types of plasma like microwave plasma, gliding arc discharge, thermal plasma, dielectric barrier discharge and corona discharge plasma has been applied for DRM. It has been reported that combining catalysts with plasma can provide a synergistic effect in increasing both activity and selectivity of the reaction through a complex interplay of highly energetic reaction intermediates created by the plasma, their interaction with the catalyst surface and the in-situ modification of catalyst properties by the plasma.

5.6.5   Minimizing Catalyst Deactivation in DRM

Catalyst deactivation over time is a critical challenge for DRM, stemming mainly from sintering of active metal phase at high temperatures and the blocking of active sites by deposition of carbonaceous species formed in the DRM reaction. Catalyst deactivation can also happen due to other reasons such as surface oxidation of active metal sites, encapsulation of active metal by reducible supports like CeO$_2$ MgO etc., poisoning by sulfur compounds etc. However, coking is the most predominant reason for loss of catalytic activity, and hence has garnered significant research interest in the recent years. Some of the strategies that can be adopted to minimize the issue of catalyst deactivation are described below.

5.6.5.1   Reaction Conditions

As discussed before, the coke forming reactions in DRM show a strong dependence on the reaction conditions. Figure 5.9 shows the equilibrium product composition from DRM as a function of temperature, wherein solid coke is considered as one of the reaction products. Equilibrium coke formation is negligible above reaction temperatures of 900 °C [221]. The highly exothermic Boudouard reaction (5.3) is inhibited above 800 °C and coke formation occurs mainly as a result of methane decomposition. The carbon species from cracking of methane is more reactive than that generated from Boudouard reaction and can be oxidized by CO$_2$ at such high temperatures. Thus, at high temperature (>800–900 °C), the rate of carbon gasification by CO$_2$ is equivalent to or faster than the rate of formation of coke precursors, leading to minimal coking [196].

Fig. 5.9 Thermodynamic equilibrium composition for dry reforming of methane (CH$_4$/CO$_2$ = 1, 1 atm) [221]

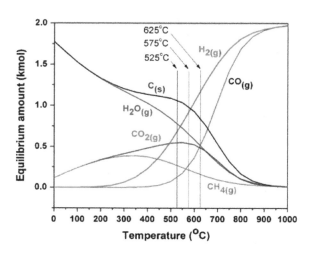

Another major factor that affects the coke formation rate is the CO_2/CH_4 ratio. Higher CO_2/CH_4 ratio naturally favours low coke formation, but affects product composition, further lowering the H_2/CO ratio. Increasing pressure increases the propensity of coke deposition; for instance, increasing pressure from 1 to 10 bars leads to an increase in the carbon deposition limit (temperature beyond which coke deposition is negligible) from 870 to 1030 °C [217]. Thus, deactivation by coking may be limited by operating in suitable regimes.

5.6.5.2 Coupling DRM with Other Reactions

Coupling DRM with Partial Oxidation of Methane (POM) [222–224] (5.16) or Steam Reforming of Methane (SRM) [225] (5.17) can reduce the coke formation tendency due to the presence of O_2 and H_2O, which are stronger oxidants than CO_2. At the same time, coupling these reactions also provide a handle in varying the H_2/CO ratio of the final product syngas. Oxidative DRM (combining DRM and POM) also results in a more favourable process thermodynamics due to the exothermic nature of POM and the strong endothermic nature of DRM [226].

$$2CH_4 + O_2 \rightarrow 2CO + 4H_2 \quad \Delta H_{298} = -36 \, kJ \, mol^{-1} \quad (5.16)$$

$$CH_4 + H_2O \rightarrow CO + 3H_2 \quad \Delta H_{298} = 228 \, kJ \, mol^{-1} \quad (5.17)$$

While coupling such reactions lead almost invariably to higher methane conversion, it can affect the hydrogen selectivity due to over oxidation of methane to CO_2. Membrane reactors have been developed to address such issues, where O_2 permeable membranes are used to provide a precisely controlled flow of additional O_2 to the DRM process, leading to high selectivity. Kawi et al. reported several perovskite membrane reactor systems to integrate POM with DRM that showed high conversion (c.a. 94% methane conversion at 725 °C), product selectivity and stable operation up to 160 h with negligible coke deposition [227, 228].

5.6.5.3 Catalyst Design

The rate of coke formation and deposition in DRM is a strong function of the catalyst structure and composition. A plethora of recent studies have focused on designing catalysts that can inherently eliminate coke formation and can be operated stably under a wide range of operating conditions.

Support Modification

A very effective way of resisting carbon formation is the incorporation of the active metal in to the support structure with high oxygen mobility. Redox oxides like CeO_2, ZrO_2 and perovskite oxides possess highly mobile oxygen species (either as lattice oxygen or surface oxygen species), that can adsorb CO_2 and activate it to accelerate the carbon oxidation process. CeO_2 can store and release lattice oxygen through a stable transition between Ce^{+4} to Ce^{+3} states. Doping redox oxides like CeO_2 in the support has shown high effectiveness in controlling coke formation by accelerating the oxygen-transport properties of the system [229, 230]. The oxygen mobility of

Fig. 5.10 Schematic diagram of the proposed mechanism of DRM reaction over the reduced $La_{0.8}Sr_{0.2}Ni_{0.8}Fe_{0.2}O_3$ perovskite catalyst [232]

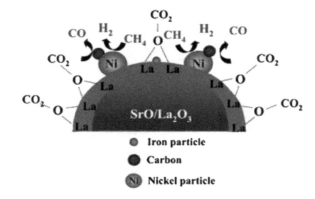

oxide supports can be further enhanced by the introduction of trace amounts of heteroatoms in the oxide structure, that creates structural defects and increases the oxygen mobility [209, 231]. For example, Sutthiumporn et al. showed that partially substituted $La_{0.8}Sr_{0.2}Ni_{0.8}M_{0.2}O_3$ perovskite (M=Bi, Cu, Cr, Co, Fe) catalysts possessed higher lattice oxygen mobility, that helped in C–H activation and also in reducing carbon formation by activating CO_2 through Lanthanum Oxycarbonate $(La_2O_2CO_3)$ formation (Fig. 5.10) [232]. In an other study [233], the Lewis acid species (Al-F) present in modified Ni/Al_2O_3 catalyst stabilized Ni species from high temperature sintering through metal-support interactions and reduce the electron density to alleviate the fast CH_4 decomposition.

Reducing Metal Particle Size
The rate of coke formation has been shown to be a strong function of the metal particle size, with coke formation being favoured on bigger ensembles. Kim et al. investigated the effect of size of nickel particles supported on alumina aerogel on coke formation in DRM and found that a minimum diameter of about 7 nm is required for the Ni particles to generate filamentous carbon [234, 235]. Hence, synthesizing catalysts with high metal dispersion and small particle size is considered an effective strategy to reduce coke formation in DRM. Several synthesis methods such as sol-gel, deposition-precipitation, microemulsion, colloidal method etc, have been explored to synthesize more dispersed metal nanoparticles. Impregnation in the presence of complex organic chelating agents like oleic acid/amine [101], PVP [236], etc. has proven to be effective in reducing metal particle size and consequently reduce coke formation in DRM. Kawi et al. reported self-assembly of transition metals into small nanoparticles coated with organic agents like oleic acid/amine, which resulted in high metal dispersion and sintering resistance [101, 113, 237–241]. Another method of controlling metal dispersion and preventing sintering is to use structured catalyst precursors with mineral type structures such as phyllosilicates [225, 242, 243], perovskites [232], hydrotalcites [244] etc. that can generate supported metal catalysts with very strong metal support interaction with lower metal mobility and sintering. Suitable promoters may be used

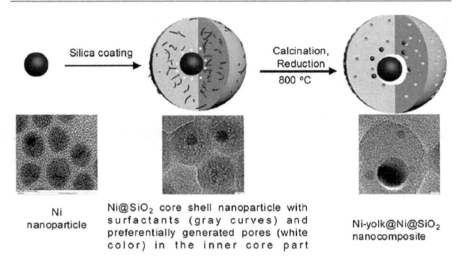

Ni nanoparticle

Ni@SiO₂ core shell nanoparticle with surfactants (gray curves) and preferentially generated pores (white color) in the inner core part

Ni-yolk@Ni@SiO₂ nanocomposite

Fig. 5.11 Schematic illustration of synthesis of Ni@SiO₂ yolk–shell structure, that shows high resistance to coke formation in DRM [247]

to enhance the metal-support interaction. Confining metal nanoparticles within the pore structures of ordered mesoporous supports like SBA-15, MCM-41 etc. also prevents metal agglomeration by physically separating the particles [245].

Another class of catalysts that have gained high interest in recent years for their high thermal stability and anti-coking properties are core@shell catalysts [246, 247]. By coating the active metal cores with thermally stable porous metal oxide shells, the particles can be segregated, preventing their agglomeration and growth under high temperatures. Several types of core@shell structured catalysts have been reported to show high stability under DRM, even under low temperature conditions with negligible coke formation (Fig. 5.11) [184, 248–257]. While SiO_2 remains the most widely used shell material for the ease of controlling the coating process by modifier Stöber process, other materials such as CeO_2 [250], ZrO_2 [258] can also provide more coke resistance by aiding in coke removal through CO_2 activation.

Bimetallic Systems

Using bimetallic catalysts can provide benefits in increasing the coke resistance of the catalyst through multiple ways, by modifying both the electronic and geometric properties of the active component [196]. Geometrically, the presence of a second metal acts to break up the metal ensembles by a 'dilution effect', which can reduce the coke formation tendency. For bimetallic catalysts involving transition metals and noble metals such as Ni–Pt, Ni–Ru etc., doping minute amounts of noble metals has shown a significant increase in activity and stability due to the inherent properties of noble metals, synergistic effect caused by alloying and an increase in metal dispersion. Depending on the method of catalyst preparation, noble metals may preferentially be enriched on the surface in the bimetallic particle, which leads to more significant improvement in performance [200].

Alloying Ni with transition metals with redox property such as Co and Fe has also shown significant improvement in coke resistance due to the higher oxygen affinity of these metals. For instance, Fan et al. studied the role of doping Co in Co$_x$Ni$_y$Mg$_{100-x-y}$O catalysts and showed that Co is enriched on Ni–Co alloy surface and enhances the chemisorption of oxygen species, thereby accelerating the gasification of coke intermediates [259]. However, it is crucial to optimize the Co/Ni ratio in the bimetallic catalyst as too high Co content may lead to oxidation of Ni and a consequent loss in activity. Similarly, adding Fe to Ni has been shown to reduce coke formation in DRM, although at a slight trade-off in activity, due to the redox nature of Fe. Kim et al. showed that under DRM conditions, there is some de-alloying of Ni and Fe and a FeO phase is formed that migrates towards the NiFe particle surface and provides oxygen species for coke removal through a redox mechanism [260].

5.7 Concluding Remarks

In this chapter, we have highlighted the technological developments of CO$_2$ to various C$_1$ valuables, with respect to heterogeneous catalysts, kinetic considerations and engineering challenges. The technologies discussed include CO$_2$ to methane, methanol, formic acid and syngas formation via hydrogenation reaction and CO$_2$ to syngas production using methane reforming process. On the common note, the role of efficient catalyst is pivotal for these technologies to be economically feasible. The nature and properties of the catalysts, such as redox or acid-base, play an important role in determining the catalytic activity as well as selectivity towards desired product. For CO$_2$ methanation reaction, being an exothermic process, the current development in the catalyst mainly focuses on low-temperature active catalysts to overcome the thermodynamic barrier and also requires appropriate reactor design to remove additional heat. Nanosized nickel particles appear to be the most suitable non-precious metal to achieve high conversion at low temperatures. Additionally, development of the catalyst is based on improving the selectivity of methane by avoiding the formation route of carbonyls and carbides which causes deactivation. Apart from active catalytic sites, designing support and addition of promoters play a crucial role in improving the performance by H$_2$ and CO$_2$ activation, metal dispersion and enhancement of metal support interaction. The fixed-bed reactors are the most widely used systems for CO$_2$ methanation. Fluidized-bed reactors have proven highly reliable for CO methanation, and other types of reactor are still under development.

Similarly, reverse water gas shift (RWGS) reaction which converts CO$_2$ to CO by combining with hydrogen is one the most straightforward processes used in industries as an intermediate process for the production of more valuable chemicals such as methanol. Additionally, RWGS is the most straightforward way for large-scale production of value-added chemical when combined with CO$_2$-FT synthesis units. To design a catalyst for RWGS reaction, it is important to consider the problems associated with the reaction such as catalyst deactivation. Some of the possible reasons for catalyst deactivation are the large amount of water formation, metal sintering and coke deposition which decreases the selectivity towards CO formation. Controlling the selectivity of RWGS requires a deep understanding of thermodynamics, reaction kinetics, and mechanism which becomes easier by the utilization of advanced techniques such as transient quantitative temporal analysis, in-situ DRIFTS, and few more. In parallel with the experimental study and simulation method involving DFT calculations, the structures of well -performing catalysts can be designed to make the RWGS process more efficient to be used industrially.

Another product from CO_2 hydrogenation reaction is methanol, which is a widely used feedstock for fuel, olefins and other chemicals. Direct conversion of CO_2 to methanol is one of the most economical and feasible ways after oil and gas. CO_2 to methanol reaction is highly exothermic and thus it can be facilitated only at high pressure and low temperature. However, this reaction suffers from low selectivity towards methanol and catalyst deactivation due to crystallization by water formed during the reaction. There are possibly two reaction routes to methanol as described in the literature: firstly via a reverse WGS for CO_2 decomposition to CO and secondly via an intermediate formate route. Cu, Pd, In, Ag are some of the active metals suitable for methanol production. Therefore, synthesis of highly-dispersed metal catalyst, with enhanced metal support interaction and increasing defect concentration in metallic nanoparticles, amount of surface steps and edges of the Cu phase helps to achieve the target yield for this process to be commercialized.

Besides methanol, synthesis of formic acid is another approach for utilizing CO_2 as C_1 source for fuels/chemicals. Heterogeneous direct synthesis of formic acid from CO_2 is also an exothermic process, so lower reaction temperature and high pressure favor the reaction towards product formation. Unlike the methanol technology, direct formic acid synthesis technology is still at conceptual research and development stage. Improvements in the yield of formic acid production remains a key factor which requires much on-going research. The key challenges that need to be addressed include the development of suitable heterogeneous solvent medium. As per the current development, Ru and Pd based catalytic systems are the most active metals for this reaction; however, the performance of active metals is greatly influenced by the choice of support. A suitable catalyst support can increase the number of available catalytic centers and minimize their agglomeration under reaction condition, thus resulting in longevity and reusability. In addition to the nature of catalyst, suitable additives and solvents have a great role in fixing gaseous reactants to make them available for the reaction to occur on the catalyst surface. DFT studies and experimental observation revealed that reaction proceeds in two steps on the metal surface i.e., abstraction of hydrogen atom by CO_2 to form a formate intermediate and this formate intermediate then takes another hydrogen atom to form formic acid. The formation of formate intermediate is observed to be the rate-determining step of the reaction.

CO_2 reforming of methane is another avenue for the fixation of CO_2 into useful fuels, and significant progress has been made in the development of catalysts and novel reactor systems that allows long-term stable operation, circumventing the inherent issues of deactivation. However, further work is still required for the translation of lab-scale research to industrial-scale implementation. In 2015, the Linde Group officially opened a dry-reforming based pilot facility at Pullach near Munich. With more research both on key catalyst development and on addressing the practical challenges of scale-up, DRM may be at the cusp of commercialization shortly.

The highlighted CO$_2$ hydrogenation and methane reforming reactions make a complex network which are inter-related, not only in terms of schematic reaction pathways (Fig. 5.12), but also in terms of the catalysts used.

For instance, for CO$_2$ methanation reaction (Route 1 in Fig. 5.12), supported Ni catalysts are one of the extensively explored catalyst and proved to be promising at slightly higher temperature (above 400 °C). Similar catalytic system can also be adopted for CO$_2$-CH$_4$ reforming reaction (route 2). However, due to the endothermic nature, the latter needs to be operated at significantly higher temperature than the former. Moreover, further research is going on to develop DRM catalyst active at lower reaction temperature. Furthermore, methane generated from scheme 1 can be used to carry out the second reaction. Vapour phase CO$_2$ hydrogenation at low temperature and high pressure and the presence of suitable catalyst generates methanol (route 3). And liquid phase CO$_2$ hydrogenation at low temperature and high pressure with suitable catalyst and additive forms formic acid (route 4). The formation of methanol can be either via formation for CO (route 5) or formate species (route 6) as an intermediate products. Thus, the reaction systems, which can catalyse CO$_2$ hydrogenation to CO via RWGS reaction, can also show promising methanol synthesis activity at its reaction conditions. Similarly, if the catalyst catalysing CO$_2$ hydrogenation to formate can be able to generate methanol from formate species with some modifications, it can open the doors for the synthesis of methanol at milder reaction conditions i.e., low temperature and high pressure. There are studies showing methanol synthesis from formic acid using H$_2$ with Cu–Zn and Cu–Al catalysts [261]. The maximum methanol yield of 36% could be achieved at 300 °C for 5 h. Further research on this direction is required to boost CO$_2$ conversion to C1 valuable chemicals.

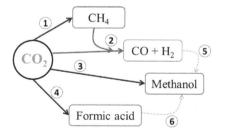

1. CO$_2$ methanation
2. CO$_2$-CH$_4$ reforming to syngas
3. CO$_2$ hydrogenation to Methanol
4. CO$_2$ hydrogenation to Formic Acid
5. Methanol synthesis via CO intermediate
6. Methanol synthesis via formate intermediate

Fig. 5.12 Network of reaction schemes showing CO$_2$ hydrogenation and methane reforming reactions

In a nutshell, the conversion of CO_2 to synthesize valuable chemicals has a tremendous potential to be commercialized on a large scale [262]. These processes require practical design of the catalyst system and an energy efficient reactor design to improve the performance regarding activity, selectivity, and stability. Finally, except CO_2 to formic acid reaction, other processes are technologically matured and can become economically viable, if the costs associated with CO_2 capture and green hydrogen production is mitigated.

References

1. Álvarez A, Bansode A, Urakawa A, Bavykina AV, Wezendonk TA, Makkee M, Gascon J, Kapteijn F (2017) Challenges in the greener production of formates/formic acid, methanol, and DME by heterogeneously catalyzed CO_2 hydrogenation processes. Chem Rev 117:9804–9838
2. Kawi S, Kathiraser Y (2015) CO_2 as an oxidant for high-temperature reactions. Front Energy Res 3:13
3. Dorner RW, Hardy DR, Williams FW, Willauer HD (2010) Heterogeneous catalytic CO_2 conversion to value-added hydrocarbons. Energy Environ Sci 3:884–890
4. Götz M, Lefebvre J, Mörs F, McDaniel Koch A, Graf F, Bajohr S, Reimert R, Kolb T (2016) Renewable power-to-gas: a technological and economic review. Renew Energy 85:1371–1390
5. Lehner, RTM, Steinmüller, H, Koppe, M (2014) Power-to-gas: technology and business models. Springer briefs in energy. Springer, Cham
6. Specht JBM, Frick V, Sturmer B, Zuberbuhler U (2016) Technical realization of power-to-gas technology (P2G): production of substitute natural gas by catalytic methanation of H_2/CO_2. In: Van Basshuysen R (ed) Natural gas and renewable methane for powertrains: future strategies for a climate-neutral mobility. Springer, Cham, pp 141–167
7. Rönsch S, Schneider J, Matthischke S, Schlüter M, Götz M, Lefebvre J, Prabhakaran P, Bajohr S (2016) Review on methanation—from fundamentals to current projects. Fuel 166:276–296
8. Barbarossa CBV, Deiana P, Vanga G (2013) CO_2 conversion to CH_4. In: de Falco M, Iaquaniello G, Centi G (eds) CO_2 a valuable source carbon. Springer, London, pp 123–145
9. Kopyscinski J, Schildhauer TJ, Biollaz SMA (2010) Production of synthetic natural gas (SNG) from coal and dry biomass—a technology review from 1950 to 2009. Fuel 89:1763–1783
10. Mazza A, Bompard E, Chicco G (2018) Applications of power to gas technologies in emerging electrical systems. Renew Sustain Energy Rev 92:794–806
11. Meylan FD, Moreau V, Erkman S (2016) Material constraints related to storage of future European renewable electricity surpluses with CO_2 methanation. Energy Policy 94:366–376
12. Napp TA, Gambhir A, Hills TP, Florin N, Fennell PS (2014) A review of the technologies, economics and policy instruments for decarbonising energy-intensive manufacturing industries. Renew Sustain Energy Rev 30:616–640
13. Sabatier P (1902) New synthesis of methane. Comptes Rendus 134:514–516
14. Mills GA, Steffgen FW (1974) Catalytic methanation. Catal Rev 8:159–210
15. Weatherbee GD, Bartholomew CH (1981) Hydrogenation of CO_2 on group VIII metals: I. Specific activity of $NiSiO_2$. J Catal 68:67–76
16. Bian, Z, Chan, YM, Yu, Y, Kawi, S (2018) Morphology dependence of catalytic properties of Ni/CeO_2 for CO_2 methanation: a kinetic and mechanism study. Catal Today (in press)

17. Visconti CG, Lietti L, Tronconi E, Forzatti P, Zennaro R, Finocchio E (2009) Fischer-tropsch synthesis on a Co/Al₂O₃ catalyst with CO₂ containing syngas. Appl Catal A 355:61–68
18. Mutschler R, Moioli E, Luo W, Gallandat N, Züttel A (2018) CO₂ hydrogenation reaction over pristine Fe Co, Ni, Cu and Al₂O₃ supported Ru: comparison and determination of the activation energies. J Catal 366:139–149
19. Solymosi F, Erdőhelyi A (1980) Hydrogenation of CO₂ to CH₄ over alumina-supported noble metals. J Mol Catal 8:471–474
20. Panagiotopoulou P (2017) Hydrogenation of CO₂ over supported noble metal catalysts. Appl Catal A 542:63–70
21. Visconti CG, Martinelli M, Falbo L, Infantes-Molina A, Lietti L, Forzatti P, Iaquaniello G, Palo E, Picutti B, Brignoli F (2017) CO₂ hydrogenation to lower olefins on a high surface area K-promoted bulk Fe-catalyst. Appl Catal B 200:530–542
22. Visconti CG, Martinelli M, Falbo L, Fratalocchi L, Lietti L (2016) CO₂ hydrogenation to hydrocarbons over Co and Fe-based Fischer-Tropsch catalysts. Catal Today 277:161–170
23. Wei W, Jinlong G (2011) Methanation of carbon dioxide: an overview. Front Chem Sci Eng 5:2–10
24. Riani P, Garbarino G, Lucchini MA, Canepa F, Busca G (2014) Unsupported versus alumina-supported Ni nanoparticles as catalysts for steam/ethanol conversion and CO₂ methanation. J Mol Catal A: Chem 383:10–16
25. Lee GD, Moon MJ, Park JH, Park SS, Hong SS (2005) Raney Ni catalysts derived from different alloy precursors part II. CO and CO₂ methanation activity. Korean J Chem Eng 22:541–546
26. Koschany F, Schlereth D, Hinrichsen O (2016) On the kinetics of the methanation of carbon dioxide on coprecipitated NiAl(O)ₓ. Appl Catal B 181:504–516
27. Mirodatos C, Praliaud H, Primet M (1987) Deactivation of nickel-based catalysts during CO methanation and disproportionation. J Catal 107:275–287
28. Agnelli, M, Kolb M, Nicot C, Mirodatos C (1991) Sintering of a Ni-based catalyst during CO hydrogenation: kinetics and modeling. In: Studies in surface science and catalysis. Elsevier, pp 605–612
29. Miao B, Ma SSK, Wang X, Su H, Chan SH (2016) Catalysis mechanisms of CO₂ and CO methanation. Catal Sci Technol 6:4048–4058
30. Czekaj I, Loviat F, Raimondi F, Wambach J, Biollaz S, Wokaun A (2007) Characterization of surface processes at the Ni-based catalyst during the methanation of biomass-derived synthesis gas: X-ray photoelectron spectroscopy (XPS). Appl Catal A 329:68–78
31. McCarty J, Wise H (1979) Hydrogenation of surface carbon on alumina-supported nickel. J Catal 57:406–416
32. Lunde PJ, Kester FL (1973) Rates of methane formation from carbon dioxide and hydrogen over a ruthenium catalyst. J Catal 30:423–429
33. Bartholomew CH (2001) Mechanisms of catalyst deactivation. Appl Catal A 212:17–60
34. Abrevaya H, Cohn M, Targos W, Robota H (1990) Structure sensitive reactions over supported ruthenium catalysts during Fischer-Tropsch synthesis. Catal Lett 7:183–195
35. Goodwin J Jr, Goa D, Erdal S, Rogan F (1986) Reactive metal volatilization from Ru/Al₂O₃ as a result of ruthenium carbonyl formation. Appl Catal 24:199–209
36. Dalla Betta R, Piken A, Shelef M (1975) Heterogeneous methanation: steady-state rate of CO hydrogenation on supported ruthenium, nickel and rhenium. J Cataly 40:173–183
37. Bowman RM, Bartholomew CH (1983) Deactivation by carbon of Ru/Al₂O₃ during CO hydrogenation. Appl Catal 7:179–187
38. Mukkavilli S, Wittmann C, Tavlarides LL (1986) Carbon deactivation of Fischer-Tropsch ruthenium catalyst. Ind Eng Chem Process Des Dev 25:487–494
39. Dalla Betta R, Shelef M (1977) Heterogeneous methanation: in situ infrared spectroscopic study of RuAl₂O₃ during the hydrogenation of CO. J Catal 48:111–119

40. Ekerdt JG, Bell AT (1979) Synthesis of hydrocarbons from CO and H_2 over silica-supported Ru: reaction rate measurements and infrared spectra of adsorbed species. J Catal 58:170–187
41. Gao J, Liu Q, Gu F, Liu B, Zhong Z, Su F (2015) Recent advances in methanation catalysts for the production of synthetic natural gas. RSC Adv 5:22759–22776
42. Wang X, Shi H, Szanyi J (2017) Controlling selectivities in CO_2 reduction through mechanistic understanding. Nat Commun 8:513
43. Lin Q, Liu XY, Jiang Y, Wang Y, Huang Y, Zhang T (2014) Crystal phase effects on the structure and performance of ruthenium nanoparticles for CO_2 hydrogenation. Catal Sci Technol 4:2058–2063
44. Aziz M, Jalil A, Triwahyono S, Ahmad A (2015) CO_2 methanation over heterogeneous catalysts: recent progress and future prospects. Green Chem 17:2647–2663
45. Zhang G, Sun T, Peng J, Wang S, Wang S (2013) A comparison of Ni/SiC and Ni/Al_2O_3 catalyzed total methanation for production of synthetic natural gas. Appl Catal A 462:75–81
46. Wang W, Wang S, Ma X, Gong J (2011) Recent advances in catalytic hydrogenation of carbon dioxide. Chem Soc Rev 40:3703–3727
47. Bartholomew CH, Vance CK (1985) Effects of support on the kinetics of carbon hydrogenation on nickel. J Catal 91:78–84
48. Yu Y, Chan YM, Bian ZF, Song FJ, Wang J, Zhong Q, Kawi S (2018) Enhanced performance and selectivity of CO_2 methanation over g-C_3N_4 assisted synthesis of Ni-CeO_2 catalyst: kinetics and DRIFTS studies. Int J Hydrogen Energy 43:15191–15204
49. Ashok J, Ang ML, Kawi S (2017) Enhanced activity of CO_2 methanation over Ni/CeO_2–ZrO_2 catalysts: influence of preparation methods. Catal Today 281:304–311
50. Kwak JH, Kovarik L, Szanyi JN (2013) Heterogeneous catalysis on atomically dispersed supported metals: CO_2 reduction on multifunctional Pd catalysts. ACS Catal 3:2094–2100
51. Aldana PU, Ocampo F, Kobl K, Louis B, Thibault-Starzyk F, Daturi M, Bazin P, Thomas S, Roger A (2013) Catalytic CO_2 valorization into CH_4 on Ni-based ceria-zirconia. Reaction mechanism by operando IR spectroscopy. Catal Today 215:201–207
52. Marwood M, Doepper R, Prairie M, Renken A (1994) Transient drift spectroscopy for the determination of the surface reaction kinetics of CO_2 methanation. Chem Eng Sci 49:4801–4809
53. Marwood M, Doepper R, Renken A (1997) In-situ surface and gas phase analysis for kinetic studies under transient conditions. The catalytic hydrogenation of CO_2. Appl Catal A 151:223–246
54. Wang X, Hong Y, Shi H, Szanyi J (2016) Kinetic modeling and transient DRIFTS–MS studies of CO_2 methanation over Ru/Al_2O_3 catalysts. J Catal 343:185–195
55. Garbarino G, Riani P, Magistri L, Busca G (2014) A study of the methanation of carbon dioxide on Ni/Al_2O_3 catalysts at atmospheric pressure. Int J Hydrogen Energy 39:11557–11565
56. Swapnesh A, Srivastava VC, Mall ID (2014) Comparative study on thermodynamic analysis of CO_2 utilization reactions. Chem Eng Technol 37:1765–1777
57. Ruterana P, Buffat P-A, Thampi K, Graetzel M (1990) The structure of ruthenium supported on titania: a catalyst for low-temperature methanation of carbon dioxide. Ultramicroscopy 34:66–72
58. Panagiotopoulou P, Verykios XE (2017) Mechanistic study of the selective methanation of CO over Ru/TiO_2 catalysts: effect of metal crystallite size on the nature of active surface species and reaction pathways. J Phys Chem C 121:5058–5068
59. Shashidhara G, Ravindram M (1992) Methanation of CO_2 over Ru–SiO_2 catalyst. React Kinet Catal Lett 46:365–372
60. Sharma S, Hu Z, Zhang P, McFarland EW, Metiu H (2011) CO_2 methanation on Ru-doped ceria. J Catal 278:297–309
61. Upham DC, Derk AR, Sharma S, Metiu H, McFarland EW (2015) CO_2 methanation by Ru-doped ceria: the role of the oxidation state of the surface. Catal Sci Technol 5:1783–1791

62. Yu Y, Bian ZF, Song FJ, Wang J, Zhong Q, Kawi S (2018) Influence of calcination temperature on activity and selectivity of Ni–CeO$_2$ and Ni–Ce$_{0.8}$Zr$_{0.2}$O$_2$ catalysts for CO$_2$ methanation. Top Catal 61:1514–1527

63. Xu J, Lin Q, Su X, Duan H, Geng H, Huang Y (2016) CO$_2$ methanation over TiO$_2$–Al$_2$O$_3$ binary oxides supported Ru catalysts. Chin J Chem Eng 24:140–145

64. Ocampo F, Louis B, Roger A-C (2009) Methanation of carbon dioxide over nickel-based Ce$_{0.72}$Zr$_{0.28}$O$_2$ mixed oxide catalysts prepared by sol–gel method. Appl Catal A Gen 369:90–96

65. Tada S, Ochieng OJ, Kikuchi R, Haneda T, Kameyama H (2014) Promotion of CO$_2$ methanation activity and CH$_4$ selectivity at low temperatures over Ru/CeO$_2$/Al$_2$O$_3$ catalysts. Int J Hydrogen Energy 39:10090–10100

66. Li K, Peng B, Peng T (2016) Recent advances in heterogeneous photocatalytic CO$_2$ conversion to solar fuels. ACS Catal 6:7485–7527

67. Fechete I, Vedrine J (2015) Nanoporous materials as new engineered catalysts for the synthesis of green fuels. Molecules 20:5638–5666

68. Zhen W, Li B, Lu G, Ma J (2015) Enhancing catalytic activity and stability for CO$_2$ methanation on Ni@ MOF-5 via control of active species dispersion. Chem Commun 51:1728–1731

69. Hwang S, Hong UG, Lee J, Baik JH, Koh DJ, Lim H, Song IK (2012) Methanation of carbon dioxide over mesoporous nickel–M–alumina (M=Fe, Zr, Ni, Y, and Mg) xerogel catalysts: effect of second metal. Catal Lett 142:860–868

70. Zhi G, Guo X, Wang Y, Jin G, Guo X (2011) Effect of La$_2$O$_3$ modification on the catalytic performance of Ni/SiC for methanation of carbon dioxide. Catal Commun 16:56–59

71. Schuurman Y, Mirodatos C, Ferreira-Aparicio P, Rodriguez-Ramos I, Guerrero-Ruiz A (2000) Bifunctional pathways in the carbon dioxide reforming of methane over MgO-promoted Ru/C catalysts. Catal Lett 66:33–37

72. Park J-N, McFarland EW (2009) A highly dispersed Pd–Mg/SiO$_2$ catalyst active for methanation of CO$_2$. J Catal 266:92–97

73. Falconer JL, Zağli AE (1980) Adsorption and methanation of carbon dioxide on a nickel/silica catalyst. J Catal 62:280–285

74. Eckle S, Anfang H-G, Behm R Jr (2010) Reaction intermediates and side products in the methanation of CO and CO$_2$ over supported Ru catalysts in H$_2$-rich reformate gases. J Phys Chem C 115:1361–1367

75. Traa Y, Weitkamp J (1999) Kinetics of the methanation of carbon dioxide over ruthenium on titania. Chem Eng Technol Ind Chem Plant Equip Process Eng Biotechnol 22:291–293

76. Vannice M (1976) The catalytic synthesis of hydrocarbons from carbon monoxide and hydrogen. Catal Rev Sci Eng 14:153–191

77. Lim JY, McGregor J, Sederman A, Dennis J (2016) Kinetic studies of CO$_2$ methanation over a Ni/γ-Al$_2$O$_3$ catalyst using a batch reactor. Chem Eng Sci 141:28–45

78. Weatherbee GD, Bartholomew CH (1982) Hydrogenation of CO$_2$ on group VIII metals: II. Kinetics and mechanism of CO$_2$ hydrogenation on nickel. J Catal 77:460–472

79. Xu J, Froment GF (1989) Methane steam reforming, methanation and water-gas shift: I. Intrinsic kinetics. AIChE J 35:88–96

80. Schlereth D, Hinrichsen O (2014) A fixed-bed reactor modeling study on the methanation of CO$_2$. Chem Eng Res Des 92:702–712

81. Falbo L, Martinelli M, Visconti CG, Lietti L, Bassano C, Deiana P (2018) Kinetics of CO$_2$ methanation on a Ru-based catalyst at process conditions relevant for power-to-gas applications. Appl Catal B 225:354–363

82. Kiewidt L, Thöming J (2015) Predicting optimal temperature profiles in single-stage fixed-bed reactors for CO$_2$-methanation. Chem Eng Sci 132:59–71

83. Seemann MC, Schildhauer TJ, Biollaz SM (2010) Fluidized bed methanation of wood-derived producer gas for the production of synthetic natural gas. Ind Eng Chem Res 49:7034–7038

84. Brooks KP, Hu J, Zhu H, Kee RJ (2007) Methanation of carbon dioxide by hydrogen reduction using the Sabatier process in microchannel reactors. Chem Eng Sci 62:1161–1170

85. Frey M, Romero T, Roger A-C, Edouard D (2016) Open cell foam catalysts for CO_2 methanation: presentation of coating procedures and in situ exothermicity reaction study by infrared thermography. Catal Today 273:83–90

86. Pastor-Pérez L, Baibars F, Le Sache E, Arellano-García H, Gu S, Reina TR (2017) CO_2 valorisation via reverse water-gas shift reaction using advanced Cs doped Fe–Cu/Al_2O_3 catalysts. J CO_2 Utiliz 21:423–428

87. Centi G, Quadrelli EA, Perathoner S (2013) Catalysis for CO_2 conversion: a key technology for rapid introduction of renewable energy in the value chain of chemical industries. Energy Environ Sci 6:1711–1731

88. Centi G, Perathoner SJCT (2009) Opportunities and prospects in the chemical recycling of carbon dioxide to fuels. Catal Today 148:191–205

89. Oshima K, Shinagawa T, Nogami Y, Manabe R, Ogo S, Sekine Y (2014) Low temperature catalytic reverse water gas shift reaction assisted by an electric field. Catal Today 232:27–32

90. Maneerung T, Hidajat K, Kawi S (2017) K-doped $LaNiO_3$ perovskite for high-temperature water-gas shift of reformate gas: Role of potassium on suppressing methanation. Int J Hydrogen Energy 42:9840–9857

91. Ang ML, Miller JT, Cui Y, Mo L, Kawi S (2016) Bimetallic Ni-Cu alloy nanoparticles supported on silica for the water-gas shift reaction: activating surface hydroxyls via enhanced CO adsorption. Catal Sci Technol 6:3394–3409

92. Saw ET, Oemar U, Ang ML, Hidajat K, Kawi S (2015) Highly active and stable bimetallic Nickel-Copper core–ceria shell catalyst for high-temperature water-gas shift reaction. ChemCatChem 7:3358–3367

93. Ang ML, Oemar U, Kathiraser Y, Saw ET, Lew CHK, Du Y, Borgna A, Kawi S (2015) High-temperature water-gas shift reaction over Ni/xK/CeO_2 catalysts: Suppression of methanation via formation of bridging carbonyls. J Catal 329:130–143

94. Nakamura J, Campbell JM, Campbell CT (1990) Kinetics and mechanism of the water-gas shift reaction catalysed by the clean and Cs-promoted Cu(110) surface: a comparison with Cu(111). J Chem Soc, Faraday Trans 86:2725–2734

95. Wambach J, Baiker A, Wokaun A (1999) CO_2 hydrogenation over metal/zirconia catalysts. Phys Chem Chem Phys 1:5071–5080

96. Saw ET, Oemar U, Ang ML, Kus H, Kawi S (2016) High-temperature water gas shift reaction on Ni–Cu/CeO_2 catalysts: effect of ceria nanocrystal size on carboxylate formation. Catal Sci Technol 6:5336–5349

97. Saw ET, Oemar U, Tan XR, Du Y, Borgna A, Hidajat K, Kawi S (2014) Bimetallic Ni-Cu catalyst supported on CeO_2 for high-temperature water-gas shift reaction: Methane suppression via enhanced CO adsorption. J Catal 314:32–46

98. Oemar U, Bian Z, Hidajat K, Kawi S (2016) Sulfur resistant $La_xCe_{1-x}Ni_{0.5}Cu_{0.5}O_3$ catalysts for an ultra-high temperature water gas shift reaction. Catal Sci Technol 6:6569–6580

99. Ang ML, Oemar U, Saw ET, Mo L, Kathiraser Y, Chia BH, Kawi S (2014) Highly active Ni/xNa/CeO_2 catalyst for the water gas shift reaction: effect of sodium on methane suppression. ACS Catal 4:3237–3248

100. Bian ZF, Li ZW, Ashok J, Kawi S (2015) A highly active and stable Ni–Mg phyllosilicate nanotubular catalyst for ultrahigh temperature water-gas shift reaction. Chem Commun 51:16324–16326

101. Mo LY, Kawi S (2014) An in situ self-assembled core-shell precursor route to prepare ultrasmall copper nanoparticles on silica catalysts. J Mater Chem A 2:7837–7844

102. Ashok J, Wai MH, Kawi S (2018) Nickel-based catalysts for high-temperature water gas shift reaction-methane suppression. ChemCatChem 10:3927–3942

103. Pati, S, Jangam, A, Zhigang, W, Dewangan, N, Ming Hui, W, Kawi, S (2018) Catalytic $Pd_{0.77}Ag_{0.23}$ alloy membrane reactor for high temperature water-gas shift reaction: methane suppression. Chem Eng J (2018) (in press)

104. Dai B, Zhou G, Ge S, Xie H, Jiao Z, Zhang G, Xiong K (2017) CO$_2$ reverse water-gas shift reaction on mesoporous M-CeO$_2$ catalysts. Can J Chem Eng 95:634–642

105. Choi S, Sang B-I, Hong J, Yoon KJ, Son J-W, Lee J-H, Kim B-K, Kim H (2017) Catalytic behavior of metal catalysts in high-temperature RWGS reaction: in-situ FT-IR experiments and first-principles calculations. Sci Rep 7:41207

106. Chen X, Su X, Duan H, Liang B, Huang Y, Zhang T (2017) Catalytic performance of the Pt/TiO$_2$ catalysts in reverse water gas shift reaction: controlled product selectivity and a mechanism study. Catal Today 281:312–318

107. Yu W, Porosoff MD, Chen JG (2012) Review of Pt-based bimetallic catalysis: from model surfaces to supported catalysts. Chem Rev 112:5780–5817

108. Zhang P, Chi M, Sharma S, McFarland E (2010) Silica encapsulated heterostructure catalyst of Pt nanoclusters on hematite nanocubes: synthesis and reactivity. J Mater Chem 20:2013–2017

109. Ro I, Sener C, Stadelman TM, Ball MR, Venegas JM, Burt SP, Hermans I, Dumesic JA, Huber GW (2016) Measurement of intrinsic catalytic activity of Pt monometallic and Pt-MoO$_x$ interfacial sites over visible light enhanced PtMoO$_x$/SiO$_2$ catalyst in reverse water gas shift reaction. J Catal 344:784–794

110. Wang LC, Tahvildar Khazaneh M, Widmann D, Behm RJ (2013) TAP reactor studies of the oxidizing capability of CO$_2$ on a Au/CeO$_2$ catalyst—a first step toward identifying a redox mechanism in the reverse water–gas shift reaction. J Catal 302:20–30

111. Yang X, Su X, Chen X, Duan H, Liang B, Liu Q, Liu X, Ren Y, Huang Y, Zhang T (2017) Promotion effects of potassium on the activity and selectivity of Pt/zeolite catalysts for reverse water gas shift reaction. Appl Catal B 216:95–105

112. Kunkes EL, Studt F, Abild-Pedersen F, Schlögl R, Behrens M (2015) Hydrogenation of CO$_2$ to methanol and CO on Cu/ZnO/Al$_2$O$_3$: is there a common intermediate or not? J Catal 328:43–48

113. Mo LY, Saw ET, Kathiraser Y, Ang ML, Kawi S (2018) Preparation of highly dispersed Cu/SiO$_2$ doped with CeO$_2$ and its application for high temperature water gas shift reaction. Int J Hydrogen Energy 43:15891–15897

114. Matsubu JC, Yang VN, Christopher P (2015) Isolated metal active site concentration and stability control catalytic CO$_2$ reduction selectivity. J Am Chem Soc 137:3076–3084

115. Kwak JH, Kovarik L, Szanyi J (2013) CO$_2$ reduction on supported Ru/Al$_2$O$_3$ catalysts: cluster size dependence of product selectivity. ACS Catal 3:2449–2455

116. Park S-W, Joo O-S, Jung K-D, Kim H, Han S-H (2001) Development of ZnO/Al$_2$O$_3$ catalyst for reverse-water-gas-shift reaction of CAMERE (carbon dioxide hydrogenation to form methanol via a reverse-water-gas-shift reaction) process. Appl Catal A 211:81–90

117. Yang L, Pastor-Pérez L, Gu S, Sepúlveda-Escribano A, Reina TR (2018) Highly efficient Ni/CeO$_2$–Al$_2$O$_3$ catalysts for CO$_2$ upgrading via reverse water-gas shift: effect of selected transition metal promoters. Appl Catal B 232:464–471

118. Lin F, Delmelle R, Vinodkumar T, Reddy BM, Wokaun A, Alxneit I (2015) Correlation between the structural characteristics, oxygen storage capacities and catalytic activities of dual-phase Zn-modified ceria nanocrystals. Catal Sci Technol 5:3556–3567

119. Silva-Calpa LDR, Zonetti PC, Rodrigues CP, Alves OC, Appel LG, de Avillez RR (2016) The Zn$_x$Zr$_{1-x}$O$_{2-y}$ solid solution on m-ZrO$_2$: creating O vacancies and improving the m-ZrO$_2$ redox properties. J Mol Catal A Chem 425:166–173

120. Zonetti PC, Letichevsky S, Gaspar AB, Sousa-Aguiar EF, Appel LG (2014) The Ni$_x$Ce$_{0.75}$Zr$_{0.25-x}$O$_2$ solid solution and the RWGS. Appl Catal A General 475:48–54

121. Hare BJ, Maiti D, Ramani S, Ramos AE, Bhethanabotla VR, Kuhn JN (2018) Thermochemical conversion of carbon dioxide by reverse water-gas shift chemical looping using supported perovskite oxides. Catal Today 323:225–232

122. Daza YA, Maiti D, Kent RA, Bhethanabotla VR, Kuhn JN (2015) Isothermal reverse water gas shift chemical looping on La$_{0.75}$Sr$_{0.25}$Co$_{(1-Y)}$Fe$_Y$O$_3$ perovskite-type oxides. Catal Today 258:691–698

123. Viana HDAL, Irvine JTS (2007) Catalytic properties of the proton conductor materials: $Sr_3CaZr_{0.5}Ta_{1.5}O_{8.75}$, $BaCe_{0.9}Y_{0.1}O_{2.95}$ and $Ba_3Ca_{1.18}Nb_{1.82}O_{8.73}$ for reverse water gas shift. Solid State Ionics 178:717–722

124. Zakowsky N, Williamson S, Irvine JTS (2005) Elaboration of CO_2 tolerance limits of $BaCe_{0.9}Y_{0.1}O_{3-\delta}$ electrolytes for fuel cells and other applications. Solid State Ionics 176:3019–3026

125. Levy RB, Boudart M (1973) Platinum-like behavior of tungsten carbide in surface catalysis. Science 181:547

126. Burghaus U (2014) Surface chemistry of CO_2—adsorption of carbon dioxide on clean surfaces at ultrahigh vacuum. Prog Surf Sci 89:161–217

127. Freund HJ, Roberts MW (1996) Surface chemistry of carbon dioxide. Surf Sci Rep 25:225–273

128. Porosoff MD, Yang X, Boscoboinik JA, Chen JG (2014) Molybdenum carbide as alternative catalysts to precious metals for highly selective reduction of CO_2 to CO. Angew Chem Int Ed 53:6705–6709

129. Posada-Pérez S, Ramírez PJ, Evans J, Viñes F, Liu P, Illas F, Rodriguez JA (2016) Highly active Au/δ-MoC and Cu/δ-MoC catalysts for the conversion of CO_2: the metal/C ratio as a key factor defining activity, selectivity, and stability. J Am Chem Soc 138:8269–8278

130. Zhang X, Zhu X, Lin L, Yao S, Zhang M, Liu X, Wang X, Li Y-W, Shi C, Ma D (2017) Highly dispersed copper over β-Mo_2C as an efficient and stable catalyst for the reverse water gas shift (RWGS) reaction. ACS Catal 7:912–918

131. Sharma S, Hilaire S, Vohs JM, Gorte RJ, Jen HW (2000) Evidence for oxidation of ceria by CO_2. J Catal 190:199–204

132. Chen CS, Wu JH, Lai TW (2010) Carbon dioxide hydrogenation on Cu nanoparticles. J Phys Chem C 114:15021–15028

133. Ginés MJL, Marchi AJ, Apesteguía CR (1997) Kinetic study of the reverse water-gas shift reaction over $CuO/ZnO/Al_2O_3$ catalysts. Appl Catal A 154:155–171

134. Fujita S-I, Usui M, Takezawa N (1992) Mechanism of the reverse water gas shift reaction over Cu/ZnO catalyst. J Catal 134:220–225

135. Ferri D, Bürgi T, Baiker A (2002) Probing boundary sites on a Pt/Al_2O_3 model catalyst by CO_2 hydrogenation and in situ ATR-IR spectroscopy of catalytic solid–liquid interfaces. Phys Chem Chem Phys 4:2667–2672

136. Goguet A, Meunier FC, Tibiletti D, Breen JP, Burch R (2004) Spectrokinetic investigation of reverse water-gas-shift reaction intermediates over a Pt/CeO_2 catalyst. J Phys Chem B 108:20240–20246

137. Arunajatesan V, Subramaniam B, Hutchenson KW, Herkes FE (2007) In situ FTIR investigations of reverse water gas shift reaction activity at supercritical conditions. Chem Eng Sci 62:5062–5069

138. Olah GA (2005) Beyond oil and gas: the methanol economy. Angew Chem Int Ed 44:2636–2639

139. Albo J, Alvarez-Guerra M, Castaño P, Irabien A (2015) Towards the electrochemical conversion of carbon dioxide into methanol. Green Chem 17:2304–2324

140. Lorenz T, Bertau M, Schmidt F, Plass L (2014) Methanol: the basic chemical and energy feedstock of the future. In Bertau M, Offermanns H, Plass L, Schmidt F, Wernicke H (Ed) Section 4.8

141. Kagan JV, Rozovskij AJ, Lin G, Slivinskij E, Lo ktev SM, Liberov LG, Bash Kirov AN (1975) Mechanism for the synthesis of methanol from carbon dioxide and hydrogen. Kinet Katal 16:809

142. Bahruji H, Esquius JR, Bowker M, Hutchings G, Armstrong RD, Jones W (2018) Solvent free synthesis of $PdZn/TiO_2$ catalysts for the Hydrogenation of CO_2 to methanol. Top Catal 61:144–153

143. Behrens M (2015) Chemical hydrogen storage by methanol: challenges for the catalytic methanol synthesis from CO_2. Recycl Catal 2:78–86

144. Sahibzada M, Metcalfe I, Chadwick D (1998) Methanol synthesis from CO/CO$_2$/H$_2$ over Cu/ZnO/Al$_2$O$_3$ at differential and finite conversions. J Catal 174:111–118
145. Lee S (1990) Methanol synthesis technology. CRC Press Inc., Boca Raton, FL
146. Baltes C, Vukojević S, Schüth F (2008) Correlations between synthesis, precursor, and catalyst structure and activity of a large set of CuO/ZnO/Al$_2$O$_3$ catalysts for methanol synthesis. J Catal 258:334–344
147. Zander S, Kunkes EL, Schuster ME, Schumann J, Weinberg G, Teschner D, Jacobsen N, Schlögl R, Behrens M (2013) The role of the oxide component in the development of copper composite catalysts for methanol synthesis. Angew Chem Int Ed 52:6536–6540
148. Schumann J, Eichelbaum M, Lunkenbein T, Thomas N, Alvarez Galvan MC, Schlögl R, Behrens M (2015) Promoting strong metal support interaction: doping ZnO for enhanced activity of Cu/ZnO: M (M=Al, Ga, Mg) catalysts. ACS Catal 5:3260–3270
149. Arena F, Barbera K, Italiano G, Bonura G, Spadaro L, Frusteri F (2007) Synthesis, characterization and activity pattern of Cu–ZnO/ZrO$_2$ catalysts in the hydrogenation of carbon dioxide to methanol. J Catal 249:185–194
150. Kühl S, Tarasov A, Zander S, Kasatkin I, Behrens M (2014) Cu-based catalyst resulting from a Cu, Zn, Al hydrotalcite-like compound: a microstructural, thermoanalytical, and in situ XAS study. Chem Eur J 20:3782–3792
151. Wu J, Saito M, Takeuchi M, Watanabe T (2001) The stability of Cu/ZnO-based catalysts in methanol synthesis from a CO$_2$-rich feed and from a CO-rich feed. Appl Catal A 218:235–240
152. Saito M, Fujitani T, Takeuchi M, Watanabe T (1996) Development of copper/zinc oxide-based multicomponent catalysts for methanol synthesis from carbon dioxide and hydrogen. Appl Catal A 138:311–318
153. Hayward JS, Smith PJ, Kondrat SA, Bowker M, Hutchings GJ (2017) The effects of secondary oxides on copper-based catalysts for green methanol synthesis. ChemCatChem 9:1655–1662
154. Bahruji H, Bowker M, Hutchings G, Dimitratos N, Wells P, Gibson E, Jones W, Brookes C, Morgan D, Lalev G (2016) Pd/ZnO catalysts for direct CO$_2$ hydrogenation to methanol. J Catal 343:133–146
155. Conant T, Karim AM, Lebarbier V, Wang Y, Girgsdies F, Schlögl R, Datye A (2008) Stability of bimetallic Pd–Zn catalysts for the steam reforming of methanol. J Catal 257:64–70
156. Bahruji H, Bowker M, Jones W, Hayward J, Esquius JR, Morgan D, Hutchings G (2017) PdZn catalysts for CO$_2$ hydrogenation to methanol using chemical vapour impregnation (CVI). Faraday Discuss 197:309–324
157. Martín O, Martín AJ, Mondelli C, Mitchell S, Segawa TF, Hauert R, Drouilly C, Curulla-Ferré D, Pérez-Ramírez J (2016) Indium oxide as a superior catalyst for methanol synthesis by CO$_2$ hydrogenation. Angew Chem Int Ed 55:6261–6265
158. Dou M, Zhang M, Chen Y, Yu Y (2018) DFT study of In$_2$O$_3$-catalyzed methanol synthesis from CO$_2$ and CO hydrogenation on the defective site. New J Chem 42:3293–3300
159. Frei MS, Capdevila-Cortada M, García-Muelas R, Mondelli C, López N, Stewart JA, Ferré DC, Pérez-Ramírez J (2018) Mechanism and microkinetics of methanol synthesis via CO$_2$ hydrogenation on indium oxide. J Catal 361:313–321
160. García-Trenco A, Regoutz A, White ER, Payne DJ, Shaffer MS, Williams CK (2018) PdIn intermetallic nanoparticles for the hydrogenation of CO$_2$ to methanol. Appl Catal B 220:9–18
161. Grabowski R, Słoczyński J, Sliwa M, Mucha D, Socha R, Lachowska M, Skrzypek J (2011) Influence of polymorphic ZrO$_2$ phases and the silver electronic state on the activity of Ag/ZrO$_2$ catalysts in the hydrogenation of CO$_2$ to methanol. ACS Catal 1:266–278
162. Tada S, Watanabe F, Kiyota K, Shimoda N, Hayashi R, Takahashi M, Nariyuki A, Igarashi A, Satokawa S (2017) Ag addition to CuO–ZrO$_2$ catalysts promotes methanol synthesis via CO$_2$ hydrogenation. J Catal 351:107–118

163. Inui T, Takeguchi T, Kohama A, Tanida K (1992) Effective conversion of carbon dioxide to gasoline. Energy Convers Manag 33:513–520
164. Marlin DS, Sarron E, Sigurbjörnsson Ó (2018) Process advantages of direct CO_2 to methanol synthesis. Front Chem 6:446
165. Lurgi (1994) 207th national meeting of the American Chemical Society, San Diego, CA, Mar 1994
166. Saito M (1998) R&D activities in Japan on methanol synthesis from CO_2 and H_2. Catal Surv Asia 2:175–184
167. Grasemann M, Laurenczy G (2012) Formic acid as a hydrogen source—recent developments and future trends. Energy Environ Sci 5:8171–8181
168. Maihom T, Wannakao S, Boekfa B, Limtrakul J (2013) Production of formic acid via hydrogenation of CO_2 over a copper-alkoxide-functionalized MOF: a mechanistic study. J Phys Chem C 117:17650–17658
169. Peng G, Sibener SJ, Schatz GC, Ceyer ST, Mavrikakis M (2012) CO_2 hydrogenation to formic acid on Ni(111). J Phys Chem C 116:3001–3006
170. Qin G, Zhang Y, Ke X, Tong X, Sun Z, Liang M, Xue S (2013) Photocatalytic reduction of carbon dioxide to formic acid, formaldehyde, and methanol using dye-sensitized TiO_2 film. Appl Catal B 129:599–605
171. Jessop PG, Ikariya T, Noyori R (1994) Homogeneous catalytic hydrogenation of supercritical carbon dioxide. Nature 368:231
172. Reutemann W, Kieczka H (2011) Formic Acid, Ullmann's encyclopedia of industrial chemistry, Wiley-VCH Verlag GmbH & Co
173. Yoshio I, Hitoshi I, Yoshiyuki S, Harukichi H (1976) Catalytic fixation of carbon dioxide to formic acid by transition-metal complexes under mild conditions. Chem Lett 5:863–864
174. Onishi N, Laurenczy G, Beller M, Himeda Y (2018) Recent progress for reversible homogeneous catalytic hydrogen storage in formic acid and in methanol. Coord Chem Rev 373:317–332
175. Schlenk W, Appenrodt J, Michael A, Thal A (1914) Über metalladditionen an mehrfache bindungen. Ber Dtsch Chem Ges 47:473–490
176. Farlow MW, Adkins H (1935) The hydrogenation of carbon dioxide and a correction of the reported synthesis of urethans. J Am Chem Soc 57:2222–2223
177. Takahashi H, Liu LH, Yashiro Y, Ioku K, Bignall G, Yamasaki N, Kori T (2006) CO_2 reduction using hydrothermal method for the selective formation of organic compounds. JMater Sci 41:1585–1589
178. Jin F, Gao Y, Jin Y, Zhang Y, Cao J, Wei Z, Smith RL Jr (2011) High-yield reduction of carbon dioxide into formic acid by zero-valent metal/metal oxide redox cycles. Energy Environ Sci 4:881–884
179. Filonenko GA, Vrijburg WL, Hensen EJM, Pidko EA (2016) On the activity of supported Au catalysts in the liquid phase hydrogenation of CO_2 to formates. J Catal 343:97–105
180. Srivastava V (2014) In situ generation of Ru nanoparticles to catalyze CO_2 hydrogenation to formic acid. Catal Lett 144:1745–1750
181. Umegaki T, Enomoto Y, Kojima Y (2016) Metallic ruthenium nanoparticles for hydrogenation of supercritical carbon dioxide. Catal Sci Technol 6:409–412
182. Stalder CJ, Chao S, Summers DP, Wrighton MS (1983) Supported palladium catalysts for the reduction of sodium bicarbonate to sodium formate in aqueous solution at room temperature and one atmosphere of hydrogen. J Am Chem Soc 105:6318–6320
183. Su J, Yang L, Lu M, Lin H (2015) Highly efficient hydrogen storage system based on ammonium bicarbonate/formate redox equilibrium over palladium nanocatalysts. Chem-SusChem 8:813–816
184. Bi Q-Y, Lin J-D, Liu Y-M, Du X-L, Wang J-Q, He H-Y, Cao Y (2014) An aqueous rechargeable formate-based hydrogen battery driven by heterogeneous Pd catalysis. Angew Chem Int Ed 53:13583–13587

185. Song H, Zhang N, Zhong C, Liu Z, Xiao M, Gai H (2017) Hydrogenation of CO$_2$ into formic acid using a palladium catalyst on chitin. New J Chem 41:9170–9177
186. Hao C, Wang S, Li M, Kang L, Ma X (2011) Hydrogenation of CO$_2$ to formic acid on supported ruthenium catalysts. Catal Today 160:184–190
187. Nguyen LTM, Park H, Banu M, Kim JY, Youn DH, Magesh G, Kim WY, Lee JS (2015) Catalytic CO$_2$ hydrogenation to formic acid over carbon nanotube-graphene supported PdNi alloy catalysts. RSC Adv 5:105560–105566
188. Thongnuam W, Maihom T, Choomwattana S, Injongkol Y, Boekfa B, Treesukol P, Limtrakul J (2018) Theoretical study of CO$_2$ hydrogenation into formic acid on Lewis acid zeolites. Phys Chem Chem Phys 20:25179–25185
189. Eseafili MD, Dinparast L (2017) A DFT study on the catalytic hydrogenation of CO$_2$ to formic acid over Ti-doped graphene nanoflake. Chem Phys Lett 682:49–54
190. Hietala J, Vuori A, Johnsson P, Pollari I, Rewtemann W, Kieczka H (2016) Formic acid. Ullmann's encyclopedia of industrial chemistry, Wiley-VCH Verlag GmbH & Co
191. Bavykina A, Goesten M, Kapteijn F, Makkee M, Gascon J (2015) Efficient production of hydrogen from formic acid using a covalent triazine framework supported molecular catalyst. ChemSusChem 8:809–812
192. Pakhare D, Spivey J (2014) A review of dry (CO$_2$) reforming of methane over noble metal catalysts. Chem Soc Rev 43:7813–7837
193. Wang S, Lu GQ, Millar GJ (1996) Carbon dioxide reforming of methane to produce synthesis gas over metal-supported catalysts: state of the art. Energy Fuels 10:896–904
194. Yabe T, Sekine Y (2018) Methane conversion using carbon dioxide as an oxidizing agent: a review. Fuel Process Technol 181:187–198
195. Jones G, Jakobsen JG, Shim SS, Kleis J, Andersson MP, Rossmeisl J, Abild-Pedersen F, Bligaard T, Helveg S, Hinnemann B, Rostrup-Nielsen JR, Chorkendorff I, Sehested J, Nørskov JK (2008) First principles calculations and experimental insight into methane steam reforming over transition metal catalysts. J Catal 259:147–160
196. Bian ZF, Das S, Wai MH, Hongmanorom P, Kawi S (2017) A review on bimetallic nickel-based catalysts for CO$_2$ reforming of methane. ChemPhysChem 18:3117–3134
197. García-Diéguez M, Pieta IS, Herrera MC, Larrubia MA, Alemany LJ (2010) Improved Pt-Ni nanocatalysts for dry reforming of methane. Appl Catal A 377:191–199
198. Elsayed NH, Roberts NRM, Joseph B, Kuhn JN (2015) Low temperature dry reforming of methane over Pt–Ni–Mg/ceria–zirconia catalysts. Appl Catal B 179:213–219
199. Li L, Zhou L, Ould-Chikh S, Anjum DH, Kanoun MB, Scaranto J, Hedhili MN, Khalid S, Laveille PV, D'Souza L, Clo A, Basset J-M (2015) Controlled surface segregation leads to efficient coke-resistant nickel/platinum bimetallic catalysts for the dry reforming of methane. ChemCatChem 7:819–829
200. Li B, Kado S, Mukainakano Y, Miyazawa T, Miyao T, Naito S, Okumura K, Kunimori K, Tomishige K (2007) Surface modification of Ni catalysts with trace Pt for oxidative steam reforming of methane. J Catal 245:144–155
201. Kawi S, Kathiraser Y, Ni J, Oemar U, Li ZW, Saw ET (2015) Progress in synthesis of highly active and stable nickel-based catalysts for carbon dioxide reforming of methane. ChemSusChem 8:3556–3575
202. Li Z, Das S, Hongmanorom P, Dewangan N, Wai MH, Kawi S (2018) Silica-based micro- and mesoporous catalysts for dry reforming of methane. Catal Sci Technol 8:2763–2778
203. Li Z, Wang Z, Kawi S (2018) Sintering and coke resistant core/yolk shell catalyst for hydrocarbon reforming. ChemCatChem (in press)
204. Li M, van Veen AC (2018) Tuning the catalytic performance of Ni-catalysed dry reforming of methane and carbon deposition via Ni–CeO$_{2-x}$ interaction. Appl Catal B 237:641–648
205. Kathiraser Y, Thitsartarn W, Sutthiumporn K, Kawi S (2013) Inverse NiAl$_2$O$_4$ on LaAlO$_3$–Al$_2$O$_3$: unique catalytic structure for stable CO$_2$ reforming of methane. J Phys Chem C 117:8120–8130

206. Sigl M, Bradford MCJ, Knözinger H, Vannice MA (1999) CO_2 reforming of methane over vanadia-promoted Rh/SiO_2 catalysts. Top Catal 8:211–222
207. Pan Y-X, Kuai P, Liu Y, Ge Q, Liu C-J (2010) Promotion effects of Ga_2O_3 on CO_2 adsorption and conversion over a SiO_2-supported Ni catalyst. Energy Environ Sci 3:1322–1325
208. Zhang G, Liu J, Xu Y, Sun Y (2018) A review of CH_4–CO_2 reforming to synthesis gas over Ni-based catalysts in recent years (2010–2017). Int J Hydrogen Energy 43:15030–15054
209. Sutthiumporn K, Kawi S (2011) Promotional effect of alkaline earth over Ni–La_2O_3 catalyst for CO_2 reforming of CH_4: role of surface oxygen species on H_2 production and carbon suppression. Int J Hydrogen Energy 36:14435–14446
210. Ni J, Chen LW, Lin JY, Kawi S (2012) Carbon deposition on borated alumina supported nano-sized Ni catalysts for dry reforming of CH_4. Nano Energy 1:674–686
211. Yu M, Zhu Y-A, Lu Y, Tong G, Zhu K, Zhou X (2015) The promoting role of Ag in Ni–CeO_2 catalyzed CH_4–CO_2 dry reforming reaction. Appl Catal B 165:43–56
212. Bitter JH, Seshan K, Lercher JA (1998) Mono and bifunctional pathways of CO_2/CH_4 reforming over Pt and Rh based catalysts. J Catal 176:93–101
213. Kathiraser Y, Oemar U, Saw ET, Li Z, Kawi S (2015) Kinetic and mechanistic aspects for CO_2 reforming of methane over Ni based catalysts. Chem Eng J 278:62–78
214. Oemar U, Kathiraser Y, Mo L, Ho XK, Kawi S (2016) CO_2 reforming of methane over highly active La-promoted Ni supported on SBA-15 catalysts: mechanism and kinetic modelling. Catal Sci Technol 6:1173–1186
215. Wang HY, Ruckenstein E (1999) CH_4/CD_4 isotope effect and the mechanism of partial oxidation of methane to synthesis gas over Rh/γ-Al_2O_3 catalyst. J Phys Chem B 103:11327–11331
216. Wei J, Iglesia E (2004) Isotopic and kinetic assessment of the mechanism of reactions of CH_4 with CO_2 or H_2O to form synthesis gas and carbon on nickel catalysts. J Catal 224:370–383
217. Usman M, Wan Daud WMA, Abbas HF (2015) Dry reforming of methane: influence of process parameters—a review. Renew Sustain Energy Rev 45:710–744
218. Chen X, Honda K, Zhang Z-G (2005) CO_2–CH_4 reforming over NiO/γ-Al_2O_3 in fixed/fluidized-bed multi-switching mode. Appl Catal A 279:263–271
219. García-García FR, Soria MA, Mateos-Pedrero C, Guerrero-Ruiz A, Rodríguez-Ramos I, Li K (2013) Dry reforming of methane using Pd-based membrane reactors fabricated from different substrates. J Membr Sci 435:218–225
220. Chung W-C, Chang M-B (2016) Review of catalysis and plasma performance on dry reforming of CH_4 and possible synergistic effects. Renew Sustain Energy Rev 62:13–31
221. Pakhare D, Shaw C, Haynes D, Shekhawat D, Spivey J (2013) Effect of reaction temperature on activity of Pt- and Ru-substituted lanthanum zirconate pyrochlores ($La_2Zr_2O_7$) for dry (CO_2) reforming of methane (DRM). J CO_2 Utiliz 1:37–42
222. Kathiraser Y, Wang Z, Ang ML, Mo L, Li Z, Oemar U, Kawi S (2017) Highly active and coke resistant Ni/SiO_2 catalysts for oxidative reforming of model biogas: effect of low ceria loading. J CO_2 Utiliz 19:284–295
223. Oemar U, Hidajat K, Kawi S (2017) High catalytic stability of Pd-Ni/Y_2O_3 formed by interfacial Cl for oxy-CO_2 reforming of CH_4. Catal Today 281:276–294
224. Oemar U, Hidajat K, Kawi S (2015) Pd-Ni catalyst over spherical nanostructured Y_2O_3 support for oxy-CO_2 reforming of methane: Role of surface oxygen mobility. Int J Hydrogen Energy 40:12227–12238
225. Ashok J, Bian Z, Wang Z, Kawi S (2018) Ni-phyllosilicate structure derived Ni–SiO_2–MgO catalysts for bi-reforming applications: acidity, basicity and thermal stability. Catal Sci Technol 8:1730–1742
226. Oemar U, Hidajat K, Kawi S (2011) Role of catalyst support over PdO–NiO catalysts on catalyst activity and stability for oxy-CO_2 reforming of methane. Appl Catal Gen 402:176–187

227. Kathiraser Y, Wang Z, Kawi S (2013) Oxidative CO$_2$ reforming of methane in La$_{0.6}$Sr$_{0.4}$Co$_{0.8}$Ga$_{0.2}$O$_{3-\delta}$ (LSCG) hollow fiber membrane reactor. Environ Sci Technol 47:14510–14517

228. Yang N-T, Kathiraser Y, Kawi S (2013) La$_{0.6}$Sr$_{0.4}$Co$_{0.8}$Ni$_{0.2}$O$_{3-\delta}$ hollow fiber membrane reactor: Integrated oxygen separation—CO$_2$ reforming of methane reaction for hydrogen production. Int J Hydrogen Energy 38:4483–4491

229. Su Y-J, Pan K-L, Chang M-B (2014) Modifying perovskite-type oxide catalyst LaNiO$_3$ with Ce for carbon dioxide reforming of methane. Int J Hydrogen Energy 39:4917–4925

230. Laosiripojana N, Assabumrungrat S (2005) Methane steam reforming over Ni/Ce–ZrO$_2$ catalyst: Influences of Ce–ZrO$_2$ support on reactivity, resistance toward carbon formation, and intrinsic reaction kinetics. Appl Catal A 290:200–211

231. Ni J, Chen LW, Lin JY, Schreyer MK, Wang Z, Kawi S (2013) High performance of Mg–La mixed oxides supported Ni catalysts for dry reforming of methane: the effect of crystal structure. Int J Hydrogen Energy 38:13631–13642

232. Sutthiumporn K, Maneerung T, Kathiraser Y, Kawi S (2012) CO$_2$ dry-reforming of methane over La$_{0.8}$Sr$_{0.2}$Ni$_{0.8}$M$_{0.2}$O$_3$ perovskite (M=Bi Co, Cr, Cu, Fe): roles of lattice oxygen on C-H activation and carbon suppression. Int J Hydrogen Energy 37:11195–11207

233. Ni J, Zhao J, Chen LW, Lin JY, Kawi S (2016) Lewis acid sites stabilized nickel catalysts for dry (CO$_2$) reforming of methane. ChemCatChem 8:3732–3739

234. Kim J-H, Suh DJ, Park T-J, Kim K-L (2000) Effect of metal particle size on coking during CO$_2$ reforming of CH$_4$ over Ni–alumina aerogel catalysts. Appl Catal A 197:191–200

235. Mo LY, Leong KKM, Kawi S (2014) A highly dispersed and anti-coking Ni-La$_2$O$_3$/SiO$_2$ catalyst for syngas production from dry carbon dioxide reforming of methane. Catal Sci Technol 4:2107–2114

236. Yang W, He D (2016) Role of poly(N-vinyl-2-pyrrolidone) in Ni dispersion for highly-dispersed Ni/SBA-15 catalyst and its catalytic performance in carbon dioxide reforming of methane. Appl Catal A 524:94–104

237. Gao XY, Ashok J, Widjaja S, Hidajat K, Kawi S (2015) Ni/SiO$_2$ catalyst prepared via Ni-aliphatic amine complexation for dry reforming of methane: Effect of carbon chain number and amine concentration. Appl Catal A 503:34–42

238. Gao X, Liu H, Hidajat K, Kawi S (2015) Anti-Coking Ni/SiO$_2$ catalyst for dry reforming of methane: role of oleylamine/oleic acid organic pair. ChemCatChem 7:4188–4196

239. Gao X, Tan Z, Hidajat K, Kawi S (2017) Highly reactive Ni–Co/SiO$_2$ bimetallic catalyst via complexation with oleylamine/oleic acid organic pair for dry reforming of methane. Catal Today 281:250–258

240. Gao XY, Hidajat K, Kawi S (2016) Facile synthesis of Ni/SiO$_2$ catalyst by sequential hydrogen/air treatment: a superior anti-coking catalyst for dry reforming of methane. J CO$_2$ Utiliz 15:146–153

241. Mo LY, Saw ET, Du YH, Borgna A, Ang ML, Kathiraser Y, Li ZW, Thitsartarn W, Lin M, Kawi S (2015) Highly dispersed supported metal catalysts prepared via in-situ self-assembled core-shell precursor route. Int J Hydrogen Energy 40:13388–13398

242. Bian Z, Kawi S (2017) Highly carbon-resistant Ni–Co/SiO$_2$ catalysts derived from phyllosilicates for dry reforming of methane. J CO$_2$ Utiliz 18:345–352

243. Bian Z, Kawi S (2018) Preparation, characterization and catalytic application of phyllosilicate: a review. Catal Today. Available online 13 December 2018. https://doi.org/10.1016/j.cattod.2018.12.030

244. Dębek R, Motak M, Duraczyska D, Launay F, Galvez ME, Grzybek T, Da Costa P (2016) Methane dry reforming over hydrotalcite-derived Ni–Mg–Al mixed oxides: the influence of Ni content on catalytic activity, selectivity and stability. Catal Sci Technol 6:6705–6715

245. Zhang S, Muratsugu S, Ishiguro N, Tada M (2013) Ceria-doped Ni/SBA-16 catalysts for dry reforming of methane. ACS Catal 3:1855–1864

246. Li Z, Li M, Bian Z, Kathiraser Y, Kawi S (2016) Design of highly stable and selective core/yolk–shell nanocatalysts—a review. Appl Catal B 188:324–341

247. Li ZW, Mo LY, Kathiraser Y, Kawi S (2014) Yolk-satellite-shell structured Ni-Yolk@Ni@SiO$_2$ nanocomposite: superb catalyst toward methane CO$_2$ reforming reaction. ACS Catal 4:1526–1536

248. Bian Z, Suryawinata IY, Kawi S (2016) Highly carbon resistant multicore-shell catalyst derived from Ni–Mg phyllosilicate nanotubes@silica for dry reforming of methane. Appl Catal B 195:1–8

249. Li Z, Kawi S (2018) Multi-Ni@Ni phyllosilicate hollow sphere for CO$_2$ reforming of CH$_4$: influence of Ni precursors on structure, sintering, and carbon resistance. Catal Sci Technol 8:1915–1922

250. Das S, Ashok J, Bian Z, Dewangan N, Wai MH, Du Y, Borgna A, Hidajat K, Kawi S (2018) Silica-ceria sandwiched Ni core-shell catalyst for low temperature dry reforming of biogas: coke resistance and mechanistic insights. Appl Catal B 230:220–236

251. Zhao Y, Li H, Li H (2018) NiCo@SiO$_2$ core-shell catalyst with high activity and long lifetime for CO$_2$ conversion through DRM reaction. Nano Energy 45:101–108

252. Han JW, Kim C, Park JS, Lee H (2014) Highly coke-resistant Ni nanoparticle catalysts with minimal sintering in dry reforming of methane. ChemSusChem 7:451–456

253. Li Z, Jiang B, Wang Z, Kawi S (2018) High carbon resistant Ni@Ni phyllosilicate@SiO$_2$ core shell hollow sphere catalysts for low temperature CH$_4$ dry reforming. J CO$_2$ Utiliz 27:238–246

254. Bian Z, Kawi S (2018) Sandwich-like Silica@Ni@Silica multicore-shell catalyst for the low-temperature dry reforming of methane: confinement effect against carbon formation. ChemCatChem 10:320–328

255. Li Z, Wang Z, Jiang B, Kawi S (2018) Sintering resistant Ni nanoparticles exclusively confined within SiO$_2$ nanotubes for CH$_4$ dry reforming. Catal Sci Technol 8:3363–3371

256. Li Z, Kathiraser Y, Ashok J, Oemar U, Kawi S (2014) Simultaneous tuning porosity and basicity of Nickel@Nickel–magnesium phyllosilicate core-shell catalysts for CO$_2$ reforming of CH$_4$. Langmuir 30:14694–14705

257. Li Z, Kathiraser Y, Kawi S (2015) Facile synthesis of high surface area yolk–shell Ni@Ni embedded SiO$_2$ via Ni phyllosilicate with enhanced performance for CO$_2$ reforming of CH$_4$. ChemCatChem 7:160–168

258. Lim Z-Y, Wu C, Wang WG, Choy K-L, Yin H (2016) Porosity effect on ZrO$_2$ hollow shells and hydrothermal stability for catalytic steam reforming of methane. J Mater Chem A 4:153–159

259. Fan X, Liu Z, Zhu Y-A, Tong G, Zhang J, Engelbrekt C, Ulstrup J, Zhu K, Zhou X (2015) Tuning the composition of metastable Co$_x$Ni$_y$Mg$_{100-x-y}$(OH)(OCH$_3$) nanoplates for optimizing robust methane dry reforming catalyst. J Catal 330:106–119

260. Kim SM, Abdala PM, Margossian T, Hosseini D, Foppa L, Armutlulu A, van Beek W, Comas-Vives A, Copéret C, Müller C (2017) Cooperativity and dynamics increase the performance of nife dry reforming catalysts. J Am Chem Soc 139:1937–1949

261. Liu J, Zeng X, Cheng M, Yun J, Li Q, Jing Z, Jin F (2012) Reduction of formic acid to methanol under hydrothermal conditions in the presence of Cu and Zn. Biores Technol 114:658–662

262. Aresta M, Dibenedetto A, Quaranta E (2016) Reaction mechanisms in carbon dioxide conversion. Springer, Berlin, Heidelberg

Use of CO_2 as Source of Carbon for Energy-Rich C_n Products

6

Jiang Xiao, Xinwen Guo and Chunshan Song

Abstract

Catalytic CO_2 conversion to clean fuels and chemicals is crucial for mitigating the climate change and reducing the dependence on nonrenewable energy resources. Converting CO_2 by hydrogenation using heterogeneous catalysts has been extensively studied in the past decades, and the products distribution can be manipulated by selecting catalysts and reaction conditions. Generally, CO_2 conversion to hydrocarbons and to alcohols are the two routes that have been explored the most, and significant advances have been made in developing efficient catalysts and understanding the thermodynamics and kinetics of the two paths. However, effective catalysts and processes are required to selectively maximize CO_2 conversion to either C_2–C_4 olefins, C_5+ hydrocarbons, or aromatics and to minimize CH_4 and CO. Catalysis for higher alcohols synthesis from CO_2 is still in the very early stage and requires more fundamental research due to the lack of understanding the possible reaction pathways and of controlling the key intermediates. This review summarizes the progresses in CO_2 conversion via heterogeneous catalysis for the two pathways in the past five years and discusses the origin of the activity and plausible reaction mechanism through a combination of computational, experimental, and analytical studies, along with suggestions for designing improved catalysts in the future.

J. Xiao · C. Song (✉)
Departments of Energy & Mineral Engineering and of Chemical Engineering, PSU-DUT Joint Center for Energy Research, EMS Energy Institute, Pennsylvania State University, 209 Academic Projects Building, University Park, PA 16802, USA
e-mail: csong@psu.edu

X. Guo · C. Song
State Key Laboratory of Fine Chemicals, School of Chemical Engineering, PSU-DUT Joint Center for Energy Research, Dalian University of Technology, Dalian 116024, China

© Springer Nature Switzerland AG 2019
M. Aresta et al. (eds.), *An Economy Based on Carbon Dioxide and Water*,
https://doi.org/10.1007/978-3-030-15868-2_6

6.1   Introduction

The average concentration of CO_2 in the atmosphere has reached 408 ppm in April 2018 and increased by 20% in the past 50 years due to the intense use of fossil fuels according to National Oceanic and Atmospheric Administration in the USA. Continuing emissions are expected to cause significant and possibly irreversible changes to the global climate [1]. Utilizing CO_2 as a carbon source to produce clean fuels and chemical feedstocks provides an avenue to mitigate CO_2 emission and reduce the dependence on fossil fuels. Due to the inertness, CO_2 conversion is energy intensive when CO_2 is a single reactant. However, the conversion is more thermodynamically favorable when combined with a co-reactant that possesses higher Gibbs energy, such as H_2 [2]. Hence, CO_2 hydrogenation (HYD) to clean fuels and chemicals is a promising approach to generating a sustainable process in which the carbon source in CO_2 is recycled efficiently [3–8].

There are multiple routes to produce energy-rich products from CO_2 hydrogenation (Fig. 6.1), and significant efforts are devoted to developing active catalysts for effective synthesis of methanol and hydrocarbons [11–17]. Table 6.1 lists some relevant CO_2 hydrogenation routes. All hydrogenation reactions are exothermic, except reverse water-gas shift (RWGS). The thermodynamic properties determine that lower reaction temperatures and higher pressures are more favorable for the

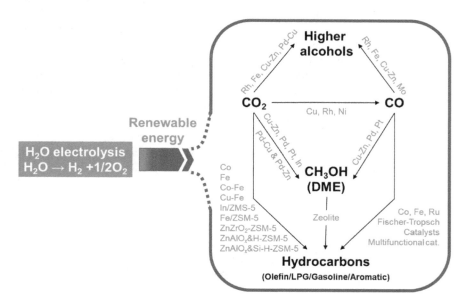

Fig. 6.1 Schematic illustration of CO_2-based hydrogenation reactions for the synthesis of value-added products

Table 6.1 Reactions involved in CO$_2$ hydrogenation to alcohols and hydrocarbons [9, 10]

Chemical equation	Enthalpy of reactions (kJ mol^{-1})	
$CO_2 + 3H_2 \leftrightarrow -(CH_2) - + 2H_2O$	$\Delta H_{298\ K} = -125.0$	(Propane)
$CO + 2H_2 \leftrightarrow -(CH_2) - + H_2O$	$\Delta H_{298\ K} = -456$	(Propane)
$nCO_2 + 3nH_2 \leftrightarrow C_nH_{2n+1}OH + (2n - 1)H_2O$	$\Delta H_{298\ K} = -86.7$	(Ethanol)
$nCO + 2nH_2 \leftrightarrow C_nH_{2n+1}OH + (n - 1)H_2O$	$\Delta H_{298\ K} = -253.6$	(Ethanol)
$CO_2 + 3H_2 \leftrightarrow CH_3OH + H_2O$	$\Delta H_{298\ K} = -49.3$	
$CO_2 + H_2 \leftrightarrow CO + H_2O$	$\Delta H_{298\ K} = 41.1$	
$CO + 2H_2 \leftrightarrow CH_3OH$	$\Delta H_{298\ K} = -90.4$	
$2CH_3OH \leftrightarrow CH_3OCH_3$	$\Delta H_{298\ K} = -37$	

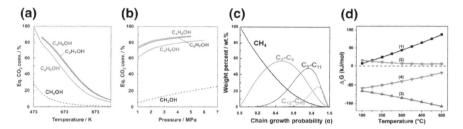

Fig. 6.2 Changes of equilibrium CO$_2$ conversion as functions of temperature (**b**) and pressure (**b**), Anderson-Schultz-Flory (ASF) production distribution of Fischer-Tropsch Synthesis (FTS), and Gibbs free energy changes at different temperatures for CO$_2$-to-CH$_3$OH (1), reverse water-gas shift (2), methanol-to-xylene (3), and CO$_2$-to-xylene (4). (3): CH$_3$OH $\leftrightarrow$ 1/8C$_8$H$_{10}$ + H$_2$O + 3/8H$_2$; (4) CO$_2$ + 21/8H$_2$ $\leftrightarrow$ 1/8C$_8$H$_{10}$ + 2H$_2$O [20]

syntheses of both alcohols and hydrocarbons. Notably, the synthesis of higher alcohols is thermodynamically more favored in comparison to that of methanol under the same reaction conditions (Fig. 6.2a, b), however, difficulty in C–C coupling and unveiled reaction pathway impairs the design of rational catalysts. In the Fischer-Tropsch Synthesis (FTS), the hydrocarbon selectivity follows the statistical Anderson-Shultz-Flory (ASF) distribution (Fig. 6.2c), wherein the chain growth probability depends on the catalysts and specific process conditions. How to break the limitation of ASF distribution and to selectively produce value-added olefin-rich light hydrocarbons (C$_2$–C$_4$) and liquid fuels (C$_5$–C$_{11}$) is still challenging. CO$_2$ hydrogenation to aromatics is an emerging research topic recently. As illustrated in Fig. 6.2d, the coupling of CO$_2$-to-methanol and methanol-to-aromatics (e.g., xylene) is feasible thermodynamically in comparison to the single methanol synthesis, indicating the potential of tandem catalysis in the effective synthesis of aromatics. Hence, efforts must be put forth to overcome the limitation of both thermodynamics and kinetics. Thermal catalytic CO$_2$ conversion attracts significant attention due to its fast kinetics, and intense studies have been conducted using heterogeneous metallic catalysts, including Fe-based catalysts for olefins and paraffins syntheses [6, 18], the metal oxide-zeolite bifunctional catalysts for the production of aromatics [19–21], and the Fe-based and modified Cu–ZnO catalysts for higher alcohol synthesis [22, 23]. Therefore, this chapter covers the progress in the thermal chemical CO_2 conversion to value-added C_n products, primarily in the past five years, and understandings of catalyst design, catalytic performance, active sites, and reaction mechanisms with an emphasis on metallic heterogeneous catalysts. Other alternative promising approaches, such as homogeneous catalysis, photocatalysis, and electrocatalysis, can be referred to corresponding review papers that are published recently [24–27].

6.2 CO_2 Hydrogenation to Hydrocarbons

6.2.1 Co-Based Catalysts

Cobalt is widely used for CO_2 hydrogenation to hydrocarbons, and key factors that could affect the activity and product distribution include the promoters, control of the phase distribution in multicomponent catalyst, and metal-support interaction. Gao and Wang investigated the effect of promoter K on activity and selectivity to C_{5+} hydrocarbons on CoCu/TiO_2 catalysts [28]. The K-promoted catalysts with the optimal composition exhibited a selectivity of C_{5+} as high as 35.1 C-mol% at CO_2 conversion of 13% at 523 K and 5.0 MPa, as presented in Table 6.2. Temperature-programmed desorption (TPD) measurements indicate that the addition of K improved the adsorption of CO_2, however, the H_2 adsorption was suppressed, thereof resulting in the enhancement in liquid fuel production.

Homs and Cabot explored the control of the phase distribution in multicomponent nanomaterials, wherein they synthesized the Co@Cu nanoparticles with narrow size and composition distribution via a facile galvanic replacement-based procedure [29]. The evolution of the Co@Cu NPs during oxidation and reduction treatments were elaborated. Results were indicative of the formation of polycrystalline CuO–Co_3O_4 nanoparticles after oxidation and recovery of the core-shell nanostructure during the posterior reduction (Fig. 6.3a). The resultant core-shell structured catalyst yielded higher C_2 compounds (i.e., ethane and ethylene, 40–65 C-mol%) than the single metallic Co/SiO_2 and physical mixed Co/SiO_2 and Cu/SiO_2 catalysts (Fig. 6.3b), the promoted chain growth of which was attributed to the synergy between the two metals in proximity to each other within the same particle domain.

However, SMSI is not always beneficial. Deo and coworkers found that the SMSI between Co and alumina support led to a loss of activity, wherein the weakened interaction gave rise to the formation of tetrahedral Co^{2+} species which was too difficult to reduce [30]. Therefore, a moderate strength between metal site and support would be more suitable, which could be adjusted by pre-calcination and reduction conditions.

Table 6.2 Summary of reaction conditions with conversion and selectivity to paraffins and olefins for selected catalysts

Catalyst	CO_2/H_2 ratio	Press./ MPa	Temp./ K	CO_2 conv./%	Hydrocarbons distri./ C-mol%			O/P (C_2– C_4)	Chain growth prob. α	References
					CH_4	C_2– C_4	C_{5+}			
K–CoCu/TiO₂	1:3	5.0	523	13.0	34.1	30.8	35.1	0.33	0.63	[28]
Fe₃O₄	1:3	2.5	623	–	87.30	9.31	0.22	0.06	–	[31]
α-Fe₂O₃	1:3	2.5	623	–	60.16	33.46	2.87	0.21	–	[31]
γ-Fe₂O₃	1:3	2.5	623	–	45.39	36.40	15.60	0.08	–	[31]
χ-Fe₅C₂	1:3	2.5	623	–	38.99	67.20	5.36	0.75	–	[31]
θ-Fe₃C	1:3	2.5	623	–	36.80	43.76	21.04	0.12	–	[31]
Fe/TiO₂ (P25)	1:3	1.1	573	16.1	34	13	28	0.02	0.39	[32]
Fe-Co/TiO₂ (P25)	1:3	1.1	573	33.3	51	11	32	0.00	0.30	[32]
Fe-Cu/TiO₂ (P25)	1:3	1.1	573	19.5	27	11	39	0.10	0.41	[32]
Fe-K/TiO₂ (P25)	1:3	1.1	573	18.0	11	8	32	0.50	0.54	[32]
Fe-Co-K/TiO₂ (P25)	1:3	1.1	573	21.2	9	6	31	4.06	0.50	[32]
Fe-Cu-K/TiO₂ (P25)	1:3	1.1	573	20.8	5	3	30	4.91	0.57	[32]
Fe/SiO₂	1:3	1.01	573	6.9	80.9	19.62	0.42	0.0062	–	[33]
Fe/TiO₂	1:3	1.01	573	11.5	51.7	41.55	6.8	0.03	–	[33]
Fe/Al₂O₃	1:3	1.01	573	22.8	43.2	48.08	8.8	0.0059	–	[33]
Fe-K/Al₂O₃	1:3	1.01	573	30.4	12.7	47.78	39.4	10.43	–	[33]
FeK/Al₂O₃	1:3	3.0	673	49.2	20.2	27.28	13.93	4.23	–	[34]
SiO₂–FeK/Al₂O₃	1:3	3.0	673	63.3	19.0	32.15	20.27	3.27	–	[34]
Fe-Co/K–Al₂O₃ (673)[b]	1:3	2.0	593	49.3	23.0	42.7	24.9	6.49	–	[35]

(continued)

Table 6.2 (continued)

Catalyst	CO$_2$/H$_2$ ratio	Press./MPa	Temp./K	CO$_2$ conv./%	Hydrocarbons distri./C-mol%			O/P (C$_2$–C$_4$)	Chain growth prob. α	References
					CH$_4$	C$_2$–C$_4$	C$_{5+}$			
Fe-Co/K–Al$_2$O$_3$ (973)[b]	1:3	2.0	593	37.6	13.8	37.3	24.4	7.60	–	[35]
Fe-Co/Al$_2$O$_3$	1:3	1.1	573	25.2	44	43	–	0.00	0.36	[36]
Fe-Co-K/Al$_2$O$_3$	1:3	1.1	573	31.0	13	69	–	5.2	0.53	[36]
Fe/Al$_2$O$_3$	1:3	1.1	573	22.0	43	39	–	0.02	0.41	[37]
Cu/Al$_2$O$_3$[c]	1:3	1.1	573	22.4	0	0	–	–	–	[37]
Fe-Cu/Al$_2$O$_3$	1:3	1.1	573	28.5	24	53	–	0.02	0.47	[37]
Fe-Cu/K/Al$_2$O$_3$	1:3	1.1	573	29.3	7	76	–	5.20	0.62	[37]
Cu-Zn-Cr/HY	1:3	5.0	673	39.9	4.2	88.3	14.1	–	–	[38]
Cu-La$_2$Zr$_2$O$_7$/HY	1:3	5.0	673	39.2	33.3	86.3	0.0	–	–	[38]
Cu-Zn-Al + HB	1:3	0.98	693	47.8	1.4	30.3	1.3	–	–	[39]
In$_2$O$_3$–ZrO$_2$/HZSM-5	1:3	3.0	613	13.1	0.55	11.26	43.39	–	–	[18]
In-Zr/SAPO-34	1:3	3.0	673	31.5	4.4[d]	92.7[d]	2.9[d]	6.54	–	[40]
In$_2$O$_3$–ZrO$_2$/SAPO-5	1:3	3.0	573	~6	3.0[d]	83[d]	14[d]	–	–	[41]
Na-Fe$_3$O$_4$/HZSM-5	1:3	3.0	593	22.0	4.0[d]	16.6[d]	79.4[d]	–	–	[6]

[a]Selectivity of C$_2$–C$_7$ hydrocarbons
[b]673 and 973 were calcination temperatures in K [35]
[c]This catalyst selectively produced CO [37]
[d]Values represent relative hydrocarbon distribution [40]

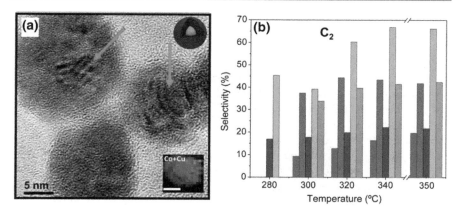

Fig. 6.3 **a** HRTEM image of Co@Cu NPs with EDS map in the inset, **b** relative selectivities related to C$_2$ products obtained from Co/SiO$_2$ (blue), Cu/SiO$_2$ (red), Co/SiO$_2$ + Cu/SiO$_2$ (purple), Co$_{0.6}$@Cu$_{0.4}$/SiO$_2$ (green), and Co$_{0.3}$@Cu$_{0.7}$/SiO$_2$ catalysts (orange). Reaction conditions: 3 MPa, 553–623 K, 3000 h^{-1} [29]

6.2.2 Fe-Based Catalysts

Fe-based bimetallic catalysts are also promising candidates for efficient CO$_2$ hydrogenation to hydrocarbons, in which the Fe sites are responsible for C–C coupling, while the other metal components account for tuning the H$_2$ adsorption property. To give a glimpse, Song et al. carried out screening tests over Fe–M bimetallic catalysts (M = Co, Cu, Ni, and Pd) [42]. Results demonstrate that Cu, Co, and Pd were promising candidates for hydrocarbon synthesis due to their synergetic effect with Fe. Ni was not favorable, because its strong hydrogenation ability led to the selective production of methane. Clearly, H$_2$ adsorption property was one of the crucial factors in determining the second metal component. In another work, Rodemerck and Baerns reported that the Cu-modified K/Fe catalysts exhibited better activity performance towards higher hydrocarbons (C$_5$–C$_{15}$) synthesis, while the Co-containing catalysts were more favorable for light hydrocarbons synthesis (C$_2$–C$_4$) [43].

(1) Active sites

Visconti and Lietti reported that the K-promoted Fe-based catalysts, prepared by fast calcination and then by carburization, exhibited a high activity in CO$_2$ hydrogenation to lower (C$_2$–C$_4$) olefins [44, 45]. The formation of hydrocarbons underwent a RWGS + Fischer-Tropsch Synthesis (FTS) pathway, and apparently CO played an important role in linking the tandem reactions. The presence of three different active sites (Type I–III) were proposed to promote this reaction pathway. RWGS occurred on Type I active site which was Fe$_3$O$_4$. Iron carbides were the Type II active sites responsible for chain growth. Catalytic Fe^0 was the Type III site which accounted for the secondary hydrogenation of olefin species. Apparently, an efficient synergy between Type I and II should lead to a better activity for lower olefin synthesis, while a slow rate of secondary hydrogenation on Type III site would be beneficial for olefin formation. Xu and Han carried out operando spectroscopic studies on the evolution of iron oxide catalysts during CO_2 hydrogenation to hydrocarbons [31]. As listed in Table 6.2, α-Fe_2O_3 and γ-Fe_2O_3 were used as catalysts, and they led to selective production of C_2–C_4 olefins (selectivity, 8.24 C-mol%, O/P = 0.21) and C_{5+} (selectivity, 15.60 C-mol%), respectively. During activation and CO_2 hydrogenation, α-Fe_2O_3 underwent: α-Fe_2O_3 → α-Fe_3O_4 → α-Fe → χ-Fe_5C_2; the evolution of γ-Fe_2O_3 was: γ-Fe_2O_3 → γ-Fe_3O_4 → γ-Fe → θ-Fe_3C. In terms of C_2–C_4 olefin production, χ-Fe_5C_2 (selectivity, 28.80 C-mol%, O/P = 0.75) outperformed θ-Fe_3C (selectivity, 4.58 C-mol%, O/P = 0.12), however, the latter species enabled a higher selectivity for C_{5+} (selectivity, 21.04 > 5.36 C-mol%). The presence of different structures in Fe carbides was speculated to account for the disparity of the selectivity.

(2) Promoters/additives

Promoters/additives, which acted as structural promoter and/or electronic promoter, were widely used in the preparation of Fe-based catalysts for improving the activity and selectivity to hydrocarbons synthesis. Their functions are diverse, and primarily include modification of the texture properties and carburization extent of Fe, tuning the capabilities of reagent adsorption and Fe carbide/Fe oxide surface ratio and valence state, as well as alteration of the state and location of promoters and their interaction with active Fe sites [46].

Alkali metals were widely used as promoters. Claeys et al. studied the effect of potassium on the structure of Fe-based catalysts and performance of CO_2 hydrogenation to hydrocarbons using novel in situ instrumentation based on XRD and magnetometry techniques [47]. As illustrated in Fig. 6.4, in situ characterization results provided straightforward evidence that the presence of K enhanced the formation of χ-Fe_5C_2 phase which was proposed as the active sites for hydrocarbon synthesis. An excessive content of K was detrimental to hydrocarbon synthesis, but barely impacted the RWGS. Prior to CO_2 hydrogenation, the carbide species were formed via carburization, the formation of which was reversible if exposed in pure H_2. However, a faster recovery of carbide species was evidenced when reintroducing CO_2.

Herskowitz et al. reported that loading 2 wt% of K promoter onto Fe–Al–O spinel-derived oxide catalysts enhanced the RWGS rate, corresponding to the increased Fe^{2+}/Fe^{3+} ratio and oxygen vacancies on the surface [48]. On Fe_5C_2 carbide, the presence of K suppressed the methanation rate, while increased the CO FTS rate to hydrocarbons. Characterization results demonstrate that K helped stabilize the surface Fe atoms in the reduced form. Combined with the activity performance on promoted carbide catalyst, it is suggested the methanation occurred on carbon vacancies, while the near-metallic iron atoms accounted for CO FTS. Song et al. observed a similar K-dependent behavior on Fe–Cu and Fe–Co

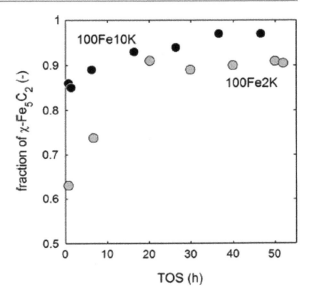

Fig. 6.4 Weight fraction of χ-Fe$_5$C$_2$ in the identified iron phases (χ-Fe$_5$C$_2$ and α-Fe) at initial reaction conditions of K-promoted Fe catalysts as a function of time-on-stream [47]

catalysts, wherein the TPD results indicate that the K addition diminished the weakly-adsorbed hydrogen but enhanced CO$_2$ adsorption capacity, which explained the significantly improved production of light olefins [49].

Apart from the promoters, additives were introduced to adjust the textural properties, tailor the structure, and tune the interaction between active Fe sites and promoters at the interface. Landau and coworkers introduced various metal oxides as additives, namely Si, Ti, Zr, Hf, Ce, and Mn, into the K-promoted Fe–Al–O spinel-derived Fe catalysts via the post-synthetic grafting method, and the catalytic performance of the resulting catalysts were evaluated in CO$_2$ hydrogenation to hydrocarbons [46]. Addition of 5–20 wt% of additives had different impacts on the performance, as presented in Fig. 6.5. Both TiO$_2$ and ZrO$_2$ evidently promoted RWGS rate and were good candidates for hydrocarbons synthesis via subsequent FTS, in which TiO$_2$-supported catalyst was selective for C$_2$–C$_4$ hydrocarbons, while ZrO$_2$ mostly benefited the production of C$_2$–C$_4$, and slightly improved the production of C$_{5+}$ hydrocarbons. These two additive NPs exhibited a strong interaction with spinel structure and iron carbide nanocrystals, resulting in the decline in the electron density of surface Fe atoms and the formation of Lewis acid sites. These sites acted as electron acceptor which greatly promoted the hydrocarbons synthesis activity by increasing the surface coverage of polar intermediate CH$^{\delta+}$ in addition to nonpolar CH$_2$ and CH$_3$. A small amount of SiO$_2$ was found to enhance CO$_2$ conversion by promoting RWGS, however, SiO$_2$ with a higher loading behaved as an inhibitor to FTS, but a promoter for methanation. Such decline in activity could be ascribed to the selective blocking of the surface-active Fe ions of the Fe carbide and their interaction with the K promoter. In striking contrast, CeO$_2$, HfO$_2$, and MnO barely affected the CO$_2$ conversion or CO

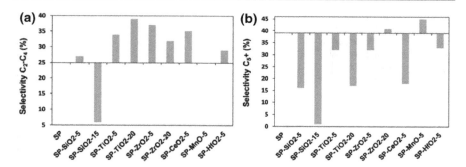

Fig. 6.5 Effects of additives on catalytic performance of K/Fe–Al–O spinel (SP) on activity performance: **a** selectivity to C_2–C_4 hydrocarbons and **b** selectivity to C_{5+} hydrocarbons [46]. Reaction conditions: 593 K, 2 MPa, WHSV$_{CO_2}$ = 3 h−1, CO_2/H_2 = 1/3

selectivity. Nor did HfO_2 and MnO have any impact on the production of methane or hydrocarbons. Although CeO_2-grafted catalyst exhibited enhancements in the selectivity to both methane and light C_2–C_4 hydrocarbons, its catalytic performance was still not comparable to that of ZrO_2 or TiO_2. Song et al. also incorporated the rare-earth metal La into the Fe-based catalysts and found that it could selectively shifted the CO production to yield more C_{5+} hydrocarbons via CO FTS, especially in the case of La- and K-copromoted catalyst [32]. Therefore, due to the difference in inherent natures, ZrO_2 and TiO_2 were better candidates if light hydrocarbons (C_2–C_4) are targeted, while La was more advantageous in chain growth for higher hydrocarbons (C_{5+}) synthesis.

(3) Support

Al_2O_3, SiO_2, and TiO_2 were widely used as support materials for Fe-based catalysts. Riedel et al. conducted comparative studies on evaluating the catalytic performance of hydrocarbon synthesis from H_2/CO_2 on γ-Al_2O_3, SiO_2, and TiO_2-supported Fe catalysts [50]. TPD results demonstrate that Al_2O_3 showed more advanced adsorption capacities of CO_2 and H_2 in comparison to the other two oxides, as listed in Table 6.3. Temperature-programmed decarburization (TPDC) profiles demonstrate that the stability of the carbides increased in the order of Fe/γ-Al_2O_3 < Fe/SiO_2 < Fe/TiO_2, wherein the hydrocarbon synthesis activity appeared to correlate to the low-temperature species. As a result, higher selectivity to C_2–C_4 was expected on Fe/γ-Al_2O_3 than that on the rest catalysts (Table 6.2). There are also articles reporting that TiO_2 is a good candidate, as the oxygen

Table 6.3 Amounts of chemisorbed CO_2 and H_2 on reduced Fe catalysts supported on different supports [50]

Catalyst[a]	CO_2 uptake/μmol g^{-1}	H_2 uptake/μmol g^{-1}
Fe/SiO_2	7.2	2.0
Fe/TiO_2	5.5	2.7
Fe/Al_2O_3	167.0	31.0

[a]Fe loading = 20 wt%

vacancies of TiO_2 could increase bridge-type adsorbed CO_2. The subsequent CO_2 dissociation to carbon species took place on the surface, resulting in the enhanced C–C coupling [32, 51].

Guo and Song investigated the effects of point of zero charge (PZC) and pore size of Al_2O_3 support on the catalytic performance [52, 53]. Both the Fe sites dispersion and particle size were dependent on the PZC value. Specifically, a gradual increase of PZC value led to an improvement of metal dispersion and a decrease of particle size. The highest CO_2 conversion (54.0%) and C_{5+} hydrocarbon selectivity (31.1 C-mol%) were achieved at PZC = 8.0, however, a further increase of PZC had negative impact on both CO_2 conversion and chain growth. In light of pore size, a suitable size range lied within 7–10 nm, in which the active Fe_2O_3 particle size could be confined to 5–8 nm, leading to higher selectivity to hydrocarbons. However, the reduction was retarded when the pore size was smaller than 7 nm, resulting in the decrease of hydrocarbon selectivity. In another work, the Al_2O_3-supported FeK catalysts were modified by SiO_2 coating, an appropriate content of which considerably improved both CO_2 conversion (63.3%) and C_{2+} hydrocarbons selectivity (72.4 C-mol%, Table 6.2) [34]. It was proposed that the SiO_2 coating enabled an improvement of surface hydrophobicity, through which not only would the negative impact of byproduct H_2O be neutralized, but also promote the hydrocarbon synthesis by boosting the RWGS. Moreover, the confirmed interactions of SiO_2-Fe and SiO_2–Al_2O_3 also played crucial roles, as these interactions weakened the metal-support interaction, but largely improved the reducbility of the Fe oxides.

6.2.3   Fe-Based Bimetallic Catalysts

(1)   Fe–Co bimetallic catalysts

The effect of calcination temperature on activity and selectivity of CO_2 hydrogenation to hydrocarbons was examined on Fe–Co/K–Al_2O_3 catalysts [35]. Both CO_2 conversion and hydrocarbon selectivity decreased with the gradual increase of calcination temperatures and the highest conversion, hydrocarbons selectivity, and olefins yield were achieved at 673 K with the values of 49.3%, 90.6 C-mol%, and 18.1%, respectively (Table 6.2). Such calcination temperature-dependent behavior was attributed to the increased crystallite size of metal oxides and stronger interaction between Fe_2O_3 and other metal oxides. However, the trend of olefin-to-paraffin ratio exhibited a volcano-like shape and was maximized at 973 K (i.e., O/P = 7.6). Higher calcination temperatures were advantageous in complete KNO_3 decomposition to K_2O, facilitating the formation of $KAlO_2$ phase via the improved interaction between Fe_2O_3 and K_2O. Therefore, the selective olefin synthesis over paraffin was achieved by suppressing the secondary hydrogenation of olefin. Song et al. prepared a series of Al_2O_3-supported Fe–Co bimetallic catalysts with various Co/(Co + Fe) atomic ratios (i.e., 0–1 at. at.$^{-1}$) and K/Fe ratios (i.e., 0–1 at. at.$^{-1}$) while fixing the total Co + Fe loading at 15 wt% [36]. A synergetic effect of combining Fe and Co was obtained on promoting olefin-rich hydrocarbon synthesis, and the optimal Fe–Co composition and K/Fe ratio were 0.17 and 0.5 at. at.$^{-1}$, respectively (Table 6.2).

(2) Fe–Cu bimetallic catalysts

Combined with experimental and computational results, Guo and Song examined bimetallic effect and the nature of the active sites of the Fe–Cu bimetallic catalysts for CO_2 hydrogenation to hydrocarbons, and the coverage of a monolayer Cu onto Fe was varied from 0 to 1 [54, 55]. Formate (HCOO*) and carboxyl (COOH*) were identified as the two intermediates involved in the initial hydrogenation of CO_2, wherein the former was more kinetically favored over the latter at all surface Cu coverages, as illustrated in Fig. 6.6. At 2/9 ML surface Cu coverage, CO* was a preferred intermediate via CO_2 dissociation, while the favorable conversion pathway shifted to HCOO* intermediate when the surface Cu coverage increased to 4/9 ML or higher. On monometallic Fe(100) surface, the hydrogenation of CO_2 occurred in the following sequence: $CO_2^* \rightarrow CO^* \rightarrow HCO^* \rightarrow HCOH^* \rightarrow CH^*$. CH* is the building block for the subsequent chain growth. However, at 4/9 ML, the reaction pathway proceeded through HCOO* and HCOOH* species and the rest remained identical. Moreover, the incorporation of Cu lowered down the activation barrier for C–C coupling, through which the C_2H_4 formation became more kinetically favorable than CH_4 via CH* hydrogenation. To verify the computational prediction, a series of Fe–Cu bimetallic catalysts were prepared, and the Cu/(Cu + Fe) varied from 0 to 1 [37]. A bimetallic promoting effect on olefin-rich hydrocarbons (C_2–C_4) synthesis was obtained at Cu/(Cu + Fe) = 0.17 at. at.$^{-1}$ (Table 6.2).

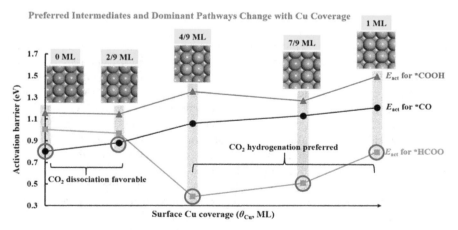

Fig. 6.6 Activation barriers for CO*, HCOO*, and COOH* formation versus surface Cu coverage (O_{Cu}, ML) on the representative surfaces [54]

6.2.4   Multifunctional Catalysts

(1)   Synthesis of paraffin and olefin

The composite catalyst with dual functionality has already been known in literature. As the vanguard of this field, Socha and coworkers reported a composite catalyst comprised of HZSM-5 zeolite (SiO$_2$/Al$_2$O$_3$ = 70) containing Cr, Zn, and Al in the atomic ratios 1:1:1, which exhibited 83 wt% selectivity to ethane from syngas conversion at 616–755 K and 100 atm [56]. A plausible reaction pathway was also proposed, which underwent CO $\rightarrow$ CH$_3$OH $\rightarrow$ CH$_3$OCH$_3$ $\rightarrow$ C$_2$H$_4$ $\rightarrow$ C$_2$H$_6$. Clearly, methanol was considered as the key building block for hydrocarbon synthesis on zeolite sites. Another pioneer work was done by Fujimoto and coworkers, in which the hybrid catalysts, consisting of the physical mixture of methanol synthesis catalysts and Y-type zeolite, gave C$_3$ and C$_4$ paraffins from syngas conversion [57]. The selectivity of C$_3$ and C$_4$ paraffins was higher than 70% with even little or no aromatic hydrocarbons under 1.0–4.1 MPa and 573–630 K. The introduction of zeolite enabled short residence time of olefin in the zeolite. A large pore size, small crystallite size, and small particle size should be essential factors to determine the activity and selectivity. Following the proposed reaction pathway for syngas, Fujiwara and coworkers prepared Cu–Zn–Cr oxide/zeolite composite catalysts for CO$_2$ hydrogenation to hydrocarbons, wherein Cu–Zn–Cr oxide was responsible for methanol synthesis, while zeolite accounted for hydrocarbon production [38]. The selective hydrocarbon production was subjected to the competition between methanol decomposition and hydrogenation, and the metallic Cu surface with low Cu surface areas benefited the hydrogenation. As shown in Table 6.2, the resulting catalyst considerably improved the selectivity toward C$_2$–C$_4$ hydrocarbons (88.3 C-mol%), in which the selectivity of C$_3$ was 40.8 C-mol%. Furthermore, an even higher selectivity to C$_2$, as high as 70.6 C-mol%, was achieved by incorporating La$_2$Zr$_2$O$_7$ into the multifunctional catalysts.

Clearly, preliminary results indicate the crucial roles of methanol and dimethyl ether (DME) as platform molecules for the subsequent hydrocarbons synthesis through cracking, isomerization, and aromatization reactions. These determining factors are also applicable to CO$_2$ hydrogenation to hydrocarbons, and the development of efficient multifunctional catalysts largely relies on the progress made in CH$_3$OH and DME syntheses from CO$_2$ hydrogenation, particularly in the CH$_3$OH synthesis as the DME can be formed through methanol dehydration. In the past decades, great efforts have been devoted to exploring the effective catalysts for CH$_3$OH synthesis, such as Cu-based catalysts (Cu/Zn, Cu/Zn/Al, and Cu/Zn/Zr) [58–62], noble catalysts (Pd and Pt) [63–66], and bimetallic catalyst with synergetic effect (Pd–Cu and Pd–Zn) [67–72]. Meanwhile, multifunctional catalysts are designed and used for the DME synthesis, in which the catalysts usually consist of two active components: one is responsible for the methanol synthesis, such as CuZn catalysts, while the other is in charge of dehydration, such as zeolites (e.g.,

HZSM-5 and ferrierite) [73–76]. These accumulative achievements offer opportunities to develop efficient multifunctional catalysts for hydrocarbon synthesis from CO_2 in tandem catalysis.

Recently, inspired by the computational prediction by Liu and coworkers [77, 78], Perez-Ramirez et al. reported the bulk In_2O_3 and ZrO_2-supported In_2O_3 catalysts with superior selectivity (i.e., 100 C-mol%) to CH_3OH under industrially relevant conditions [79, 80]. They reported that both In_2O_3 and In_2O_3/ZrO_2 catalysts exhibit 100 C-mol% of CH_3OH in the whole temperature range studied, and they claimed advantageous CH_3OH synthesis activity over commercial CuZnAl catalyst at higher temperatures (e.g., >540 K), where RWGS should dominate as a result of its endothermicity. ZrO_2-supported In_2O_3 catalyst seems to show excellent stability for 1000 h on stream, while the commercial catalyst deactivated rapidly in the first 100 h on stream. More research is in progress on the multifunctional catalysts for hydrocarbon synthesis from CO_2 by incorporating In_2O_3 as the sites responsible for methanol synthesis, as discussed below.

Sun and coworkers reported a bifunctional catalyst comprising of reducible In_2O_3 and HZSM-5 zeolites which selectively yielded gasoline-range hydrocarbons (e.g., 78.6% in hydrocarbons) with very low methane selectivity (e.g., 1%) at 613 K and 3.0 MPa, as shown in Fig. 6.7a and Table 6.2 [18]. The reaction pathway appear to be similar to the case of syngas conversion, wherein CO_2 hydrogenation to methanol took place on the oxygen vacancies on In_2O_3 surface, while subsequent C–C coupling occurred inside zeolite pores, specifically on the acidic sites via the hydrocarbon-pool mechanism, as illustrated in Fig. 6.7b. Characterization results demonstrate that the activity and selectivity were determined by the proximity of these two components. Moreover, the excellent performance in an industry-relevant test with pellet catalyst provided promising prospects for industrial application in the future. As a follow-up work, the same group prepared another bifunctional catalyst comprising of In_2O_3/ZrO_2 and SAPO-34, in which the latter zeolite was employed because of its superior activity in methanol-to-olefin reaction [40]. As

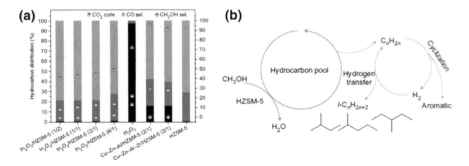

Fig. 6.7 **a** CO_2 hydrogenation over various bifunctional catalysts that contained Cu-based catalysts or In_2O_3 and HZSM-5 with different mass ratios. Reaction conditions: 613 K, 3.0 MPa, 9000 mL g-cat^{-1} h^{-1}, CO_2/H_2 = 1:3. **b** Schematic of the hydrocarbon-pool mechanism for CH_3OH conversion into hydrocarbons inside HZSM-5 [18]

listed in Table 6.2, through the control of oxygen vacancies on the oxide surface and the manner of integrating the components, the authors successfully achieved superior selectivity to $C_2^=-C_4^=$ olefins (as high as 80% in hydrocarbons), as well as C_2-C_4 selectivity around 93% with more than 35% CO_2 conversion at 673 K and 3.0 MPa. A similar bifunctional catalyst composite, In_2O_3/ZrO_2-SAPO-34, was prepared, and the catalyst performed well in light olefin production (ethylene and propylene) with a selectivity as high as 80–90% in hydrocarbons, as well as a CO_2 conversion of ∼20% [81]. This composite catalyst outperformed the conventional Fe or Co catalysts (typically <50% selectivity), which promote the conversion via RWGS + FTS pathway. A similar effort was also reported by Guo and Song with selective production of C_2-C_4 hydrocarbons (83%) on bifunctional In_2O_3/ZrO_2-SAPO-5 composite catalysts at 573 K and 3 MPa [41]. Furthermore, the catalyst exhibited good stability with 50 h on stream without obvious activity decay.

Ge and Xu prepared the multifunctional catalysts comprising of Na-promoted Fe_3O_4 and HZSM-5 zeolite [6]. The resulting catalysts exhibited a high selectivity (i.e., 78 C-mol%) of gasoline-range hydrocarbons (C_5-C_{11}) with a CO_2 conversion of 22%, while the methane selectivity was dramatically suppressed to only 4 C-mol %. They believe that the prominent activity originated from the proximity and synergy between three active sites, namely Fe_3O_4, Fe_5C_2, and acid sites, which cooperatively catalyzed the tandem reactions. More importantly, the multifunctional catalyst showed an excellent stability with 1000 h on stream. These results are promising for further development of our understanding of CO_2 activation and subsequent hydrogenation, shedding lights upon potential industrial applications in the future.

(2) Synthesis of aromatics

Aromatics are important for the production of chemicals and polymers, constituting about 1/3 of the market for commodity petrochemicals [19]. However, the production of aromatics relies heavily on petroleum, the depletion of which makes it desirable to seek alternative and non-traditional carbon sources. In this context, CO_2 hydrogenation to aromatics is promising and becomes an emerging field most recently. However, challenges other than C–C coupling impair the efforts for developing active catalysts, including the high unsaturation degree and complex structure of aromatics, as well as high H_2 content during reaction.

As an example of recent work on CO_2 conversion to aromatics through CO_2–CH_3OH-aromatics pathway, Guo and Song reported a tandem bifunctional catalyst consisting of ZnO/ZrO_2 and ZSM-5 for aromatics synthesis from CO_2 hydrogenation, in which CO_2–to–methanol and methanol-to-aromatics (MTA) took place, respectively, on these different active sites [19]. As shown in Table 6.4, the optimal aromatics selectivity was achieved at 70%, comprising of benzene, toluene, xylene, and heavy aromatics (C_9, and C_{10}). Comparative studies unveiled the key factors to manipulate the product distribution: (i) an appropriate temperature range for CH_3OH synthesis was prerequisite, as reverse water-gas shift (RWGS) dominated at high temperatures; (ii) strong acidic sites on ZSM-5 was crucial for the

Table 6.4 Summary of reaction conditions with conversion and selectivity to aromatics for selected catalysts[a]

Cat.	Temp/K	Press./ MPa	W/F/h g mol[-1]	CO_2 conv./%	CO selec./ C-mol%	Other C-containing product distribution[b]/%		References
						CH$_4$	Aromatics	
Zn/ZrO$_2$-Z5-300	613	3.0	8.28	9.1	42.5	0.6	70.0	[19]
ZnAlO$_x$ & H-ZSM-5	593	3.0	11.2	9.1	57.4	0.4	73.9	[21]

[a]$CO_2/H_2 = 1:3$
[b]Other carbon-containing products include C_2–C_4 paraffins and C_2–C_4 olefins, as well as C_{5+} hydrocarbons

MTA; (iii) a suitable contact time was necessary, as a longer contact time shifted the selectivity from benzene, toluene, and xylene to heavy aromatics through alkylation.

In another report, Li and coworkers explored the similar bifunctional catalyst, but with particular interests in understanding the roles of integration manner and H_2O effect [20]. As presented in Fig. 6.8a, an intimately spatial contact between ZnZrO and H-ZSM-5 was indispensable to boost the synthesis of aromatics, because such intimacy facilitated the migration of the intermediate formed on the surface of ZnZrO to the pore structure of H-ZSM-5. An appropriate amount of H_2O was beneficial to the formation of aromatics (Fig. 6.8b). Characterization results indicate that both the H_2O and ethylene adsorption primarily took place on the weak acidic sites of H-ZSM-5, whereas the presence of H_2O significantly suppressed the ethylene adsorption, thereby promoting its conversion to aromatics on the strong acidic sites (Fig. 6.8c).

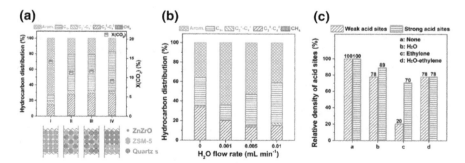

Fig. 6.8 a Effect of integration manner on catalytic performance (Cat, 1.0 g, 593 K, 4.0 MPa, 1200 mL g[-1] h[-1]). **b** Ethylene conversion over H-ZSM-5 with co-feeding different content of H_2O (Cat, 0.5 g, 593 K, 4.0 MPa, 1200 mL g[-1] h[-1]). **c** Normalized density of weak and strong acidic sites for H-ZSM-5 with different treatments: a, the pristine H-ZSM-5; b, with pre-adsorption of H_2O at room temperature; c, with pre-adsorption of ethylene; d, with adsorption of ethylene after H_2O adsorption [20]

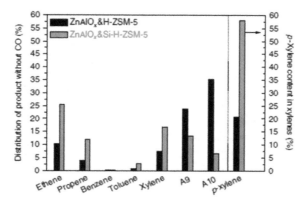

Fig. 6.9 Comparisons of catalytic behaviors over ZnAlO$_x$ & H-ZSM-5 and ZnAlO$_x$ & Si-H-ZSM-5. Reaction conditions: space velocity = 6000 mL g^{-1} h^{-1}, 3.0 MPa, 593 K, CO$_2$/H$_2$ = 1:3. A9 and A10 represent the aromatics containing 9 and 10 carbon atoms, respectively [21]

Liu et al. reported a composite catalyst with multi-components catalyst design concept, in which the catalyst comprised of nano-scaled spinel structural ZnAlO$_x$ oxide and H-ZSM-5 [21]. The importance of the intimate contact between ZnAlO$_x$ and H-ZSM-5 was corroborated, as the grinding-prepared catalyst exhibited superior selectivity to aromatics among all catalysts integrated by different manners, and the optimal aromatics selectivity was achieved at 73.9% with extremely low CH$_4$ selectivity of 0.4% among the CO-free carbon products (Table 6.4). More importantly, by incorporating the tetraethoxysilane (TEOS)-modified Si-H-ZSM-5, the catalyst produced more toluene (2.7%) and xylenes (16.8%), even with 58.1% p-xylene, 25.3% ethylene, and 11.9% propylene, which are important commodity hydrocarbon chemicals in industry (Fig. 6.9). H$_2$-TPR results suggest that Zn^{2+} in the spinel structure of ZnAlO$_x$ played a crucial role in the activation of CO$_2$ hydrogenation. The Bronsted acidic sites on the H-ZSM-5 was detrimental to the synthesis of aromatics through promoting the hydrogenation of unsaturated hydrocarbons to paraffins [82]. However, this undesired reaction was significantly suppressed when ZnAlO$_x$ was introduced. The 2,6-di-tert-butyl-pyridine absorption (DTBPy)-based FT-IR spectra of the catalysts revealed that the ZnAlO$_x$ shielded the external Bronsted acidic sites of H-ZSM-5 after the grinding and pressing under high pressures, thereof mitigating their negative impact.

6.2.5 Reaction Pathways and Mechanisms of C–C Coupling

In general, two reaction routes were proposed for CO$_2$ hydrogenation to hydrocarbons, including modified FTS route and methanol-mediated route. The CO$_2$-FTS route was similar to the reaction pathway for FTS (Fig. 6.13 in the next section), except that the RWGS and CO activation were the initial steps. Due to the important role of CO in this route, two active sites were essential to account for RWGS and FTS, respectively. As aforementioned in Sect. 6.2.2, Fe oxide and Fe carbides were identified as corresponding active sites on Fe-based catalysts. With the formed CO on Fe oxide sites, the light hydrocarbons were produced subsequently on χ-Fe$_5$C$_2$ carbide site, while the higher hydrocarbons proceeded on θ-Fe$_3$C carbide site [31]. This reaction route was developed on the multifunctional Na-Fe$_3$O$_4$/Zeolite catalyst. As shown in Fig. 6.10a, in addition to the above types of sites, zeolite was incorporated as the third sites for boosting the acid-catalyzed reactions, leading to the production of gasoline-range isoparaffins and aromatics [6].

In methanol-mediated route, methanol was considered as the key building unit for the hydrocarbon synthesis from syngas, through which the reaction underwent CH$_3$OH $\rightarrow$ CH$_3$OCH$_3$, followed by the formation of C$_2$H$_4$ with the production of H$_2$O [56]. Therefore, bifunctional catalyst systems are required to complete the tandem reactions, including methanol synthesis and subsequent conversion (Fig. 6.10b). A good example was the composite between In$_2$O$_3$ and HZSM-5. The former In$_2$O$_3$ was the newly emerging vacancy-based catalyst that exhibited superior selectivity of CH$_3$OH, while the latter HZSM-5 was responsible for chain growth [18]. An indirect methanol-mediated route was also proposed by Fujiwara and coworkers, in which a characteristic two-stage reactor was designed [39]. The first stage, catalyzed by Cu–Zn–Al oxide, was for RWGS, while methanol synthesis and conversion were combined in the second stage and catalyzed by the composite catalyst comprising of Cu–Zn–Al oxide and HB (Table 6.2). A higher selectivity of C$_{2+}$ hydrocarbons was largely dependent on the concentration of CO in the feed gas of the second stage, and a cold trap of H$_2$O removal was indispensable to preserve the activity of the composite catalyst.

There is some similarity between CO and CO$_2$ hydrogenation over the Fe-based catalysts, and the C–C coupling in both cases is relevant. Pham et al. studied the C–C coupling mechanism on the terraced-like χ-Fe$_5$C$_2$ (510) phase, as it exhibited the lower surface energy and was evidenced in HRTEM images and XRD patterns [84]. As depicted in Fig. 6.11a, CO insertion and carbide mechanisms were proposed and examined, and the overall barrier of the former pathway was 0.47 eV

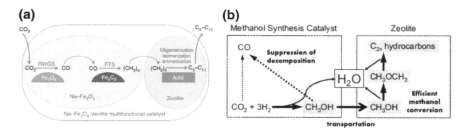

Fig. 6.10 **a** Reaction scheme for CO$_2$ hydrogenation to gasoline-range hydrocarbons on Na-Fe$_3$O$_4$/Zeolite multifunctional catalyst, including RWGS $\rightarrow$ CO to α-olefins via FTS $\rightarrow$ gasoline-range hydrocarbons via acid-catalyzed reactions [18], **b** reaction steps of CO$_2$ hydrogenation over composite catalyst (favorable paths are shown in bold lines) [83]

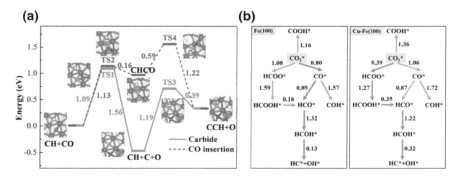

Fig. 6.11 a Energies and structures for the CH + CO → CCH + O formation pathways on χ-Fe$_5$C$_2$ (510) surface in terms of the carbide mechanism (red-solid line) and CO-insertion mechanism (blue-dash line). Zero-point energies are included. Blue: Fe atoms; gray: C atoms; green: C atoms involved in reactions; while: H atoms; red: O atoms [84], **b** reaction networks examined to identify energetically favorable C$_1$ species from CO$_2$ hydrogenation on Fe(100) and Cu–Fe (100) surface at 4/9 ML Cu coverage. Activation barriers are given in eV [55]. (The networks connected with red arrows represent the preferred path for CO$_2$ conversion to CH*.)

higher than that of the latter pathway. Therefore, the carbide mechanism was thought to be dominant on χ-Fe$_5$C$_2$ (510) surface, and C–C coupling proceeded by the C* + CH* and CH* + CH*. Recently, Nie et al. studied the C–C coupling on Fe–Cu bimetallic catalysts, and CH* was proposed as the building unit for chain growth (Fig. 6.11b) [55]. On the Fe(100) facet, CH* was formed through CO$_2^*$ dissociation to CO*, and CH$_4$ synthesis was favored via CH* hydrogenation with lower barrier. Unlike the Fe (100) alone facet, the adsorbed CO$_2^*$ proceeded to HCOO* via hydrogenation, followed by further hydrogenation and dissociation on Cu-doped Fe (100) facet. In the following step, C–C coupling of CH* species was more favorable in comparison to CH* hydrogenation, leading to the improvement of chain growth for hydrocarbons synthesis. These new results suggest that the incorporation of Cu enabled the alteration of the reaction pathway by boosting the CO$_2^*$ hydrogenation without going through CO* and lower the barrier of C–C coupling. Similar reaction mechanisms were also reported on other Fe facets [85–87].

In CO$_2$ hydrogenation to aromatics, the *operando* diffuse reflectance FT-IR study identified the presence of formate and methoxy species on the surface of ZnZr/H-ZSM-5 [20] and ZnAlO$_x$/H-ZSM-5 [21], and their correlation with the subsequent surface reaction was also evidenced. Therefore, combined with the functionality of the bifunctional catalysts, it was proposed that the surface formate species formed on either ZnZrO or ZnAlO$_x$ was hydrogenated to methoxy species; then, both surface CH$_3$OH and DME, formed via the dissociation of methoxy species, migrated to H-ZSM-5 to synthesize olefin intermediates; finally, the transformation of olefins to aromatics occurred in the micropores of H-ZSM-5. Due to the similarity with the syngas-to-aromatics, a hydrocarbon-pool mechanism was also speculated. As presented in Fig. 6.12, the aromatization proceeded primarily via the arene-based cycles [19].

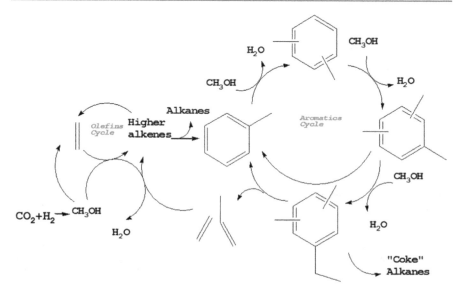

Fig. 6.12 Speculated hydrocarbon-pool mechanism for the aromatization of CO_2 over the bifunctional $ZnZrO_2/ZSM-5$ [19]

6.3 CO₂ Hydrogenation to Higher Alcohols

6.3.1 Plausible Reaction Pathways

CO_2 conversion to higher alcohols may be related with higher alcohols synthesis (HAS) from syngas. In the case of syngas, a so-called "CO insertion" mechanism was generally accepted [88]. As illustrated in Fig. 6.13, CO undergoes dissociation and hydrogenation to form CH_x species, followed by the insertion of nondissociated CO into CH_x, ultimately leading to form higher alcohols. C–C coupling reaction and CO insertion on the surface are considered as key steps. In the past decades, great efforts have been put forth to develop efficient catalysts for direct syngas conversion to higher alcohols, which involves a huge variety of elements and different combinations. Generally, the following four catalyst systems are distinguished: (1) Rh-based catalysts, (2) modified Mo-based catalysts, (3) modified methanol synthesis catalysts, and (4) modified Fischer-Tropsch Synthesis (FTS) catalysts [89]. However, each of them suffers from some drawbacks, which impairs the effort for improvement, as listed in Table 6.5.

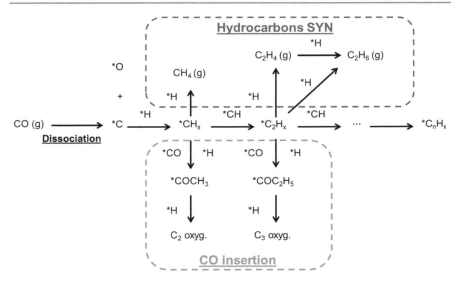

Fig. 6.13 Schematic illustration of higher alcohol synthesis from CO hydrogenation [88]

Table 6.5 Comparison of different catalyst systems in HAS from syngas [89]

Catalyst system	Features	Reaction mechanism	Issues
Rh-based catalysts	• High number of Rh-MO$_x$ interface sites • CO insertion into the Rh-CH$_x$ bond	• CO insertion	• Low conversion • High cost of noble metals
Mo-based catalysts	• Alkali doping • High resistance against sulfur poisoning	• CO insertion	• Sulfur formed with significant amount • Costly posttreatment
Methanol synthesis catalysts	• Cu/Zn catalysts • Alkali metal doping • Basic sites for C–C and C–O bond formation	• Condensation of lower alcohol molecules for higher alcohols • CO insertion	• CH$_3$OH as major product • Low CH$_3$CH$_2$OH selectivity
Modified FTS catalysts	• Active metals, Co, Fe, Ni, or Ru supported on Al$_2$O$_3$, SiO$_2$, or carbon nanotubes • Promoters, Cu, Mo, Mn, Pd, and La, alkali metals • Synergistic effect of an FT-active metal for CO dissociation and a promoter metal for nondissociation of CO	• CO insertion	• High selectivity to hydrocarbons, especially CH$_4$ • Anderson-Schulz-Flory distribution limitation

6.3.2 Early Exploration on HAS from CO_2 Hydrogenation

In an early work, Arakawa et al. investigated the effect of various metal oxide promoters on ethanol synthesis from CO_2 hydrogenation over Rh/SiO_2 catalysts at 513 K and 5 MPa, as listed in Table 6.6 [90]. Among the metals tested, Li salts showed the most prominent ethanol selectivity, which was 15.5 C-mol%. In situ FT-IR analysis suggests that the reaction proceeded via CO intermediates. Later, the same group studied the ethanol synthesis over $Rh-Fe/SiO_2$ catalysts, wherein they reported the correlation of ethanol selectivity with Fe^0 composition on the surface, as well as with the amount of bridged-CO adsorbed species by in situ FT-IR [91]. An ethanol selectivity of 16.0 C-mol% and a CO_2 conversion of 26.7% were obtained at 533 K and 5 MPa (Table 6.6). Then, the same group shifted from Rh-Fe-based catalysts to Co/SiO_2-based catalysts, and developed a Na salt-promoted $Ir/Co/SiO_2$ catalysts, with the ethanol selectivity up to 8 C-mol% at 493 K and 2.1 MPa (Table 6.6) [92]. In another early test, Souma and coworkers investigated the Fe catalysts in alcohol synthesis from CO_2 hydrogenation at 523 K and 1 MPa [93]. An ethanol selectivity of 11.6 C-mol% and a propanol selectivity of 4.5 C-mol% were obtained when CO/H_2 was the feed gas. However, no alcohols were produced, except CH_3OH, when the feed gas was switched to CO_2/H_2. With above early attempts, the CO insertion mechanism may still be mostly valid in HAS from CO_2 hydrogenation, however, these catalysts suffered from low activity and selectivity.

6.3.3 Recent Progress in Fe-, Cu- and Noble Metal-Based Catalysts

Considerable attention has been given to the modified Fe- and Cu-based catalysts recently. Worayingyong et al. evaluated the activity performance of K- and Mn-promoted Fe catalysts supported on N-functionalized carbon nanotubes (NCNTs) in CO_2 hydrogenation, and the effect of K and Mn promoters were addressed [94]. Apart from the hydrocarbons, the K-promoted Fe/NCNT and Mn/Fe/CNT catalysts exhibited an activity toward ethanol synthesis. The role of K was relevant to weaken the C–O bond, resulting in the formation of CH_x species responsible for chain growth. Mn acted as a structural promoter, hampering the iron active sites from agglomeration during the reaction. The production of ethene appeared to display an adverse trend with the ethanol synthesis, implying the contribution of ethene to ethanol formation. The proposed reaction was analogous to the CO insertion mechanism, except that the "OH" group, formed through CO dissociation and subsequent reaction with adsorbed H, was inserted into R-CH-CH$_2$ instead of "CO". Bimetallic catalysts are also exploited with the combination of two components with different roles.

With respect to the modification of methanol synthesis catalyst, namely Cu–Zn catalysts, adding modifiers into Cu–Zn is proved to be an effective way to improve the alcohol synthesis activity. The reaction pathway in this case shall undergo the

Table 6.6 Summary of reaction conditions with conversion and selectivity to alcohols for selected catalysts

Catalyst	CO_2/H_2 ratio	Flow		Press./MPa	Temp./K	CO_2 conv./%	Alcohol selec./C-mol%			References
		$GHSV/h^{-1}$	W/F/g-cat h mol^{-1}				MeOH	EtOH	$C_{2+}OH$	
Li/Rh/SiO₂	1:3	–	3.73	5.0	513	7.0	5.2	15.5	0.0	[90]
Sr/Rh/SiO₂	1:3	–	3.73	5.0	513	2.8	7.6	2.5	0.0	[90]
Ag/Rh/SiO₂	1:3	–	3.73	5.0	513	2.1	7.4	1.8	0.0	[90]
Fe/SiO₂	1:3	–	3.73	5.0	533	12.7	6.2	1.2	0.0	[22]
Rh-Fe/SiO₂	1:3	–	3.73	5.0	533	26.7	29.4	16.0	0.0	[22]
Ir/Co–Li₂O/SiO₂	1:3	2000	–	2.1	493	5.6	16.3	2.5	0.0	[92]
Ir/Co–Na₂O/SiO₂	1:3	2000	–	2.1	493	7.6	7.8	7.9	0.0	[92]
Ir/Co–K₂O/SiO₂	1:3	2000	–	2.1	493	7.8	5.6	6.1	0.0	[92]
CuZnK[a]	1:3	5000	–	6.0	573	25.1	6.95	–	3.96	[23]
CuZnFeK[a]	1:3	5000	–	6.0	573	45.1	4.78	–	31.89	[23]
CuZnK + CuFeCoK[a]	1:3	5000	–	6.0	623	32.4	5.28	–	6.54	[95]
Pt/Co₃O₄-p	1:3		3.73	2.0	473	22.4	17.6	–	2.3	[96]
Pd NPs/P25[b]	1:3	–	–	3.2	473	–	49.3	50.7	–	[97]
Pd₁.₅Cu NPs/P25[b]	1:3	–	–	3.2	473	–	19.7	80.3	–	[97]
Pd₁Cu NPs/P25[b]	1:3	–	–	3.2	473	–	22.5	77.5	–	[97]
Pd₂Cu NPs/P25[b]	1:3	–	–	3.2	473	–	8.0	92.0	–	[97]
Pd₂Cu NPs/SiO₂[b]	1:3	–	–	3.2	473	–	26.2	73.8	–	[97]
Pd₂Cu NPs/CeO₂[b]	1:3	–	–	3.2	473	–	30.4	69.6	–	[97]
Pd₂Cu NPs/Al₂O₃[b]	1:3	–	–	3.2	473	–	16.9	83.1	–	[97]

[a]Alcohol selectivity was based on wt%. Others were evaluated by C-mol%

[b]Impregnation method was referred to [97]. Activity performance was collected by batch reactor, while others were collected based on fixed-bed flow reactor

$$2CO_2 \longrightarrow 2^\cdot CO_2 \xrightarrow{H_2} 2^\cdot HOCO \longrightarrow 2^\cdot CO + 2^\cdot HO \xrightarrow{\frac{3}{2}H_2} {}^\cdot CHO + {}^\cdot CO + 2H_2O \xrightarrow{H_2}$$

$$^\cdot CH_2OH + {}^\cdot CO + 2H_2O \longrightarrow {}^\cdot CH_2 + {}^\cdot OH + {}^\cdot CO + 2H_2O \xrightarrow{H_2} {}^\cdot CH_3 + {}^\cdot CO + 3H_2O \longrightarrow$$

$$^\cdot CH_3CO + 3H_2O \xrightarrow{H_2} {}^\cdot CH_3CH_2O + 3H_2O \xrightarrow{\frac{1}{2}H_2} CH_3CH_2OH + 3H_2O$$

Scheme 6.1 Conceivable reaction mechanism of CO_2 hydrogenation to EtOH on Pd–Cu NPs/P25 [97]

RWGS first, followed by CO hydrogenation to alcohols. Chen et al. evaluated the activity performance of Cu–Zn–Fe–K using a one-stage catalyst bed [23]. The highest alcohol selectivity of 36.7 wt% with a high ethanol/methanol ratio of 6.76 was obtained when the Fe/(Cu + Zn + Fe) ratio was 0.2 (Table 6.6). The same group further reported that the higher alcohol synthesis from CO_2 hydrogenation can be achieved by a two-stage bed catalyst combination system comprising of two catalysts that are responsible for RWGS and CO hydrogenation, respectively [95]. The first stage was K-promoted Cu-Zn catalyst, while Co and K-promoted Cu-Fe catalyst accounted for the second page. The optimal catalyst loading volumes for the first and second stage were 1.5 and 4.5 mL, respectively, at which the ROH selectivity and relative $C_{2+}OH$ selectivity within the alcohol were 11.8 and 55.3 wt%, respectively (Table 6.6).

Meanwhile, other metal-based catalysts were also developed for higher alcohol synthesis from CO_2 hydrogenation. Pt NPs were supported on Co_3O_4 with two different morphologies, namely nanoplate (Co_3O_4-p) and nanorod (Co_3O_4-r) [96]. Pt/Co_3O_4-p was more easily reduced, through which parts of Co_3O_4 were reduced to CoO and Co^0 under the reaction conditions, forming new active sites Pt–Co/Co_3O_{4-x}. This catalyst yielded the most $R_{2+}OH$ of 0.56 mol kg-cat^{-1} h^{-1} at 473 K and 2.0 MPa, which could be attributed to the synergic effect of Pt, Co NPs, and oxygen vacancies of Co_3O_{4-x} on the improvement of H_2 and CO_2 adsorption. Huang and coworkers prepared Pd-rich Pd_2Cu alloy NPs and supported them onto commercial P25 (Degussa) [97]. The resulting catalyst exhibited a superior ethanol selectivity as high as 92.0 C-mol% and a high turnover frequency of 359.0 h^{-1} at 473 K and 3.2 MPa (Table 6.6). A "CO insertion" mechanism was also suggested, and the P25-supported Pd_2Cu NPs can boost CO* hydrogenation to $^*CHO$, which was proposed as the rate-determining step (RDS), as illustrated in Scheme 6.1.

6.4 Summary and Future Perspective

Several routes have been explored for CO_2 hydrogenation to clean fuels and chemicals. For the syntheses of hydrocarbons (olefins and paraffins) and alcohols, numerous efforts have been focused on Fe-based catalysts and Cu–ZnO-based catalysts, respectively. The emerging new oxide-based (such as In_2O_3-based) catalysts enable the development of a new type of catalytic materials with high selectivity to methanol and for the hydrocarbon synthesis from CO_2 via a methanol-mediated route. This route has also been applied to the CO_2-to-aromatics over metal oxide-zeolite bifunctional catalysts.

Significant progress has been made in new bimetallic catalysts (such as Fe–Co and Fe–Cu), new metal-oxide hybrid catalysts (Fe_3O_4–FeC_n/ZSM-5), new binary oxide-based multi-components or hybrid catalysts (such as In_2O_3–ZrO_2/ZSM-5 and ZnO–ZrO_2/ZSM-5 hybrids) for CO_2 conversion to energy-rich C_n products in the past five years. Based on the achievements that were reported so far, a future direction of CO_2 hydrogenation to C_n products should focus on synthesizing catalysts that are active, selective, water-resistant, and cost-effective in the range of lower temperatures such as 373–573 K where hydrogenation is more thermodynamically favored. Furthermore, the stability is important in catalyst evaluation, especially in the large-scale processes. Hence, long-term catalytic performance should be examined, as well as the solutions to prevent deactivation.

Some recent endeavors have been devoted to applying novel materials and approaches in CO_2 hydrogenation. The development of novel catalytic materials stems from the idea to tailor and control the catalyst structure from nanoscale, including metal-organic frameworks (MOF)/zeolitic-imidazolate frameworks (ZIF)-derived catalysts with top-down or bottom-up strategies, quantum dots, intermetallic compounds, and graphene- and carbide-based catalysts. Energy efficiency is of great importance in utilizing CO as untraditional carbon source because the integration of efficient catalyst and reasonable energy consumption can ultimately guarantee the environmental benefits. For this purpose, exploration revolving around reactor innovation is in progress, such as microreactor, membrane reactor, and magnetic field-assisted reactor. Continued endeavors will be focusing on these two directions in the near future. However, the prerequisite is whether we can unveil the reaction mechanism such as the identification of active sites by using the state-of-the-art techniques and the establishment of descriptors from DFT calculations for mechanistic pathways. Apparently, the integration of theoretical and experimental studies would provide avenues to achieve the goal.

For the higher alcohols synthesis, key questions that remain to be answered include (i) the mechanism for cleaving one C–O bond while maintaining the other or preserving one C–O* during C–C chain growth; (ii) the active sites responsible for C–C chain growth; (iii) is CO insertion or CH insertion important? (iv) how to tune the surface H/C ratio of chemisorbed species to benefit higher alcohol synthesis rather than higher hydrocarbons. Seeking other alternative reaction pathways through different mediated intermediates, instead of CO and/or CH insertion, is also worth consideration.

More laboratory research and computational study as well as pilot-scale research and development are required for establishing efficient processes for CO_2 utilization in order to promote sustainable development and mitigate climate change due to CO_2 emissions. Meanwhile, investigations on developing more efficient and cost-effective ways to produce H_2 from water using renewable energy is also indispensable for CO_2 hydrogenation with environmental and economic viability in the near future.

Acknowledgements This work was supported in part by the Pennsylvania State University through the EMS Energy Institute, the Institutes of Energy and the Environment and the Joint Center for Energy Research established between Penn State and Dalian University of Technology.

References

1. Pontzen F, Liebner W, Gronemann V, Rothaemel M, Ahlers B (2011) Catal Today 171: 242–250
2. Song C (2006) Catal Today 115:2–32
3. Gusain R, Kumar P, Sharma OP, Jain SL, Khatri OP (2016) Appl Catal B-Environ 181: 352–362
4. Xia S, Meng Y, Zhou X, Xue J, Pan G, Ni Z (2016) Appl. Catal B-Environ 187:122–133
5. Gutiérrez-Guerra N, Moreno-López L, Serrano-Ruiz JC, Valverde JL, de Lucas-Consuegra A (2016) Appl Catal B-Environ 188:272–282
6. Wei J, Ge Q, Yao R, Wen Z, Fang C, Guo L, Xu H, Sun J (2017) Nat Commun 8:15174
7. Alvarez A, Bansode A, Urakawa A, Bavykina AV, Wezendonk TA, Makkee M, Gascon J, Kapteijn F (2017) Chem Rev 117:9804–9838
8. Aresta M, Dibenedetto A, Angelini A (2013) J CO_2 Util 3–4:65–73
9. He X (2017) Int J Oil Gas Coal Eng 5:145–152
10. Swapnesh A, Srivastava VC, Mall ID (2014) Chem Eng Technol 37:1765–1777
11. Grabow LC, Marvrikakis M (2011) ACS Catal 1:365–384
12. Wang W, Wang S, Ma X, Gong J (2011) Chem Soc Rev 40:3703–3727
13. Ansari MB, Park S-E (2012) Energy Environ Sci 5:9419–9437
14. Li MM-J, Tsang SCE (2018) Catal. Sci Technol 8:3450–3464
15. Li W, Wang H, Jiang X, Zhu J, Liu Z, Guo X, Song C (2018) RSC Adv 8:7651–7669
16. Yang H, Zhang C, Gao P, Wang H, Li X, Zhong L, Wei W, Sun Y (2017) Catal. Sci Technol 7:4580–4598
17. Wang WH, Himeda Y, Muckerman JT, Manbeck GF, Fujita E (2015) Chem Rev 115: 12936–12973
18. Gao P, Li S, Bu X, Dang S, Liu Z, Wang H, Zhong L, Qiu M, Yang C, Cai J, Wei W, Sun Y (2017) Nat Chem 9:1019–1024
19. Zhang X, Zhang A, Jiang X, Zhu J, Liu J, Li J, Zhang G, Song C, Guo X (2019) J CO_2 Util 29:140–145
20. Li Z, Qu Y, Wang J, Liu H, Li M, Miao S, Li C Joule (2018)
21. Ni Y, Chen Z, Fu Y, Liu Y, Zhu W, Liu Z (2018) Nat Commun 9:3457
22. Kusama H, Okabe K, Syama K, Arakawa H (1997) Energy 22:343–348
23. Li S, Guo H, Luo C, Zhang H, Xiong L, Chen X, Ma L (2013) Catal Lett 143:345–355
24. Wu J, Huang Y, Ye W, Li Y (2017) Adv Sci (Weinh) 4:1700194
25. Qiao J, Liu Y, Hong F, Zhang J (2014) Chem Soc Rev 43:631–675
26. Francke R, Schille B, Roemelt M (2018) Chem Rev 118:4631–4701
27. Zhao G, Huang X, Wang X, Wang X (2017) J Mater Chem A 5:21625–21649
28. Shi Z, Yang H, Gao P, Li X, Zhong L, Wang H, Liu H, Wei W, Sun Y (2017) Catal Today 311:65–73
29. Nafria R, Genc A, Ibanez M, Arbiol J, de la Piscina PR, Homs N, Cabot A (2016) Langmuir 32:2267–2276
30. Das T, Sengupta S, Deo G (2013) Reac Kinet Mech Cat 110:147–162
31. Zhang Y, Fu D, Liu X, Zhang Z, Zhang C, Shi B, Xu J, Han Y-F (2018) ChemCatChem 10:1272–1276
32. Boreriboon N, Jiang X, Song C, Prasassarakich P (2018) J CO_2 Util 25:330–337
33. Riedel T, Claeys M, Schulz H, Schaub G, Nam S-S, Jun K-W, Choi M-J, Kishan G, Lee K-W (1999) Appl Catal A-Gen 186:201–213
34. Ding F, Zhang A, Liu M, Guo X, Song C (2014) RSC Adv 4:8930

35. Numpilai T, Witoon T, Chanlek N, Limphirat W, Bonura G, Chareonpanich M, Limtrakul J (2017) Appl Catal A-Gen 547:219–229
36. Satthawong R, Koizumi N, Song C, Prasassarakich P (2013) J CO$_2$ Util 3–4:102–106
37. Wang W, Jiang X, Wang X, Song C (2018) Ind Eng Chem Res 57:4535–4542
38. Fujiwara M, Kieffer R, Ando H, Souma Y (1995) Appl Catal A-Gen 121:113–124
39. Fujiwara M, Sakurai H, Shiokawa K, Iizuka Y (2015) Catal Today 242:255–260
40. Gao P, Dang S, Li S, Bu X, Liu Z, Qiu M, Yang C, Wang H, Zhong L, Han Y, Liu Q, Wei W, Sun Y (2017) ACS Catal 8:571–578
41. Wang J, Zhang A, Jiang X, Song C, Guo X (2018) J CO$_2$ Util 27:81–88
42. Satthawong R, Koizumi N, Song C, Prasassarakich P (2013) Top Catal 57:588–594
43. Rodemerck U, Holeňa M, Wagner E, Smejkal Q, Barkschat A, Baerns M (2013) ChemCatChem 5:1948–1955
44. Visconti CG, Martinelli M, Falbo L, Infantes-Molina A, Lietti L, Forzatti P, Iaquaniello G, Palo E, Picutti B, Brignoli F (2017) Appl Catal B-Environ 200:530–542
45. Visconti CG, Martinelli M, Falbo L, Fratalocchi L, Lietti L (2016) Catal Today 277:161–170
46. Samanta A, Landau MV, Vidruk-Nehemya R, Herskowitz M (2017) Catal Sci Technol 7:4048–4063
47. Fischer N, Henkel R, Hettel B, Iglesias M, Schaub G, Claeys M (2015) Catal Lett 146:509–517
48. Amoyal M, Vidruk-Nehemya R, Landau MV, Herskowitz M (2017) J Catal 348:29–39
49. Satthawong R, Koizumi N, Song C, Prasassarakich P (2015) Catal Today 251:34–40
50. Riedel T, Claeys M, Shulz H, Schaub G, Nam S-S, Jun K-W, Choi M-J, Kishan G, Lee K (1999) Appl Catal A-Gen 186:201–213
51. Xiao J, Mao D, Guo X, Yu J (2015) Appl Surf Sci 338:146–153
52. Xie T, Wang J, Ding F, Zhang A, Li W, Guo X, Song C (2017) J CO$_2$ Util 19:202–208
53. Ding F, Zhang A, Liu M, Zuo Y, Li K, Guo X, Song C (2014) Ind Eng Chem Res 53:17563–17569
54. Nie X, Wang H, Janik MJ, Guo X, Song C (2016) J Phys Chem C 120:9364–9373
55. Nie X, Wang H, Janik MJ, Chen Y, Guo X, Song C (2017) J Phys Chem C 121:13164–13174
56. Chang CD, Miale JN, Socha RF (1984) J Catal 90:84–87
57. Fujimoto K, Saima H, Tominaga H (1988) Ind Eng Chem Res 27:920–926
58. Arena F, Mezzatesta G, Zafarana G, Trunfio G, Frusteri F, Spadaro L (2013) J Catal 300:141–151
59. Arena F, Mezzatesta G, Zafarana G, Trunfio G, Frusteri F, Spadaro L (2013) Catal Today 210:39–46
60. Liao G, Chen S, Quan X, Yu H, Zhao H (2012) J Mater Chem 22:2721–2726
61. Li MM-J, Zeng Z, Liao F, Hong X, Tsang SCE (2016) J Catal 343:157–167
62. Schumann J, Eichelbaum M, Lunkenbein T, Thomas N, Álvarez Galván MC, Schlögl R, Behrens M (2015) ACS Catal 5:3260–3270
63. Matsumura Y, Shen W-J, Ichihashi Y, Okumura M (2001) J Catal 197:267–272
64. Zhou X, Qu J, Xu F, Hu J, Foord JS, Zeng Z, Hong X, Tsang SC (2013) Chem Commun (Camb) 49:1747–1749
65. Wang X, Shi H, Kwak JH, Szanyi J (2015) ACS Catal 5:6337–6349
66. Koizumi N, Jiang X, Kugai J, Song C (2012) Catal Today 194:16–24
67. Jiang X, Koizumi N, Guo X, Song C (2015) Appl Catal B-Environ 170:173–185
68. Nie X, Jiang X, Wang H, Luo W, Janik MJ, Chen Y, Guo X, Song C (2018) ACS Catal 8:4873–4892
69. Jiang X, Jiao Y, Moran C, Nie X, Gong Y, Guo X, Walton KS, Song C (2018) Catal Commun 118:10–14
70. Jiang X, Wang X, Nie X, Koizumi N, Guo X, Song C (2018) Catal Today 316:62–70
71. Jiang X, Nie X, Wang X, Wang H, Koizumi N, Chen Y, Guo X, Song C (2019) J Catal 369:21–32

72. Bahruji H, Bowker M, Hutchings G, Dimitratos N, Wells P, Gibson E, Jones W, Brookes C, Morgan D, Lalev G (2016) J Catal 343:133–146
73. Bonura G, Migliori M, Frusteri L, Cannilla C, Catizzone E, Giordano G, Frusteri F (2018) J CO_2 Util 24:398–406
74. Bonura G, Cannilla C, Frusteri L, Mezzapica A, Frusteri F (2017) Catal Today 281:337–344
75. Zhou X, Su T, Jiang Y, Qin Z, Ji H, Guo Z (2016) Chem Eng Sci 153:10–20
76. Frusteri F, Bonura G, Cannilla C, Drago Ferrante G, Aloise A, Catizzone E, Migliori M, Giordano G (2015) Appl Catal B-Environ 176–177:522–531
77. Ye J, Liu C, Mei D, Ge Q (2013) ACS Catal 3:1296–1306
78. Ye J, Liu C, Ge Q (2012) J Phys Chem C 116:7817–7825
79. Martin O, Martín AJ, Mondelli C, Mitchell S, Segawa TF, Hauert R, Drouilly C, Curulla-Ferré D, Pérez-Ramírez J (2016) Angew Chem Int Ed 55:1–6
80. Frei MS, Capdevila-Cortada M, García-Muelas R, Mondelli C, López N, Stewart JA, Curulla Ferré D, Pérez-Ramírez J (2018) J Catal 361:313–321
81. Gao J, Jia C, Liu B (2017) Catal. Sci Technol 7:5602–5607
82. Cheng K, Gu B, Liu X, Kang J, Zhang Q, Wang Y (2016) Angew Chem Int Ed 55:4725–4728
83. Fujiwara M, Satake T, Shiokawa K, Sakurai H (2015) Appl Catal B-Environ 179:37–43
84. Pham TH, Qi Y, Yang J, Duan X, Qian G, Zhou X, Chen D, Yuan W (2015) ACS Catal 5:2203–2208
85. Cheng J, Hu P, Ellis P, French S, Kelly G, Lok CM (2008) J Phys Chem C 112:6082–6086
86. Li H-J, Chang C-C, Ho J-J (2011) J Phys Chem C 115:11045–11055
87. Lee S-B, Kim J-S, Lee W-Y, Lee K-W, Choi M-J Studies in surface science and catalysis. In: Chang J-S, Park S-E, Kyu-Wan L (eds) Proceedings of 7th the international conference on carbon dioxide utilization, Elsevier, 2004, pp 8–73
88. Chuang SSC, Stevens RW, Khatri R (2005) Top Catal 32:225–232
89. Muhler M, Kaluza S Syngas to methanol and ethanol. In: J Sa (Ed.) Fuel production with heterogeneous catalysis. Taylor & Francis Group, LLC, 2014, pp 169–192
90. Kusama H, Okabe K, Sayama K, Arakawa H (1996) Catal Today 28:261–266
91. Kusama H, Okabe K, Sayama K, Arakawa H (1997) Energy 22:343–348
92. Okabe K, Yamada H, Hanaoka T, Matsuzaki T, Arakawa H, Abe Y (2001) Chem Lett 30:904–905
93. Ando H, Matsumura Y, Souma Y (2000) Appl Organomet Chem 14:831–835
94. Kangvansura P, Chew LM, Saengsui W, Santawaja P, Poo-arporn Y, Muhler M, Schulz H, Worayingyong A (2016) Catal Today 275:59–65
95. Guo H, Li S, Peng F, Zhang H, Xiong L, Huang C, Wang C, Chen X (2014) Catal Lett 145:620–630
96. Ouyang B, Xiong S, Zhang Y, Liu B, Li J (2017) Appl Catal A-Gen 543:189–195
97. Bai S, Shao Q, Wang P, Dai Q, Wang X, Huang X (2017) J Am Chem Soc 139:6827–68300

Electrochemical and Photoelectrochemical Transformations of Aqueous CO$_2$

Aubrey R. Paris, Jessica J. Frick, Danrui Ni, Michael R. Smith and Andrew B. Bocarsly

Abstract
This chapter covers the electrochemical and photoelectrochemical conversion of CO$_2$ in aqueous media. It is divided into sections that consider heterogeneous electrocatalysts on metal electrodes, homogeneous catalysts interacting with metal surfaces, light-driven semiconductor electrodes, and hybrid systems that combine heterogeneous interfaces with surface-confined molecular components.

7.1 Introduction

It has been suggested that CO$_2$ may be a "drop in" chemical feedstock replacement for petroleum and natural gas. To fulfill that role, it must be possible to produce both fuels and bulk organic starting materials from CO$_2$. One major challenge to adopting this vision is that CO$_2$ represents a highly oxidized form of carbon, whereas fossil fuels are primarily composed of highly reduced carbon compounds. Thus, converting CO$_2$ into a "replacement fossil resource" involves the development of a rich and cost-effective reduction chemistry. Cathodic electrochemical processes provide a viable route to achieving that goal.

To avoid further consumption of fossil fuels and the generation of the undesirable greenhouse gas CO$_2$, the transformations of interest must be driven by an alternate energy source. While any alternative energy resource that allows for efficient production of electricity is viable in this regard, solar energy-derived electricity provides a very appealing mimic of the geobiological processes that gave rise to our fossil fuel reserves. To that end, two different electrochemical config-

A. R. Paris · J. J. Frick · D. Ni · M. R. Smith · A. B. Bocarsly (✉)
Department of Chemistry, Princeton University, Princeton, NJ 08544, USA
e-mail: bocarsly@princeton.edu

© Springer Nature Switzerland AG 2019
M. Aresta et al. (eds.), *An Economy Based on Carbon Dioxide and Water*,
https://doi.org/10.1007/978-3-030-15868-2_7

urations are conceivable: (1) a classic metal electrode-based electrochemical cell coupled to a photovoltaic panel or (2) a photoelectrochemical cell consisting of one or two semiconductor electrodes directly immersed in an electrolyte as shown in Fig. 7.1. Both systems can be said to perform *artificial photosynthesis*, although that term has traditionally been applied to systems that specifically split H_2O to generate H_2 and O_2. More recently the term *solar fuels* has been used to describe light-driven electrochemistry, be it water splitting or CO_2 reduction. However, this term ignores the fact that a major motivation for electrochemically reducing CO_2 is the formation of feedstocks for organic synthesis. Thus, a more creative term that

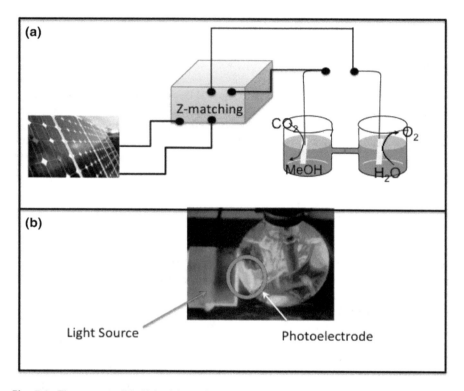

Fig. 7.1 The two possible light-driven electrochemical devices for the conversion of CO_2 to organic products. **a** A photovoltaic system coupled to an electrochemical cell composed of two half cells employing metallic electrodes. This system requires coupling electronics (z-matching) to align the impedance of the two devices so that both are operating at or near their maximum power efficiency points. **b** A photoelectrochemical cell (PEC) composed of a photocathode and a metallic anode in this example. The particular cell shown utilizes a p-GaP photocathode and was demonstrated to convert CO_2 to methanol with 96% Faradaic efficiency under illumination with 450 nm light. The "one pot" design of a PEC minimizes the need for external electronics, but forces a compromise on the photoelectrode material since it must meet both the optical and electrochemical demands of the system

seems to capture the overall intent of studies in this area is *reverse combustion*, coined by Biello [1] and supported by the equation:

$$m\text{CO}_2 + n\text{H}_2\text{O} \xrightarrow{\text{electrons}} \text{C}_x\text{H}_y\text{O}_z \qquad (7.1)$$

The debate about whether an electrochemical cell/photovoltaic system is preferred over a photoelectrochemical cell has raged for several decades, which suggests there is no true answer to this question. This is the case, in part, because the answer lies in the specifics of the system's application, location, and quality of illumination. On the one hand, the two-component system allows for independent optimization of the electrolyzer and the photovoltaic panel. However, infrastructure must be provided to couple the two systems such that they both operate near their maximum power. Use of a photoelectrochemical cell avoids infrastructure overhead but does so at the cost of not allowing optimization of both the optical and charge transfer properties of the device. Given the competitive nature of these two approaches, this chapter considers the chemistry of both metal electrode-based electrochemical cells and semiconductor-based photoelectrochemical cells. One distinction that bears noting from the outset is that semiconductor-based devices are subject to photoinduced degradation, which often limits the practical operation of such systems. While stability of metal electrodes can be a challenge, corrosion in pure metal systems is typically easier to handle than photodecomposition in semiconductors; this has hindered commercialization of large-scale photoelectrochemical cells.

While a commercial electrolyzer is a two-electrode device (Fig. 7.1a) in which all internal resistances must be minimized by careful design, research-scale electrochemical cells typically add a third electrode (actually an electrochemical half-cell) which allows measurement of the working electrode potential with minimal error due to cell impedance. Since the focus of this chapter will be the reduction process given by (7.1), the working electrode of interest is the cathode. Aqueous CO_2 reduction potentials are often measured against a Ag/AgCl/4 M KCl reference half-cell, although data are often reported with respect to the reversible hydrogen electrode (RHE, a pH-corrected value for the reduction of H^+ to H_2). In some cases, the saturated calomel electrode (SCE, $\text{Hg}_2\text{Cl}_{2(s)} + 2\text{ e}^- + $ (sat'd) KCl $\Leftrightarrow$ $2\text{ Hg}^0_{(l)} + $ (sat'd) KCl) is employed. Values for these reference systems as measured against a standard hydrogen electrode (SHE) are available in standard redox potential tables; however, for simplicity's sake, all systems discussed here will be referenced to Ag/AgCl.

Recall that the potential of the working electrode and voltage dropped across the two electrodes of an electrolyzer are different values. The total cell voltage not only accounts for the potential of the cathode, but also the potential of the reaction occurring at the anode and the total resistive losses of the electrochemical cell.

The operation of a semiconductor-based photoelectrochemical (PEC) cell (Fig. 7.1b) introduces further complexity. Unlike the electrochemical cell, where the potential applied to the working electrode is dropped across the electrolyte side (double layer) of the electrode-electrolyte interface, the potential in a PEC cell falls across both the double layer and the semiconductor's space charge region. The theoretical and operational implications of this physics are well-described by White et al. [2] for the interested reader.

The important aspect of the semiconductor-electrolyte interface to consider during the present discussion is that the potential of a metal electrode is set by an external power supply, whereas the potential of a semiconducting electrode is set by both the material employed as the semiconductor and the composition of the electrolyte. Changing the applied potential of an ideal semiconductor will impact the efficiency of charge photogeneration and transport across the semiconductor-electrolyte interface. However, it will not vary the electrode potential of the electrons transiting the interface. Thus, the CO_2 reduction products observed in a photoelectrochemical experiment are expected to be more dependent on the electrode material than the potential at which the electrode is held. In contrast, the applied potential at a metal electrode is a critical parameter influencing product distribution. The basis for these statements is the physics of an *ideal* semiconductor-electrolyte interface. When dealing with *real* interfaces, however, examples of potential-dependent product distributions have been noted. The source of this phenomenon is still under investigation.

An important takeaway from this discussion is that regardless of whether the CO_2-reducing electrode is semiconducting or metallic, the thermodynamics of the reduction process have little to do with the observed products; rather, the kinetics of interfacial charge transfer tend to control the observed chemistry. As such, the electrocatalytic nature of the electrode-electrolyte interface is a major factor in the development of CO_2 electrochemistry. While not the only factor (i.e., the activity and flux of protons at the electrode as well as the kinetics and thermodynamics of proton coupled electron transfer (PCET) must also be considered), it sufficiently dominates the current understanding of CO_2 electrochemistry that it is the focus of this chapter.

Accordingly, several electrocatalytic routes are imaginable. The electrode surface itself may be catalytic. For example, we have noted that while a p-$CuFeO_2$ photocathode is effective in the conversion of aqueous CO_2 to formate [3], the p-$CuRhO_2$ analog only reduces H_2O to H_2 when administered in an aqueous CO_2 electrolyte [4]. If an electrode surface does not exhibit catalytic properties, introduction of an independent heterogeneous catalyst onto the electrode surface may be effective. This was observed with In-based systems. Although an In electrode is not particularly catalytic toward CO_2 reduction, addition of a thin layer of In hydroxide onto the surface creates an interface that is highly efficient for the conversion of aqueous CO_2 to formate [5, 6].

Alternatively, an electrode interface that is poorly catalytic can be accommodated by using a homogeneous catalyst. For example, the complexes [$ClRe(CO_3)$ bpy] [7] and [$BrMn(CO_3)bpy$] [8] can be reduced at a glassy carbon electrode to generate an electrocatalyst that converts CO_2 to CO via a mediated charge transfer mechanism. One-electron reduction of the complex removes the chloride ligand, freeing a coordination site for CO_2. A second charge transfer leads to the two-electron reduction of CO_2 to CO. Another homogeneous catalyst that has received attention is protonated pyridine, which has been shown to catalyze the reduction of CO_2 to methanol at various electrodes. In this case, the mechanism is less well understood. Regardless, an aspect of CO_2 chemistry that must always be considered is the need for H^+, even when the product is CO. Therefore, CO_2 electrochemistry always competes with the reduction of H^+ or H_2O to H_2. Whether a homogeneous or heterogeneous electrocatalyst is employed, the system must inhibit H_2 evolution. Thus, the role of the electrocatalyst is not only to deliver H^+ to the reductive carbon center (i.e., PCET), but it must also obstruct putative hydrogen atom coupling processes.

Given the mechanistic and operational requirements outlined here, this chapter is divided into sections that consider heterogeneous electrocatalysts on metal electrodes, homogeneous catalysts interacting with metal surfaces, light-driven semiconductor electrodes, and hybrid systems that combine heterogeneous interfaces with surface-confined molecular components.

7.2 Electrochemical CO_2 Reduction

7.2.1 Heterogeneous Catalytic Systems

Heterogeneous electrocatalysts, solid state materials that comprise the surface of a working electrode in an electrochemical cell, have been studied widely for applications in CO_2 reduction. The CO_2-reducing abilities of single-metal electrodes have been known for decades [9–11]. More recent work in this area has focused on manipulating catalyst morphology, electrode topology, and surface composition (i.e., oxide(s), etc.) to elicit improved electrochemical responses toward CO_2. Over the past five years, one of the more popular goals for electrocatalyst improvement has been to increase either the Faradaic efficiency (η_F) or selectivity of the reduction reaction:

$$\eta_F = \frac{(\text{moles of specific product}) \times (n)}{(Q/F)} \tag{7.2}$$

In this equation, n is the number of electrons stoichiometrically required to make the selected product, Q is the quantity of charge passed in coulombs, and F is Faraday's constant (96,486 C/mol). The Faradaic efficiencies of all products generated by an electrochemical system should sum to 100%, accounting for all the charge passed by the system. For many metal-based electrodes active in aqueous CO_2 reduction, some portion of the charge passed contributes to the competing H_2 evolution reaction, which consumes electrons by reducing H^+ or H_2O to H_2 [12]. It is therefore necessary to reduce competition from H_2 evolution and, instead, increase Faradaic efficiencies for one or more carbon-containing reduction products.

7.2.1.1 Copper

The vast majority of heterogeneous CO_2 electroreduction systems explored in recent years use Cu electrodes as catalysts due to Cu's unique ability to transform CO_2 into multi-carbon products [13]. Though the reactivity of Cu electrodes has been known since the late 1970s, in 2012, Kuhl et al. reported the use of a Cu foil electrode to achieve 16 distinct reduction products, each containing one to three carbon atoms, at varying efficiencies [14]. This and similar reports launched a massive effort to understand and optimize Cu-based systems by improving selectivity, lowering overpotential, and creating large quantities of multi-carbon products, such as ethylene or ethanol. A popular strategy for achieving these goals and increasing Faradaic efficiencies for carbon-containing products has been morphological tuning of the Cu catalyst.

Of all morphological variations, nanostructuring has proven to be the surface-altering method of choice. Cu nanoparticles (7 nm) have been shown to generate methane selectively at 80% Faradaic efficiency using an applied potential of nearly -1.5 V versus Ag/AgCl, representing an improvement in both efficiency and selectivity compared to Cu foil [15]. Other morphologies have facilitated Faradaic efficiency increases for two-carbon products. For instance, by altering the pore sizes of Cu foam electrodes, Dutta et al. preferentially generated two-carbon products over one-carbon products at -0.9 V versus Ag/AgCl. Using a pore diameter of 50–100 μm, the authors calculated a total Faradaic efficiency of 50% for ethane and ethylene and only 25% for CO; the remainder of the charge passed contributed to H_2 evolution [16]. Similarly, Ren and colleagues showed that adjusting the thickness of Cu_2O thin films, which are reduced to Cu(0) during electrolysis, impacts the Faradaic efficiencies for ethylene and ethanol, their two major carbon-containing products, which reach maxima at 39 and 16%, respectively [17]. In some cases, thin film thickness has proven more consequential than crystal orientation in influencing product efficiencies [18].

Besides improving Faradaic efficiencies for products, morphology tuning has been used to adjust *selectivity* for one product—or one type of product—over others, which is significant due to the wide range of products possible using Cu electrodes. For example, nanostructured Cu catalysts have been shown in several instances to alter CO_2 reduction product distribution compared to simple Cu foils [15, 16, 19]. The term "selectivity" can also describe explicitly favoring CO_2 reduction over H_2 evolution; this outcome was observed by Loiudice and coworkers when tuning the size of cube-shaped Cu nanocrystals operated at -1.3 V versus Ag/AgCl. Using an optimal 44-nm edge length, Faradaic efficiencies of 41 and $\sim 20\%$ were achieved for ethylene and methane, respectively, and many minor carbon-containing products were generated, as well [20]. Selectivity for highly reduced products (i.e., reduction by six or more electrons) over CO, formate, or H_2 has been attained using nanoparticles of varying sizes. By increasing the diameter of their Cu nanoparticles, Reske et al. achieved higher methane and ethylene Faradaic efficiencies compared to CO and H_2, which they attributed to the availability of high-coordination active sites on progressively larger nanoparticles (Fig. 7.2) [21]. In sum, nanostructuring trends have manifested in morphologically

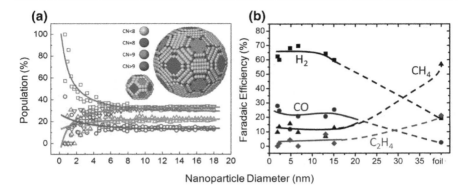

Fig. 7.2 Correlation between diameter, surface atom coordination numbers, and CO_2 reduction activity for Cu nanoparticles. **a** Percent coverage by high-coordination number surfaces increases with nanoparticle diameter, as shown graphically as well as pictorially (via comparison of 2.2- and 6.9-nm spheres). **b** Faradaic efficiencies for H_2 (squares), CO (circles), methane (triangles), and ethylene (diamonds) also depend on nanoparticle size. Adapted with permission from [21]. Copyright 2014 American Chemical Society

distinct systems exhibiting notable selectivity for one- versus multi-carbon reduction products [22, 23].

Aside from morphology, both synthetic (i.e., during catalyst preparation) and electrochemical conditions have been implicated as factors affecting selectivity during CO_2 reduction. In terms of catalyst synthetic techniques, many researchers have noted that oxide-derived Cu catalysts, synthesized by reducing Cu oxide precursors [24], exhibit improved efficiencies for multi-carbon products. One such example is Ma et al.'s report of achieving, amongst other compounds, propanol at around 10% Faradaic efficiency using oxide-derived Cu nanowires operated at −1.3 V versus Ag/AgCl [25]. To determine the origin of this enhanced electrocatalysis, Huang and colleagues tested both oxide-derived Cu electrodes and single-crystal variants and noted a correlation between the onset potentials for and electrogenerated quantities of CO and ethylene. This led them to propose that oxide-derived Cu enhances dimerization of surface-adsorbed CO, thereby resulting in higher Faradaic efficiencies for ethylene and related multi-carbon species [26].

Unsurprisingly, electrochemical conditions have proven just as consequential in dictating product distribution and selectivity. Electrode potential, for example, is known to influence whether one- or multi-carbon products will be generated on Cu electrodes. Oxide-derived Cu nanowires have been shown to switch from producing almost entirely CO (61.8%), H_2 (23.5%), and formate ($\sim 8\%$) at low potentials to generating significant quantities of two-carbon products ($\sim 25\%$) at high potentials [27]. Furthermore, electrolyte identity, concentration, and buffer capacity (implicating local pH effects) have all been reported to impact product selectivity and rates of formation on Cu-based electrodes [28–31]. Overall, progress toward achieving high Faradaic efficiencies for multi-carbon compounds using Cu electrode systems has been summarized by Gattrell et al. [32]. Figure 7.3 depicts some

Fig. 7.3 Summary of recent Cu-based electrode systems capable of generating multi-carbon products at a total Faradaic efficiency of 50% or higher. Two- (red), three- (yellow), and four-carbon (blue) products have been produced using a range of Cu morphologies and applied potentials

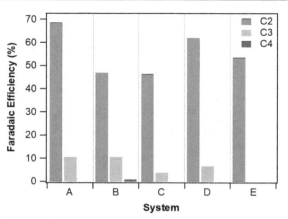

System	Structure	Applied Potential (V vs. Ag/AgCl)	Reference
A	Cu(711)	−1.6	Hori 2002 (13)
B	Cu₂O-Cu	−1.8	Lee 2015 (33)
C	Cu nanocubes	−1.3	Loiudice 2016 (20)
D	Oxide-derived Cu	−1.2	Lum 2017 (34)
E	Cu₂O nanoneedles	−1.4	De Luna 2018 (35)

of the most effective Cu catalysts established in the past five years, where systems were selected based on ability to generate a total Faradaic efficiency of 50% or more for all multi-carbon products.

Hori et al.'s 2002 report of electrocatalysis using a series of Cu single crystals has become an early benchmark for highly reduced products from CO_2; specifically, an applied potential of −1.6 V versus Ag/AgCl at the Cu(711) surface provided the best Faradaic efficiencies of 68.6 and 10.4% for all two- and three-carbon products, respectively [13]. Since then, several systems have broken the 40% Faradaic efficiency threshold for two-carbon products. Lee and coworkers employed a considerably negative potential (−1.8 V vs. Ag/AgCl) with a biphasic Cu_2O-Cu electrode to achieve 10.6% total Faradaic efficiency for three-carbon products and even witnessed trace quantities of the four-carbon molecule butane [33]. Cu nanocubes [20], oxide-derived Cu [34], and Cu_2O nanoneedles [35] have also proven effective at generating primarily two-carbon products from CO_2 at high Faradaic efficiencies.

7.2.1.2 Transition Metals

Despite the popularity of Cu electrode systems, progress has been made in reducing CO_2 using alternative transition metals, as well. As a general rule, transition metal electrodes either generate CO or are incapable of reducing CO_2, instead favoring the H_2 evolution reaction [36], though recent work has indicated that most transition metals can produce small quantities of other one-carbon compounds [37]. Nonetheless, prolific CO-generating electrodes have been the predominant subjects of current transition metal literature.

One such transition metal electrocatalyst is Ag, whose product distribution in CO_2 reduction is highly dependent on the electrochemical environment (e.g., local pH) [38]. Like Cu, however, the morphologies of Ag catalysts have also attracted attention for their ability to influence CO_2 electrochemistry at the electrode surface. Nanostructuring, in particular, has been shown to increase CO Faradaic efficiencies compared to values achieved using Ag foil [39, 40]. This efficacy has been attributed to the higher percentage of reactive crystal faces found on nanostructured surfaces compared to polycrystalline catalysts. For example, Peng et al. witnessed 96.7% Faradaic efficiency for CO using (110)/(100)-oriented Ag nanoparticles at −0.9 V versus Ag/AgCl applied potential [41]. At −1.1 V versus Ag/AgCl, Liu and coworkers achieved a similar Faradaic efficiency (96.8%) with an electrocatalyst comprised of triangular, (100)-dominant Ag nanoplates [42]. As depicted in Fig. 7.4, different crystal faces require different activation energies to attain critical surface-bound intermediates, impacting the overpotentials needed to produce CO [43].

Nanostructuring of Au electrodes yields similar results to those reported for Ag systems. Applying just −0.7 V versus Ag/AgCl, Feng and colleagues achieved nearly 100% Faradaic efficiency for CO using a catalyst comprised of Au nanoparticles supported on carbon nanotubes. The authors found that CO_2

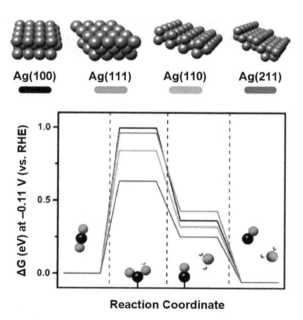

Fig. 7.4 Free energy diagram for CO_2 reduction to CO on flat (i.e., (100)/black and (111)/red) and edge (i.e., (110)/green and (211)/blue) Ag surfaces, which are portrayed at the top with gray spheres. The molecular configuration of each reaction step is depicted at the bottom of the free energy diagram, where black, red, and white spheres represent C, O, and H atoms, respectively. Adapted with permission from [43]. Copyright 2015 American Chemical Society

reduction activity was linearly correlated with the density of catalyst grain boundaries [44]. Similarly, ultrathin Au nanowires (2 nm wide) attained 94% Faradaic efficiency for CO at a remarkably low potential of −0.56 V versus Ag/AgCl. This activity was attributed to the catalyst's high density of reactive edge sites which purportedly exhibit weak CO binding, thereby facilitating product release [45]. Calculations suggest that low-coordinate sites favor H_2 evolution over CO_2 reduction. This has been observed experimentally by comparing CO_2 reduction selectivity with increasing Au nanoparticle size (i.e., increasing coordination number) [46].

Despite their prevalence, developments in transition metal-mediated CO generation have not been limited to coinage metal catalysts. Pd nanoparticles (3.7 nm) supported on carbon black were shown to generate CO at 91.2% Faradaic efficiency at −1.1 V versus Ag/AgCl [47]. At −1.3 V versus Ag/AgCl, electrochemically synthesized, nanostructured Zn dendrites achieved 79% Faradaic efficiency for the same product [48]. In both cases, the morphology-tuned catalysts facilitated CO_2 electroreduction more efficiently than their bulk electrode analogs. Cobalt oxide (Co_3O_4) thin films have proven effective in CO_2 reduction to formate (64.3%) at −0.84 V versus Ag/AgCl [49]. This system is an outlier, however, as transition metal catalysts that generate non-CO products remain elusive.

7.2.1.3 Heavy Post-transition Metals

While transition metal-based electrodes are predominantly known for reducing CO_2 to CO, the heavy post-transition metals In, Sn, Pb, and Bi are effective formate generators. Despite their close proximity to one another on the periodic table, however, these electrodes achieve formate via different mechanisms. Using infrared spectroelectrochemistry, Bocarsly and coworkers determined that In and Sn electrodes exhibit metastable surface oxides that serve as the catalytically active species during CO_2 reduction to formate. These surface oxides play a direct role in the reduction mechanism, allowing for CO_2 binding as a bicarbonate/carbonate species [5, 6, 50, 51]. In contrast, Pb's surface oxide does not interact directly with CO_2 and instead influences the electrochemistry indirectly by buffering the solution at the interface. Bi-mediated CO_2 reduction does not appear to be dependent on a surface oxide in any capacity [52].

Of the heavy post-transition metals, oxidized Sn has been the subject of most recent studies. Keeping with the trend, researchers have successfully improved formate Faradaic efficiencies by using nanostructured SnO_2 electrodes, such as Li et al.'s SnO_2 nanosheets on carbon cloth whose Faradaic efficiency for formate is 87% [53]. SnO_2 nanowires, exhibiting high grain boundary density [54], and graphene-bound nanocrystals, with tunable diameters for optimal electroreduction [55], were reported to generate formate at 78 and 93% Faradaic efficiency, respectively, though the latter system required a sizeable applied potential of −1.8 V versus Ag/AgCl. SnO_2 electrocatalysts have also been studied widely as core-shell structures, in which a SnO_2 shell covers a support material. Core materials have ranged from Cu nanoparticles to Ag-Sn alloys, but Faradaic efficiencies for formate have been shown to depend on thickness [56] and other qualities of the SnO_2 shell. For instance, Luc and coworkers found that CO_2 reduction efficacy was largely dictated by the number of oxygen vacancies on the metal oxide (101) surface [57].

It should be noted that morphology has been explored using systems comprised of other heavy post-transition metals. For instance, White et al. studied In and In oxide nanoparticles and found that $\sim 100\%$ Faradaic efficiencies for formate could be achieved at lower applied potentials than those required for bulk In electrodes [6].

7.2.1.4 Alloys and Mixed-Metal Systems

Recent studies have started to expand the repertoire of CO_2-reducing electrocatalysts to include alloys and mixed-metal systems. As portrayed by Hansen et al. (Fig. 7.5), the presence of two (or more) different metal atoms at the surface of an electrode could alter the binding configuration of intermediates, lower activation barriers for energetically taxing reaction steps, or otherwise alter the stability of

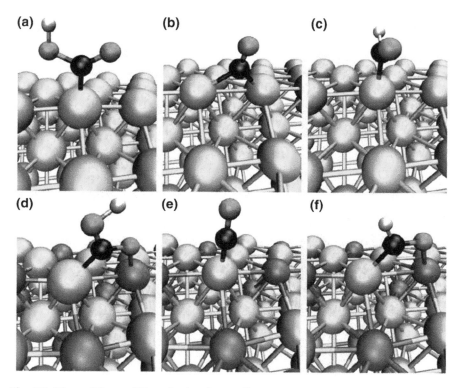

Fig. 7.5 Three different CO_2 reduction intermediates (a/d = COOH, b/e = CO, c/f = CHO) adsorbed at the surface of Au(211) (top) versus $Au_3Zn(211)$ (bottom). Alloyed surfaces such as Au_3Zn introduce different binding configurations compared to single metals. Yellow spheres correspond to Au, purple to Zn, black to C, red to O, and white to H. Reproduced from [58] with permission from the PCCP Owner Societies

reactants, intermediates, or products during CO_2 reduction [58]. As such, bimetallic surfaces have started to be tested as CO_2 electrocatalysts. Likely due to the highly reduced, multi-carbon products derived from Cu electrodes, alloy electrocatalysis research has focused on systems that include Cu as one of two metal components. For example, a phase-separated (i.e., not alloyed) Cu-Pd system was reported to achieve Faradaic efficiencies of 48% for ethylene and $\sim 15\%$ for ethanol, alongside CO at 20%, when held at a potential of -0.96 V versus Ag/AgCl [59].

Interestingly, however, alloying Cu with other metals usually eliminates Cu's characteristic ability to generate multi-carbon products, instead producing CO at >80% Faradaic efficiency. In general, atomic-level ordering within these two-metal systems has been shown to improve selectivity for CO_2 reduction to CO over H_2 evolution [59, 60]. Some of the best CO Faradaic efficiencies for two-metal systems have been achieved by alloying Cu with heavy post-transition metals. Specifically, electrodeposition of Sn [61] or In [62] onto Cu eliminates hydrocarbon formation in the potential range of -0.8 to -0.9 V versus Ag/AgCl and yields CO at Faradaic efficiencies of 90 and 95%, respectively.

When considering alloy and mixed-metal systems that do not include a Cu component, fewer catalysts have been reported. These systems are summarized in Fig. 7.6. For example, Kortlever et al. performed CO_2 electroreduction on carbon black-supported $Pd_{70}Pt_{30}$ nanoparticles and achieved 88% Faradaic efficiency for formate at only -0.6 V versus Ag/AgCl. Perhaps more interesting than this high Faradaic efficiency is the fact that neither of the alloy's constituent metals can facilitate CO_2 reduction to formate [63].

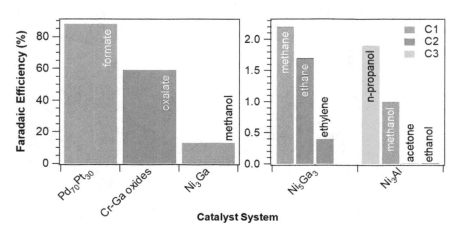

Fig. 7.6 Summary of recent non-Cu-containing bimetallic catalysts capable of reducing CO_2 to products other than CO. Systems are slowly being discovered that surpass one-carbon product (blue) generation, as a handful of two- (red) and three-carbon (yellow) products have now been reported. Catalyst system references: $Pd_{70}Pt_{30}$ [63], Cr-Ga oxides [64], Ni_3Ga [69], Ni_5Ga_3 [68], Ni_3Al [70]

Another system exhibiting impressive stability and unexpected chemical selectivity is the mixed-metal Cr-Ga system comprised of metal oxides [64]. Thin films of this catalyst supported on glassy carbon were shown to generate the two-carbon product oxalate at 59% Faradaic efficiency with an applied potential of -1.48 V versus Ag/AgCl. The film itself remained electrocatalytic for a period of at least ten days [64]. This system is particularly interesting because the applied potential and aqueous electrochemical environment eliminate the possibility of passing through a CO_2^- intermediate, a mechanistic route which had long been accepted as the sole means of achieving the electrochemical transformation of CO_2 to oxalate [65, 66]. As such, the Cr–Ga system provides the first demonstration of CO_2 reduction to oxalate in aqueous solution while simultaneously exhibiting a competitive Faradaic efficiency for a two-carbon product.

Two-carbon products have also been achieved using Ni–Ga intermetallic catalysts, which were first implicated as CO_2 reduction candidates in non-electrochemical, high-temperature CO_2 hydrogenation studies [67]. As a thin film on carbon electrodes, Ni_5Ga_3 has been shown to produce low efficiencies for methane ($\sim 2.2\%$), ethane ($\sim 1.7\%$), and ethylene (trace) at -1.1 V versus Ag/AgCl. Onset potentials for two-carbon products were more than 250 mV more positive than those required on polycrystalline Cu [68]. The Ni–Ga on carbon system has been shown to demonstrate extreme morphological dependence, switching from one dominant carbon-containing product to another depending on electrode surface morphology. Strategic, combinatorial use of different Ni_3Ga morphologies was shown to result in methanol Faradaic efficiencies up to 13.2% [69]. These observations regarding Ni–Ga have inspired examination of similar alloys for electrocatalytic ability. An example is the Ni_3Al intermetallic, whose thin film was used to generate propanol at 1.9% Faradaic efficiency, among other minor liquid products, as well as CO at 33%. Ni_3Al thin films represent one of the first non-Cu-containing systems to facilitate CO_2 electroreduction to three-carbon products [70].

7.2.2 Homogeneous Catalytic Systems

Homogeneous electrocatalysts are molecules dissolved in the electrolyte that interact with the working electrode to facilitate CO_2 reduction. Homogeneous catalyst systems have shown promise in reducing CO_2 to CO, as well as formate [71, 72], oxalate, and various alcohols [73]. These catalysts are often, though not exclusively, comprised of one or more metal centers bonded to various chemical groups, or ligands. Researchers interested in modern design of CO_2-reducing electrocatalysts have tended to focus on the catalytic implications of ligand manipulation [74].

7.2.2.1 Tricarbonyl Bipyridyl Complexes and N-Heterocyclic Carbenes

Re and, more recently, Mn tricarbonyl bipyridyl complexes (i.e., fac-[M(bpy) $(CO)_3X$] (X=Br, Cl)) have been shown to electrochemically reduce CO_2. Since their discovery in 1984 by Hawecker et al., Re complexes have been extensively studied as CO_2-reducing electrocatalysts [7, 75–77]. Mn analogs of fac-[Re(bpy)(CO)$_3X$] have been the focus of several recent studies due to the elemental abundance of Mn (compared to Re) and the observation that these two chemically similar systems have several mechanistic distinctions as CO_2 electrocatalysts. Bourrez et al. first demonstrated that fac-[Mn(bpy)(CO)$_3$Br] could reduce CO_2 to CO with comparable selectivity and significantly lower overpotential (400 mV) compared to Re tricarbonyl bipyridyl systems [78]. Unlike the Re systems, experiments using Mn-based catalysts have required a proton source in order to catalytically reduce CO_2. While H_2O is typically a suitable proton source, recent investigations have shown that other Brönsted acids such as methanol, phenol, and trifluoroethanol can serve as viable alternatives [79].

The CO_2 reduction mechanism of these systems has been studied extensively using cyclic voltammetry and bulk electrolysis techniques. fac-[Mn(bpy-R) $(CO)_3$Br] (R=H, tBu, Me) typically exhibits two non-reversible, one-electron reduction waves separated by 200–300 mV. The first wave has been associated with a charge transfer that induces dehalogenation of the initial fac-[Mn(bpy-R) $(CO)_3$Br], yielding a coordinately unsaturated species that follows a bifurcated reaction pathway. Most of the one-electron reduction product undergoes a second reduction which results in coordination of CO_2 to the metal center. However, a significant fraction of the coordinately unsaturated species dimerizes to form fac-[Mn(bpy-R)(CO)$_3$]$_2$ [78–80].

Sampson et al. have shown that adding sterically taxing mesityl (Mes) substituents at the 6 and 6′ bipyridine positions leads to significant changes in the mechanism of active catalyst formation. The second reduction wave disappears, and the complex instead exhibits a quasi-reversible, two-electron reduction wave at −1.40 V versus Ag/AgCl (Fig. 7.7). This change in redox behavior is attributed to the mesityl substituents hindering formation of the Mn–Mn dimer complex commonly observed for other fac-[Mn(bpy-R)(CO)$_3$Br] analogs. Instead, the sterically hindered fac-[Mn(Mes-bpy)(CO)$_3$Br] undergoes rapid dehalogenation and subsequent reduction to produce fac-[Mn(Mes-bpy)(CO)$_3$]$^-$ at a potential 300 mV more positive than fac-[Mn(bpy-R)(CO)$_3$Br]. The mesityl version of the catalyst was determined to selectively reduce CO_2 to CO with a Faradaic efficiency of 76% at −1.80 V versus Ag/AgCl [80].

Further research in Mn bipyridyl catalyst design has sought to understand the mechanism of CO_2 reduction from both electronic and proximity-assisted stabilization effects of the Mn–CO adduct [81–83]. Advances in proximity-assisted stabilization were achieved by Agarwal et al. with the inclusion of a hydrogen bond-donating phenol at the 6-position of the bipyridyl ligand (Fig. 7.8) [81]. This group within the fac-[Mn(HOPh-bpy)(CO)$_3$Br] complex is closely positioned to the CO_2 binding site, allowing for proton-assisted stabilization of the rate-determining

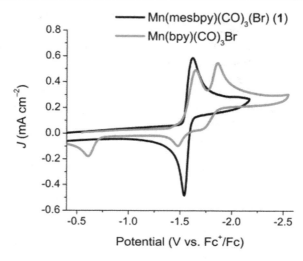

Fig. 7.7 Cyclic voltammograms reported for *fac*-[Mn(Mes-bpy)(CO)₃Br] (black) and [Mn(bpy)(CO)₃Br] (red). Experiments were performed under N_2 atmosphere using 1.0 mM catalyst in acetonitrile with 0.1 M TBAPF₆ as the supporting electrolyte. Glassy carbon, Pt wire, and Ag wire were used as the working, counter, and pseudo-reference electrodes, respectively, with addition of Fc as an internal reference. Scan rate = 100 mV/s. Adapted with permission from [80]. Copyright 2014 American Chemical Society

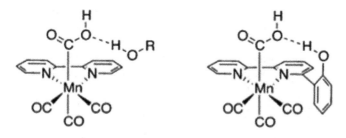

Added Brönsted acid:
Intermolecular H-bond

H-bond donating group on bipyridine:
Intramolecular H-bond

Fig. 7.8 CO_2 stabilization in *fac*-[Mn(HOPh-bpy)(CO)₃Br] is achieved using intramolecular hydrogen bonding (right), while *fac*-[M(bpy)(CO)₃Br] elicits stabilization using a Brönsted acid (left). Adapted with permission from [81]. Copyright 2015 American Chemical Society

C–O bond cleavage step and resulting in a seven times larger current enhancement compared to *fac*-[Mn(bpy)(CO)₃Br].

Related investigations by Franco et al. expanded on this work by determining that selectivity for CO versus formate using *fac*-[Mn(pdbpy)(CO)₃Br] (pdbpy = 4-phenyl-6-(phenyl-2,6-diol)-2,2′-bipyridine) and *fac*-[Mn(ptbpy)(CO)₃Br] (ptbpy = 4-phenyl-6-(phenyl-3,4,5-triol)-2,2′-bipyridine) depends on acid strength [82].

Recent investigations on the electronic effects of Mn–CO stabilization were performed by Tignor et al., who determined that inclusion of electron-withdrawing and electron-donating groups on the bipyridyl scaffold results in generation of catalytically *inactive* and catalytically *active* species, respectively [83].

In addition to the many studies concerning Mn tricarbonyl bipyridyl complexes, modifications to the coordinating atoms of *N*-heterocyclic ligands have been extensively explored. In 2014, Agarwal et al. reported the first example of *fac*-Mn(I) tricarbonyl *N*-heterocyclic carbene (NHC) complexes serving as electrocatalysts for CO_2 reduction [84]. A [Mn(N–C)(CO)$_3$Br] ((N–C)=*N*-methyl-*N*′-2-pyridylimidazolium) catalyst was shown to reduce CO_2 to CO with a Faradaic efficiency of 34.6% at significantly more positive potentials than *fac*-[Mn(bpy)(CO)$_3$Br] (−1.42 V and −1.70 V vs. Ag/AgCl, respectively). Additionally, lower reduction potentials, enhanced currents, and improved product selectivities [85–87] have been achieved for metal–NHC complexes via inclusion of π-acidic aryl [88] and electron-withdrawing substituents [89] on the NHC framework.

The photodecomposition of Mn(I) tricarbonyl *N*-heterocyclic complexes under visible light has been widely documented [90–102] and presents a host of new mechanistic challenges. Rosa et al. suggested that photochemical dissociation of the halide ligand generates a dimerized *fac*-[Mn(bpy-R)(CO)$_3$]$_2$ species [96]. Because of this, researchers have studied the effect of ligand substitution at the halogen position of *fac*-[Mn(bpy-R)(CO)$_3$Br]. Machan et al. determined through both experimental and computational studies that the strong-field nature of the cyanide ligand on *fac*-[Mn(bpy)(CO)$_3$(CN)] inhibits dimerization and forms the catalytically active [Mn(bpy)(CO)$_3$]$^-$. This anionic species is highly selective for reducing CO_2 to CO, achieving average Faradaic efficiencies of 98% for CO and 1% for H_2 at −1.90 V versus Ag/AgCl [103].

Researchers have also replaced the halogen ligand in metal–NHC complexes with strong-field options, such as cyanide and thiocyanate, in order to determine photodecomposition rates and CO_2-reducing capabilities. It was shown that cyanide-incorporated metal–NHC complexes persisted up to five times longer than the halogenated species as a result of this strong-field nature. However, metal–NHC complexes containing either cyanide or thiocyanate ligands exhibited poor Faradaic efficiencies for CO of 20–29.7 and 27%, respectively [104]. Although investigations into ligand substitution are generally geared toward inhibiting complex dimerization, Kuo et al. recently reported a cyano-bridged di-manganese complex, {[Mn(bpy)(CO)$_3$]$_2$(μ-CN)}ClO$_4$, which is both electrocatalytically and photocatalytically active toward CO_2 reduction at a potential 700 mV lower than the corresponding *fac*-[Mn(bpy)(CO)$_3$(CN)] monomer. Furthermore, the cyano-bridged dimer was found to be more resistant to destructive photodimerization compared to *fac*-[Mn(bpy)(CO)$_3$Br] [105].

7.2.2.2 Ni(Cyclam) Complexes

A common theme amongst the homogeneous electrocatalysts discussed thus far is that most metal complexes of bipyridyl and *N*-heterocyclic carbenes are catalytically active in solvent mixtures consisting of acetonitrile or dimethylformamide and 5–10%

water. In contrast, $[Ni(cyclam)]^{2+}$ (cyclam = 1,4,8,11-tetraazacyclotetradecane) has attracted interest due to its selectivity in producing CO in entirely aqueous conditions. $[Ni(cyclam)]^{2+}$ has been shown to produce CO with 99% Faradaic efficiency on a hanging Hg electrode at -1.2 V versus Ag/AgCl and 90% Faradaic efficiency on a glassy carbon electrode at -1.41 V versus Ag/AgCl [106]. The one-electron-reduced $[Ni(cyclam)]^{+}$ is a catalytically active species when adsorbed onto a Hg electrode [107]. Both Hg and glassy carbon electrodes have commonly been used in conjunction with $[Ni(cyclam)]^{2+}$ systems due to their large negative potential window in an aqueous electrolyte. Structurally, $[Ni(cyclam)]^{2+}$ predominantly exists as the *trans*-III and *trans*-I isomers (85 and 15%, respectively) in aqueous solution. Computational studies have indicated that the less-abundant *trans*-I species is the better catalyst in situations where the catalytic intermediate is not adsorbed to the electrode [108]. Accordingly, the *trans*-III species is a better catalyst in the presence of Hg electrodes because of conformational changes to the catalytic intermediate when adsorbed to the electrode surface [109].

Most recent studies of $[Ni(cyclam)]^{2+}$ electrocatalysts have focused on elucidating mechanistic details [108–110] and optimizing catalytic performance [106, 111, 112]. Notably, common H_2-producing side reactions do not occur in aqueous $[Ni(cyclam)]^{2+}$ systems containing CO_2 at pH = 5. Recent investigations by Schneider et al. have provided mechanistic insight into the catalytic selectivity for both CO and H_2 based on modifications to solution pH, as well as the catalytic implications of structural components [110].

7.2.2.3 Fe Porphyrin Complexes

Fe complexes have garnered significant interest because of their CO_2 reduction versatility; specifically, both Fe porphyrin complexes [113–122] and Fe carbonyl clusters [123–126] are capable of reducing CO_2. Seminal work from Hammouche et al. reported the first use of an Fe(0) porphyrin complex, Fe(0)TPP (TTP = tetraphenylporphyrin), as a homogeneous electrocatalyst for CO_2 reduction when coupled with a Lewis or Brönsted acid [113, 114]. The complex was found to produce CO with a Faradaic efficiency of 98% at -1.63 V versus Ag/AgCl. Ambre et al. demonstrated that the inclusion of methoxy substituents at the *ortho* and *meta* positions of the TPP ligand scaffold increased both the Faradaic efficiency and selectivity for CO production [115]. This work was later expanded by Costentin et al., who added positively charged trimethylanilium (TMA) and negatively charged sulfonate (PSULF) groups to Fe(0)TPP. Complexes with the TMA ligand in the *para* and *ortho* positions exhibited high Faradaic efficiencies for CO (92% and 93%) at -1.46 and -1.14 V versus Ag/AgCl, respectively. On the contrary, complexes with PSULF ligands in the para position showed no change in the CO_2 reduction ability compared to Fe(0)TPP [116, 117].

Costentin et al. have also shown that adding electron-withdrawing or electron-donating groups to the TPP scaffold can electronically influence catalytic activity through inductive effects [118, 119]. In a series of experiments where hexafluorinated phenyl groups were added to the porphyrin ligand, it was observed

Table 7.1 Standard CO$_2$ reduction potentials reported for Fe(0)TPP and its analogs [118] compared to Fe(0)TPP [113]

Catalyst	E° (V vs. Ag/AgCl)
FeF20TPP	−1.32
FeF10TPP	−1.47
FeF5TPP	−1.56
FeMeO8TPP	−1.92
Fe(0)TPP	−1.63

that increasing the number of fluorinated phenyl groups resulted in the complex's standard reduction potential for CO$_2$ becoming more positive compared to Fe(0) TPP. Additionally, complexes containing methoxy groups on the porphyrin scaffold exhibited a negative shift in standard reduction potential compared to Fe(0)TPP (Table 7.1). It has since been rationalized that electron-withdrawing groups inductively stabilize the Fe(0)–CO adduct intermediate, favoring a more positive overpotential [118]. Linear sweep voltammograms of the TMA, PSULF, and various fluorine-substituted Fe(0)TPP complexes are shown in Fig. 7.9.

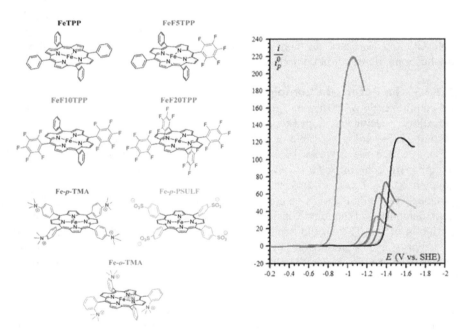

Fig. 7.9 Linear sweep voltammograms obtained for Fe-o-TMA (light blue), FeF20TPP (magenta), FeF10TPP (red), Fe-p-TMA (purple), FeF5TPP (green), FeTPP (black), and Fe-p-PSULF (orange) in the potential domain of the catalytic CO$_2$ reduction wave. Experiments were performed under 1 atm CO$_2$ using 1.0 mM catalyst in DMF with 0.1 M n-Bu$_4$NPF$_6$, 0.1 M H$_2$O, and 3 M PhOH. Glassy carbon, Pt wire, and SCE were used as the working, counter, and reference electrodes, respectively. Scan rate = 0.1 V/s. Adapted with permission from [117]. Copyright 2016 American Chemical Society

Homogeneous organometallic systems comprise a highly multidisciplinary subset of electrocatalysts that are promising CO_2 remediation solutions. An in-depth understanding of homogenous electrocatalysis requires knowledge of many subfields, such as synthesis, catalysis, electronic structure, electrochemistry, and theory. The vast array of research possibilities as a result of these subfields suggests that advancements in homogeneous electrocatalysis will be strengthened by a diverse collection of scientific perspectives.

7.2.3 Hybrid Catalytic Systems

A variety of electrocatalytic CO_2-reducing systems exist in which a molecular catalyst has been immobilized on an electrode surface. Here, such chemically modified electrode interfaces are referred to as hybrid systems.

7.2.3.1 Organic Coatings

Perhaps the simplest hybrid catalysts include layers of organic molecules coated onto solid phases; in these systems, the organic coating frequently improves the CO_2 reduction ability of the material on which it is anchored. For example, Au nanoparticles, which are effective CO_2-to-CO converters, exhibit improved Faradaic efficiencies or current densities when organic functionalities are added to their surfaces. When modified with N-heterocyclic carbenes [127] or polyethylene glycol (PEG) [128], Au nanoparticles reach 83 and 100% Faradaic efficiency for CO, respectively, at no more than −0.78 V versus Ag/AgCl. While the carbene functionalities are said to affect the CO_2 reduction mechanism, PEG is assumed to increase the concentration of CO_2 reactant at the Au surface. Similar rationales have been invoked for treating N-doped carbon nanotubes with polyethylenimine. Namely, the organic coating stabilizes the reaction intermediate while acting as a CO_2-concentrating agent, resulting in ∼80% Faradaic efficiency for formate, albeit at high applied potentials (−1.76 V vs. Ag/AgCl) [129].

7.2.3.2 Porphyrinoids

Porphyrins and their derivatives, collectively making up the "porphyrinoids," have also been immobilized on solid electrodes by methods ranging from covalent tethering to electropolymerization. The basic structures of the porphyrinoids discussed here, the majority of which generate CO from CO_2, are depicted in Fig. 7.10. Fe porphyrins derivatized with hydroxy-substituted phenyls, anchored via pyrene molecules onto carbon nanotubes, selectively generated CO at 97% Faradaic efficiency at an applied potential of −1.24 V versus Ag/AgCl [130]. Co protoporphyrins electropolymerized onto glassy carbon reached slightly lower CO Faradaic efficiencies (84%) at similar potentials (−1.2 V) [131]. Interestingly, the same Co protoporphyrins, when applied to pyrolytic graphite by dip-coating and operated at −1 V versus Ag/AgCl, generated small amounts of methane (up to 2.5%) alongside smaller CO quantities (∼50%). The distribution of CO versus methane was found to be highly pH-dependent [132].

Porphyrin	Protoporphyrin	Phthalocyanine

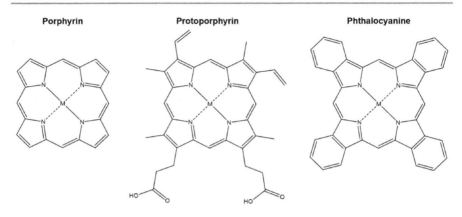

Fig. 7.10 Chemical structures of the protoporphyrins discussed here; each can be further derivatized with a variety of functional groups. M represents a metal center, which is typically required to achieve CO_2 electroreduction

Within the porphyrinoid class, the most effective CO producers have consistently been functionalized Co phthalocyanines immobilized on carbon. For instance, Morlanés et al. used a perfluorinated variant deposited onto carbon cloth to achieve 93% Faradaic efficiency for CO at approximately −1 V versus Ag/AgCl [133], while Zhang and coworkers' cyano-modified analog reached >95% Faradaic efficiency at 170 mV lower overpotential [134].

To achieve major products other than CO, more complicated strategies combining porphyrins and organic coatings have been lightly explored. As an example, Weng et al. anchored Co porphyrin onto polytetrafluoroethylene-treated carbon paper and, applying −1.18 V versus Ag/AgCl, achieved Faradaic efficiencies for methane and ethylene of ∼30 and ∼15%, respectively, compared to ∼10% for CO. These are the highest hydrocarbon Faradaic efficiencies achieved to-date using hybrid electrocatalysts [135].

7.2.3.3 Metal-Organic Frameworks (MOFs)

A recent trend in hybrid catalysis for electrochemical CO_2 reduction has been the use of porous framework complexes that incorporate metals into their otherwise-organic assemblies. These metal-organic frameworks (MOFs) [136], as well as metal-modified covalent-organic frameworks (COFs), have garnered interest due to the theory that their porous structures could help facilitate CO_2 concentration near the catalytically active metal centers.

Like many other hybrid electrodes, the majority of MOF catalysts reduce CO_2 to CO, and Co porphyrins have been employed as typical subunits to comprise MOF structures. When Co(II) ions were embedded in a classical COF structure—thereby forming the COF-366-Co catalyst—and the catalyst was subsequently applied to a carbon support, Faradaic efficiencies of 90% were attained at −0.88 V versus Ag/AgCl. Rates of reaction could be increased by seven times when COF-366-Co was grown directly onto a glassy carbon electrode. Modifying the framework structure to achieve different pore sizes also impacted reaction efficiency, but in all cases electrocatalyst stability was said to improve compared to Co porphyrin used as an electrocatalyst on its own [137]. It has been shown using the related MOF $Al_2(OH)_2TCPP$-Co that the Co(II) center is reduced to Co(I) during electrochemistry [138], much like the electropolymerized Co protoporphyrin catalyst discussed previously [131].

The MOF structure ZIF-8 has also been popularized in CO_2 reduction studies because its Zn(II) centers can be easily replaced with other CO_2-reducing metals. For example, the ion-exchange process shown in Fig. 7.11 was used to incorporate Ni(II) into the ZIF-8 framework, creating an electrocatalyst that achieved 71.9% Faradaic efficiency for CO at −1.1 V versus Ag/AgCl [139]. In a different study, ZIF-8 was subjected to pyrolysis to integrate FeN_4 centers; when added to a carbon paper electrode, this catalyst surpassed the Ni-ZIF-8 catalyst by reaching 90% Faradaic efficiency at only −0.7 V, representing a 400 mV lower overpotential. In this case, the authors claimed that isolated FeN_4 sites specifically favor CO_2 reduction over H_2 evolution, especially when compared to Fe nanoparticles [140].

Examples of MOF electrocatalysts capable of generating more-reduced products from CO_2 are very limited. However, one successful case was reported by Albo et al., who examined a handful of Cu-containing MOFs and aerogels exhibiting nano- and microscale pores, respectively [141]. While each of their catalysts, deposited onto carbon paper, could produce some quantity of alcohols, the MOF called HKUST-1 was most effective. At −0.9 V versus Ag/AgCl, HKUST-1 electrocatalysts resulted in 10.3 and 5.6% Faradaic efficiencies for ethanol and methanol, respectively. For comparison, the second best-performing electrocatalyst examined in that study, a Cu-containing aerogel, reached only 6.5 and 3.4%

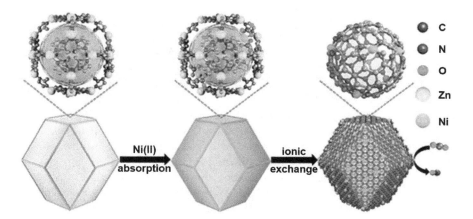

Fig. 7.11 Rhombododecahedral ZIF-8 MOF structures can be subjected to various treatments, such as ion exchange, to incorporate new metal centers that behave as single-metal CO_2 reduction electrocatalysts. Adapted with permission from [139]. Copyright 2017 American Chemical Society

efficiencies for the same products at potentials 350 mV more negative. The superiority of HKUST-1 was attributed to its accessible Cu centers, which had open coordination sites, coupled with the material's high surface area [141]. In any case, the HKUST-1 system remains a standout amongst MOF electrocatalysts due to its unique ability to generate alcohols from CO_2.

7.3 Photoelectrochemical CO_2 Reduction

7.3.1 Heterogeneous Catalytic Systems

When CO_2 reduction occurs at the interface of a light-absorbing, semiconducting material and a liquid electrolyte, the catalyst system is characterized as photoelectrochemical (PEC) and heterogeneous. Some advantages of PEC heterogeneous systems are similar to electrochemical heterogeneous systems and include small catalyst quantity requirements, high concentration of reactive sites, and facile separation of the catalyst from the reaction mixture. However, PEC systems have the added benefit of using light in addition to electricity, introducing the possibility of lowering reaction overpotentials.

7.3.1.1 Binary Semiconductors

As discussed in prior sections, elemental Cu is well known for its unique ability to electrochemically reduce CO_2 to hydrocarbon products [36]. Accordingly, researchers have used Cu as a building block for semiconductors when studying PEC CO_2 reduction. Both CuO and Cu_2O are p-type semiconductors with bandgaps of 1.4 and 2.2 eV, respectively [142]. The first report of CuO PEC behavior in aqueous media was in 1977 [7, 143] while the first PEC CO_2 reduction study involving Cu_2O was published in 1989 [144]. Since this time, a multitude of work has been performed with copper oxides, hereafter referred to as Cu_xO, for PEC reduction of CO_2. Here we provide a sampling of the results published in this field within the last five years. The reader is directed to Jánáky et al. for a more complete history of Cu_xO-mediated PEC CO_2 reduction [145].

A recent trend in Cu_xO photoelectrochemistry has involved studying the effects of Cu_xO morphology on PEC CO_2 reduction activity. Different Cu_xO morphologies can be achieved by modifying the photoelectrode synthesis and/or assembly method. Ba et al. highlighted the effect of Cu_xO morphology by preparing what they have referred to as "nanobelt array" and "stone-like" p-Cu_2O films, seen in Fig. 7.12 [146]. To classify these different morphologies, scanning electron microscopy (SEM) and high-resolution transmission electron microscopy (TEM) were employed. Figure 7.12a–d shows SEM and TEM images of the nanobelt arrays, while images of the stone-like morphology are pictured in Fig. 7.12e–f. The major hydrocarbon product achieved during PEC CO_2 reduction (-2.0 V vs. Ag/AgCl; 0.1 M $KHCO_3$; 435–450 nm irradiation) using the nanobelt and stone-like photoelectrodes was ethylene at Faradaic efficiencies of 32.69 and

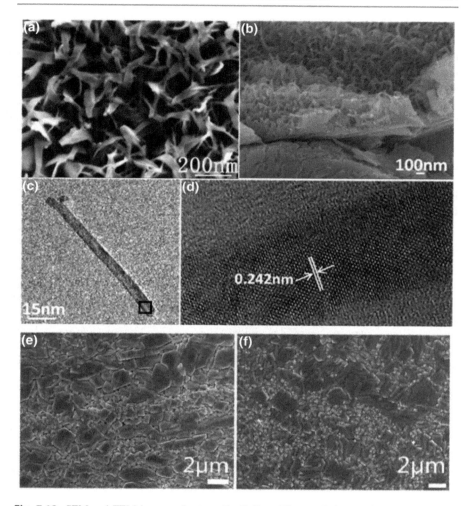

Fig. 7.12 SEM and TEM images of *p*-type Cu_2O films. The nanobelt array is shown from the **a** top view and **b** cross section, while a single nanobelt is examined at **c** low and **d** high magnification (i.e., the square box in panel **c**). Stone-like Cu_2O films of **e** *n*-type and **f** *p*-type are also shown. Adapted with permission from [146]. Copyright 2014 American Chemical Society

30.60%, respectively. However, further experimentation prompted the authors to conclude that the stone-like morphology has higher activity for PEC CO_2 reduction overall, which they attributed to this morphology's thicker Cu_2O layer relative to the nanobelt array; a similar rationale was conceived by Kanan et al. in 2012 [147].

The use of core-shell nanorod photoelectrodes is another common strategy for eliciting enhanced PEC CO_2 reduction efficiencies. In 2013, Rajeshwar et al. reported a CuO core/Cu_2O shell nanorod array that produced methanol at 95% Faradaic efficiency at −0.43 V versus Ag/AgCl; this represents an *underpotential* of approximately 800 meV relative to the thermodynamic requirement for CO_2 reduction to methanol [148]. The authors attribute this enhanced PEC performance to the energetically aligned valence and conduction band edges of the CuO/Cu_2O nanorods, facilitating generation of photoexcited electrons from both the CuO core and Cu_2O shell. The authors also mention morphological advantages of their nanorod arrays, which include large surface areas compared to other thin film morphologies.

A 2016 study by Janáky et al. also utilized this core-shell morphological approach by combining carbon nanotubes (CNT) and Cu_2O [149]. Bare Cu_2O photoelectrodes were tested alongside the CNT/Cu_2O hybrid for PEC CO_2 reduction. The overall product distributions achieved using both photoelectrodes were identical, resulting in methanol, ethanol, and formic acid as the only carbon-containing products. However, the CNT/Cu_2O hybrids exhibited a five-fold higher electrical conductivity and displayed superior performance in photoelectrolysis stability tests. It must be noted that although the synthetic assemblies are similar for the CNT/Cu_2O hybrids and the previously discussed CuO/Cu_2O hybrids, the underlying reasons for the systems' advantages are fundamentally different. Unlike the CuO/Cu_2O hybrid, the enhanced performance of the CNT/Cu_2O hybrid was attributed, in part, to the fast transport of photogenerated holes from the Cu_2O nanoparticles to the highly conductive CNT. The enhanced charge transfer rates are evident when comparing the photoactivity of the bare Cu_2O photoelectrode (red) to the CNT/Cu_2O hybrid (black) in Fig. 7.13. At the start of the chopped-light, current versus time experiment, the ratio of photocurrent from the CNT/Cu_2O hybrid to bare Cu_2O is ~ 1.9, and after four hours the ratio increases to 4.3—attesting to the hybrid's higher conductivity and photoelectrocatalytic longevity.

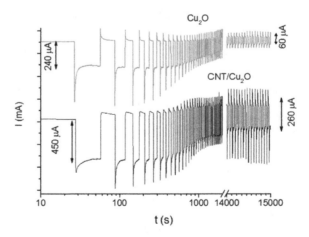

Fig. 7.13 Chronoamperometry data for bare Cu_2O (red) and hybrid CNT/Cu_2O (black). The CNT/Cu_2O photocurrent is both larger and longer-lasting compared to bare Cu_2O. Data were collected at 0.05 V vs. Ag/AgCl in CO_2-saturated, 0.1 M Na_2SO_4 under chopped simulated solar irradiation. Adapted with permission from [149]. Copyright 2016 The Royal Society of Chemistry

Early research on PEC CO_2 reduction was dominated by III-V binary semi-conductors like p-InP [150] and p-GaP [151]. The past decade, however, has witnessed an increase in the use of metal oxides for PEC CO_2 reduction, such as the Cu_xO photoabsorbers discussed above. Other binary oxides, including TiO_2, ZrO_2, and Co_3O_4, have also been explored in various morphologies to fine-tune their PEC CO_2 reduction performance. In and of themselves, TiO_2 and ZrO_2 are poor photocatalysts for CO_2 reduction due to either unfavorable CO_2 adsorption and conduction band energetics [152] or a massively insulating bandgap [153]. Nonetheless, Zanoni et al. created a photoelectrode comprised of these two wide-band gap metal oxides, and the system reduced CO_2 under illumination to produce methanol and ethanol at concentrations of 485 and 268 μ, respectively, at only -0.3 V versus Ag/AgCl (0.1 M Na_2SO_4; pH = 4; UV-vis 120 mW/cm^2) [154]. The authors attributed the PEC success resulting from pairing the oxides to effective CO_2 adsorption sites contributed by ZrO_2, which subsequently enables the TiO_2 to shuttle photoexcited electrons to CO_2 molecules for reduction [155, 156].

Similar in photoelectrode construction to the CNT/Cu_2O hybrids discussed previously are the Cu-Co_3O_4 nanotubes (NTs) reported by Zhao et al. [157], representing the combination of a photoabsorbing metal oxide and a highly conductive counterpart. However, in this case, the Co_3O_4 semiconductor forms the nanotube core and the Cu acts as a co-catalyst on the surface of the NTs (Fig. 7.14) in order to facilitate the flow of electrons into the electrolyte for CO_2 reduction. With or without the Cu co-catalyst, the photoelectrode exhibited an impressive selectivity of nearly 100% for the reduction of CO_2 to formate. After eight hours of photoelectrolysis, bare Co_3O_4 NTs produced 4.34 mM/cm^2 formate, while addition of Cu increased this yield to 6.75 mM/cm^2.

Coupled with binary oxides, the II-IV semiconductor ZnTe has been examined for its utility in PEC CO_2 reduction. In 2014 [158] and 2015 [159], Lee et al.

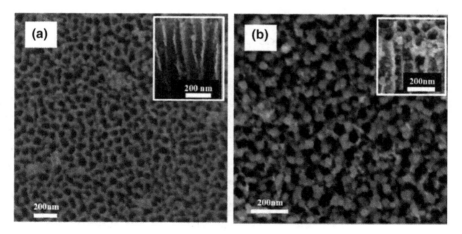

Fig. 7.14 SEM images of **a** Co_3O_4 NTs and **b** Cu/Co_3O_4 NTs. Insets portray SEM images from a side view. Adapted with permission from [157]. Copyright 2015 American Chemical Society

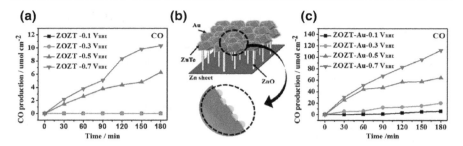

Fig. 7.15 **a** CO production at various applied potentials using ZnTe/ZnO nanowires (ZOZT). **b** Schematic of Au nanoparticles deposited onto the photoelectrode, resulting in the composite ZOZT-Au. **c** CO production at various applied potentials using the ZOZT-Au photoelectrode. Addition of the Au co-catalyst increases CO production by approximately ten times. Adapted with permission from [159]. Copyright 2015 The Royal Society of Chemistry

released two studies on PEC CO_2 reduction using photoelectrodes comprised of ZnTe/ZnO heterostructures. Their initial study reported PEC CO_2 reduction selective for CO production (-1.37 V versus Ag/AgCl; 0.5 M $KHCO_3$; medium-pressure Hg lamp, 490 mW/cm^2); however, the competing H_2 evolution reaction remained the dominant charge transfer process. These authors demonstrated that the addition of Au nanoparticles to the electrode surface dramatically improved the yield of CO from 8.7 to 64.9% Faradaic efficiency under 100 mW/cm^2 irradiation at -1.37 V versus Ag/AgCl. A schematic illustration of the optimized photoelectrode composite is shown in Fig. 7.15 along with CO production versus time plots for bare and Au-deposited photoelectrodes. The authors attributed this enhancement following addition of Au NPs not only to their highly conductive behavior—shuttling electrons from the photoabsorber to the electrolyte —but also to the CO_2 reduction reaction sites provided by the NPs themselves. However, the authors did not provide experimental data to support this claim.

7.3.1.2 Ternary Semiconductors

Within the last five years, PEC CO_2 reduction studies using ternary semiconductors have primarily focused on ternary metal oxide photoelectrodes, paying particular attention to the delafossite material p-$CuFeO_2$. In 2013, Bocarsly et al. reported PEC CO_2 reduction to formate at 400 mV underpotential with maximum PEC conversion efficiency achieved at -0.8 V versus Ag/AgCl (0.1 M $NaHCO_3$; pH = 6.8; 470 nm irradiation) [160]. However, PEC CO_2 reduction on the bare $CuFeO_2$ photoelectrode could not compete with H_2 evolution. Further, the material showed signs of slow degradation via self-reduction.

In 2015, Park et al. improved the $CuFeO_2$ photoelectrode by adding p-type CuO [161]. The mixed $CuFeO_2$/CuO photoelectrode shifted the selectivity to almost 100% Faradaic efficiency for CO_2 reduction to formate; furthermore, the photoelectrode remained stable for over one week while continuing to produce formate. Figure 7.16 depicts time-profiled changes in potential (red and blue solid lines),

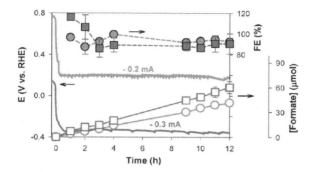

Fig. 7.16 Time-profiled changes in potential (red and blue solid lines), formate production (open symbols), and Faradaic efficiency (closed symbols) achieved using CuFeO$_2$/CuO under constant current (red circles = − 0.2 mA; blue squares = − 0.3 mA). Conditions: CO$_2$-purged, 0.1 M bicarbonate solution under simulated light (100 mW/cm^2). Adapted with permission from [161]. Copyright 2015 The Royal Society of Chemistry

formate production (open symbols), and Faradaic efficiency (closed symbols) for the CuFeO$_2$/CuO system under constant current (red circles = −0.2 mA; blue squares = −0.3 mA). The enhanced efficiency for PEC CO$_2$ reduction and photo-electrode stability is attributed to the heterojunction formed within the CuFeO$_2$/CuO structure, which is capable of absorbing a broad range of the solar spectrum and facilitating charge transfer at the interface [148, 162].

In addition, this PEC CO$_2$ reduction system also evolved O$_2$ from water when coupled with a Pt electrode, a finding that was confirmed via experimentation with H$_2^{18}$O. In 2017, Park et al. published a subsequent study on the CuFeO$_2$/CuO system expanding on the photocatalytic reduction of CO$_2$ to formate while driving O$_2$ evolution under simulated sunlight at ∼1% efficiency [163].

In 2015, a new p-type photoelectrode, Cu$_3$Nb$_2$O$_8$, was reported by Ohno et al. to photoelectrochemically reduce CO$_2$ to CO with a Faradaic efficiency of 9% at −0.20 V versus Ag/AgCl [164]. Self-reduction of the photoelectrode was identified using X-ray photoelectron spectroscopy (XPS), which indicated the presence of Cu (II), Cu(I), and Cu(0) on electrodes post-electrolysis. However, thermal annealing in air was reported to recover the photoelectrode. The authors proposed expanding on their work by incorporating a protective layer or depositing a metal co-catalyst on the surface of the Cu$_3$Nb$_2$O$_8$ photoelectrode, but no further studies on this system have been published at this time.

7.3.1.3 Additional Semiconductor Systems

The classic, elemental semiconductor p-Si, though primarily reported as a photo-cathode for H$_2$ evolution [165], has been shown to photoelectrochemically reduce CO$_2$ into CO and formate. Park et al. studied PEC CO$_2$ reduction efficiencies achieved using Sn-coupled p-Si arrays while altering the photoabsorber morphol-ogy [166]. p-Si photoelectrodes are usually planar, and the metallic co-catalyst is then deposited on this flat surface [167]. The authors compared this typical photoelectrode fabrication to a nanowire array, hypothesizing that the Sn-coupled *p*-Si nanowire array would orthogonalize and efficiently trap incident light (Fig. 7.17) [168]. Enhanced photocurrent was observed using the nanoarrays, and the amount of formate produced was double that achieved with the planar analogs under the same light intensity (100 mW/cm^2).

Graphitic carbon nitride (g-C$_3$N$_4$; a two-dimensional conjugated polymer displayed in Fig. 7.18) has attracted considerable attention as a low-cost, metal-free, visible-light-responsive photocatalyst for energy conversion. Within the last decade, this semiconductor has shown promising activity for photocatalytic CO$_2$ reduction [169]. In 2016, Ohno et al. expanded on preliminary photocatalytic studies by reporting PEC CO$_2$ reduction using a *p*-type, B-doped g-C$_3$N$_4$ photoelectrode (BCN$_x$) [170]. In this case, B-doping is necessary to shift the naturally *n*-type g-C$_3$N$_4$ to *p*-type, making it suitable for photoreduction processes [171]. The authors loaded various co-catalysts onto the BCN$_x$ photoelectrodes to increase the overall quantity of PEC CO$_2$ reduction products. Figure 7.19 shows a bar graph of

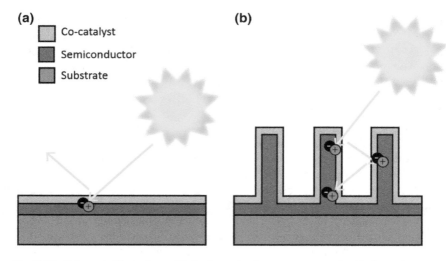

(a)

Co-catalyst

Semiconductor

Substrate

(b)

Fig. 7.17 Schematic illustrations of light absorption between **a** conventionally layered arrays and **b** nanowire arrays; the latter facilitate light orthogonalization, resulting in more effective utilization of solar rays

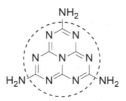

Fig. 7.18 Tris-s-triazine, the repeating subunit of the two-dimensional polymer g-C$_3$N$_4$

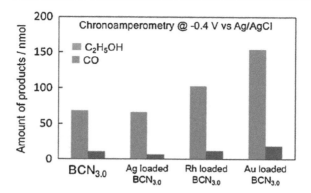

Fig. 7.19 Amount of photoelectrochemically generated CO_2 reduction products achieved using bare $BCN_{3.0}$ and co-catalyst-loaded $BCN_{3.0}$. Adapted with permission from [170]. Copyright 2016 Elsevier

the ethanol and CO products generated by the bare BCN_x photoelectrode compared to Ag-, Rh-, and Au-loaded analogs. An overall increase in product generation was observed by adding the metals, and the main photoreduction product, ethanol, was confirmed by $^{13}CO_2$ reduction.

7.3.2 Homogeneous Catalytic Systems

A major, yet typically unavoidable, concern associated with PEC CO_2 reduction is competition with the more kinetically favorable process of water reduction [160, 172, 173]. One promising strategy to combat this problem involves combining semiconductor photoelectrodes with molecular CO_2 reduction catalysts. The rationale for this strategy is that molecular catalysts can lower the activation barrier for the CO_2 reduction pathway, allowing this pathway to outcompete water reduction in aqueous media. Combining a solid-phase, semiconducting photo-electrode with a water-soluble molecular catalyst creates a homogeneous PEC CO_2 reduction system.

In 2016, the II-VI binary species CdTe was reported by Woo et al. to perform enhanced PEC CO_2 reduction to formic acid when pyridine was added to the electrolyte [174]. Without the molecular catalyst in solution, the CdTe photoelectrode produced formic acid at -0.5 V versus Ag/AgCl with a Faradaic efficiency of 43.6%; however, addition of 10 mM pyridine increased the Faradic efficiency to 60.7%. Figure 7.20 tracks formic acid production over a six-hour period when different concentrations of pyridine were supplied in the electrolyte (0.1 M $NaHCO_3$). For this system, 10 mM pyridine proved to be the optimal catalytic concentration. This finding is consistent with Bocarsly's earlier results suggesting that 10 mM pyridine enhanced aqueous CO_2 reduction at p-GaP [175–177] p-GasAs [178], and p-FeS_2 photocathodes [179].

Fig. 7.20 PEC formic acid production over the course of six hours by CdTe in conjunction with various concentrations of pyridine. Adapted with permission from [174]. Copyright 2014 The Royal Society of Chemistry

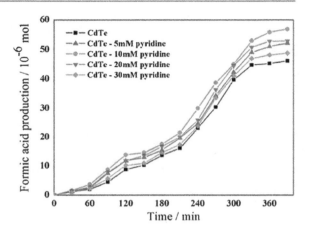

The classic binary semiconductor TiO_2 was reported by Sundararajan et al. in 2014 to photoelectrochemically reduce CO_2 in conjunction with an organic molecular catalyst; the reduction was selective for CO [180]. It was necessary that the TiO_2 be alloyed with 50 wt% Cu to achieve p-type photoelectrode behavior [181]. In aqueous solution, the highest CO selectivity recorded was 27% at -1.5 V versus Ag/AgCl in the presence of 20 mM pyridine.

Work by Cronin et al. also reported a TiO_2 system capable of photoelectro-chemically reducing CO_2; however, their study employed n-type TiO_2 to form a p-n junction with the III-V semiconductor p-GaP [182]. Figure 7.21 shows the increase in photocurrent and decrease in overpotential for the system with increasing thickness of the n-TiO_2 layer. The authors reported that PEC CO_2 reduction to methanol occurred with or without the presence of pyridine; however, addition of 10 mM pyridine resulted in a 300% increase in methanol yield. The authors attributed this dramatic rise to pyridine forming an inner-sphere-type electron transfer system, which lowers the energy barrier for this CO_2 transformation [175, 183].

With the aid of pyridine, chalcopyrite p-CuInS$_2$ was reported by Yuan et al. to produce methanol under PEC CO_2 reduction conditions [184, 185]. In their 2013 study, the authors reported a thin-film fabrication of p-CuInS$_2$ whose performance remained stable up to 11 h. At just 20 mV overpotential, the photoelectrode converted CO_2 to methanol at 97% Faradaic efficiency when coupled with 10 mM pyridine. In 2016, the authors reported a hybrid system comprised of CuInS$_2$/-graphene thin film photoelectrodes that yielded 95% Faradaic efficiency for methanol; however, upon closer inspection the more recently reported system is somewhat less impressive than the first report. Both studies employed 0.1 M Na$_2$SO$_4$ electrolyte solution (pH = 5.2) with 10 mM pyridine under simulated sunlight irradiation (100 mW/cm^2). The slight difference in experimental conditions involves the applied potential: the 2013 study used -0.44 V versus Ag/AgCl, while the 2016 study used -0.49 V versus Ag/AgCl, representing a -50 mV difference.

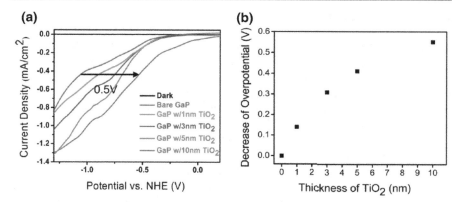

Fig. 7.21 **a** Current-potential curves recorded for GaP photoabsorbers coated with various thicknesses of TiO_2 in 0.5 M NaCl, 10 mM pyridine under 532-nm illumination. **b** Decrease in overpotential plotted as a function of TiO_2 thickness. Adapted with permission from [182]. Copyright 2014 American Chemical Society

Figure 7.22 shows the methanol concentration versus time plots for each study side-by-side. At 10 h of photoelectrolysis, the 2013 system had produced almost 1.2 mM methanol (~ 0.12 mM/hour), while the $CuInS_2$/graphene hybrid had produced just ~ 0.9 mM (~ 0.09 mM/hour) at an applied potential 50 meV more negative than the original system. The reason for incorporating graphene is its high electrical conductivity, which was anticipated to improve charge transfer. Nonetheless, both studies lack a critical $^{13}CO_2$ labeling experiment to fully support their claim of PEC methanol production from CO_2.

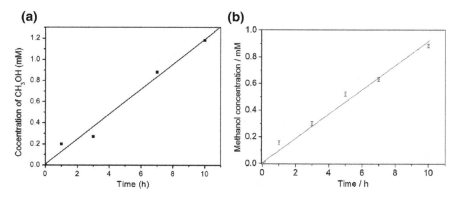

Fig. 7.22 **a** Initial study by Yuan et al. (2013) showing a concentration versus time plot for the production of methanol using p-$CuInS_2$ thin films. **b** Subsequent study by Yuan et al. (2016) presenting the same information acquired using a $CuInS_2$/graphene hybrid photoelectrode. No improvements in methanol production were made by altering the photoelectrode. Adapted with permission from [185]. Copyright 2016 Elsevier

7.3.3 Hybrid Catalytic Systems

In addition to heterogeneous and homogeneous systems for PEC CO_2 reduction, hybrid systems—semiconductor materials coupled with transition metal complexes, functional organic molecules, or conducting polymers—have been studied [186]. Polymers or metal complexes stabilized on semiconductor surfaces are commonly intended to facilitate electron transfer from the semiconductor to CO_2 molecules in the electrolyte [187, 188]. Engineering these hybrid electrodes has allowed researchers to optimize or tune reactivity, selectivity, and Faradaic efficiency in CO_2 reduction experiments [189].

In fabricating these hybrid electrode systems, different methods have been invoked to modify the semiconductor surfaces. Drop-casting and immersion have often been used for immobilizing transition metal complexes [187, 188, 190], while various polymerization techniques have been employed to stabilize larger molecules on the semiconductor surface [189, 191].

7.3.3.1 Metal Oxides

Semiconductor metal oxides have attracted much attention in PEC CO_2 reduction due to their impressive performance and high stability in aqueous electrolyte [192]. Hybrid systems consisting of a metal oxide and molecular co-catalyst have been shown to further improve the activity and selectivity compared to the naked electrode. An example of such a system was reported in 2014 by Bachmeier et al., who assembled a photocathode comprised of a porous p-type NiO thin film, a visible light-responsive organic dye (P1), and carbon monoxide dehydrogenase (CODH) enzymes [187]. Under visible light irradiation (>420 nm), electrons flowed from p-NiO to P1 and then CODH, where they were used to reduce CO_2 to CO (Fig. 7.23). In this case, P1 served as a photosensitizer at the semiconductor surface and CODH behaved as a reversible catalyst. The visible light-driven selective reduction of CO_2 to CO was carried out at −0.48 V versus Ag/AgCl, well below the reaction's thermodynamic reduction potential. Additionally, the hybrid structure increased the stability and selectivity of the photocathode. The authors suggested that the enzymes, having well-defined and stable active sites, served as depolarizers which overcame the kinetic burden of catalysis, thereby allowing for mechanistic study of the system.

In 2017, researchers reported a hybrid photocathode composed of p-CuGaO$_2$ and a Ru(II)–Re(I) supramolecular photocatalyst ([(dmb)(L1)Ru(BL)Re(CO)$_3$Br]$^{2+}$ (RuRe; BL = 1,2-bis(4′-methyl-[2,2′-bipyridine]-4-yl)ethane; L1 = 4,4′-bis (methyl-phosphonate)-2,2′-bipyridine)), hereafter referred to as RuRe/CuGaO$_2$ [188]. To obtain the hybridized structure, the electrode was soaked in a RuRe/acetonitrile solution overnight, resulting in covalent attachment of RuRe to the electrode surface via phosphonate bonds (Fig. 7.24). Using this photoelectrode, a Faradaic efficiency of 49% was obtained for CO at −0.3 V versus Ag/AgCl under visible light (>460 nm) in 0.05 M, CO_2-saturated NaHCO$_3$ (pH = 6.6). Controls performed using bare CuGaO$_2$ or electrodes modified with only one of the model

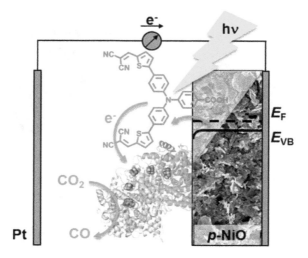

Fig. 7.23 Scheme depicting a PEC cell for selective reduction of CO_2 to CO at p-type NiO. Light absorption by the organic dye P1 (red) is followed by electron transfer to CODH, which is co-adsorbed on the NiO surface and carries out CO_2 reduction; hole injection into the NiO valence band regenerates the P1 ground state. The porous nature of the NiO surface is conveyed by the SEM image. The Fermi level and valence band position of NiO are denoted as E_F and E_{VB}. Adapted with permission from [187]. Copyright 2014 American Chemical Society

mononuclear complexes (i.e., $[Ru(dmb)_2(L1)]^{2+}$ *or* $[Re(CO)_3Br(L1)])$ demonstrated that the RuRe/CuGaO$_2$ hybrid was needed to generate CO.

Recently, a hybrid photocathode consisting of a ruthenium complex (*trans*(Cl)-$[Ru(bpyX_2)-(CO)_2Cl_2]$ (bpyX$_2$ = 2,2′-bipyridine with X substituents in the 4,4′-positions; X = H, CO_2H, CO_2CH_3, $CO_2C_3H_6$-1H-pyrrolyl)) and a multi-heterojunction semiconductor thin film electrode (TiO$_2$/N,Zn-Fe$_2$O$_3$/Cr$_2$O$_3$, where

Fig. 7.24 CO_2 reduction scheme facilitated by the RuRe/CuGaO$_2$ hybrid photocathode. Adapted with permission from [188]. Copyright 2017 The Royal Society of Chemistry

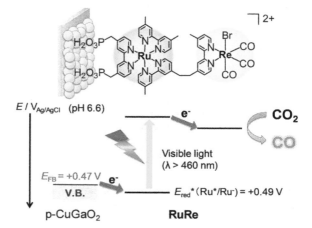

(a)   **(b)**

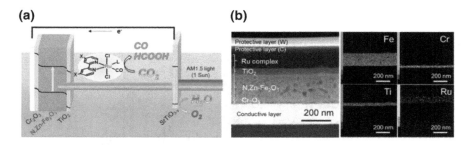

Fig. 7.25 **a** Schematic for PEC reduction of CO_2 using a two-electrode configuration comprised of Ru(MeCN)CO$_2$C3Py-P/TiO$_2$/N,Zn-Fe$_2$O$_3$/Cr$_2$O$_3$ and SrTiO$_{3-x}$. **b** Cross-sectional STEM image and STEM-EDX elemental maps for the Ru(MeCN)CO$_2$C3Py-P/TiO$_2$/N,Zn-Fe$_2$O$_3$/Cr$_2$O$_3$ electrode. Adapted with permission from [190]. Copyright 2018 American Chemical Society

N represents doping from N$_2$ gas present in the atmosphere during synthesis) was developed [190]. The multilayer design (Fig. 7.25) improved the photocurrent of the system, because the TiO$_2$ overlayer protected the unstable Fe$_2$O$_3$ surface and improved electron transfer from N,Zn-Fe$_2$O$_3$ to the Ru complex (localized on the electrode surface) by forming a *p-n* junction. Furthermore, the Cr$_2$O$_3$ underlayer improved hole transfer to the conductive substrate. Multiple Ru complexes were drop-casted onto the electrode surface and tested as CO$_2$ reduction catalysts, and it was noted that stability of the complexes was improved by introducing an electrical network of polypyrrole chains. One of the polymeric Ru complexes, Ru(MeCN) CO$_2$C3Py-P, produced a stable photocurrent that generated formate (Faradaic efficiency = 63%), CO (30%), and a small amount of H$_2$ (7%) under 1 sun irradiation (100 mW/cm^2) at −0.49 V versus Ag/AgCl in CO$_2$-saturated, 0.1 M KHCO$_3$ (pH = 6.6). Following a control experiment using ^{13}CO$_2$ reactant, this study was suggested to be the first example of stoichiometric, selective CO$_2$ reduction in an aqueous environment using a *p*-Fe$_2$O$_3$-based photocathode.

7.3.3.2 Metal Chalcogenides

Metal chalcogenide semiconductors have also been fabricated into hybrid structures for PEC CO$_2$ reduction. For instance, in 2015, a polypyrrole-coated *p*-ZnTe electrode (PPy/ZnTe) was prepared by electropolymerization, and electron transfer from ZnTe to PPy was confirmed [189]. Under CO$_2$-saturated conditions, the composite PPy/ZnTe photoelectrode yielded Faradaic efficiencies of 37.2% for formate and 13.8% for CO at –0.2 V versus Ag/AgCl in 0.1 M KHCO$_3$ (pH = 6.6) under visible light irradiation (>420 nm). Surface-bound PPy, which provided active sites for CO$_2$ conversion to formate and hindered H$_2$ evolution, was shown to improve Faradaic efficiency, selectivity, and product generation rate.

In a 2015 study, a CdTe quantum dot-modified electrode capable of photoelectrochemically reducing CO$_2$ was assembled layer-by-layer using an electric field [191]. These polycation/QD assemblies (polycation = poly diallyldimethylammonium (PDDA) and poly(2-trimethylammonium)ethyl methacrylate

(PMAEMA); QD = CdTe quantum dots) were deposited on an indium tin oxide (ITO) substrate, and it was shown that choice of polycation influenced the products of PEC CO_2 reduction. Specifically, at −0.45 V versus Ag/AgCl under irradiation (500 W Xe lamp) and in 0.1 M $NaClO_4$ (pH = 4), the PDDA/QD composite generated CO and methanol, while the PMAEMA/QD composite was more selective toward production of formaldehyde.

7.4 Miscellaneous CO_2 Reduction Systems

Other successful systems for electrochemical or photoelectrochemical reduction of CO_2 exist that do not necessarily fit into the above categories. These systems are discussed here.

7.4.1 Photoanode-Driven Systems

7.4.1.1 *n*-TiO_2

For decades, variations on *n*-type TiO_2 semiconductors have been studied as dark cathodes and photoanodes due to their high stability [192]. Relevant to this review, *n*-type TiO_2 can be used as a dark cathode, driven by an external bias, to directly reduce CO_2, or an indirect route involving H_2O oxidation can be employed. In this case, TiO_2 serves as a H_2O-oxidizing photoanode, and the electrons derived from this process drive CO_2 reduction at the opposite electrode.

Recently, several studies have been carried out using these *n*-TiO_2, photoanode-driven systems. One example is work performed by the Cen group featuring Pt-modified TiO_2 nanotube arrays coupled with various cathodes [193–195]. Multicarbon products such as ethanol and acetic acid were obtained under near-UV irradiation (320–410 nm, 10 mW/cm^2) at a bias of +2.0 V versus Ag/AgCl. Different pH values were maintained for the anolyte and catholyte solutions. In 2016, Gong et al. reported a strategy using Cu_2O as a dark cathode and TiO_2 nanorods as a photoanode to conduct CO_2 reduction in an aqueous environment. This system boasted 92.6% selectivity for carbon-containing products, including Faradaic efficiencies of 54.6, 30.0, and 2.8% for methane, CO, and methanol, respectively [196]. Importantly, use of the photoanode improved the stability of the Cu_2O dark cathode.

7.4.1.2 Other Photoanode Systems

Many other *n*-type semiconductors have been employed as photoanode candidates for CO_2 reduction. A photoanode-driven PEC system consisting of a WO_3 thin film photoanode and either a Cu or Sn/SnO_x cathode for CO_2 reduction has been investigated under visible light irradiation [197]. When matching WO_3 with a Cu cathode (0.5 M $KHCO_3$, pH = 7.5), a total Faradaic efficiency of 71.6% was recorded for all carbon-containing products (including 67% for methane alone) at

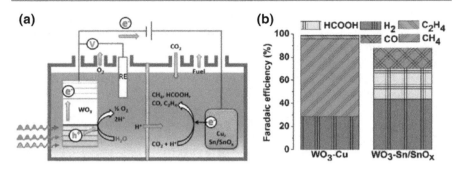

Fig. 7.26 **a** Depiction of a PEC CO_2 reduction system utilizing a WO_3 photoanode and Cu or Sn/SnO$_x$ cathode under visible light irradiation. RE denotes use of a Ag/AgCl reference electrode. **b** CO_2 reduction products achieved using WO_3–Cu and WO_3–Sn/SnO$_x$ systems under visible light irradiation (> 420 nm). Adapted with permission from [197]. Copyright 2014 The Royal Society of Chemistry

+0.65 V versus Ag/AgCl. Replacing Cu with Sn/SnO$_x$ yielded a total Faradaic efficiency for CO and formate of 44.3% at +0.8 V versus Ag/AgCl (Fig. 7.26).

An n-type 3C(cubic)-SiC photoanode was also studied alongside Pt and Ag cathodes in 0.1 M KHCO$_3$ (pH = 6.8) at +1.0 V versus Ag/AgCl under irradiation (Xe lamp, 100 mW/cm^2) [198]. In this system, the rate of CO_2 reduction to CO was higher when Ag was selected as the CO_2-reducing component. In 2016, a Ni-coated, n-type Si electrode was matched with a nanoporous Ag cathode for CO_2 conversion to CO [199]. At an external bias of +2.0 V versus Ag/AgCl under visible light irradiation (>400 nm), the PEC cell delivered a current density of 10 mA/cm^2 with a CO Faradaic efficiency of about 70% in 0.5 M KHCO$_3$. The system remained stable for up to three hours during photoelectrolysis. It has been suggested that, compared with common p-type cathode-driven systems, photoanode-driven PEC cells could exhibit superior long-term stability. Furthermore, these systems exhibit high versatility since they can incorporate a variety of CO_2-reducing electrocatalysts as the cathode [197].

7.4.2 Diamond

Diamond has attracted much attention from chemists and physicists due to its extreme physical properties and interesting behavior [200–202]. More recently, however, diamond has become a candidate material for CO_2 reduction catalysis. In 2015, efficient electrochemical reduction of CO_2 to acetate was realized on N-doped nanodiamond [203]. In this study, doped nanodiamond particles grown on a Si rod yielded Faradaic efficiencies of 80% for acetate and 15% for formate at −1.0 V versus Ag/AgCl in 0.5 M NaHCO$_3$ solution. The electrode's high overpotential for H$_2$ evolution as well as its N-doping (in which N-sp^{3}C species were highly active for CO_2 reduction) contributed to this high selectivity and good performance.

PEC CO_2 reduction has also been carried out on diamond materials. In 2016, a B-doped diamond semiconductor thin film electrode was shown to achieve efficient PEC CO_2 reduction following surface modification with Ag nanoparticles [204]. A Faradaic efficiency of 72.5% for CO was obtained using the optimal Ag nanoparticle-modified, oxygen-terminated, lightly boron-doped diamond electrode at -1.6 V versus Ag/AgCl under 222-nm irradiation (7 W) in 0.025 M Na_2SO_4. The performance and stability of this photocathode were ascribed to in situ photoactivation of the B-doped diamond surface during the PEC reaction.

In summary, diamond materials have demonstrated attractive properties and great potential in electrocatalytic and photoelectrocatalytic CO_2 reduction, much like the full range of heterogeneous, homogeneous, electrochemical, and photoelectrochemical systems discussed here. Ongoing work in catalyst discovery, optimization, and understanding will lend invaluable insight into the real-world possibility of transforming CO_2 into useful products using chemistry tools.

Acknowledgements The electrochemistry sections in this work were supported in part by the US National Science Foundation under Grant No. CHE-1800400. Additionally, ARP and JJF acknowledge funding from the National Science Foundation Graduate Research Fellowship Program under Grant No. DGE-1148900 and Grant No. DGE-1656466. The photoelectrochemistry sections in this work were supported in part by the US Department of Energy, Basic Energy Sciences under Grant No. DE-SC0002133 Any opinions, findings, and conclusions or recommendations expressed in this material are those of the authors and do not necessarily reflect the views of the National Science Foundation or the Department of Energy.

References

1. Biello D (2010) Reverse combustion: can CO_2 be turned back into fuel? (Video). Sci Am, 23 Sept 2010
2. White JL, Baruch MF, Pander JE III, Hu Y, Fortmeyer IC, Park JE, Zhang T, Liao K, Gu J, Yan Y et al (2015) Light-driven heterogeneous reduction of carbon dioxide: photocatalysts and photoelectrodes. Chem Rev 115(23):12888–12935
3. Gu J, Wuttig A, Krizan JW, Hu Y, Detweiler ZM, Cava RJ, Bocarsly AB (2013) Mg-doped $CuFeO_2$ photocathodes for photoelectrochemical reduction of carbon dioxide. J Phys Chem C 117(24):12415–12422
4. Gu J, Yan Y, Krizan JW, Gibson QD, Detweiler ZM, Cava RJ, Bocarsly AB (2014) *P*-type $CuRhO_2$ as a self-healing photoelectrode for water reduction under visible light. J Am Chem Soc 136(3):830–833
5. Detweiler ZM, White JL, Bernasek SL, Bocarsly AB (2014) Anodized indium metal electrodes for enhanced carbon dioxide reduction in aqueous electrolyte. Langmuir 30(25):7593–7600
6. White JL, Bocarsly AB (2016) Enhanced carbon dioxide reduction activity on indium-based nanoparticles. J Electrochem Soc 163(6):H410–H416
7. Hawecker J, Lehn J-M, Ziessel R (1984) Electrocatalytic reduction of carbon dioxide mediated by Re(Bipy)(CO)$_3$Cl (Bipy = 2,2'-Bipyridine). J Chem Soc Chem Commun, 328–330
8. Bourrez M, Molton F, Chardon-Noblat S, Deronzier A (2011) [Mn(Bipyridyl)(CO)$_3$Br]: an abundant metal carbonyl complex as efficient electrocatalyst for CO_2 reduction. Angewandte Chemie Int Edn 50(42):9903–9906

9. Hori Y, Murata A, Takahashi R, Suzuki S (1988) Enhanced formation of ethylene and alcohols at ambient temperature and pressure in electrochemical reduction of carbon dioxide at a copper electrode. J Chem Soc, Chem Commun 1:17–19
10. Hori Y, Murata A, Takahashi R (1989) Formation of hydrocarbons in the electrochemical reduction of carbon dioxide at a copper electrode in aqueous solution. J Chem Soc Faraday Trans 1 Phys Chem Condens Phases 85(8):2309–2326
11. Hori Y, Wakebe H, Tsukamoto T, Koga O (1994) Electrocatalytic process of CO selectivity in electrochemical reduction of CO_2 at metal electrodes in aqueous media. Electrochim Acta 39(11–12):1833–1839
12. Ooka H, Figueiredo MC, Koper MTM (2017) Competition between hydrogen evolution and carbon dioxide reduction on copper electrodes in mildly acidic media. Langmuir 33 (37):9307–9313
13. Hori Y, Takahashi I, Koga O, Hoshi N (2002) Selective formation of C2 compounds from electrochemical reduction of CO_2 at a series of copper single crystal electrodes. J Phys Chem B 106(1):15–17
14. Kuhl KP, Cave ER, Abram DN, Jaramillo TF (2012) New Insights into the electrochemical reduction of carbon dioxide on metallic copper surfaces. Energy Environ Sci 5(5):7050
15. Manthiram K, Beberwyck BJ, Alivisatos AP (2014) Enhanced electrochemical methanation of carbon dioxide with a dispersible nanoscale copper catalyst. J Am Chem Soc 136 (38):13319–13325
16. Dutta A, Rahaman M, Luedi NC, Mohos M, Broekmann P (2016) Morphology matters: tuning the product distribution of CO_2 electroreduction on oxide-derived Cu foam catalysts. ACS Catal 6(6):3804–3814
17. Ren D, Deng Y, Handoko AD, Chen CS, Malkhandi S, Yeo BS (2015) Selective electrochemical reduction of carbon dioxide to ethylene and ethanol on copper(I) oxide catalysts. ACS Catal 5(5):2814–2821
18. Kas R, Kortlever R, Milbrat A, Koper MTM, Mul G, Baltrusaitis J (2014) Electrochemical CO_2 reduction on Cu_2O-derived copper nanoparticles: controlling the catalytic selectivity of hydrocarbons. Phys Chem Chem Phys 16(24):12194–12201
19. Sen S, Liu D, Palmore GTR (2014) Electrochemical reduction of CO_2 at copper nanofoams. ACS Catal 4(9):3091–3095
20. Loiudice A, Lobaccaro P, Kamali EA, Thao T, Huang BH, Ager JW, Buonsanti R (2016) Tailoring copper nanocrystals towards C_2 products in electrochemical CO_2 reduction. Angewandte Chemie Int Edn 55(19):5789–5792
21. Reske R, Mistry H, Behafarid F, Roldan Cuenya B, Strasser P (2014) Particle size effects in the catalytic electroreduction of CO_2 on Cu nanoparticles. J Am Chem Soc 136(19):6978–6986
22. Roberts FS, Kuhl KP, Nilsson A (2015) High selectivity for ethylene from carbon dioxide reduction over copper nanocube electrocatalysts. Angew Chem 127(17):5268–5271
23. Li Y, Cui F, Ross MB, Kim D, Sun Y, Yang P (2017) Structure-sensitive CO_2 electroreduction to hydrocarbons on ultrathin 5-fold twinned copper nanowires. Nano Lett 17(2):1312–1317
24. Verdaguer-Casadevall A, Li CW, Johansson TP, Scott SB, McKeown JT, Kumar M, Stephens IEL, Kanan MW, Chorkendorff I (2015) Probing the active surface sites for CO reduction on oxide-derived copper electrocatalysts. J Am Chem Soc 137(31):9808–9811
25. Ma M, Djanashvili K, Smith WA (2016) Controllable hydrocarbon formation from the electrochemical reduction of CO_2 over Cu nanowire arrays. Angewandte Chemie Int Edn 55 (23):6680–6684
26. Huang Y, Handoko AD, Hirunsit P, Yeo BS (2017) Electrochemical reduction of CO_2 using copper single-crystal surfaces: effects of CO* coverage on the selective formation of ethylene. ACS Catal 7(3):1749–1756
27. Raciti D, Livi KJ, Wang C (2015) Highly dense Cu nanowires for low-overpotential CO_2 reduction. Nano Lett 15(10):6829–6835

28. Kas R, Kortlever R, Yilmaz H, Koper MTM, Mul G (2015) Manipulating the hydrocarbon selectivity of copper nanoparticles in CO_2 electroreduction by process conditions. ChemElectroChem 2(3):354–358
29. Varela AS, Kroschel M, Reier T, Strasser P (2016) Controlling the selectivity of CO_2 electroreduction on copper: the effect of the electrolyte concentration and the importance of the local pH. Catal Today 260:8–13
30. Varela AS, Ju W, Reier T, Strasser P (2016) Tuning the catalytic activity and selectivity of Cu for CO_2 electroreduction in the presence of halides. ACS Catal 6(4):2136–2144
31. Resasco J, Lum Y, Clark E, Zeledon JZ, Bell AT (2018) Effects of anion identity and concentration on electrochemical reduction of CO_2. ChemElectroChem 5(7):1064–1072
32. Gattrell M, Gupta N, Co A (2006) A review of the aqueous electrochemical reduction of CO_2 to hydrocarbons at copper. J Electroanal Chem 594(1):1–19
33. Lee S, Kim D, Lee J (2015) Electrocatalytic production of C3-C4 compounds by conversion of CO_2 on a chloride-induced Bi-phasic Cu_2O-Cu catalyst. Angew Chem 127(49):14914–14918
34. Lum Y, Yue B, Lobaccaro P, Bell AT, Ager JW (2017) Optimizing C-C coupling on oxide-derived copper catalysts for electrochemical CO_2 reduction. J Phys Chem C 121(26):14191–14203
35. De Luna P, Quintero-Bermudez R, Dinh C-T, Ross MB, Bushuyev OS, Todorović P, Regier T, Kelley SO, Yang P, Sargent EH (2018) Catalyst electro-redeposition controls morphology and oxidation state for selective carbon dioxide reduction. Nat Catal 1(2):103–110
36. Hori Y (2008) Electrochemical CO_2 reduction on metal electrodes. In: Gamboa-Aldeco ME (ed) Modern aspects of electrochemistry. Springer, New York, vol 42, pp 89–189
37. Kuhl KP, Hatsukade T, Cave ER, Abram DN, Kibsgaard J, Jaramillo TF (2014) Electrocatalytic conversion of carbon dioxide to methane and methanol on transition metal surfaces. J Am Chem Soc 136(40):14107–14113
38. Singh MR, Kwon Y, Lum Y, Ager JW, Bell AT (2016) Hydrolysis of electrolyte cations enhances the electrochemical reduction of CO_2 over Ag and Cu. J Am Chem Soc 138(39):13006–13012
39. Kim C, Jeon HS, Eom T, Jee MS, Kim H, Friend CM, Min BK, Hwang YJ (2015) Achieving selective and efficient electrocatalytic activity for CO_2 reduction using immobilized silver nanoparticles. J Am Chem Soc 137(43):13844–13850
40. Ma M, Trześniewski BJ, Xie J, Smith WA (2016) Selective and efficient reduction of carbon dioxide to carbon monoxide on oxide-derived nanostructured silver electrocatalysts. Angew Chem 128(33):9900–9904
41. Peng X, Karakalos SG, Mustain WE (2018) Preferentially oriented Ag nanocrystals with extremely high activity and faradaic efficiency for CO_2 electrochemical reduction to CO. ACS Appl Mater Interfaces 10(2):1734–1742
42. Liu S, Tao H, Zeng L, Liu Q, Xu Z, Liu Q, Luo J-L (2017) Shape-dependent electrocatalytic reduction of CO_2 to CO on triangular silver nanoplates. J Am Chem Soc 139(6):2160–2163
43. Rosen J, Hutchings GS, Lu Q, Rivera S, Zhou Y, Vlachos DG, Jiao F (2015) Mechanistic insights into the electrochemical reduction of CO_2 to CO on nanostructured Ag surfaces. ACS Catal 5(7):4293–4299
44. Feng X, Jiang K, Fan S, Kanan MW (2015) Grain-boundary-dependent CO_2 electroreduction activity. J Am Chem Soc 137(14):4606–4609
45. Zhu W, Zhang Y-J, Zhang H, Lv H, Li Q, Michalsky R, Peterson AA, Sun S (2014) Active and selective conversion of CO_2 to CO on ultrathin au nanowires. J Am Chem Soc 136(46):16132–16135
46. Mistry H, Reske R, Zeng Z, Zhao Z-J, Greeley J, Strasser P, Cuenya BR (2014) Exceptional size-dependent activity enhancement in the electroreduction of CO_2 over Au nanoparticles. J Am Chem Soc 136(47):16473–16476

47. Gao D, Zhou H, Wang J, Miao S, Yang F, Wang G, Wang J, Bao X (2015) Size-dependent electrocatalytic reduction of CO_2 over Pd nanoparticles. J Am Chem Soc 137(13):4288–4291
48. Rosen J, Hutchings GS, Lu Q, Forest RV, Moore A, Jiao F (2015) Electrodeposited Zn dendrites with enhanced CO selectivity for electrocatalytic CO_2 reduction. ACS Catal 5 (8):4586–4591
49. Gao S, Jiao X, Sun Z, Zhang W, Sun Y, Wang C, Hu Q, Zu X, Yang F, Yang S et al (2016) Ultrathin Co_3O_4 layers realizing optimized CO_2 electroreduction to formate. Angewandte Chemie Int Edn 55(2):698–702
50. Baruch MF, Pander JE, White JL, Bocarsly AB (2015) Mechanistic insights into the reduction of CO_2 on tin electrodes using in situ ATR-IR spectroscopy. ACS Catal 5 (5):3148–3156
51. Detweiler ZM, Wulfsberg SM, Frith MG, Bocarsly AB, Bernasek SL (2016) The oxidation and surface speciation of indium and indium oxides exposed to atmospheric oxidants. Surf Sci 648:188–195
52. Pander JE, Baruch MF, Bocarsly AB (2016) Probing the mechanism of aqueous CO_2 reduction on post-transition-metal electrodes using ATR-IR spectroelectrochemistry. ACS Catal 6(11):7824–7833
53. Li F, Chen L, Knowles GP, MacFarlane DR, Zhang J (2017) Hierarchical mesoporous SnO_2 nanosheets on carbon cloth: a robust and flexible electrocatalyst for CO_2 reduction with high efficiency and selectivity. Angewandte Chemie Int Edn 56(2):505–509
54. Kumar B, Atla V, Brian JP, Kumari S, Nguyen TQ, Sunkara M, Spurgeon JM (2017) Reduced SnO_2 porous nanowires with a high density of grain boundaries as catalysts for efficient electrochemical CO_2-into-HCOOH conversion. Angewandte Chemie Int Edn 56 (13):3645–3649
55. Zhang S, Kang P, Meyer TJ (2014) Nanostructured tin catalysts for selective electrochemical reduction of carbon dioxide to formate. J Am Chem Soc 136(5):1734–1737
56. Li Q, Fu J, Zhu W, Chen Z, Shen B, Wu L, Xi Z, Wang T, Lu G, Zhu J et al (2017) Tuning Sn-catalysis for electrochemical reduction of CO_2 to CO via the core/shell Cu/SnO_2 structure. J Am Chem Soc 139(12):4290–4293
57. Luc W, Collins C, Wang S, Xin H, He K, Kang Y, Jiao F (2017) Ag–Sn bimetallic catalyst with a core-shell structure for CO_2 reduction. J Am Chem Soc 139(5):1885–1893
58. Hansen HA, Shi C, Lausche AC, Peterson AA, Nørskov JK (2016) Bifunctional alloys for the electroreduction of CO_2 and CO. Phys Chem Chem Phys 18(13):9194–9201
59. Ma S, Sadakiyo M, Heima M, Luo R, Haasch RT, Gold JI, Yamauchi M, Kenis PJA (2017) Electroreduction of carbon dioxide to hydrocarbons using bimetallic Cu–Pd catalysts with different mixing patterns. J Am Chem Soc 139(1):47–50
60. Kim D, Xie C, Becknell N, Yu Y, Karamad M, Chan K, Crumlin EJ, Nørskov JK, Yang P (2017) Electrochemical activation of CO_2 through atomic ordering transformations of AuCu nanoparticles. J Am Chem Soc 139(24):8329–8336
61. Sarfraz S, Garcia-Esparza AT, Jedidi A, Cavallo L, Takanabe K (2016) Cu–Sn bimetallic catalyst for selective aqueous electroreduction of CO_2 to CO. ACS Catal 6(5):2842–2851
62. Rasul S, Anjum DH, Jedidi A, Minenkov Y, Cavallo L, Takanabe K (2015) A highly selective copper-indium bimetallic electrocatalyst for the electrochemical reduction of aqueous CO_2 to CO. Angewandte Chemie Int Edn 54(7):2146–2150
63. Kortlever R, Peters I, Koper S, Koper MTM (2015) Electrochemical CO_2 reduction to formic acid at low overpotential and with high faradaic efficiency on carbon-supported bimetallic Pd–Pt nanoparticles. ACS Catal 5(7):3916–3923
64. Paris AR, Bocarsly AB (2018, submitted) High-efficiency conversion of CO_2 to oxalate in water is possible using a Cr-Ga oxide electrocatalyst. Nat Chem. ACS Catal 2019(9):2324–2333.
65. Amatore C, Saveant JM (1981) Mechanism and kinetic characteristics of the electrochemical reduction of carbon dioxide in media of low proton availability. J Am Chem Soc 103 (17):5021–5023

66. Gennaro A, Isse AA, Severin M-G, Vianello E, Bhugun I, Savéant J-M (1996) Mechanism of the electrochemical reduction of carbon dioxide at inert electrodes in media of low proton availability. J Chem Soc Faraday Trans 92(20):3963–3968

67. Studt F, Sharafutdinov I, Abild-Pedersen F, Elkjær CF, Hummelshøj JS, Dahl S, Chorkendorff I, Nørskov JK (2014) Discovery of a Ni-Ga catalyst for carbon dioxide reduction to methanol. Nat Chem 6(4):320–324

68. Torelli DA, Francis SA, Crompton JC, Javier A, Thompson JR, Brunschwig BS, Soriaga MP, Lewis NS (2016) Nickel–gallium-catalyzed electrochemical reduction of CO_2 to highly reduced products at low overpotentials. ACS Catal 6(3):2100–2104

69. Paris AR, Chu AT, O'Brien CB, Frick JJ, Francis SA, Bocarsly AB (2018) Tuning the products of CO_2 electroreduction on a Ni_3Ga catalyst using carbon solid supports. J Electrochem Soc 165(7):H385–H392

70. Paris AR, Bocarsly AB (2017) Ni–Al films on glassy carbon electrodes generate an array of oxygenated organics from CO_2. ACS Catal 7(10):6815–6820

71. Ghosh D, Kobayashi K, Kajiwara T, Kitagawa S, Tanaka K (2017) Catalytic hydride transfer to CO_2 using Ru-NAD-type complexes under electrochemical conditions. Inorg Chem 56 (18):11066–11073

72. Min S, Rasul S, Li H, Grills DC, Takanabe K, Li L-J, Huang K-W (2016) Electrocatalytic reduction of carbon dioxide with a well-defined PN3 − Ru pincer complex. ChemPlusChem 81(2):166–171

73. Boston DJ, Pachón YMF, Lezna RO, de Tacconi NR, MacDonnell FM (2014) Electrocatalytic and photocatalytic conversion of CO_2 to methanol using ruthenium complexes with internal pyridyl cocatalysts. Inorg Chem 53(13):6544–6553

74. Francke R, Schille B, Roemelt M (2018) Homogeneously catalyzed electroreduction of carbon dioxide—methods, mechanisms, and catalysts. Chem Rev 118(9):4631–4701

75. Machan CW, Chabolla SA, Yin J, Gilson MK, Tezcan FA, Kubiak CP (2014) Supramolecular assembly promotes the electrocatalytic reduction of carbon dioxide by Re (I) bipyridine catalysts at a lower overpotential. J Am Chem Soc 136(41):14598–14607

76. Sung S, Kumar D, Gil-Sepulcre M, Nippe M (2017) Electrocatalytic CO_2 reduction by imidazolium-functionalized molecular catalysts. J Am Chem Soc 139(40):13993–13996

77. Clark ML, Cheung PL, Lessio M, Carter EA, Kubiak CP (2018) Kinetic and mechanistic effects of Bipyridine (Bpy) substituent, Labile Ligand, and Brønsted acid on electrocatalytic CO_2 reduction by Re(Bpy) complexes. ACS Catal 8(3):2021–2029

78. Bourrez M, Molton F, Chardon-Noblat S, Deronzier A (2011) [Mn(Bipyridyl)(CO)$_3$Br]: an abundant metal carbonyl complex as efficient electrocatalyst for CO_2 reduction. Angewandte Chemie Int Edn 50(42):9903–9906

79. Smieja JM, Sampson MD, Grice KA, Benson EE, Froehlich JD, Kubiak CP (2013) Manganese as a substitute for rhenium in CO_2 reduction catalysts: the importance of acids. Inorg Chem 52(5):2484–2491

80. Sampson MD, Nguyen AD, Grice KA, Moore CE, Rheingold AL, Kubiak CP (2014) Manganese catalysts with bulky bipyridine ligands for the electrocatalytic reduction of carbon dioxide: eliminating dimerization and altering catalysis. J Am Chem Soc 136 (14):5460–5471

81. Agarwal J, Shaw TW, Schaefer HF, Bocarsly AB (2015) Design of a catalytic active site for electrochemical CO_2 reduction with Mn(I)-tricarbonyl species. Inorg Chem 54(11):5285–5294

82. Franco F, Cometto C, Nencini L, Barolo C, Sordello F, Minero C, Fiedler J, Robert M, Gobetto R, Nervi C. Local proton source in electrocatalytic CO_2 reduction with [Mn(Bpy–R) (CO)$_3$Br] complexes. Chem Eur J 23(20):4782–4793

83. Tignor SE, Kuo H-Y, Lee TS, Scholes GD, Bocarsly AB (2018, Submitted) manganese based catalysts with varying ligand substituents for the electrochemical reduction of CO_2 to CO. Organometallics

84. Agarwal J, Shaw TW, Stanton CJ, Majetich GF, Bocarsly AB, Schaefer HF. NHC-containing manganese(I) electrocatalysts for the two-electron reduction of CO_2. Angewandte Chemie Int Edn 53(20):5152–5155

85. Kang P, Chen Z, Nayak A, Zhang S, Meyer TJ (2014) Single catalyst electrocatalytic reduction of co_2 in water to H_2 + CO syngas mixtures with water oxidation to O_2. Energy Environ Sci 7(12):4007–4012

86. Sheng M, Jiang N, Gustafson S, You B, Ess DH, Sun Y (2015) A nickel complex with a biscarbene pincer-type ligand shows high electrocatalytic reduction of CO_2 over H_2O. Dalton Trans 44(37):16247–16250

87. Cope JD, Liyanage NP, Kelley PJ, Denny JA, Valente EJ, Webster CE, Delcamp JH, Hollis TK (2017) Electrocatalytic reduction of CO_2 with CCC-NHC pincer nickel complexes. Chem Commun 53(68):9442–9445

88. Stanton CJ, Vandezande JE, Majetich GF, Schaefer HF, Agarwal J (2016) Mn-NHC electrocatalysts: increasing π acidity lowers the reduction potential and increases the turnover frequency for CO_2 reduction. Inorg Chem 55(19):9509–9512

89. Liyanage NP, Dulaney HA, Huckaba AJ, Jurss JW, Delcamp JH (2016) Electrocatalytic reduction of CO_2 to CO with Re-Pyridyl-NHCs: proton source influence on rates and product selectivities. Inorg Chem 55(12):6085–6094

90. Carrington SJ, Chakraborty I, Bernard JML, Mascharak PK (2014) Synthesis and characterization of a "Turn-On" PhotoCORM for trackable CO delivery to biological targets. ACS Med Chem Lett 5(12):1324–1328

91. Yempally V, Kyran SJ, Raju RK, Fan WY, Brothers EN, Darensbourg DJ, Bengali AA (2014) Thermal and photochemical reactivity of manganese tricarbonyl and tetracarbonyl complexes with a bulky diazabutadiene ligand. Inorg Chem 53(8):4081–4088

92. Takeda H, Koizumi H, Okamoto K, Ishitani O (2014) Photocatalytic CO_2 reduction using a Mn complex as a catalyst. Chem Commun 50(12):1491–1493

93. Stor GJ, Morrison SL, Stufkens DJ, Oskam A (1994) The remarkable photochemistry of fac-XMn(CO)$_3$(alpha-diimine) (X = Halide): formation of $Mn_2(CO)_6$(alpha-diimine)$_2$ via the mer isomer and photocatalytic substitution of X$^-$ in the presence of PR$_3$. Organometallics 13(7):2641–2650

94. Stor GJ, Stufkens DJ, Vernooijs P, Baerends EJ, Fraanje J, Goubitz K (1995) X-ray structure of fac-IMn(CO)$_3$(Bpy) and electronic structures and transitions of the complexes fac-XMn (CO)$_3$(Bpy) (X = Cl, I) and mer-ClMn(CO)$_3$(Bpy). Inorg Chem 34(6):1588–1594

95. Staal LH, Oskam A, Vrieze K (1979) The syntheses and coordination properties of M (CO)$_3$X(DAB) (M = Mn, Re; X = Cl, Br, I; DAB = 1,4-Diazabutadiene). J Organomet Chem 170(2):235–245

96. Rosa A, Ricciardi G, Baerends EJ, Stufkens DJ (1996) Metal-to-ligand charge transfer (MLCT) photochemistry of fac-Mn(Cl)(CO)3(H-DAB): a density functional study. J Phys Chem 100(38):15346–15357

97. Kottelat E, Ruggi A, Zobi F (2016) Red-light activated PhotoCORMs of Mn(I) species bearing electron deficient 2,2′-Azopyridines. Dalton Trans 45(16):6920–6927

98. Kleverlaan CJ, Hartl F, Stufkens DJ (1997) Real-time fourier transform IR (FTIR) spectroscopy in organometallic chemistry: mechanistic aspects of the fac to mer photoisomerization of fac-[Mn(Br)(CO)$_3$(R-DAB)]. J Photochem Photobiol A Chem 103 (3):231–237

99. Govender P, Pai S, Schatzschneider U, Smith GS (2013) Next generation PhotoCORMs: polynuclear tricarbonylmanganese(I)-functionalized polypyridyl metallodendrimers. Inorg Chem 52(9):5470–5478

100. Gonzalez MA, Yim MA, Cheng S, Moyes A, Hobbs AJ, Mascharak PK (2012) Manganese carbonyls bearing tripodal polypyridine ligands as photoactive carbon monoxide-releasing molecules. Inorg Chem 51(1):601–608

101. Fei H, Sampson MD, Lee Y, Kubiak CP, Cohen SM (2015) Photocatalytic CO_2 reduction to formate using a Mn(I) molecular catalyst in a robust metal-organic framework. Inorg Chem 54(14):6821–6828

102. Amsterdam W (1996) Alkyl-dependent photochemistry of $Mn(R)(CO)_3(R'\text{-DAB})$ (R = Me, Bz; R' = iPr, pTol): homolysis of the Mn-R bond for R = Bz and release of CO for R = Me. Inorg Chim Acta 15

103. Machan CW, Stanton CJ, Vandezande JE, Majetich GF, Schaefer HF, Kubiak CP, Agarwal J (2015) Electrocatalytic reduction of carbon dioxide by $Mn(CN)(2,2'\text{-Bipyridine})(CO)_3$: CN coordination alters mechanism. Inorg Chem 54(17):8849–8856

104. Agarwal J, Iii CJS, Shaw TW, Vandezande JE, Majetich GF, Bocarsly AB, Iii HFS (2015) Exploring the effect of axial ligand substitution (X = Br, NCS, CN) on the photodecomposition and electrochemical activity of $[MnX(N–C)(CO)_3]$ complexes. Dalton Trans 44 (5):2122–2131

105. Kuo H-Y, Lee TS, Chu AT, Tignor SE, Scholes GD, Bocarsly AB (2018, Submitted) A Cyanide-Bridged Di-Manganese carbonyl complex that photochemically reduces CO_2 to CO. Dalton Trans

106. Froehlich JD, Kubiak CP (2012) Homogeneous CO_2 reduction by Ni(Cyclam) at a glassy carbon electrode. Inorg Chem 51(7):3932–3934

107. Beley M, Collin JP, Ruppert R, Sauvage JP (1986) Electrocatalytic reduction of carbon dioxide by nickel $Cyclam^{2+}$ in water: study of the factors affecting the efficiency and the selectivity of the process. J Am Chem Soc 108(24):7461–7467

108. Song J, Klein EL, Neese F, Ye S (2014) The mechanism of homogeneous CO_2 reduction by Ni(Cyclam): product selectivity, concerted proton-electron transfer and C-O bond cleavage. Inorg Chem 53(14):7500–7507

109. Wu Y, Rudshteyn B, Zhanaidarova A, Froehlich JD, Ding W, Kubiak CP, Batista VS (2017) Electrode-ligand interactions dramatically enhance CO_2 conversion to CO by the [Ni (Cyclam)]$(PF_6)_2$ catalyst. ACS Catal 7(8):5282–5288

110. Schneider J, Jia H, Kobiro K, Cabelli DE, Muckerman JT, Fujita E (2012) Nickel(II) macrocycles: highly efficient electrocatalysts for the selective reduction of CO_2 to CO. Energy Environ Sci 5(11):9502–9510

111. Neri G, Aldous IM, Walsh JJ, Hardwick LJ, Cowan AJ (2016) A highly active nickel electrocatalyst shows excellent selectivity for CO_2 reduction in acidic media. Chem Sci 7 (2):1521–1526

112. Froehlich JD, Kubiak CP (2015) The homogeneous reduction of CO_2 by [Ni(Cyclam)]$^+$: increased catalytic rates with the addition of a CO scavenger. J Am Chem Soc 137 (10):3565–3573

113. Hammouche M, Lexa D, Savéant JM, Momenteau M (1988) Catalysis of the electrochemical reduction of carbon dioxide by Iron(0) porphyrins. J Electroanal Chem Interfacial Electrochem 249(1):347–351

114. Costentin C, Drouet S, Passard G, Robert M, Savéant J-M (2013) Proton-coupled electron transfer cleavage of heavy-atom bonds in electrocatalytic processes. Cleavage of a C–O bond in the catalyzed electrochemical reduction of CO_2. J Am Chem Soc 135(24):9023–9031

115. Ambre RB, Daniel Q, Fan T, Chen H, Zhang B, Wang L, Ahlquist MSG, Duan L, Sun L (2016) Molecular engineering for efficient and selective iron porphyrin catalysts for electrochemical reduction of CO_2 to CO. Chem Commun 52(100):14478–14481

116. Costentin C, Robert M, Savéant J-M, Tatin A (2015) Efficient and selective molecular catalyst for the CO_2-to-CO electrochemical conversion in water. PNAS 112(22):6882–6886

117. Azcarate I, Costentin C, Robert M, Savéant J-M (2016) Through-space charge interaction substituent effects in molecular catalysis leading to the design of the most efficient catalyst of CO_2-to-CO electrochemical conversion. J Am Chem Soc 138(51):16639–16644

118. Azcarate I, Costentin C, Robert M, Savéant J-M (2016) Dissection of electronic substituent effects in multielectron–multistep molecular catalysis. Electrochemical CO_2-to-CO conversion catalyzed by iron porphyrins. J Phys Chem C 120(51):28951–28960

119. Costentin C, Passard G, Robert M, Savéant J-M (2014) Ultraefficient homogeneous catalyst for the CO_2-to-CO electrochemical conversion. Proc Natl Acad Sci U S A 111(42):14990–14994

120. Mohamed EA, Zahran ZN, Naruta Y (2015) Efficient electrocatalytic CO_2 reduction with a molecular cofacial iron porphyrin dimer. Chem Commun 51(95):16900–16903

121. Zahran ZN, Mohamed EA, Naruta Y (2016) Bio-inspired cofacial Fe porphyrin dimers for efficient electrocatalytic CO_2 to CO conversion: overpotential tuning by substituents at the porphyrin rings. Sci Rep 6

122. Fukuzumi S, Lee Y-M, Ahn HS, Nam W (2018) Mechanisms of catalytic reduction of CO_2 with heme and nonheme metal complexes. Chem Sci 9(28):6017–6034

123. Loewen ND, Thompson EJ, Kagan M, Banales CL, Myers TW, Fettinger JC, Berben LA (2016) A pendant proton shuttle on $[Fe_4N(CO)_{12}]^-$ alters product selectivity in formate vs. H_2 production via the hydride $[H–Fe_4N(CO)_{12}]^-$. Chem Sci 7(4):2728–2735

124. Taheri A, Thompson EJ, Fettinger JC, Berben LA (2015) An iron electrocatalyst for selective reduction of CO_2 to formate in water: including thermochemical insights. ACS Catal 5(12):7140–7151

125. Taheri A, Carr CR, Berben LA (2018) Electrochemical methods for assessing kinetic factors in the reduction of CO_2 to formate: implications for improving electrocatalyst design. ACS Catal 8(7):5787–5793

126. Taheri A, Loewen ND, Cluff DB, Berben LA (2018) Considering a possible role for $[H-Fe_4N(CO)_{12}]^{2-}$ in selective electrocatalytic CO_2 reduction to formate by $[Fe_4N(CO)_{12}]^-$. Organometallics 37(7):1087–1091

127. Cao Z, Kim D, Hong D, Yu Y, Xu J, Lin S, Wen X, Nichols EM, Jeong K, Reimer JA et al (2016) A molecular surface functionalization approach to tuning nanoparticle electrocatalysts for carbon dioxide reduction. J Am Chem Soc 138(26):8120–8125

128. Chung MW, Cha IY, Ha MG, Na Y, Hwang J, Ham HC, Kim H-J, Henkensmeier D, Yoo SJ, Kim JY et al (2018) Enhanced CO_2 reduction activity of polyethylene glycol-modified Au nanoparticles prepared via liquid medium sputtering. Appl Catal B Environ 237:673–680

129. Zhang S, Kang P, Ubnoske S, Brennaman MK, Song N, House RL, Glass JT, Meyer TJ (2014) Polyethylenimine-enhanced electrocatalytic reduction of CO_2 to formate at nitrogen-doped carbon nanomaterials. J Am Chem Soc 136(22):7845–7848

130. Maurin A, Robert M (2016) Noncovalent immobilization of a molecular iron-based electrocatalyst on carbon electrodes for selective, efficient CO_2-to-CO conversion in water. J Am Chem Soc 138(8):2492–2495

131. Pander JE, Fogg A, Bocarsly AB (2016) Utilization of electropolymerized films of cobalt porphyrin for the reduction of carbon dioxide in aqueous media. ChemCatChem 8 (22):3536–3545

132. Shen J, Kortlever R, Kas R, Birdja YY, Diaz-Morales O, Kwon Y, Ledezma-Yanez I, Schouten KJP, Mul G, Koper MTM (2015) Electrocatalytic reduction of carbon dioxide to carbon monoxide and methane at an immobilized cobalt protoporphyrin. Nat Commun 6(1)

133. Morlanés N, Takanabe K, Rodionov V (2016) Simultaneous reduction of CO_2 and splitting of H_2O by a single immobilized cobalt phthalocyanine electrocatalyst. ACS Catal 6 (5):3092–3095

134. Zhang X, Wu Z, Zhang X, Li L, Li Y, Xu H, Li X, Yu X, Zhang Z, Liang Y et al (2017) Highly selective and active CO_2 reduction electrocatalysts based on cobalt phthalocyanine/carbon nanotube hybrid structures. Nat Commun 8:14675

135. Weng Z, Jiang J, Wu Y, Wu Z, Guo X, Materna KL, Liu W, Batista VS, Brudvig GW, Wang H (2016) Electrochemical CO_2 reduction to hydrocarbons on a heterogeneous molecular Cu catalyst in aqueous solution. J Am Chem Soc 138(26):8076–8079

136. Farrusseng D, Aguado S, Pinel C (2009) Metal-organic frameworks: opportunities for catalysis. Angewandte Chemie Int Edn 48(41):7502–7513
137. Lin S, Diercks CS, Zhang Y-B, Kornienko N, Nichols EM, Zhao Y, Paris AR, Kim D, Yang P, Yaghi OM et al (2015) Covalent organic frameworks comprising cobalt porphyrins for catalytic CO_2 reduction in water. Science 349(6253):1209–1213
138. Kornienko N, Zhao Y, Kley CS, Zhu C, Kim D, Lin S, Chang CJ, Yaghi OM, Yang P (2015) Metal-organic frameworks for electrocatalytic reduction of carbon dioxide. J Am Chem Soc 137(44):14129–14135
139. Zhao C, Dai X, Yao T, Chen W, Wang X, Wang J, Yang J, Wei S, Wu Y, Li Y (2017) Ionic exchange of metal-organic frameworks to access single nickel sites for efficient electroreduction of CO_2. J Am Chem Soc 139(24):8078–8081
140. Huan TN, Ranjbar N, Rousse G, Sougrati M, Zitolo A, Mougel V, Jaouen F, Fontecave M (2017) Electrochemical reduction of CO_2 catalyzed by Fe-N-C materials: a structure-selectivity study. ACS Catal 7(3):1520–1525
141. Albo J, Vallejo D, Beobide G, Castillo O, Castaño P, Irabien A (2017) Copper-based metal-organic porous materials for CO_2 electrocatalytic reduction to alcohols. Chemsuschem 10(6):1100–1109
142. Ghijsen J, Tjeng LH, van Elp J, Eskes H, Westerink J, Sawatzky GA, Czyzyk MT (1988) Electronic structure of Cu_2O and CuO. Phys Rev B 38(16):11322–11330
143. Hardee KI, Bard AJX (1977) Photoelectrochemical behavior of several polycrystalline metal oxide electrodes in aqueous solutions. J Electrochem Soc 124(2):10
144. Tennakone K, Jayatissa AH, Punchihewa S (1989) Selective photoreduction of carbon dioxide to methanol with hydrous cuprous oxide. J Photochem Photobiol A Chem 49 (3):369–375
145. Janáky C, Hursán D, Endrődi B, Chanmanee W, Roy D, Liu D, de Tacconi NR, Dennis BH, Rajeshwar K (2016) Electro- and photoreduction of carbon dioxide: the twain shall meet at copper oxide/copper interfaces. ACS Energy Lett 1(2):332–338
146. Ba X, Yan L-L, Huang S, Yu J, Xia X-J, Yu Y (2014) New way for CO_2 reduction under visible light by a combination of a Cu electrode and semiconductor thin film: Cu_2O conduction type and morphology effect. J Phys Chem C 118(42):24467–24478
147. Li CW, Kanan MW (2012) CO_2 reduction at low overpotential on Cu electrodes resulting from the reduction of thick Cu_2O films. J Am Chem Soc 134(17):7231–7234
148. Ghadimkhani G, de Tacconi NR, Chanmanee W, Janaky C, Rajeshwar K (2013) Efficient solar photoelectrosynthesis of methanol from carbon dioxide using hybrid $CuO-Cu_2O$ semiconductor nanorod arrays. Chem Commun 49(13):1297
149. Kecsenovity E, Endrődi B, Pápa Z, Hernádi K, Rajeshwar K, Janáky C (2016) Decoration of ultra-long carbon nanotubes with Cu_2O nanocrystals: a hybrid platform for enhanced photoelectrochemical CO_2 reduction. J Mater Chem A 4(8):3139–3147
150. Parkinson BA, Weaver PF (1984) Photoelectrochemical pumping of enzymatic CO_2 reduction. Nature 309(5964):148–149
151. Halmann M (1978) Photoelectrochemical reduction of aqueous carbon dioxide on p-type gallium phosphide in liquid junction solar cells. Nature 275(5676):115–116
152. Kočí K, Obalová L, Matějová L, Plachá D, Lacný Z, Jirkovský J, Šolcová O (2009) Effect of TiO_2 particle size on the photocatalytic reduction of CO_2. Appl Catal B Environ 89(3):494–502
153. Lo C-C, Hung C-H, Yuan C-S, Wu J-F (2007) Photoreduction of carbon dioxide with H_2 and H_2O over TiO_2 and ZrO_2 in a circulated photocatalytic reactor. Solar Energy Mater Solar Cells 91(19):1765–1774
154. Perini JAL, Cardoso JC, de Brito JF, Zanoni MVB (2018) Contribution of thin films of ZrO_2 on TiO_2 nanotubes electrodes applied in the photoelectrocatalytic CO_2 conversion. J CO_2 Utilization 25:254–263
155. Morterra C, Orio L (1990) Surface characterization of zirconium oxide. II. The interaction with carbon dioxide at ambient temperature. Mater Chem Phys 24(3):247–268

156. Bachiller-Baeza B, Rodriguez-Ramos I, Guerrero-Ruiz A (1998) Interaction of carbon dioxide with the surface of zirconia polymorphs. Langmuir 14(13):3556–3564
157. Shen Q, Chen Z, Huang X, Liu M, Zhao G (2015) High-yield and selective photoelectrocatalytic reduction of CO_2 to formate by metallic copper decorated Co_3O_4 nanotube arrays. Environ Sci Technol 49(9):5828–5835
158. Jang J-W, Cho S, Magesh G, Jang YJ, Kim JY, Kim WY, Seo JK, Kim S, Lee K-H, Lee JS (2014) Aqueous-solution route to zinc telluride films for application to CO_2 reduction. Angewandte Chemie Int Edn 53(23):5852–5857
159. Jang YJ, Jang J-W, Lee J, Kim JH, Kumagai H, Lee J, Minegishi T, Kubota J, Domen K, Lee JS (2015) Selective CO production by Au coupled ZnTe/ZnO in the photoelectrochemical CO_2 reduction system. Energy Environ Sci 8(12):3597–3604
160. Gu J, Wuttig A, Krizan JW, Hu Y, Detweiler ZM, Cava RJ, Bocarsly AB (2013) Mg-doped $CuFeO_2$ photocathodes for photoelectrochemical reduction of carbon dioxide. J Phys Chem C 117(24):12415–12422
161. Kang U, Choi SK, Ham DJ, Ji SM, Choi W, Han DS, Abdel-Wahab A, Park H (2015) Photosynthesis of formate from CO_2 and water at 1% energy efficiency via copper iron oxide catalysis. Energy Environ Sci 8(9):2638–2643
162. Jeong HW, Jeon TH, Jang JS, Choi W, Park H (2013) Strategic modification of $BiVO_4$ for improving photoelectrochemical water oxidation performance. J Phys Chem C 117 (18):9104–9112
163. Kang U, Park H (2017) A facile synthesis of $CuFeO_2$ and CuO composite photocatalyst films for the production of liquid formate from CO_2 and water over a month. J Mater Chem A 5(5):2123–2131
164. Kamimura S, Murakami N, Tsubota T, Ohno T (2015) Fabrication and characterization of a p-type $Cu_3Nb_2O_8$ photocathode toward photoelectrochemical reduction of carbon dioxide. Appl Catal B Environ 174–175:471–476
165. Boettcher SW, Warren EL, Putnam MC, Santori EA, Turner-Evans D, Kelzenberg MD, Walter MG, McKone JR, Brunschwig BS, Atwater HA et al (2011) Photoelectrochemical hydrogen evolution using Si microwire arrays. J Am Chem Soc 133(5):1216–1219
166. Choi SK, Kang U, Lee S, Ham DJ, Ji SM, Park H (2014) Sn-coupled p-Si nanowire arrays for solar formate production from CO_2. Adv Energy Mater 4(11):1301614
167. Hinogami R, Nakamura Y, Yae S, Nakato Y (1998) An approach to ideal semiconductor electrodes for efficient photoelectrochemical reduction of carbon dioxide by modification with small metal particles. J Phys Chem B 102(6):974–980
168. Kuang Y, Di Vece M, Rath JK, van Dijk L, Schropp REI (2013) Elongated nanostructures for radial junction solar cells, vol 76
169. Ohno T, Murakami N, Koyanagi T, Yang Y (2014) Photocatalytic reduction of CO_2 over a hybrid photocatalyst composed of WO_3 and graphitic carbon nitride (g-C_3N_4) under visible light. J CO_2 Utilization 6:17–25
170. Sagara N, Kamimura S, Tsubota T, Ohno T (2016) Photoelectrochemical CO_2 reduction by a p-type boron-doped g-C_3N_4 electrode under visible light. Appl Catal B Environ 192:193–198
171. Wang Y, Li H, Yao J, Wang X, Antonietti M (2011) Synthesis of boron doped polymeric carbon nitride solids and their use as metal-free catalysts for aliphatic C-H bond oxidation. Chem Sci 2(3):446–450
172. Zhang Y, Sethuraman V, Michalsky R, Peterson AA. Competition between CO_2 reduction and H_2 evolution on transition-metal electrocatalysts
173. Tinnemans AHA, Koster TPM, Thewissen DHMW, Mackor A. Tetraaza-macrocyclic cobalt (II) and nickel(II) complexes as electron-transfer agents in the photo(electro)chemical and electrochemical reduction of carbon dioxide. Recueil des Travaux Chimiques des Pays-Bas 103(10):288–295

174. Jeon JH, Mareeswaran PM, Choi CH, Woo SI (2014) Synergism between CdTe semiconductor and pyridine—photoenhanced electrocatalysis for CO_2 reduction to formic acid. RSC Adv 4(6):3016–3019

175. Barton Cole E, Lakkaraju PS, Rampulla DM, Morris AJ, Abelev E, Bocarsly AB (2010) Using a one-electron shuttle for the multielectron reduction of CO_2 to methanol: kinetic, mechanistic, and structural insights. J Am Chem Soc 132(33):11539–11551

176. Barton EE, Rampulla DM, Bocarsly AB (2008) Selective solar-driven reduction of CO_2 to methanol using a catalyzed p-GaP based photoelectrochemical cell. J Am Chem Soc 130 (20):6342–6344

177. Cole EB, Bocarsly AB (2010) Photochemical, electrochemical, and photoelectrochemical reduction of carbon dioxide. In: Carbon dioxide as chemical feedstock. Wiley-Blackwell, pp 291–316

178. Keets K, Morris A, Zeitler E, Lakkaraju P, Bocarsly A (2010) Catalytic conversion of carbon dioxide to methanol and higher order alcohols at a photoelectrochemical interface. In: Proceedings of SPIE—the international society for optical engineering 7770

179. Bocarsly AB, Gibson QD, Morris AJ, L'Esperance RP, Detweiler ZM, Lakkaraju PS, Zeitler EL, Shaw TW (2012) Comparative study of imidazole and pyridine catalyzed reduction of carbon dioxide at illuminated iron pyrite electrodes. ACS Catal 2(8):1684–1692

180. Ganesh I, Kumar PP, Annapoorna I, Sumliner JM, Ramakrishna M, Hebalkar NY, Padmanabham G, Sundararajan G (2014) Preparation and characterization of Cu-doped TiO_2 materials for electrochemical, photoelectrochemical, and photocatalytic applications. Appl Surf Sci 293:229–247

181. Bak T, Nowotny J, Rekas M, Sorrell CC (2002) Photo-electrochemical hydrogen generation from water using solar energy. Materials-related aspects. Int J Hydrogen Energy 27(10):991–1022

182. Zeng G, Qiu J, Li Z, Pavaskar P, Cronin SB (2014) CO_2 reduction to methanol on TiO_2-passivated GaP photocatalysts. ACS Catal 4(10):3512–3516

183. Yan Y, Zeitler EL, Gu J, Hu Y, Bocarsly AB (2013) Electrochemistry of aqueous pyridinium: exploration of a key aspect of electrocatalytic reduction of CO_2 to methanol. J Am Chem Soc 135(38):14020–14023

184. Yuan J, Hao C (2013) Solar-driven photoelectrochemical reduction of carbon dioxide to methanol at $CuInS_2$ thin film photocathode. Solar Energy Mater Solar Cells 108:170–174

185. Yuan J, Wang P, Hao C, Yu G (2016) Photoelectrochemical reduction of carbon dioxide at $CuInS_2$/graphene hybrid thin film electrode. Electrochim Acta 193:1–6

186. Zhang N, Long R, Gao C, Xiong Y (2018) Recent progress on advanced design for photoelectrochemical reduction of CO_2 to fuels. Sci China Mater 61(6):771–805

187. Bachmeier A, Hall S, Ragsdale SW, Armstrong FA (2014) Selective visible-light-driven CO_2 reduction on a p-Type dye-sensitized NiO photocathode. J Am Chem Soc 136 (39):13518–13521

188. Kumagai H, Sahara G, Maeda K, Higashi M, Abe R, Ishitani O (2017) Hybrid photocathode consisting of a $CuGaO_2$ p-type semiconductor and a Ru(II)–Re(I) supramolecular photocatalyst: non-biased visible-light-driven CO_2 reduction with water oxidation. Chem Sci 8(6):4242–4249

189. Hye Won D, Chung J, Hyeon Park S, Kim E-H, Ihl Woo S (2015) Photoelectrochemical production of useful fuels from carbon dioxide on a polypyrrole-coated p-ZnTe photocathode under visible light irradiation. J Mater Chem A 3(3):1089–1095

190. Sekizawa K, Sato S, Arai T, Morikawa T (2018) Solar-driven photocatalytic CO_2 reduction in water utilizing a ruthenium complex catalyst on p-type Fe_2O_3 with a multiheterojunction. ACS Catal 8(2):1405–1416

191. Guzmán D, Isaacs M, Osorio-Román I, García M, Astudillo J, Ohlbaum M (2015) Photoelectrochemical reduction of carbon dioxide on quantum-dot-modified electrodes by electric field directed layer-by-layer assembly methodology. ACS Appl Mater Interfaces 7 (36):19865–19869

192. White JL, Baruch MF, Pander JE, Hu Y, Fortmeyer IC, Park JE, Zhang T, Liao K, Gu J, Yan Y et al (2015) Light-driven heterogeneous reduction of carbon dioxide: photocatalysts and photoelectrodes. Chem Rev 115(23):12888–12935

193. Cheng J, Zhang M, Wu G, Wang X, Zhou J, Cen K (2014) Photoelectrocatalytic reduction of CO_2 into chemicals using Pt-modified reduced graphene oxide combined with Pt-modified TiO_2 nanotubes. Environ Sci Technol 48(12):7076–7084

194. Cheng J, Zhang M, Wu G, Wang X, Zhou J, Cen K (2015) Optimizing CO_2 reduction conditions to increase carbon atom conversion using a Pt-RGO‖Pt-TNT photoelectrochemical cell. Solar Energy Mater Solar Cells 132:606–614

195. Cheng J, Zhang M, Liu J, Zhou J, Cen K (2015) A Cu foam cathode used as a Pt–RGO catalyst matrix to improve CO_2 reduction in a photoelectrocatalytic cell with a TiO_2 photoanode. J Mater Chem A 3(24):12947–12957

196. Chang X, Wang T, Zhang P, Wei Y, Zhao J, Gong J. Stable aqueous photoelectrochemical CO_2 reduction by a Cu_2O dark cathode with improved selectivity for carbonaceous products. Angewandte Chemie Int Edn 55(31):8840–8845

197. Magesh G, Kim ES, Kang HJ, Banu M, Kim JY, Kim JH, Lee JS (2014) A versatile photoanode-driven photoelectrochemical system for conversion of CO_2 to fuels with high faradaic efficiencies at low bias potentials. J Mater Chem A 2(7):2044

198. Song JT, Iwasaki T, Hatano M (2015) Photoelectrochemical CO_2 reduction on 3C-SiC photoanode in aqueous solution. Jpn J Appl Phys 54(4S):04DR05

199. Zhang Y, Luc W, Hutchings GS, Jiao F (2016) Photoelectrochemical carbon dioxide reduction using a nanoporous Ag cathode. ACS Appl Mater Interfaces 8(37):24652–24658

200. May PW (2000) Diamond thin films: a 21st-century material. Philos Trans R Soc Lond A Math Phys Eng Sci 358(1766):473–495

201. Ekimov EA, Sidorov VA, Bauer ED, Mel'nik NN, Curro NJ, Thompson JD, Stishov SM (2004) Superconductivity in diamond. Nature 428(6982):542–545

202. Balasubramanian G, Neumann P, Twitchen D, Markham M, Kolesov R, Mizuochi N, Isoya J, Achard J, Beck J, Tissler J et al (2009) Ultralong spin coherence time in isotopically engineered diamond. Nat Mater 8(5):383–387

203. Liu Y, Chen S, Quan X, Yu H (2015) Efficient electrochemical reduction of carbon dioxide to acetate on nitrogen-doped nanodiamond. J Am Chem Soc 137(36):11631–11636

204. Roy N, Hirano Y, Kuriyama H, Sudhagar P, Suzuki N, Katsumata K, Nakata K, Kondo T, Yuasa M, Serizawa I et al (2016) Boron-doped diamond semiconductor electrodes: efficient photoelectrochemical CO_2 reduction through surface modification. Sci Rep 6(1)

Plasma-Based CO$_2$ Conversion

8

Annemie Bogaerts and Ramses Snoeckx

Abstract

In this chapter, we will explain why plasma is promising for CO$_2$ conversion. First, we will give a brief introduction on plasma technology (Sect. 8.1), and highlight its unique feature for CO$_2$ conversion (Sect. 8.2). Next, we will briefly illustrate the most common types of plasma reactors, explaining why some plasma types exhibit better energy efficiency than others (Sect. 8.3). In Sect. 8.4, we will present the state-of-the-art on plasma-based CO$_2$ conversion, for pure CO$_2$ splitting and the combined conversion of CO$_2$ with either CH$_4$, H$_2$O or H$_2$, for different types of plasma reactors. To put plasma technology in a broader perspective of emerging technologies for CO$_2$ conversion, we will discuss in Sect. 8.5 its inherent promising characteristics for this application. Finally, in Sect. 8.6 we will summarize the state-of-the-art and the current limitations, and elaborate on future research directions needed to bring plasma-based CO$_2$ conversion into real application.

A. Bogaerts (✉)
Research Group PLASMANT, Department of Chemistry, University of Antwerp, Universiteitsplein 1, 2610 Antwerp, Belgium
e-mail: annemie.bogaerts@uantwerpen.be

R. Snoeckx
Physical Science and Engineering Division (PSE), Clean Combustion Research Center (CCRC), King Abdullah University of Science and Technology (KAUST), Thuwal 23955, Saudi Arabia
e-mail: ramses.snoeckx@kaust.edu.sa

© Springer Nature Switzerland AG 2019
M. Aresta et al. (eds.), *An Economy Based on Carbon Dioxide and Water*,
https://doi.org/10.1007/978-3-030-15868-2_8

8.1 Plasma, the Fourth State of Matter

One of the emerging technologies for CO_2 conversion is based on plasma technology. Plasma is an ionized gas, consisting of molecules, but also electrons, various types of ions, radicals, excited species, and photons. This reactive cocktail makes plasma useful for a wide range of applications, including materials technology and microelectronics (for thin film deposition, surface modification, etching,...), medical applications, light sources, lasers, plasma displays, as well as for environmental and energy applications, such as air pollution control and gas conversion [1]. This chapter focuses on the application of CO_2 conversion into value-added chemicals and fuels.

Plasma is also called the "fourth state of matter", next to solid, liquid, and gas. It is formed by increasing the temperature of a gas, or by inserting another form of energy, like electrical energy. More than 99% of the visible matter in the universe is in plasma state, e.g., our Sun and the other stars, as well as the interstellar matter. Closer to Earth, Saint Elmo's fire, lightning, red sprites, the Aurora Borealis and Australis are also natural plasmas.

Beside natural plasmas, plasmas are also created on Earth, and we can distinguish two main groups, i.e., (i) high-temperature plasmas used for fusion research, which are typically fully ionized, and (ii) so-called low-temperature gas discharge plasmas, which are partially ionized (typical ionization degree of 10^{-4} to 10^{-6}). The latter are used for CO_2 conversion, and will be discussed in this chapter.

Plasma can be either in thermal equilibrium or not. Plasma contains many different species, as mentioned above, and they can all exhibit a different temperature (or average energy). When the temperature of all species is the same, the plasma is said to be in 'thermal equilibrium'. Alternatively, when the temperature of the (light) electrons is much higher than for the other (heavy) species, the plasma is called 'non-thermal plasma'.

Non-thermal plasmas are much more suitable for CO_2 conversion. Indeed, in thermal plasmas, the maximum energy efficiency is limited to the thermodynamic equilibrium efficiency [2], while non-thermal plasmas can overcome these thermal equilibrium values. Indeed, as mentioned above, non-thermal plasmas are created by applying (electrical) energy to a gas. This energy will selectively heat the electrons because of their small mass, and subsequently, the energetic electrons will activate the gas molecules (e.g., CO_2), causing excitation, ionization and dissociation. The excited species, ions and/or radicals are very reactive, and will quickly react into new molecules. Hence, non-thermal plasma allows the activation of stable gas molecules, such as CO_2, in an energy-efficient way, because the gas does not have to be heated as a whole for the chemical conversions to take place. The thermal non-equilibrium between the highly energetic electrons (typically with energy of a few eV, i.e., several 10,000 K) and the gas molecules (virtually at room temperature up to a maximum of a few 1000 K) allows energy-intensive or thermodynamically unfavorable chemical reactions, such as CO_2 splitting or dry reforming of methane (DRM), to be carried out at mild operating conditions (typically atmospheric pressure and near room temperature).

Besides the higher temperature of the electrons, the vibrationally excited molecules also typically exhibit a higher temperature than the other plasma species, i.e., neutral gas molecules and ions, and this is one of the underlying reasons for the energy-efficient CO_2 conversion, as will be explained below [3].

The fact that the gas itself must not be heated as a whole makes that plasma can nearly instantaneously be switched on and off, i.e., so-called "turn-key" process. Conversion and product yields stabilize typically within 30 min. Furthermore, the power consumption can easily be scaled, making plasma a suitable technology for utilizing excess fluctuating renewable electricity (e.g. wind or solar energy), to store it in a chemical form. Hence, plasma technology might provide a solution for the current challenges on efficient storage and transport of renewable electricity, i.e., peak shaving and grid stabilization.

In the next section, we will discuss why plasma can provide good energy efficiency for CO_2 conversion. In Sect. 8.3, we will describe the different kinds of plasma reactors used for CO_2 conversion, with their advantages and disadvantages, and we will outline the promising combination of plasma with catalysts (i.e., plasma catalysis). Section 8.4 will present the state-of-the-art on plasma-based CO_2 conversion for different plasma reactor types, both for pure CO_2 splitting and for the combined conversion of CO_2 with either CH_4, H_2O or H_2. In Sect. 8.5, we will summarize the main advantages of plasma-based CO_2 conversion with respect to other emerging technologies. Finally, Sect. 8.6 will discuss the future research directions for plasma-based CO_2 conversion.

8.2 Plasma and Its Unique Feature for Energy-Efficient CO_2 Conversion

Besides thermal and non-thermal plasmas, mentioned in previous section, there is a transitional type of plasma, so-called "warm plasma", which operates at conditions in between thermal and non-thermal plasmas, and which seems very promising for CO_2 conversion. It allows a high power (yielding a high electron density) as well as thermal non-equilibrium, to selectively populate the vibrationally excited states (see below) [4]. Its temperature can reach up to 2000–3000 K, which can also be beneficial for the chemical kinetics in the plasma.

For energy-efficient CO_2 conversion, the crucial parameter is the electron energy distribution. If most of the electron energy is around 1 eV, the electrons will mainly give rise to vibrational excitation of CO_2, as discussed below, which provides the most energy-efficient channel for CO_2 dissociation [3, 5]. The higher gas temperature of warm plasma might also be an unwanted effect, as revealed by modeling [6–8], to create the strongest thermal non-equilibrium.

To understand better why different discharge types (as described in the next section) exhibit different energy efficiency, we need to explain the different channels in which electrons can transfer their energy to CO_2 molecules, i.e., through excitation, ionization and dissociation. Figure 8.1 illustrates the fraction of electron energy transferred to these different channels, as a function of the reduced electric field (E/n) [2, 3, 9]. The reduced electric field is defined as the ratio of the electric field over the gas number density in the plasma. It is a very important characteristic, which defines the differences between different plasma types. A dielectric barrier discharge (DBD; see Sect. 8.3.1), which is a typical non-thermal plasma, has a reduced electric field above 100-200 Td (Townsend; 1 Td = 10^{-21} V m^2), whereas microwave (MW) and gliding arc (GA) discharges, which are defined as warm plasmas; see Sects. 8.3.2 and 8.3.3, typically operate around 50 Td.

Figure 8.1 clearly illustrates that the reduced electric field determines the fraction of electron energy going to different channels. Above 200 Td, 70–80% of the electron energy goes into electronic excitation, about 5% is used for dissociation and for ionization (increasing with E/n), while only 10% is used for vibrational excitation (decreasing with E/n). On the other hand, at 50 Td only 10% goes into electronic excitation and 90% of the energy is used for vibrational excitation.

The fraction of electron energy going into vibrational excitation is crucial, because the vibrational levels of CO_2 provide the most energy-efficient dissociation path of CO_2, as illustrated in Fig. 8.2. Direct electron impact dissociation proceeds through a dissociative electronically excited state, for which at least 7 eV electron energy is needed (see Fig. 8.2), i.e., much more than the theoretical value needed for C = O bond breaking (5.5 eV). On the other hand, electron impact vibrational excitation of the lowest vibrational levels, followed by vibrational-vibrational (VV) collisions (so-called "ladder-climbing") gradually populates the higher

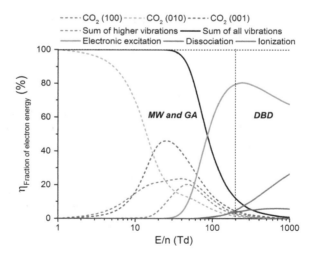

Fig. 8.1 Fraction of electron energy transferred to different channels of excitation, ionization and dissociation of CO_2, as a function of the reduced electric field (E/n), calculated from the corresponding cross sections of the electron impact reactions. The E/n regions characteristic for MW and GA plasma and for DBD plasma are indicated. Adopted from [2] with permission

Fig. 8.2 Schematic diagram of some CO₂ electronic and vibrational levels, illustrating that much more energy is needed for direct electronic excitation–dissociation than for stepwise vibrational excitation, i.e. so-called ladder climbing. Adopted from [9] with permission

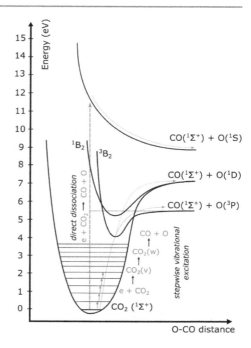

vibrational levels, which eventually lead to dissociation (see Fig. 8.2). This is a much more efficient dissociation pathway, because only the minimum energy of 5.5 eV for bond breaking is required [9].

8.3   Plasma Reactor Types for CO₂ Conversion

As mentioned in the Introduction, plasma is created by applying electrical energy to a gas. In its most simple form, this electrical energy is provided by applying a potential difference between two electrodes. This potential difference can be direct current (DC), alternating current (AC), ranging from 50 Hz over kHz to MHz (radio-frequency; RF), or pulsed. Furthermore, the electrical energy can be provided in other ways as well, e.g., by a coil (inductively coupled plasma; ICP) or as microwaves (MW). The gas pressure in the plasma can vary from a few Torr up to several atm.

   Three types of plasma reactors are most often investigated for CO₂ conversion, i.e., dielectric barrier discharges (DBD), microwave (MW) plasmas and gliding arc (GA) discharges. We will briefly present their characteristic features and typical operating conditions, to explain their strengths and limitations. Besides these three major types of plasma reactors, other plasma types are also being explored for CO₂ conversion, as will also be briefly discussed below. Finally, we will explain the working principles of plasma catalysis, for the selective production of value-added chemicals.

8.3.1 Dielectric Barrier Discharge (DBD)

A DBD is created by applying an AC potential difference between two electrodes, of which at least one is covered by a dielectric barrier. The role of the latter is to limit the charge transported between both electrodes, and thus to prevent the formation of a thermal plasma, which is less efficient for CO_2 conversion (as explained in Sect. 8.1). The most common design for CO_2 conversion is based on two concentric cylindrical electrodes (cf. Figure 8.3a): the inner electrode is surrounded by a dielectric tube, covered by a mesh or foil electrode. One of the electrodes is powered, while the other is grounded. The gas flows through the gap between inner electrode and dielectric tube (with typical dimensions of a few millimeter).

As a DBD typically operates at atmospheric pressure, it is most suitable for industrial applications. It can easily be scaled up, due to its simple design, as demonstrated already long time ago for ozone synthesis, by placing a large number of DBD reactors in parallel [11].

The major limitation of a DBD plasma, however, is its limited energy efficiency for CO_2 conversion (typically around 10% [2]; see next section). The reason is the high reduced electric field (as explained in previous section), creating electrons with too high energy for efficient vibrational excitation of CO_2 molecules, so that this most energy-efficient CO_2 dissociation channel is of minor importance in a DBD (see Fig. 8.2).

We can enhance the energy efficiency in a DBD by introducing a packing of dielectric material in the gap. The applied electrical potential difference will then

Fig. 8.3 Schematic illustration of the three plasma reactors most often used for CO_2 conversion, i.e., dielectric barrier discharge (DBD) (**a**), microwave (MW) plasma (**b**), and gliding arc (GA) discharge, in classical configuration (**c**) and cylindrical geometry, called gliding arc plasmatron (GAP) (**d**). Adopted from [10] with permission

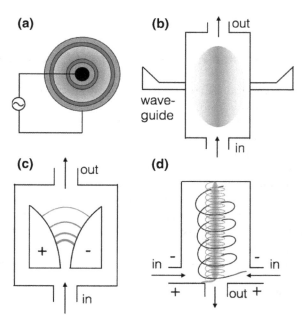

cause polarization of the dielectric packing beads, enhancing the electric field near their contact points, and thus the electron energy [12]. This gives rise to more CO$_2$ conversion (albeit not through the most energy-efficient pathway) for the same applied power. In addition, a packed bed DBD is very suitable for plasma catalysis, as discussed in Sect. 8.3.5 below. Nevertheless, the CO$_2$ conversion efficiency is not always enhanced in a packed-bed DBD [13, 14]. The reason is the competing effect of reduced residence time in the smaller discharge volume (when compared at the same gas flow rate) and the loss of electrons and reactive plasma species at the surface of the packing materials.

8.3.2 Microwave (MW) Plasma

A microwave (MW) plasma is a typical warm plasma, created by applying electromagnetic radiation (with frequency between 300 MHz and 10 GHz) to a gas, without using electrodes. Different types of MW plasmas can be distinguished, depending on their configuration, but for CO$_2$ conversion the so-called surface wave discharges are most often used, in which the gas flows through a quartz tube (transparent to MW radiation), intersecting with a rectangular waveguide, to initiate the discharge (see Fig. 8.3b).

MW plasmas can operate at pressures ranging from 10 mTorr up to atmospheric pressure. The low pressure regime can provide the most energy-efficient CO$_2$ conversion, with maximum values reported up to 90% for supersonic gas flow and pressures around 100–200 Torr [15]. Indeed, these conditions yield typical reduced electric fields around 50 Td, causing electron energies around 1 eV, which give rise to most vibrational excitation of CO$_2$, being the most energy-efficient CO$_2$ dissociation pathway, as explained above [2, 3, 9].

Besides VV relaxation, which typically gradually populates the higher vibrational levels, the vibrational levels can also get lost, by so-called vibrational-translational (VT) relaxation, i.e., upon collision with ground state molecules. Computer models have indicated that this VT relaxation becomes especially important at high gas temperatures [7], resulting in a vibrational distribution function (VDF) close to thermal equilibrium with the gas temperature. MW plasmas at atmospheric pressure exhibit a quite high gas temperature (order of several 1000 K), yielding a VDF close to thermal [7]. Model calculations revealed that deviation from a thermal distribution can be realized by increasing the power density, reducing the pressure and the gas temperature [7]. However, it is not straightforward to realize a low gas temperature at atmospheric pressure. One possible solution might be to apply pulsed power, so that the gas temperature would drop in between the applied pulses, although recent model calculations showed that the vibrational temperature drops faster than the gas temperature, so this might not be a good solution. Another possibility might be to apply a supersonic gas flow, as demonstrated by Asisov et al. [15], so that the gas does not have enough time for being heated. On the other hand, model calculations revealed that the gas might not have a sufficiently long residence time in the plasma, thereby limiting the conversion [16].

8.3.3   Gliding Arc (GA) Discharge

A gliding arc (GA) discharge is also a typical warm plasma. A classical (two-dimensional) GA discharge is created between two flat diverging electrodes (see Fig. 8.3c). The arc is initiated at the shortest interelectrode distance, and it "glides" towards larger interelectrode distance due to the gas flow, until it extinguishes. Subsequently, a new arc is created at the shortest inter-electrode distance.

The classical GA plasma, however, exhibits some disadvantages. The flat 2D electrode geometry, makes a classical GA plasma less compatible with industrial systems. Moreover, a considerable fraction of the gas does not pass through the active plasma region, so that the gas conversion is quite limited. Finally, a high gas flow rate is required for the arc to glide along the electrodes, limiting the gas residence time, and thus the conversion.

Due to these limitations of the classical GA plasma, other types of (three-dimensional) GA discharges have been designed, such as a gliding arc plasmatron (GAP) and a rotating GA, which both operate between cylindrical electrodes. The operating principle of the GAP is illustrated in Fig. 8.3d [17]. The cylindrical reactor body functions as powered electrode (cathode), while the reactor outlet is at anode potential (grounded). The gas enters tangentially between both electrodes, and when the outlet diameter is smaller than the reactor diameter, it flows in an outer vortex towards the upper part of the reactor, creating an insulating and cooling effect, thus protecting the reactor walls from the warm plasma arc in the center. Once the gas reaches the upper end of the reactor, it will start flowing downwards, but with a smaller vortex, because it has lost some speed due to friction and inertia. This so-called reverse vortex causes mixing of the gas with the plasma arc, resulting in enhanced conversion compared to a classical GA. Indeed, the arc is initiated at the shortest interelectrode distance, but after stabilization (at ca. 1 ms), it rotates around the reactor axis, so that the inner gas vortex passes through the arc. Nevertheless, computer modeling has revealed that the fraction of gas passing through the arc is still rather limited, thereby limiting the overall CO_2 conversion [18, 19].

As the GA discharge operates at atmospheric pressure, it is highly suitable for industrial implementation. Its energy efficiency is fairly good, i.e., around 30% for CO_2 splitting [18] and 60% for DRM [19]. This is again attributed to the reduced electric field and electron energy being most favorable for vibrational excitation of CO_2, thus promoting the vibrational pathway of CO_2 dissociation, like in a MW plasma. On the other hand, the gas temperature is also quite high (typically a few 1000 K), again causing a lot of VT relaxation, and thus a nearly thermal VDF, limiting the energy efficiency [20, 21]. We believe the energy efficiency could be further enhanced by reducing the gas temperature, so that the non-equilibrium behavior of a GA plasma can be better exploited.

8.3.4 Other Plasma Types Used for CO_2 Conversion

Other plasma types, such as nanosecond (ns)-pulsed discharges [22], spark discharges [23], corona discharges [24] and atmospheric pressure glow discharges (APGDs) [25] are also gaining increasing interest for CO_2 conversion.

Ns-pulsed discharges are created by applying ns-pulses, causing strong non-equilibrium with very high plasma densities at a relatively low power consumption, due to the short pulse duration. The short pulses might also allow good control of the electron energy, so that more energy might be directed towards the desired dissociation channels.

Spark discharges are based on streamers between two electrodes, which develop into highly energetic spark channels, extinguishing and reigniting periodically, even without pulsed power supply.

Corona discharges are created near sharp electrode edges. We can distinguish between negative or positive corona discharges, depending on whether a negative or positive voltage is applied at the electrode. Corona discharges exhibit a strong electric field near the sharp electrode, while the charged particles are dragged to the other electrode by a weak electric field. They are not very efficient for CO_2 conversion, due to a too high reduced electric field, like for DBDs.

Finally, "APGD" stands for a collection of several plasma types operating at atmospheric pressure, including miniaturized DC glow discharges, microhollow cathode DC discharges, RF discharges and DBDs. When they operate in a homogeneous glow mode, their gas temperature is around 900 K, hence considerably lower than for GA and MW plasmas. This is beneficial to create thermal non-equilibrium. Furthermore, they exhibit a typical electron temperature around 2 eV, thus suitable for vibrational excitation of CO_2. These two aspects make them quite promising for CO_2 conversion, as shown in next section.

8.3.5 Plasma Catalysis

As explained above, plasma contains a cocktail of chemical species (electrons, various types of molecules, atoms, radicals, ions and excited species), which easily form various new molecules. Hence, plasma is very reactive, but not really selective for the production of specific compounds. In so-called plasma catalysis, the high reactivity of plasma is combined with the selectivity of a catalyst [26–28]. Most plasma catalysis research is performed in packed bed DBDs, as the packing beads can have catalytic properties, or can be covered by a catalytic material.

In so-called one-stage plasma catalysis, the catalyst is placed inside the plasma, like in a packed bed DBD, while in two-stage plasma catalysis, the catalyst is placed after the plasma reactor. The first option can give rise to more synergy between plasma and catalyst (see below), because all plasma components, such as short-lived excited species, radicals, photons, and electrons, as well as electric fields, can interact with the catalyst. On the other hand, in two-stage plasma catalysis, only long-lived species, who survive after leaving the plasma, can interact with the catalyst. Nevertheless, two-stage plasma catalysis also has some benefits: it is easier to combine with other plasma types, such as MW and GA discharges, which have too high gas temperatures for most catalysts. Furthermore, the catalyst material would interact with the microwaves in a MW plasma, and the geometry of (for instance) GA reactors might not easily allow catalyst integration in the plasma, although it is not impossible, as demonstrated with a spouted catalyst bed [29]. As mentioned above, packed bed DBD reactors have been mostly used for plasma catalysis up to now, but the combination with MW and GA plasmas (and other types of warm plasmas, such as APGD) may provide other opportunities, such as thermal activation of catalysts.

In spite of the great potential of plasma catalysis to improve the product selectivity, the underlying mechanisms are still far from understood.

On the one hand, the plasma can affect the catalyst and catalysis mechanisms in various ways, i.e.:

(a) Changes in the physicochemical properties of the catalyst, e.g., higher adsorption probability [30], higher surface area [31], change in oxidation state [32–34], reduced coke formation, preventing catalyst deactivation [33], and a change in the work function due to charge accumulation at the catalyst surface [35], affecting the catalytic activity [36].

(b) The formation of hot spots [37], which can modify the local plasma chemistry [38] and even thermally activate the catalyst [39], or deactivate the catalyst due to plasma-induced damage [40].

(c) Lower activation barriers, due to the presence of short-lived active species, e.g., radicals and vibrationally excited species [32].

(d) Activation by photon irradiation [41], although this effect is subject to discussion, because other studies reported no effect [42]. Indeed, the UV light produced by the plasma is probably not intensive enough [41, 43]: the UV dose in typical photocatalytic processes should be in the order of several mW/cm^2, whereas in typical (air) plasmas it is only in the order of $\mu W/cm^2$ [43]. However, photocatalysts might still be activated by other (energetic) plasma species, like ions, metastables or electrons with suitable energy [37, 44].

(e) Changes in the reaction pathways, because of the reactive species present in plasma, i.e., radicals, ions, electrons, vibrationally and electronically excited species, which can give rise to different types of reactions at the catalyst surface. For instance, next to the Langmuir-Hinshelwood mechanism, where two adsorbed species undergo a chemical reaction, plasma catalysis also allows the Eley-Rideal mechanism to take place, where a plasma-produced radical reacts with an adsorbed species [28, 45].

On the other hand, the catalysts can also affect the plasma behavior, causing:

(a) Enhancement of the local electric field in the plasma, in a packed-bed DBD reactor, where the catalyst is mostly present in a structured packing (e.g., pellets, beads, honeycomb, or due to the porosity at the catalyst surface [12, 41, 46]. This enhanced electric field gives rise to higher electron energies [38].
(b) Changes in the discharge type from streamers inside the plasma to streamers along the catalyst surface [34, 47, 48]. This might result in more intense plasma around the contact points [37, 48], affecting the plasma chemistry.
(c) Formation of microdischarges in the catalyst pores [46, 49−51], which can give rise to a strong electric field inside the pores, again affecting the plasma chemistry, and it also enlarges the catalyst surface area available for plasma species.
(d) Adsorption of plasma species on the catalyst surface, affecting the residence time and thus the concentration of species in the plasma [37], as well as the formation of new reactive species at the catalyst surface. The adsorption will increase with the porosity of the catalyst surface [52].

Figure 8.4 schematically illustrates some of these plasma-catalyst interactions. Two types of effects can be distinguished, i.e., physical and chemical effects. The physical effects, such as enhanced electric field, are mainly responsible for gaining a better energy efficiency, while the chemical effects can lead to improved selectivity towards value-added products. In case of CO_2 splitting, CO and O_2 are virtually the only products, so the added value of the catalyst is mainly to increase the energy efficiency, although the conversion might also be improved by chemical effects, e.g., enhanced dissociative chemisorption due to catalyst acid/basic sites. In case of

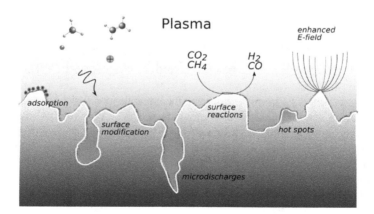

Fig. 8.4 Schematic illustration of some plasma-catalyst interaction mechanisms. Adopted from [53] with permission

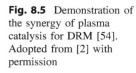

Fig. 8.5 Demonstration of the synergy of plasma catalysis for DRM [54]. Adopted from [2] with permission

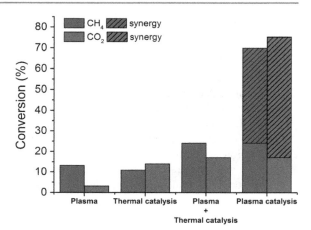

adding a co-reactant (e.g., CH_4, H_2O, H_2), the catalyst can of course also modify the selectivity towards value-added products.

Plasma catalysis sometimes gives better performance than the sum of the performances of plasma only and catalyst only, as illustrated in Fig. 8.5 for DRM. However, this so-called synergy is not always realized. Up to now, typically commercial (thermal) catalysts are being used, which in our opinion limit the real potential of plasma catalysis in selectively producing the desired products. More dedicated research is needed to effectively design catalysts tailored to the plasma environment, making profit of the typical plasma conditions, such as vibrationally and electronically excited species, ions, electrons and the electric field. To fully exploit the possible synergy of plasma catalysis, more insights are needed in the plasma-catalyst interactions, which can be obtained by combined experiments and computer modelling.

8.4 Overview of Plasma-Based CO_2 Conversion

In this section, we present the state-of-the-art of plasma-based CO_2 conversion, summarizing the main results based on a thorough and critical assessment that we performed recently [2], aiming to illustrate the advantages and disadvantages of the various plasma reactors described in previous section. We will first discuss pure CO_2 splitting into CO and O_2, followed by the mixture of CO_2 with other gases acting as hydrogen source, i.e., CH_4 (dry reforming of methane, DRM), H_2O (artificial photosynthesis) and H_2 (CO_2 hydrogenation). The aim of adding a hydrogen source is to produce value-added chemicals and/or fuels, such as syngas, hydrocarbons and oxygenates.

We recently defined a target efficiency of 60% for plasma technology to be competitive with classical and other emerging technologies [2]. This is based on two criteria. First, we believe that electrolysis is the main competitor of plasma technology, as it can also rely on all kinds of renewable electricity (see further). Thus, plasma technology should become competitive with electrochemical water splitting, which reaches commercial energy efficiencies of 65–75%. Thus, plasma technology should aim for a similar energy efficiency (see details in [2]). The second criterion is based on comparing plasma technology with other emerging technologies making directly use of solar energy, e.g., solar thermochemical conversion, for which a solar-to-fuel conversion efficiency of 20% is considered industrially competitive [55]. Assuming a solar panel efficiency of 25%, the energy efficiency for plasma-based CO$_2$ conversion needs to be 60–80%, to yield a competitive solar-to-fuel efficiency of 15–20%.

However, this efficiency target of 60% is based on syngas production, but the conversion of syngas into value-added chemicals and fuels through the Fischer-Tropsch process or for methanol synthesis (and subsequently methanol or ethanol to olefin synthesis) is also an energy-intensive process. Thus, if plasma (catalysis) can directly produce these value-added compounds with sufficient yields, the target efficiency can be drastically reduced. Because a solar-to-methanol conversion efficiency of 7.1% is already economically feasible [55], an energy efficiency of 30% for plasma-based conversion (instead of the above 60%) would already make plasma technology competitive with other technologies. This direct oxidative pathway is one of the advantages of plasma technology, but as mentioned above, major research efforts are required to better understand the underlying processes, especially in terms of catalyst design, to improve the yield of the desired compounds. Therefore, we use the 60% efficiency target in this section for the production of syngas, which is still the major reaction product in plasma technology.

Before we discuss the results obtained in literature, we first present the definitions for the key performance indicators in plasma technology, i.e., conversion, energy efficiency and energy cost:

The (absolute) conversion is based on the molar flow rates of the reactants, e.g. CO$_2$, CH$_4$, H$_2$O or H$_2$:

$$\chi_{abs,reactant_i} = \frac{\dot{n}_{reactant_i,inlet} - \dot{n}_{reactant_i,outlet}}{\dot{n}_{reactant_i,inlet}} \qquad (8.1)$$

where $\dot{n}$ is the molar flow rate of reactant species i.

The effective conversion takes the dilution into account when using mixtures of CO$_2$ with a H-source:

$$\chi_{eff,reactant_i} = \chi_{abs,reactant_i} \cdot \frac{\dot{n}_{reactant_i,inlet}}{\sum_i \dot{n}_{reactant,inlet}} \qquad (8.2)$$

The total conversion is the sum of the effective conversions:

$$\chi_{\text{Total}} = \sum_i \left(\frac{\dot{n}_{\text{reactant}_i,\text{inlet}}}{\sum_i \dot{n}_{\text{reactant}_i,\text{inlet}}} \cdot \chi_{\text{abs,reactant}_i} \right) = \sum_i \chi_{\text{eff,reactant}_i} \qquad (8.3)$$

The energy efficiency indicates how efficient is the conversion process compared to the standard reaction enthalpy, based on the specific energy input (SEI):

$$\eta = \frac{\chi_{\text{Total}} \cdot \Delta H^0_{298\,\text{K}}\left(\text{kJ mol}^{-1}\right)}{\text{SEI}\left(\text{kJ mol}^{-1}\right)} = \frac{\chi_{\text{Total}} \cdot \Delta H^0_{298\,\text{K}}\left(\text{eV molecule}^{-1}\right)}{\text{SEI}\left(\text{eV molecule}^{-1}\right)} \qquad (8.4)$$

$\Delta H^0_{298\,\text{K}}$ is 283 kJ/mol (or 2.93 eV/molecule) for pure CO_2 splitting and 247 kJ/mol (or 2.56 eV/molecule) for DRM into syngas.

The SEI is defined as the plasma power divided by the gas flow rate, which are the dominant factors for determining conversion and energy efficiency. It can be expressed in J/cm^3 (or kJ/L):

$$\text{SEI}\left(\frac{\text{J}}{\text{cm}^3}\right) = \text{SEI}\left(\frac{\text{kJ}}{\text{L}}\right) = \frac{\text{Power}(\text{kW})}{\text{Flow rate}\left(\frac{\text{L}}{\text{min}}\right)} \cdot 60\left(\frac{\text{s}}{\text{min}}\right) \qquad (8.5)$$

But also in electron volt per molecule:

$$\text{SEI}\left(\frac{\text{eV}}{\text{molec}}\right) = \text{SEI}\left(\frac{\text{kJ}}{\text{L}}\right) \cdot \frac{6.24 \times 10^{21}\left(\text{eV kJ}^{-1}\right) \cdot 24.5\left(\text{L mol}^{-1}\right)}{6.022 \times 10^{23}\left(\text{molec mol}^{-1}\right)} \qquad (8.6)$$

Note that the value of 24.5 L mol^{-1} *is only valid for* 298 K *and* 1 atm.

Finally, the energy cost is the amount of energy consumed by the process (generally expressed as kJ/converted mol or eV/converted molecule):

$$\text{EC}\left(\text{kJ mol}^{-1}_{\text{conv}}\right) = \frac{\text{SEI}\left(\text{kJ L}^{-1}\right) \cdot 24.5\left(\text{L mol}^{-1}\right)}{\chi_{\text{Total}}} \qquad (8.7)$$

$$\text{EC}\left(\text{eV molec}^{-1}_{\text{conv}}\right) = \text{EC}\left(\text{kJ mol}^{-1}_{\text{conv}}\right) \cdot \frac{6.24 \times 10^{21}\left(\text{eV kJ}^{-1}\right) \cdot}{6.022 \times 10^{23}\left(\text{molec mol}^{-1}\right)} \qquad (8.8)$$

8.4.1 Pure CO_2 Splitting

Many different plasma types have been applied for CO_2 splitting, but most research has been performed with DBD, MW and GA plasmas. Figure 8.6 summarizes the energy efficiency as a function of the CO_2 conversion, grouped per plasma type, for

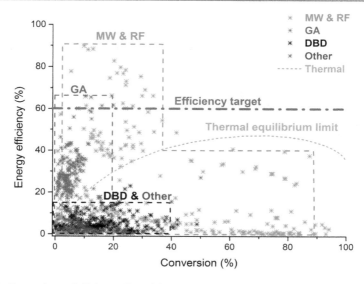

Fig. 8.6 Comparison of all data collected from literature for CO_2 splitting in the different plasma types, showing the energy efficiency as a function of the conversion. The thermal equilibrium limit and the 60% efficiency target are also indicated. Adopted from [2] with permission

all data reported in literature, and critically analysed in [2]. In addition, it indicates both the thermal equilibrium limit and the efficiency target of 60% (see above).

Many papers focused on CO_2 conversion by DBD plasmas, with the most detailed systematic studies, applied to a wide range of conditions, being carried out by Aerts et al. [56], Paulussen et al. [57], Yu et al. [58] and Ozkan et al. [59]. It is apparent from Fig. 8.6 that a DBD plasma can provide reasonable conversions up to 40%, but the energy efficiency is only around 5–10%, with maximum values up to 15% and some exceptions of 20% by applying the power in so-called burst mode [59, 60]. This is still at least a factor 3–4 away from the 60% efficiency target. Moreover, a high conversion is typically accompanied by a low energy efficiency, and vice versa, as demonstrated in [2, 56].

Inserting a packing in a DBD (packed bed DBD; see Sect. 8.3.1) can sometimes improve the performance. Van Laer et al. [61] and Mei et al. [62] reported a simultaneous increase in both conversion and energy efficiency by a factor 2, for ZrO_2 and for TiO_2 and $BaTiO_3$ packing, respectively, attributed to the enhanced electric field upon polarization of the packing beads. In addition, for TiO_2 and $BaTiO_3$ packing, Mei et al. also reported a (photo)catalytic effect, i.e., the formation of electron-hole pairs due to the plasma electrons, which cause oxygen vacancies, thus acting as active sites for adsorption and activation of the reactants [62].

However, a packed bed DBD not always yields better performance [2, 13, 14, 61–63], as also discussed in Sect. 8.3.1. Figure 8.7 illustrates CO_2 conversions and energy efficiencies obtained in a DBD plasma, with and without dielectric packing [13]. The conversion is only about 5–20%, with corresponding energy efficiency of 1–3%. These values are quite low, although they are somewhat better in a packed bed DBD, due to the electric field enhancement, as explained above. However, this is not the case for all conditions. Indeed, a packing reduces the residence time at constant flow rate, due to a smaller plasma volume, which will reduce the conversion, as explained in Sect. 8.3.1. Depending on the conditions (e.g., applied power and flow rate, discharge gap, packing material and bead size), either the positive effect of electric field enhancement or the negative effect of smaller plasma volume might be dominant, resulting in either a higher or lower CO_2 conversion and energy efficiency. Indeed, Fig. 8.7 shows that when we compare at

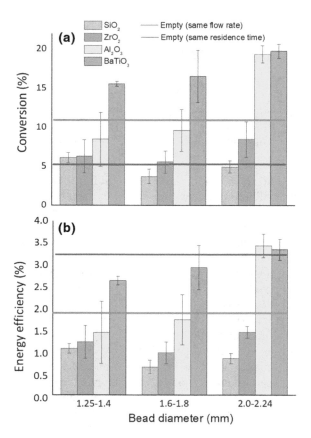

Fig. 8.7 CO_2 conversion (**a**) and energy efficiency (**b**) in a DBD, with and without dielectric packing, for four different packing materials (see legend) and three different bead sizes (x-axis), in case of a DBD with Al_2O_3 dielectric barrier, 4.5 mm gap, stainless steel outer electrode, applied frequency of 23.5 kHz, 10 W input power and 50 mL/min gas flow rate [13]. The error bars are defined based on 12 gas chromatography measurements. Comparison is also made with the results of an unpacked reactor, at the same flow rate (red line) and the same residence time (but much higher flow rate of 192 mL/min; blue line). Adopted from [10] with permission

the same flow rate (cf. red horizontal line), some packing materials (e.g., Al$_2$O$_3$ and BaTiO$_3$) and bead sizes (typically the larger beads) yield higher conversion and energy efficiency than in the unpacked reactor, while others yield lower results. When we compare with the unpacked reactor at the same residence time (and thus much higher flow rate, cf. blue horizontal line in Fig. 8.7), the conversions are typically higher, because only the enhanced electric field comes into play, but the energy efficiency is lower, because the latter is defined by the SEI, and thus by the flow rate [see (8.4) and (8.5) above], i.e., a higher flow rate results in a lower energy efficiency for the same conversion.

It is important to note that the results of Fig. 8.7 do not represent the best performance collected from literature for (packed bed) DBD reactors (cf. also Fig. 8.6). We plotted them to illustrate that inserting a packing does not always yield better performance, and that the results greatly depend on the packing material and geometry, and on the reactor geometry [13]. We can conclude that, in spite of the inherent strengths of DBDs (in terms of scalability and ease of operation), their low energy efficiency makes it questionable whether they would ever become suitable for pure CO$_2$ splitting.

The first types of plasmas that were extensively studied for CO$_2$ splitting, both theoretically and experimentally, already back in the 1980s [3, 15, 64], are MW (and RF) discharges. They were found to reach very high energy efficiencies, attributed to rather high electron densities in combination with low reduced electric field, thus promoting the vibrational kinetics, being the most energy efficient CO$_2$ dissociation pathway (see Sect. 8.2 above). Rusanov et al. reported energy efficiencies up to 80% with conversions around 20% in a MW plasma at reduced pressure and subsonic flow [64], while Asisov et al. obtained record values for the energy efficiency of 90%, along with a conversion of 10%, for reduced pressure and supersonic flow [15]. These values are clearly above the 60% efficiency target defined for pure CO$_2$ splitting (see Fig. 8.6).

Nevertheless, these excellent results have not yet been reproduced since then, although more recently, van Rooij, Bongers and colleagues also obtained quite high energy efficiencies around 40–50%, for conversions of 10–20% [65, 66], and Silva et al. reached conversions of 50–80%, albeit at lower energy efficiencies of 10% [67]. However, most of these results were reached at reduced pressure (typically up to a few 100 mbar). At atmospheric pressure, Spencer et al. obtained energy efficiencies up to 20% for conversions of 10%, or conversions up to 50% for energy efficiencies of 5% [68].

Figure 8.8 summarizes the best results obtained for MW plasmas, at either reduced pressure (open symbols) or atmospheric pressure (closed symbols), clearly pointing out the superior results at reduced pressure. The best energy efficiency in the more recent work is around the thermodynamic equilibrium value of 45–50%. Keeping in mind that the gas temperatures in this case is clearly above 1000 K, we doubt whether vibrational excitation followed by ladder climbing (i.e., the most energy-efficient dissociation pathway; cf. Section 8.2 above) is the dominant dissociation mechanism in this case. Indeed, the conversion appears to be mainly due to thermal dissociation, as is also confirmed by computer modelling, where at

Fig. 8.8 Summary of the best results published in literature for energy efficiency versus CO_2 conversion in a MW plasma, both at reduced pressure (open symbols) and atmospheric pressure (full symbols)

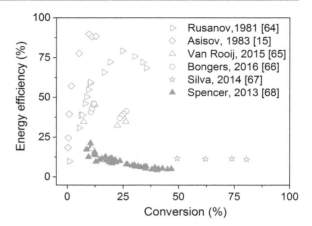

atmospheric pressure and corresponding high temperature, the VDF is nearly thermal [7]. A strong non-equilibrium VDF with pronounced overpopulation of the higher vibrational levels, needed to promote the vibrational kinetics, could only be reached at low pressure, low gas temperature, but still high power density, as revealed by modeling [7].

As mentioned in Sect. 8.2 above, a low gas temperature at atmospheric pressure is not straightforward to realize. Applying a supersonic gas flow [15] might provide a solution, if the gas has not enough time to be heated, but recent model calculations showed that this supersonic flow also limits the gas residence time in the plasma, and thus the conversion [16]. Another possibility might be to apply pulsed power, so that the gas temperature can drop in between the applied pulses, but recent modeling revealed that the vibrational temperature drops faster than the gas temperature, hence not generating the desired effect. Finally, applying a higher power density could also provide a solution, e.g., by reducing the dimensions of the plasma reactor, so that the same power is applied over a smaller plasma volume. However, this will reduce the gas throughput, unless several reactors could be placed in parallel.

Both DBD and MW discharges thus clearly have their distinctive advantages and disadvantages. DBDs operate at atmospheric pressure, but do not produce significant vibrational excitation, yielding the most energy efficient dissociation process, while MW plasmas do give rise to vibrational excitation, but the VDF is too thermal, so the vibrational dissociation pathway cannot be fully exploited, except at low pressure. The GA plasma, on the other hand, operates at atmospheric pressure and gives rise to significant vibrational excitation, so it tries to combine the best of both worlds [17].

As illustrated in Fig. 8.6, energy efficiencies above the target efficiency of 60% have already been reported for CO_2 splitting in GA discharges, although most values are around 20–40% [17, 18, 20, 69–71], pointing again towards thermal equilibrium conversion. It is indeed predicted by computer modeling [20, 21, 72] that GA plasmas are also often too close to thermal. Thus, the non-equilibrium character of the GA—making full potential of the vibrational kinetics pathway—should also be further exploited to improve the energy efficiency, like for MW discharges. Operating GA plasmas at lower gas temperatures would again be the key solution, but is not straightforward to realize.

Moreover, GA plasmas have only limited conversion around 10% at maximum, with some exceptions up to 20%. The reason is the limited fraction of gas that passes through the active arc plasma [21, 73]. We believe that smart reactor design, in order to enhance the processed gas fraction, is required to enhance the conversion. To illustrate this, we plot in Fig. 8.9 the energy efficiency as a function of CO_2 conversion in a GAP, for different gas outlet diameters. As explained in Sect. 8.3.3 above, the gas enters tangentially in a GAP and when the outlet diameter is smaller than the reactor diameter, it first flows in an outer vortex towards the upper end of the reactor, followed by a reverse vortex with smaller diameter towards the outlet (see Fig. 8.3d above). Ideally, this reverse vortex gas flow passes exactly through the active arc in the middle of the reactor, to obtain maximum conversion. It is clear from Fig. 8.9 that the conversion in the case of the smallest outlet diameter is larger than for the case with outlet diameter equal to the reactor diameter, where this reverse vortex gas flow is almost absent. Nevertheless, the fraction of gas passing through the arc is still quite limited, even in the case with smallest outlet diameter, and this limits the CO_2 conversion, not only in a classical GA (as explained in Sect. 8.3.3 above), but also in the improved design of GAP. Enhancing this gas fraction should be possible by modifying the reactor setup and/or gas flow configuration, but more research, including fluid dynamics simulations, are needed to achieve this. For a classical GA, model calculations demonstrated that such an improvement in treated gas fractions could be realized by a higher relative velocity between arc and gas flow [20, 71].

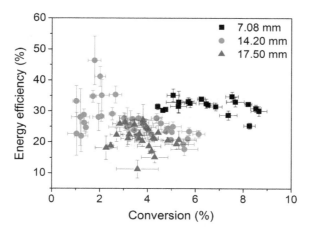

Fig. 8.9 Energy efficiency versus CO_2 conversion in a GAP, for three different configurations with different anode diameters (cf. legend) and fixed cathode diameter of 17.50 mm, and different combinations of power and gas flow rate [18]. Adopted from [10] with permission

Other plasma types, such as corona discharges, glow discharges, non-self-sustained discharges, capillary discharges, and ns-pulsed discharges, have also been tested for CO_2 conversion, but their performance was found to be similar to the results obtained for DBDs, with maximum conversions up to 40% and energy efficiencies below 15% [2]. Only the so-called non-self-sustained discharge presented by Andreev et al. [74] could reach a conversion of 50% with an energy efficiency of almost 30%. This could be attributed to a precise control of the mean electron energy, by changing the reduced electric field. The highest energy efficiency was reached at reduced electric fields around 20 Td, favouring vibrational excitation (see Figs. 8.1 and 8.2), but just like for most MW plasmas, this discharge was operating at reduced pressures (1550 Pa), limiting its industrial applicability.

Finally, modeling revealed that the conversion and energy efficiency of CO_2 splitting can be improved by removing the produced O_2 molecules from the gas mixture [20, 71], to avoid the back-reaction, i.e., the recombination with CO with O_2 into CO_2. Removing O_2 from the gas mixture might be realized by scavenging materials or chemicals [75] or by membranes, as demonstrated by Mori et al. for a hybrid reactor composed of a DBD with solid oxide electrolysis cell (SOEC) [76]. However, more research is required to fully explore all possibilities.

8.4.2 CO_2 + CH_4: Dry Reforming of Methane

Plasma-based DRM is gaining increasing interest [2, 3, 77–80]. Figure 8.10 summarizes the energy cost versus total conversion, collected from all results published up to now in literature [2]. Unlike Fig. 8.6 for pure CO_2 splitting, we do not plot the energy efficiency because the latter can only be determined if the theoretical reaction enthalpies of all possible reactions are accounted for [see (8.4) above], or if the higher (or lower) heating values of all reactions products are taken into account. However, papers in literature typically only report the selectivity (or yield) towards CO and H_2, and sometimes the light hydrocarbons, which makes it impossible to deduce the true energy efficiency. The energy cost, on the other hand, can be calculated without knowledge of all reaction products [see (8.7) above], and that's the reason why we have plotted the latter in Fig. 8.10. To allow easy comparison with Fig. 8.6 for pure CO_2 splitting, we have reversed the y-axis in Fig. 8.10.

Again, we have also plotted the thermal equilibrium limit and efficiency target in Fig. 8.10. We assume the same 60% efficiency target, keeping in mind that syngas is the main product in DRM. This equals an energy cost of 4.27 eV/molec for general stoichiometric DRM [2]. Nevertheless, when directly forming products with higher added-value, the efficiency target can be drastically reduced, because the energy-intensive step of processing syngas into the desired products can be avoided.

Again most of the research on DRM has been performed with DBDs, due to their simple design and ease of operation, especially also in combination with (catalytic) packing materials. For plasma catalytic DRM, a wide range of materials have been used, with Ni being the most common active phase and alumina the most common

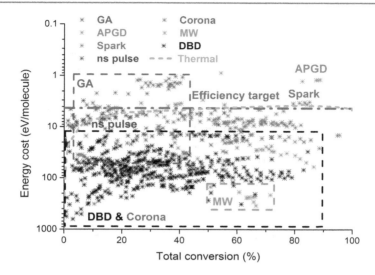

Fig. 8.10 Comparison of all data collected from literature for DRM in the different plasma types, showing the energy cost as a function of the conversion. The thermal equilibrium limit and the target energy cost of 4.27 eV/molec for the production of syngas (corresponding to 60% efficiency target) are also indicated. Note that the y-axis is reversed to be more comparable with Fig. 8.6. Adopted from [2] with permission

support. A detailed overview of all catalysts used, along with the corresponding references, can be found in [2]. A Zeolite Na-ZSM-5 catalyst in an AC packed bed DBD yielded the highest (total) conversion of 37%, with energy cost of 24 eV/molec [81]. A packed bed DBD in pulsed operation mode gave even better results with BZT (BaZr0.75T0.25O3) and BFN (BaFe0.5Nb0.5O3) catalysts, reaching total conversions around 45–60% at an energy cost of 13–16 eV/molec [82].

As mentioned above, syngas appears the major product formed, even when using catalysts. Clearly more research is needed in catalyst design to selectively produce higher value compounds. The CH$_4$/CO$_2$ ratio in the feed gas plays an important role in determining the conversion and energy cost, as well as the product distribution and syngas ratio. At high CH$_4$/CO$_2$ ratios, carbon deposition becomes predominant, causing detrimental effects, while at low ratios, part of the H atoms get lost into H$_2$O formation.

Based on all DBD results from literature, we can conclude that a DBD can give rise to reasonable conversions (up to 60% for non-packed DBDs, and 80% for packed bed DBDs; see Fig. 8.10), but the energy costs are around 20–100 eV/molec, which correspond to an energy efficiency of 12.8–2.6%, in case when only syngas would be formed. This is at least a factor 5 away from the target (i.e., 4.27 eV/molec energy cost or 60% energy efficiency). Model predictions revealed that the lowest achievable energy cost, upon careful selection of the operating conditions, would be 16.9 eV/molec [83]. This is still far too high to be competitive with other technologies, unless suitable catalysts could be found for the direct production of value-added compounds with high enough yields, which would drastically lower the efficiency target.

Some first results with DBDs and suitable catalysts have already been reported, showing the selective production of oxygenates. When using copper or nickel electrodes instead of stainless steel, Scapinello et al. [84] reported an enhanced selectivity towards carboxylic acid formation, which was attributed to a chemical catalytic effect of the metals, i.e., hydrogenation of chemisorbed CO_2. The latter process indeed has a rather high barrier in the gas phase, and seems to play a key role in the synthesis of these carboxylic acids [84].

Another example is shown in Fig. 8.11 which illustrates the conversion of CO_2 and CH_4 (top), the selectivity of gaseous products (middle) and liquid products (bottom), for catalyst alone (using a Cu/Al_2O_3 catalyst), DBD plasma alone, and DBD plasma + catalyst, at 30 °C and atmospheric pressure, for different catalyst materials [85]. When using only catalysts, no conversion was obtained at this low temperature. Plasma catalysis yields a slightly lower conversion of CO_2 and CH_4 than plasma alone, which might be explained by the change in discharge behavior or the shorter gas residence time in the plasma reactor upon packing, as also mentioned above. This also results in a slightly lower total energy efficiency, from 12.4% (plasma alone) to 8.7, 11.7 and 11.1% for Cu, Au or Pt catalyst, respectively [85].

The selectivities of the gaseous products were similar for plasma alone and plasma catalysis, with H_2, CO and C_2H_6 being the major products formed (middle panel in Fig. 8.11). As far as the liquid compounds are concerned, acetic acid, methanol and ethanol, and some fractions of acetone were reported as the major products in the plasma-only case, with a total selectivity of 59% (bottom panel in Fig. 8.11). In the case of plasma catalysis, these products were also the major ones, but the distribution of different liquid products can to some extent be tuned. For instance, the Cu/Al_2O_3 catalyst could enhance the selectivity of acetic acid to 40%, while HCHO could be detected when using Pt/Al_2O_3 and Au/Al_2O_3 catalysts, while it was not observed in the plasma-only case. Nevertheless, it must be realized that the total fraction of liquids was only ca. 1% [85]. Indeed, our knowledge of selecting appropriate catalysts for directly converting CO_2 and CH_4 into value-added oxygenates by plasma catalysis is still very limited. More dedicated research is clearly needed for designing catalysts tailored to the plasma environment, with high selectivity towards targeted value-added compounds.

Other plasma types are also being investigated for DRM, as is clear from Fig. 8.10. It is, however, surprising that only a few experiments on DRM have been reported with MW plasmas, in view of the good results obtained for pure CO_2 splitting (cf. previous section). The few studies reported show high conversions but also high energy costs (see Fig. 8.10).

On the other hand, quite promising results have been obtained already for GA plasmas. The energy costs are much lower than for DBDs, while much higher gas feeds can be processed. Figure 8.12a shows the effective CO_2 and CH_4 conversion in a GAP, for different CH_4 fractions in the mixture, for a total gas flow rate of 10 L/min and a plasma power of 500 W, corresponding to an SEI of 3 kJ/L or 0.75 eV/molec [19]. The effective CO_2 and CH_4 conversion, and hence also the

Fig. 8.11 Conversion of
CO$_2$ and CH$_4$ (top),
selectivity of gaseous
products (middle) and liquid
products (bottom), for catalyst
only, plasma only, and plasma
catalysis for three different
catalyst materials, in a DBD
at 30 °C and atmospheric
pressure [85]. Adopted from
[10] with permission

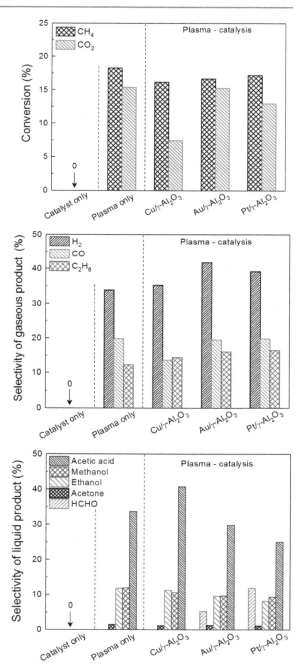

total conversion, increase with the CH_4 fraction. Note that the absolute conversions (not plotted here) are still higher, especially for CH_4, i.e., up to 60% [19], because the effective conversions account for the gas fraction in the mixture [see (8.2) above]. The energy efficiency also rises with CH_4 fraction (see Fig. 8.12b), up to values above 60%, corresponding to an energy cost of ca. 10 kJ/L or 2.6 eV/molec [19]. These are very promising results, which might be further improved upon higher CH_4 fractions in the mixture. The limiting factor, as also revealed by model calculations, is the fraction of gas passing through the arc plasma. Increasing the latter, by smart reactor design, will further enhance the conversion, and therefore also the energy efficiency.

Fig. 8.12 Conversion of CO_2 and CH_4 and total conversion (**a**), energy efficiency (**b**) and product selectivity (**c**), as a function of CH_4 fraction in the mixture, for a GAP at constant total gas flow rate of 10 L/min and plasma power of 500 W, yielding an SEI of 3 kJ/L or 0.75 eV/molec [19]. Adopted from [10] with permission

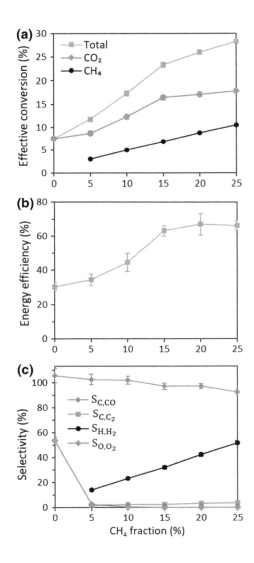

Figure 8.12c also depicts the product selectivity, obtained without catalyst. The C-based selectivity for CO is ca. 100% (or even slightly above 100% at low CH$_4$ fraction, but this is attributed to uncertainties in the measurements). Upon higher CH$_4$ fraction, the CO selectivity slightly drops due to the formation of some C$_2$ hydrocarbons (mainly C$_2$H$_2$, as also revealed from plasma chemistry modeling [19]). The O-based selectivity of O$_2$ is about 60% in pure CO$_2$, but it drops significantly upon addition of (5%) CH$_4$. This indicates that the O atoms formed from CO$_2$ splitting mainly yield O$_2$ and CO in pure CO$_2$, but they are converted into other compounds (e.g., oxygenates, but also H$_2$O) upon addition of CH$_4$. The H-based selectivity of H$_2$ rises with increasing CH$_4$ fraction, while the remaining H atoms give rise to higher hydrocarbons and H$_2$O. In general, CO and H$_2$ (i.e., the syngas components) are the major compounds formed, but this could change when applying a catalyst.

Similar results have been reported for other GA plasmas, with maximum conversions of 30–50% and energy costs as low as 1–2 eV/molec (see Fig. 8.10), and mainly forming syngas [19, 86–89]. The GA thus easily reaches the energy cost target of 4.27 eV/molec (or energy efficiency target of 60%) for syngas production by DRM, and it clearly exhibits non-equilibrium character, as the results are obviously above the thermal equilibrium limit in Fig. 8.10.

The best reported results for GA plasmas were obtained for a rotating GA reactor, with a total conversion of 39% and an energy cost of 1 eV/molec [86]. Furthermore, a GA plasma with Ni-based catalysts in a heat-insulated reactor, operating in a mixture of CO$_2$/CH$_4$/O$_2$ (2/3/1.8), also-called oxidative DRM, gave a dramatic rise in energy efficiency (up to 86%) with absolute CH$_4$ conversion of 92% and absolute CO$_2$ conversion of 23% [87]. A similar setup recently yielded an energy efficiency of 79% at an absolute CH$_4$ conversion of 99% and an absolute CO$_2$ conversion of 79% [89]. The latter study reported on an interesting concept of combining this GA-based plasma-catalytic reforming with water electrolysis, being both renewable electricity-driven approaches, to produce high-quality syngas from CH$_4$, CO$_2$ and H$_2$O. In this concept, pure O$_2$ produced from water electrolysis can directly be used by the GA plasma for oxidative DRM, without the need for prior air separation like in conventional processes. If we consider an energy efficiency of 80% for water electrolysis, this combined system would yield an overall energy efficiency of 79%, producing high-quality syngas without the need for post-treatment, and featuring the ideal stoichiometric number of 2, with concentration of nearly 95%, and a desired CO$_2$ fraction of 1.9% for methanol synthesis [89]. This concept is highly interesting for the large-scale energy storage of renewable electricity via electricity-to-fuel conversion, but will need further investigation to test the feasibility.

Finally, other plasma types have also been investigated for DRM, and especially ns-pulsed discharges, spark discharges and atmospheric pressure glow discharges (APGDs) indicated very promising results (see Fig. 8.10). Ns-pulsed discharges yield total conversions around 40–60%, with energy costs of about 3–10 eV/molec [22], in contrast to pulsed power DBD and corona discharges, which operate in the microsecond pulse regime. This indicates that the nanosecond timescale is essential to create the necessary strong non-equilibrium [90]. C_2H_2 is the major hydrocarbon formed, like in a GA (see above), but different from a DBD where mainly C_2H_6 is formed (cf. Figure 8.11; middle panel above).

Spark discharges and APGDs yield conversions up to 80–90%, with minimum energy costs around 3 eV/molec (for spark discharges [91]) and even down to 1.2 eV/molec (for APGDs [92]). Especially the APGD seems quite promising for DRM, which can be attributed to its high electron density, proper electron temperature for vibrational excitation and limited gas temperature, as explained in Sect. 8.3.4 above. However, it is clear from Fig. 8.10 that not all APGDs yield such good results, and most often the results are only comparable to DBD and corona discharges. Furthermore, only a limited number of results have been reported up to now for APGDs, as well as for ns-pulsed and spark discharges, and more research is certainly required to gain better insight into their underlying mechanisms and to further exploit their possibilities.

8.4.3 CO_2 + H_2O: Artificial Photosynthesis

Only a limited number of studies have been reported on the combined conversion of CO_2 and H_2O (so-called artificial photosynthesis). Some papers reported the use of a DBD, both without and with (ferroelectric or $Ni/\gamma Al_2O_3$ catalytic) packing, MW, GA, surface discharge and corona discharge. Figure 8.13 summarizes the energy cost as a function of the conversion, for all results obtained up to now in literature,

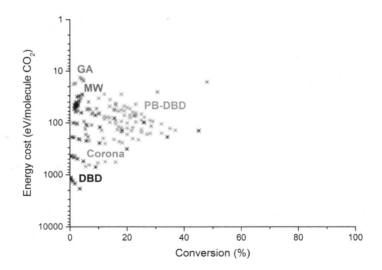

Fig. 8.13 Comparison of all data collected from literature for the artificial photosynthesis (CO_2 + H_2O) in the different plasma types, showing the energy cost per converted CO_2 molecule as a function of the conversion. PB-DBD = packed bed DBD. Some of the data has been recalculated from the original references, to take (among others) dilution effects into account

again based on the extensive critical assessment performed in [2]. We could not define an efficiency target and thermal equilibrium limit in this case, as this is only possible for a specific overall reaction towards specified products, which was typically not reported in literature.

Without catalysts, even small amounts of H$_2$O addition (1–2%) give rise to a significant drop in CO$_2$ conversion, and thus a higher energy cost, compared to pure CO$_2$ splitting. We therefore believe that DBDs, which already have high energy costs, will not be suitable for the combined CO$_2$/H$_2$O conversion, unless dedicated catalysts can be found (see below). MW and GA plasmas are far more efficient, due to their vibrational-induced dissociation pathway (cf. Section 8.2 above). Moreover, they operate at somewhat higher temperatures (order of 1000 K or more), allowing more H$_2$O vapour addition. However, the H$_2$O molecules might quench the CO$_2$ vibrational levels, thus reducing the most energy efficient conversion process, and increasing the energy cost. This was confirmed in a GA [17, 69], but on the other hand, in a MW plasma, conversions up to 50%, at an energy cost around 20 eV/molec were obtained [93]. However, this MW set-up was again operating at low pressures (30–60 Torr), which might explain why it was less prone to quenching of the CO$_2$ vibrational levels. In addition, H$_2$O might have a cooling effect, possibly increasing the thermal non-equilibrium and thus the energy efficiency.

The main products formed by the combined CO$_2$/H$_2$O conversion are again H$_2$ and CO, like in DRM, as well as O$_2$, although the production of hydrogen peroxide, oxalic acid, formic acid, methane, dimethyl ether, methanol, ethanol, acetylene, propadiene and even carbon nanofibres has been reported; see detailed overview in [2]. Hence, the combined CO$_2$/H$_2$O conversion might allow the direct formation of oxygenates, although significant yields can again only be obtained when suitable catalysts can be designed. Indeed, without catalyst, too many steps are involved in generating oxygenates in the plasma, and all of them involve H atoms, which will faster recombine with OH into H$_2$O or with O$_2$ into HO$_2$, and further H$_2$O.

This is illustrated in the reaction scheme of Fig. 8.14, obtained from computer simulations for a DBD [94]. The main reactive species formed in the CO$_2$/H$_2$O mixture are OH, CO, O and H. The O and H atoms recombine in a few steps into H$_2$O, explaining why no oxygenated products were formed, and also why the H$_2$O conversion is typically limited. Furthermore, the OH radicals quickly recombine with CO into CO$_2$, as revealed by the model, which explains the limited CO$_2$ conversion upon H$_2$O addition [94].

The need for catalysts is thus clear, so that plasma-generated CO and H$_2$ can selectively react into oxygenates. We believe that the critical point will be the arrival and binding (by either physi- or chemisorption) of the reactants to the catalyst surface, which should be faster than the recombination rate of OH with H into H$_2$O. Eliasson et al. [95] applied a CuO/ZnO/Al$_2$O$_3$ catalyst in a CO$_2$/H$_2$ DBD, which gave a rise in methanol yield and selectivity by more than a factor ten. Similar effects might be expected for a CO$_2$/H$_2$O DBD, because the CO/CO$_2$/H$_2$ mixture is also formed in this case. Mahammadunnisa et al. [96] reported a higher conversion and syngas ratio than without catalyst, as well as the formation of other products, including CH$_4$, CH$_3$OH, C$_2$H$_2$, and propadiene, upon addition of a

Fig. 8.14 Reaction scheme to illustrate the main pathways for CO_2 and H_2O conversion and their interactions. The arrow lines represent the formation rates of the species, with full green lines being formation rates over 10^{17} cm^{-3} s^{-1}, orange dashed lines between 10^{17} and 10^{16} cm^{-3} s^{-1} and red dotted lines between 10^{16} and 10^{15} cm^{-3} s^{-1}. Adopted from [94] with permission

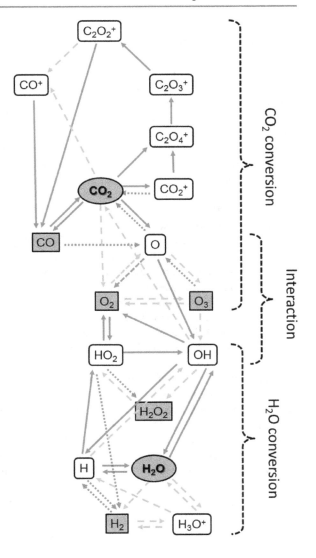

NiO/γ-Al_2O_3 catalyst (in unreduced form), while the reduced Ni catalyst also produced carbon nanofibers. Likewise, Chen et al. [93] reported a higher CO_2 conversion (up to 48%), with energy cost of 14.5 eV/molec, upon adding NiO/TiO_2 catalysts in a MW plasma (see best data point in Fig. 8.13). The authors explained this by CO_2 adsorption at oxygen vacancies on the catalyst surface, reducing the threshold for dissociative electron attachment into CO, adsorbed O atoms at the vacancies and electrons. The adsorbed O atoms may subsequently recombine with gas phase O atoms or OH radicals into O_2 (and H atoms). The catalyst thus seems to be able to tune the gas phase O/OH into O_2, by means of adsorbed O atoms at the vacancies, before they recombine again into CO_2 and H_2O, which is indeed considered to be the limiting step in CO_2/H_2O conversion, as revealed by computer modelling [94] (see above). However, the catalyst was only found to enhance the CO_2 conversion and reduce the energy cost, but not to improve the selectivity towards value-added products, because no oxygenated products were detected in this setup.

CO_2/H_2O plasmas, even without catalysts, seem to be able to deliver an easily controllable H_2/CO ratio, with a rich hydrogen content (even up to 8.6 according to model predictions), when sufficient amounts of H_2O can be added [94]. Thus, they might still be suitable to create value-added chemicals in a two-step process, combined with Fischer-Tropsch or methanol synthesis. Finally, the direct production of sufficient amounts of H_2O_2, to be used as a disinfectant, seems possible as well [94].

It should, however, be noted that a CO_2/H_2O plasma might form an explosive mixture, due to the presence of O_2, together with CO, H_2 and an ignition source. For research purposes this might not be a problem, due to the low volumes and conversions, but when using larger volumes and conversions, on pilot or industrial scale, the explosion risk will increase, and this will enhance the capital and operating costs to ensure safe operations. One possibility could be to dilute this mixture with an inert gas, such as argon or helium, but this would require an additional separation (for the products) and recuperation step (for the inert gas), thus also increasing the cost. In addition, some of the input energy would be used for electron impact excitation and ionization of the inert gas, hence reducing the energy efficiency of the conversion process.

8.4.4 $CO_2 + H_2$: Hydrogenation of CO_2

Also for plasma-based CO_2 hydrogenation, i.e., using H_2 as co-reactant, not much research has been performed yet. The type of plasma reactors that have been applied are DBD (with and without packing), MW and RF plasmas, as well as a surface discharge.

In Fig. 8.15 we plot the energy cost as a function of conversion for all data available in literature [2]. The conversion is about 2–3 times lower (and thus the energy costs the same factor higher) than for DRM and pure CO_2 splitting. This lower conversion (compared to DRM) can be explained by the low concentrations of CH_2 and CH_3 radicals, which help in CO_2 conversion, as revealed by computer modeling [98]. Moreover, the same limitation as for the CO_2/H_2O mixture applies here as well, i.e., the CO_2 conversion is limited by the formation of CHO (CO + H + M → CHO + M), which reacts back to CO_2 (CHO + O → CO_2 + H). Hence, plasma-based CO_2 hydrogenation appears not to be very successful yet, at least without catalyst (see below).

From literature it is clear that CO_2 hydrogenation typically yields CO and H_2O as main products, with some secondary products in much smaller—almost negligible—amounts. The most important secondary products are CH_4 and methanol,

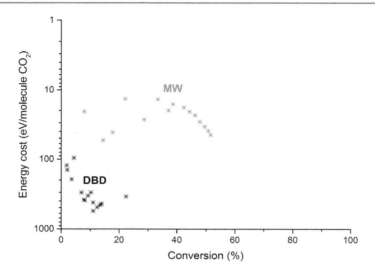

Fig. 8.15 Comparison of all data collected from literature for CO_2 hydrogenation in the different plasma types, showing the energy cost per converted CO_2 molecule as a function of the conversion. Some of the data has been recalculated from the original references, to take (among others) dilution effects into account

but the production of ethanol, formaldehyde, formic acid, dimethyl ether, acetylene, ethylene and ethane has been reported as well. We can conclude that CO_2 hydrogenation without catalytic packing resembles the combined CO_2/H_2O conversion, which is in fact quite logical, because both mixtures yield the same reactive species. In spite of the low amounts in which these products were formed, the fact that they were formed indicates that plasma-based CO_2 hydrogenation has potential to directly produce value-added compounds, again if suitable catalysts can be developed.

Most papers in literature indeed report the use of catalysts to enhance the performance of CO_2 hydrogenation, as least for DBD reactors. Eliasson et al. [95] was able to enhance the methanol yield and selectivity in a CO_2/H_2 DBD by more than an order of magnitude, as mentioned in previous section, by using a $CuO/ZnO/Al_2O_3$ catalyst. Jwa et al. [97] and Nizio et al. [98] presented a Ni-based plasma-catalytic set-up, where the plasma accounts for dissociation of the adsorbed CO into C (i.e., the rate determining step), whereas the catalyst can reduce C with H into CH_4. In addition, Wang et al. [99] presented very interesting results on highly selective methanol synthesis using a water-cooled DBD, in combination with $Cu/\gamma\text{-}Al_2O_3$ or $Pt/\gamma\text{-}Al_2O_3$ catalysts. The a $Cu/\gamma\text{-}Al_2O_3$ catalyst yielded a maximum methanol yield of 11.3% and methanol selectivity of 53.7%, for a CO_2 conversion of 21.2% in plasma catalysis, whereas no methanol was formed at ambient conditions without plasma [99]. Other catalysts might also be promising for CO_2 hydrogenation, such as Ni-zeolite catalysts for methanation [97], a Rh_{10}/Se catalyst for ethanol formation [100], a Ni-Ga catalyst for methanol formation [101], and $CuO/ZnO/ZrO_2$, Cu/ZnO-based catalysts promoted with Pd and Ga, or Pd/ZnO and Pd/SiO_2 upon addition of Ga [102]. Finally, also multicomponent systems ($Cu/ZnO/ZrO_2/Al_2O_3/SiO_2$), showing good performance for methanol formation from $CO/CO_2/H_2$ mixtures [100], might be promising, because this mixture is indeed automatically generated during plasma-based conversion of CO_2/H_2. From model calculations we expect that the catalyst should allow the plasma-generated C atoms (or CO) and H_2 to react into methane (or oxygenates), just like for the combined CO_2/H_2O conversion (see previous section).

8.5 Plasma Technology in a Broader Perspective of Emerging Technologies for CO_2 Conversion

We recently made a detailed assessment of plasma-based CO_2 conversion (both pure and in combination with a H-source, i.e., CH_4, H_2O or H_2), and we compared the performance with other (emerging) technologies, i.e., electrochemical, solar thermochemical, photochemical, biochemical and catalytic conversion (see details in [2]). From this comparison we can conclude that plasma technology is promising for CO_2 conversion, for the following reasons:

(1) *Process versatility*: Different types of reactions can be performed in plasma, i.e., pure CO_2 splitting, but also the combined CO_2 conversion with CH_4, H_2 or H_2O, as H-sources. Nevertheless, the combination with H_2 or H_2O only appears promising if suitable catalysts can be designed (cf. previous section). In addition, when suitable catalysts can be developed, plasma catalysis might yield the direct production of higher hydrocarbons or oxygenates in a one-step process, without additional Fisher-Tropsch or methanol synthesis and subsequent methanol/ethanol-to-olefins synthesis. Nevertheless, clearly more research is needed, especially in the design of suitable catalysts tailored to the plasma environment. In general, we may conclude that plasma technology can be used at many different locations, independent of the feed gas available, and even allowing various feed gas compositions. Indeed, in previous work we showed that large amounts of N_2, which might be present in CO_2 effluent gases, can even enhance the CO_2 conversion [103].

(2) *Plasma does not make use of rare earth metals*, which is currently a limitation for various other technologies.

(3) *Plasma is a turnkey process*: as plasma requires no pre-heating or long stabilization times, and no cool-down times, it can be turned on and off very quickly. Moreover, the CO_2 conversion is instantaneous upon plasma ignition (i.e., from the first second). Thus, plasma technology is very suitable for converting fluctuating renewable energy into fuels or chemical building blocks.

(4) **Plasma technology has low investment and operating costs**: no expensive equipment is needed, and the plasma reactors can be used in a modular setting, with virtually no economy of scale, making plasma technology useful for local on-demand production of fuels from renewable energy.

(5) **Various types of (renewable) energy** can be applied to plasma, while other emerging technologies (e.g., solar thermochemical, photochemical and bio-chemical conversion) are limited to solar energy. This increases the application versatility of plasma technology, because it can be operated independent of the availability of solar radiation.

8.6 Conclusion and Future Research Directions

This chapter discusses the possibilities and limitations of plasma-based CO_2 conversion, using a variety of plasma reactors. We can conclude that DBD plasmas exhibit too low energy efficiency for industrial implementation. This is certainly true for CO_2 splitting in absence of a catalyst (Sect. 8.4.1). The energy efficiency in this case is only about 5–10%, hence far below the target of 60%, which we defined to be competitive with classical and other emerging technologies [2] (see Fig. 8.6 in Sect. 8.4.1). On the other hand, GA and MW plasmas yield energy efficiencies for CO_2 splitting clearly above the thermal equilibrium conversion and approaching the defined target of 60%. This is attributed to the important role of the vibrational kinetics for energy-efficient CO_2 dissociation. Hence, GA and MW plasma are highly promising for CO_2 conversion.

When looking at DRM, again DBD plasmas perform far below the defined efficiency target of 60% (or above the target energy cost of 4.27 eV/molec), so they cannot yet be competitive with existing technologies (see Fig. 8.10 in Sect. 8.4.2). Surprisingly, very limited results have been published in literature on MW plasmas for DRM, while GA discharges seem very promising, with energy efficiencies close to or above the efficiency target. The same applies to some other types of plasmas that have recently been investigated, including ns-pulsed discharges, spark discharges and APGDs, although more research is needed to fully explore their capabilities.

It must be realized, however, that the efficiency target of 60% for DRM was only defined for syngas production. This is indeed the major product in plasma-based DRM, but in principle, plasma can directly produce value-added compounds, such as oxygenates or higher hydrocarbons, when suitable catalysts can be found. This gives plasma technology a clear advantage, and it would drastically reduce the efficiency target to compete with other technologies, which need two steps for this purpose. Thus, even DBD plasmas might become viable in this case, especially due to their simple design for easy implementation of catalyst and even upscaling. Nevertheless, to realize this, the precise underlying mechanisms of plasma catalysis (which may be different from thermal mechanisms [104]) must first be better understood, to allow the design of catalysts tailored to the plasma environment. This conclusion does not only apply to DRM, but also to artificial photosynthesis (CO$_2$/H$_2$O) and CO$_2$ hydrogenation (CO$_2$/H$_2$), as discussed in Sects. 8.4.3 and 8.4.4.

In general, plasma technology has great advantages, certainly combined with renewable electricity, due to its flexibility, in terms of (i) feed gas (i.e., allowing both CO$_2$ splitting and mixtures with any H-source), (ii) energy source (i.e., all types of renewable electricity, as well as nuclear power), and (iii) operation (i.e., modular upscaling possible and turnkey process).

Nevertheless, more research is crucial, especially in terms of (i) plasma operation, to tune the reduced electric field, plasma power and gas temperature for optimal energy efficiency, (ii) plasma reactor and gas flow design, to improve the conversion, and (iii) catalyst design, tailored to the plasma environment.

Although promising results have been obtained in terms of energy efficiency, the latter largely depends on the plasma reactor and operating conditions. For energy-efficient CO$_2$ conversion, a reduced electric field around 5–100 Td is needed, in combination with high plasma power for sufficient vibrational excitation, which yields the most energy-efficient dissociation pathway. Finally, this should be combined with a low gas temperature, to reduce VT relaxation, i.e., the vibrational losses upon collision with other gas molecules. In other words, a strong thermal non-equilibrium should be targeted.

To enhance the conversion, efforts should be directed to increasing the fraction of gas treated by the plasma, especially in GA plasma reactors; cf. section 8.3.3). Finally and probably most importantly, the product yield/selectivity must be improved. Indeed, the major disadvantage of plasma-based CO$_2$ conversion is, in our opinion, the need for post-reaction separation, because the CO$_2$ conversion is far from 100%, and many different products can be formed, due to the reactive and non-selective character of the plasma chemistry.

In a recent paper, the potential of MW plasma for CO$_2$ splitting into CO was evaluated, and the authors estimated a CO mean selling price of 1.2 kUS\$/ton [105]. Knowing that the bulk CO price is estimated around 228 US\$/ton, this shows that plasma-based CO$_2$ splitting into CO needs further improvement [105]. It is worth to mention that the largest energy cost was found to be the separation of CO and O$_2$. Hence, the authors concluded that improvements in the conversion would be more effective than improvements in energy efficiency. However, if plasma technology would be able to directly produced liquid compounds (e.g., oxygenates), the post-reaction separation would not be so critical, as the liquid compounds are easier to separate. Therefore, this would make the business case for plasma technology much more optimistic. Nevertheless, as discussed above, more research is needed towards the smart design of catalysts tailored to the plasma chemistry.

Finally, we want to stress that plasma technology might not be chemically selective in producing value-added compounds, but it can selectively populate the vibrational levels of CO$_2$, without activating the other degrees of freedom, i.e., specifically without heating the gas. This type of selectivity induces thermal non-equilibrium, and explains the good energy efficiency compared to thermal conversion, at least for some plasma types (e.g., MW and GA plasmas, and maybe APGDs). We believe that this characteristic feature distinguishes plasma technology from other technologies, and should be further exploited, to bring plasma-based CO_2 conversion into commercial application.

References

1. Bogaerts A, Neyts E, Gijbels R, Van der Mullen J (2002) Gas discharge plasmas and their applications. Spectrochim Acta B Atom Spectrosc 57:609–658
2. Snoeckx R, Bogaerts A (2017) Plasma technology—a novel solution for CO_2 conversion? Chem Soc Rev 46:5805–5863
3. Fridman A (2008) Plasma chemistry. Cambridge University Press, Cambridge
4. Fridman A, Chirokov A, Gutsol A (2005) Non-thermal atmospheric pressure discharges. J Phys D Appl Phys 38:R1–R24
5. Kozák T, Bogaerts A (2014) Splitting of CO_2 by vibrational excitation in non-equilibrium plasmas: a reaction kinetics model. Plasma Sources Sci Technol 23:045004
6. Kozák T, Bogaerts A (2015) Evaluation of the energy efficiency of CO_2 conversion in microwave discharges using a reaction kinetics model. Plasma Sources Sci Technol 24:015024
7. Berthelot A, Bogaerts A (2017) Modeling of CO_2 splitting in a microwave plasma: how to improve the conversion and energy efficiency? J Phys Chem C 121:8236–8251
8. Bogaerts A, Berthelot A, Heijkers S, St Kolev, Snoeckx R, Sun SR, Trenchev G, Van Laer K, Wang W (2017) CO_2 conversion by plasma technology: insights from modeling the plasma chemistry and plasma reactor design. Plasma Sources Sci Technol 26:063001
9. Bogaerts A, Kozák T, Van Laer K, Snoeckx R (2015) Plasma-based conversion of CO_2: current status and future challenges. Faraday Discuss 183:217–232
10. Bogaerts A, Neyts E (2018) Plasma technology: an emerging technology for energy storage. ACS Energy Lett 3:1013–1027
11. Kogelschatz U (2003) Dielectric-barrier discharges: their history, discharge physics, and industrial applications. Plasma Chem Plasma Process 23:1–46
12. Van Laer K, Bogaerts A (2016) Fluid modeling of a packed bed dielectric barrier discharge plasma reactor. Plasma Sources Sci Technol 25:015002
13. Michielsen I, Uytdenhouwen Y, Pype J, Michielsen B, Mertens J, Reniers R, Meynen V, Bogaerts A (2017) CO_2 dissociation in a packed bed DBD reactor: first steps towards a better understanding of plasma catalysis. Chem Eng J 326:477–488
14. Uytdenhouwen Y, Van Alphen S, Michielsen I, Meynen V, Cool P, Bogaerts A (2018) A packed-bed DBD micro plasma reactor for CO_2 dissociation: does size matter? Chem Eng J 348:557–568
15. Asisov R, Vakar AK, Jivotov VK, Krotov MF, Zinoviev OA, Potapkin BV, Rusanov AA, Rusanov VD, Fridman AA (1983) Non-equilibrium plasma-chemical process of CO_2 decomposition in a supersonic microwave discharge. Proc USSR Acad Sci 271:94–97
16. Vermeiren V, Bogaerts A. Paper in preparation
17. Nunnally T, Gutsol K, Rabinovich A, Fridman A, Gutsol A, Kemoun A (2011) Dissociation of CO_2 in a low current gliding arc plasmatron. J Phys D Appl Phys 44:274009
18. Ramakers M, Trenchev G, Heijkers S, Wang W, Bogaerts A (2017) Gliding arc plasmatron: providing a novel method for CO_2 conversion. Chemsuschem 10:2642–2652
19. Cleiren E, Heijkers S, Ramakers M, Bogaerts A (2017) Dry reforming of methane in a gliding arc plasmatron: towards a better understanding of the plasma chemistry. Chemsuschem 10:4025–4036

20. Sun SR, Wang HX, Mei DH, Tu X, Bogaerts A (2017) CO$_2$ conversion in a gliding arc plasma: performance improvement based on chemical reaction modeling. J CO$_2$ Utilization 17:220–234
21. Heijkers S, Bogaerts A (2017) CO$_2$ conversion in a gliding arc plasmatron: elucidating the chemistry through kinetic modelling. J Phys Chem C 121:22644–22655
22. Scapinello M, Martini LM, Dilecce G, Tosi P (2016) Conversion of CH$_4$/CO$_2$ by a nanosecond repetitively pulsed discharge. J Phys D Appl Phys 49:75602
23. Zhu B, Li X, Liu J, Zhu X, Zhu A (2015) Kinetics study on carbon dioxide reforming of methane in kilohertz spark-discharge plasma. Chem Eng J 264:445–452
24. Indarto A, Choi J, Lee H, Song H (2006) Effect of additive gases on methane conversion using gliding arc discharge. Energy 31:2986–2995
25. Trenchev G, Nikiforov A, Wang W, Kolev S, Bogaerts A (2019) Atmospheric pressure glow discharge for CO$_2$ conversion: model-based exploration of the optimum reactor configuration. Chem Eng J 362:830–841
26. Neyts EC, Ostrikov K, Sunkara MK, Bogaerts A (2015) Plasma catalysis: synergistic effects at the nanoscale. Chem Rev 115:13408–13446
27. Whitehead JC (2016) Plasma-catalysis: the known knowns, the known unknowns and the unknown unknowns. J Phys D Appl Phys 49:243001
28. Tu X, Whitehead JC, Nozaki T (eds) (2018) Plasma-catalysis: fundamentals and applications. Springer, to be published
29. Lee H, Sekiguchi H (2011) Plasma–catalytic hybrid system using spouted bed with a gliding arc discharge: CH$_4$ reforming as a model reaction. J Phys D Appl Phys 44:274008
30. Blin-Simiand N, Tardivaux P, Risacher A, Jorand F, Pasquiers S (2005) Removal of 2-heptanone by dielectric barrier discharges—the effect of a catalyst support. Plasma Process Polym 2:256–262
31. Hong JP, Chu W, Chernavskii PA, Khodakov AY (2010) Cobalt species and cobalt-support interaction in glow discharge plasma-assisted Fischer-Tropsch catalysts. J Catal 273:9–17
32. Demidyuk V, Whitehead JC (2007) Influence of temperature on gas-phase toluene decomposition in plasma-catalytic system. Plasma Chem Plasma Process 27:85–94
33. Shang S, Liu G, Chai X, Tao X, Li X, Bai M, Chu W, Dai X, Zhao Y, Yin Y (2009) Research on Ni/γ-Al$_2$O$_3$ catalyst for CO$_2$ reforming of CH$_4$ prepared by atmospheric pressure glow discharge plasma jet. Catal Today 148:268–274
34. Tu X, Gallon HJ, Twigg MV, Gorry PA, Whitehead JC (2011) Dry reforming of methane over a Ni/Al$_2$O$_3$ catalyst in a coaxial dielectric barrier discharge reactor. J Phys D Appl Phys 44:274007
35. Liu C-J, Mallison R, Lobban L (1998) Nonoxidative methane conversion to acetylene over zeolite in a low temperature plasma. J Catal 179:326–334
36. Poppe J, Völkening S, Schaak A, Schütz E, Janek J, Imbihl R (1999) Electrochemical promotion of catalytic CO oxidation on Pt/YSZ catalysts under low pressure conditions. Phys Chem Chem Phys 1:5241–5249
37. van Durme J, Dewulf J, Leys C, Van Langenhove H (2008) Combining non-thermal plasma with heterogeneous catalysis in waste gas treatment: a review. Appl Catal B: Environ 78:324–333
38. Liu CJ, Wang JX, Yu KL, Eliasson B, Xia Q, Xue B (2002) Floating double probe characteristics of non-thermal plasmas in the presence of zeolite. J Electrostat 54:149–158
39. Kim HH, Ogata A, Futamura S (2006) Effect of different catalysts on the decomposition of VOCs using flow-type plasma-driven catalysis. IEEE Trans Plasma Sci 34:984–995
40. Löfberg A, Essakhi A, Paul S, Swesi Y, Zanota M-L, Meille V, Pitault I, Supiot P, Mutel B, Le Courtois V, Bordes-Richard E (2011) Use of catalytic oxidation and dehydrogenation of hydrocarbons reactions to highlight improvement of heat transfer in catalytic metallic foams. Chem Eng J 176–177:49–56

41. Guaitella O, Thevenet F, Puzenat E, Guillard C, Rousseau A (2008) C_2H_2 oxidation by plasma/TiO$_2$ combination: Influence of the porosity, and photocatalytic mechanisms under plasma exposure. Appl Cat B: Environ 80:296–305

42. Kim HH, Ogata A, Futamura S (2008) Oxygen partial pressure-dependent behavior of various catalysts for the total oxidation of VOCs using cycled system of adsorption and oxygen plasma. Appl Catal B: Environ 79:356–367

43. Kim HH, Ogata A (2011) Nonthermal plasma activates catalyst: from current understanding and future prospects. Eur Phys J Appl Phys 55:13806

44. Mei D, Zhu X, He Y, Yan JD, Tu X (2015) Plasma-assisted conversion of CO_2 in a dielectric barrier discharge reactor: understanding the effect of packing materials. Plasma Sources Sci Technol 24:015011

45. Shirazi M, Neyts EC, Bogaerts A (2017) DFT study of Ni-catalyzed plasma dry reforming of methane. Appl Cat B: Environ 205:605–614

46. Zhang Q-Z, Bogaerts A (2018) Propagation of a plasma streamer in catalyst pores. Plasma Sources Sci Technol 27:035009

47. Kim HH, Kim J-H, Ogata A (2009) Microscopic observation of discharge plasma on the surface of zeolites supported metal nanoparticles. J Phys D Appl Phys 42:135210

48. Wang W, Kim H-H, Van Laer K, Bogaerts A (2018) Streamer propagation in a packed bed plasma reactor for plasma catalysis applications. Chem Eng J 334:2467–2479

49. Holzer F, Kopinke FD, Roland U (2005) Influence of ferroelectric materials and catalysts on the performance of non-thermal plasma (NTP) for the removal of air pollutants. Plasma Chem Plasma Proc 25:595–611

50. Hensel K, Martisovits V, Machala Z, Janda M, Lestinsky M, Tardiveau P, Mizuno A (2007) Electrical and optical properties of AC microdischarges in porous ceramics. Plasma Process Polym 4:682–693

51. Zhang Y-R, Van Laer K, Neyts EC, Bogaerts A (2016) Can plasma be formed in catalyst pores? A modeling investigation. Appl Cat B: Environm 185:56–67

52. Rousseau A, Guaitella O, Röpcke J, Gatilova LV, Tolmachev YA (2004) Combination of a pulsed microwave plasma with a catalyst for acetylene oxidation. Appl Phys Lett 85:2199–2201

53. Neyts EC, Bogaerts A (2014) Understanding plasma catalysis through modelling and simulation—a review. J Phys D Appl Phys 47:224010

54. Zhang A, Zhu A, Guo J, Xu Y, Shi C (2010) Conversion of greenhouse gases into syngas via combined effects of discharge activation and catalysis. Chem Eng J 156:601–606

55. Kim J, Henao CA, Johnson TA, Dedrick DE, Miller JE, Stechel EB, Maravelias CT (2011) Methanol production from CO_2 using solar-thermal energy: process development and techno-economic analysis. Energy Environ Sci 4:3122

56. Aerts R, Somers W, Bogaerts A (2015) Carbon dioxide splitting in a dielectric barrier discharge plasma: a combined experimental and computational study. Chemsuschem 8:702–716

57. Paulussen S, Verheyde B, Tu X, De Bie C, Martens T, Petrovic D, Bogaerts A, Sels B (2010) Conversion of carbon dioxide to value-added chemicals in atmospheric pressure dielectric barrier discharges. Plasma Sources Sci Technol 19:034015

58. Yu Q, Kong M, Liu T, Fei J, Zheng X (2012) Characteristics of the decomposition of CO_2 in a dielectric packed-bed plasma reactor. Plasma Chem Plasma Process 32:153–163

59. Ozkan A, Bogaerts A, Reniers F (2017) Routes to increase the conversion and the energy efficiency in the splitting of CO_2 by a dielectric barrier discharge. J Phys D Appl Phys 50:084004

60. Ozkan A, Dufour T, Silva T, Britun N, Snyders R, Reniers F, Bogaerts A (2016) DBD in burst mode: solution for more efficient CO_2 conversion? Plasma Sources Sci Technol 25:055005

61. Van Laer K, Bogaerts A (2015) Improving the conversion and energy efficiency of carbon dioxide splitting in a zirconia-packed dielectric barrier discharge reactor. Energy Technol 3:1038–1044

62. Mei D, Zhu X, Wu C, Ashford B, Williams PT, Tu X (2016) Plasma-photocatalytic conversion of CO_2 at low temperatures: understanding the synergistic effect of plasma-catalysis. Appl Catal B: Environ 182:525–532

63. Duan X, Hu Z, Li Y, Wang B (2015) Effect of dielectric packing materials on the decomposition of carbon dioxide using DBD microplasma reactor. AlChe 61:898–903

64. Rusanov VD, Fridman AA, Sholin GV (1981) The physics of a chemically active plasma with nonequilibrium vibrational excitation of molecules. Uspekhi Fiz Nauk 134:185–235

65. van Rooij GJ, van den Bekerom DCM, den Harder N, Minea T, Berden G, Bongers WW, Engeln R, Graswinckel MF, Zoethout E, van de Sanden MCM (2015) Taming microwave plasma to beat thermodynamics in CO_2 dissociation. Faraday Discuss 183:233–248

66. Bongers W, Bouwmeester H, Wolf B, Peeters F, Welzel S, van den Bekerom D, den Harder N, Goede A, Graswinckel M, Groen PW, Kopecki J, Leins M, van Rooij G, Schulz A, Walker M, van de Sanden R (2017) Plasma-driven dissociation of CO_2 for fuel synthesis. Plasma Process Polym 14:e1600126

67. Silva T, Britun N, Godfroid T, Snyders R (2014) Optical characterization of a microwave pulsed discharge used for dissociation of CO_2. Plasma Sources Sci Technol 23:025009

68. Spencer LF, Gallimore AD (2013) CO_2 dissociation in an atmospheric pressure plasma/catalyst system: a study of efficiency. Plasma Sources Sci Technol 22:015019

69. Indarto A, Yang DR, Choi JW, Lee H, Song HK (2007) Gliding arc plasma processing of CO_2 conversion. J Hazard Mater 146:309–315

70. Liu JL, Park HW, Chung WJ, Park DW (2016) High-efficient conversion of CO_2 in AC-pulsed tornado gliding arc plasma. Plasma Chem Plasma Process 36:437–449

71. Wang W, Mei D, Tu X, Bogaerts A (2017) Gliding arc plasma for CO_2 conversion: better insights by a combined experimental and modelling approach. Chem Eng J 330:11–25

72. Wang W, Berthelot A, Berthelot A, Kolev S, Tu X (2016) CO_2 conversion in a gliding arc plasma: 1D cylindrical discharge model. Plasma Sources Sci Technol 25:065012

73. Trenchev G, Kolev S, Wang W, Ramakers M, Bogaerts A (2017) CO_2 conversion in a gliding arc plasmatron: multi-dimensional modeling for improved efficiency. J Phys Chem C 121:24470–24479

74. Andreev SN, Zakharov VV, Ochkin VN, Savinov SY (2004) Plasma-chemical CO_2 decomposition in a non-self-sustained discharge with a controlled electronic component of plasma. Spectrochim Acta A: Mol Biomol Spectrosc 60:3361–3369

75. Aerts R, Snoeckx R, Bogaerts A (2014) In-situ chemical trapping of oxygen after the splitting of carbon dioxide by plasma. Plasma Process Polym 11:985–992

76. Mori S, Matsuura N, Tun LL, Suzuki M (2016) Direct synthesis of carbon nanotubes from only CO_2 by a hybrid reactor of dielectric barrier discharge and solid oxide electrolyser cell. Plasma Chem Plasma Process 36:231–239

77. Lavoie JM (2014) Review on dry reforming of methane, a potentially more environmentally-friendly approach to the increasing natural gas exploitation. Front Chem 2:1–17

78. Tao X, Bai M, Li X, Long H, Shang S, Yin Y, Dai X (2011) CH_4-CO_2 reforming by plasma—challenges and opportunities. Prog Energy Combust Sci 37:113–124

79. Lebouvier A, Iwarere SA, D'Argenlieu P, Ramjugernath D, Fulcheri L (2013) Assessment of carbon dioxide dissociation as a new route for syngas production: a comparative review and potential of plasma-based technologies. Energy Fuels 27:2712–2722

80. Istadi I, Amin NAS (2006) Co-generation of synthesis gas and C2+ hydrocarbons from methane and carbon dioxide in a hybrid catalytic-plasma reactor: a review. Fuel 85:577–592

81. Krawczyk K, Młotek M, Ulejczyk B (2014) Methane conversion with carbon dioxide in plasma-catalytic system. Fuel 117:608–617

82. Chung W, Pan K, Lee H, Chang M (2014) Dry reforming of methane with dielectric barrier discharge and ferroelectric packed-bed reactors. Energy Fuels 28:7621–7631

83. Snoeckx R, Zeng YX, Tu X, Bogaerts A (2015) Plasma-based dry reforming: improving the conversion and energy efficiency in a dielectric barrier discharge. RSC Adv 5:29799–29808

84. Scapinello M, Martini LM, Tosi T (2014) CO_2 hydrogenation by CH_4 in a dielectric barrier discharge: catalytic effect of Ni and Cu. Plasma Process Polym 11:624–628

85. Wang L, Yi Y, Wu C, Guo H, Tu X (2017) One-step reforming of CO_2 and CH_4 into high-value liquid chemicals and fuels at room temperature by plasma-driven catalysis. Angew Chem Int Ed 129:13867–13871

86. Wu W, Yan J, Zhang H, Zhang M, Du C, Li X (2014) Study of the dry methane reforming process using a rotating gliding arc reactor. Int J Hydrogen Energy 39:17656–17670

87. Li K, Liu JL, Li XS, Zhu X, Zhu AM (2016) Warm plasma catalytic reforming of biogas in a heat-insulated reactor: dramatic energy efficiency and catalyst auto-reduction. Chem Eng J 288:671–679

88. Tu X, Whitehead JC (2014) Plasma dry reforming of methane in an atmospheric pressure AC gliding arc discharge: Co-generation of syngas and carbon nanomaterials. Int J Hydrogen Energy 39:9658–9669

89. Li K, Liu J-L, Li X-S, Lian H-Y, Zhu X, Bogaerts A, Zhu A-M (2018) Novel power-to-syngas concept for plasma catalytic reforming coupled with water electrolysis. Chem Eng J (submitted)

90. Ghorbanzadeh AM, Norouzi S, Mohammadi T (2005) High energy efficiency in syngas and hydrocarbon production from dissociation of CH_4-CO_2 mixture in a non-equilibrium pulsed plasma. J Phys D Appl Phys 38:3804–3811

91. Chung WC, Chang MB (2016) Review of catalysis and plasma performance on dry reforming of CH_4 and possible synergistic effects. Renew Sustain Energy Rev 62:13–31

92. Li X, Bai M, Tao X, Shang S, Dai X, Yin Y (2009) CO_2 reforming of CH_4 by atmospheric pressure glow discharge plasma: a high conversion ability. Int J Hydrogen Energy 34:308–313

93. Chen G, Britun N, Godfroid T, Georgieva V, Snyders V, Delplancke-Ogletree M-P (2017) An overview of CO_2 conversion in a microwave discharge: the role of plasma-catalysis. J Phys D Appl Phys 50:084001

94. Snoeckx R, Ozkan A, Reniers F, Bogaerts A (2017) The quest for value-added products from carbon dioxide and water in a dielectric barrier discharge plasma: a chemical kinetics study. Chemsuschem 10:409–424

95. Eliasson B, Kogelschatz U, Xue B, Zhou L-M (1998) Hydrogenation of carbon dioxide to methanol with a discharge-activated catalyst. Ind Eng Chem Res 37:3350–3357

96. Mahammadunnisa S, Reddy EL, Ray D, Subrahmanyam C, Whitehead JC (2013) CO_2 reduction to syngas and carbon nanofibres by plasma-assisted in situ decomposition of water. Int J Greenhouse Gas Control 16:361–363

97. Jwa E, Lee SB, Lee HW, Mok YS (2013) Plasma-assisted catalytic methanation of CO and CO_2 over Ni–zeolite catalysts Fuel Process Technol 108:89–93

98. De Bie C, van Dijk J, Bogaerts A (2016) CO_2 hydrogenation in a dielectric barrier discharge plasma revealed. J Phys Chem C 120:25210–25224

99. Wang L, Yi Y, Guo H, Tu X (2018) Atmospheric pressure and room temperature synthesis of methanol through plasma-catalytic hydrogenation of CO_2. ACS Catal 8:90–100

100. Centi G, Perathoner S (2009) Opportunities and prospects in the chemical recycling of carbon dioxide to fuels. Catal Today 148:191–205

101. Studt F, Sharafutdinov I, Abild-Pedersen F, Elkjær CF, Hummelshøj JS, Dahl S, Chorkendorff I, Nørskov JK (2014) Discovery of a Ni-Ga catalyst for carbon dioxide reduction to methanol. Nat Chem 6:320–324

102. Jadhav SG, Vaidya PD, Bhanage BM, Joshi JB (2014) Catalytic carbon dioxide hydrogenation to methanol: a review of recent studies. Chem Eng Res Des 92:2557–2567

103. Snoeckx R, Heijkers S, Van Wesenbeeck K, Lenaerts S, Bogaerts A (2016) CO$_2$ conversion in a dielectric barrier discharge plasma: N$_2$ in the mix as helping hand of problematic impurity? Energy Environ Sci 9:999–1011
104. Kim J, Abbott MS, Go DB, Hicks JC (2016) Enhancing C-H bond activation of methane via temperature-controlled, catalyst-plasma interactions. ACS Energy Lett 1:94–99
105. van Rooij GJ, Akse HN, Bongers WA, van de Sanden MCM (2018) Plasma for electrification of chemical industry: a case study on CO$_2$ reduction. Plasma Phys Control Fusion 60:014019

Bioelectrochemical Syntheses

9

9

Suman Bajracharya, Nabin Aryal, Heleen De Wever
and Deepak Pant

Abstract

Bioelectrosynthesis from CO_2 offers the prospect to reuse CO_2 emissions as a feedstock and generate fuels and value-added chemicals from CO_2 and its derivatives working in water. The technology has environmental advantages due to its sustainability, renewability and environmentally friendly qualities. The future potential of these systems can be associated to the framework of CO_2 biorefineries, the power-to-gas concept, or biogas upgrading, thus helping to step-up in the desired global transition from fossil fuel-based to electricity-based economy.

9.1 Introduction

Bioelectrochemical systems (BESs) encompass the electrode-based oxidation and reduction reactions catalyzed by biological agents such as microorganisms and/or enzymes to convert the chemical energy stored in biodegradable materials to

S. Bajracharya
Water Desalination and Reuse Center, King Abdullah University
of Science and Technology, Thuwal 23955-6900, Saudi Arabia

N. Aryal
Biological and Chemical Engineering, Aarhus University, Hangovej 2,
DK-8200 Aarhus N, Denmark

N. Aryal
Danish Gas Technology Centre, Dr Neergaards Vej 5B, DK-2970 Horsholm, Denmark

H. De Wever · D. Pant (✉)
Separation and Conversion Technology, Flemish Institute for Technological
Research (VITO), Boeretang 200, 2400 Mol, Belgium
e-mail: deepak.pant@vito.be; pantonline@gmail.com

© Springer Nature Switzerland AG 2019
M. Aresta et al. (eds.), *An Economy Based on Carbon Dioxide and Water*,
https://doi.org/10.1007/978-3-030-15868-2_9

electric current or to produce value-added chemicals from low value chemicals with an external voltage supply [19, 102, 138]. BES reactors comprise a combination of either a biotic anode (bioanode) with chemical cathode or a bioanode with biotic cathode (biocathode) or an abiotic anode with biocathode in a configuration with or without an ion exchange membrane as a separator for the anode-cathode compartments. When microbial cells are involved in any of the electrode-based redox reaction the term 'microbial' is used whereas the term 'biological/enzymatic' is used when enzymes, proteins or any other biological agents are involved in the redox reaction [71, 123]. The conversion of low-value waste into electricity, fuels and chemicals using biocatalysts especially microbes that are rejuvenating and adapting themselves to the required conversion activity is the unique feature of bioelectrochemical systems [82, 108].

In principle, biodegradable materials are oxidized at the anode employing a biological agent as a catalyst and generate electrons, which flow (current) towards the cathode due to the difference in potential along an external circuit where they are consumed in reduction reaction, producing electricity. These systems are called biological/microbial fuel cells. In the opposite case, by applying an external voltage, the electrons produced at the bioanode are used in the reduction of water to hydrogen at the cathode in systems called Microbial Electrolysis Cells (MECs); or, other oxidized components, such as metal ions, CO_2, or organic chemicals are reduced to value-added chemicals using microbes or enzymes at cathode in the process called Bioelectrosynthesis. Bioelectrosynthesis embraces a synergistic association of biological metabolism into the electrochemical process. The concept of bioelectrosynthesis is to synthesize organic compounds by involving microbes or enzymes as biocatalysts in electrode-based redox reactions more commonly in cathode-based reduction reactions. An external potential is applied to provide sufficient reduction potentials at the cathode to initiate different reduction reactions to produce value-added chemicals. Bioelectrosynthesis is an innovative standpoint of BESs which utilizes the external input of electrical energy to drive the biosynthesis of value-added products using biocatalyst at cathodic microenvironment.

9.2   Bioelectrosynthesis from CO_2

The conversion of captured CO_2 from the waste streams in today's and tomorrow's chemical industry back into chemical feedstocks could be an efficient way to close the carbon cycle within a chemical site and simultaneously could limit the reliance on fossil feedstocks. A new approach that has emerged in recent years is to introduce biological CO_2 fixation metabolism in the electrochemical process of CO_2 reduction. The biological catalysts comprise microbial cells/communities, predominantly the chemolithoautotrophs and methanogens that use CO_2 as electron acceptor, and also the enzymes derived from these microbes. Bioelectrosynthesis from CO_2 here, means the electricity driven cathodic process for the production of

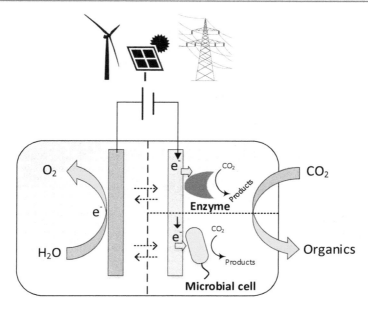

Fig. 9.1 Conceptual representation of bioelectrosynthesis of organic products from CO_2

value-added compounds from CO_2 using biological catalysts (Fig. 9.1). A combined approach of electrochemical hydrogen production at cathode (hydrogen evolution reaction, HER) and microbial CO_2 reduction is also gaining attention as a hybrid biosystem for the bioelectrosynthesis of value-added products from CO_2 [17, 80, 97].

The use of microbes as biocatalyst in the electricity-driven production of chemical compounds by reducing CO_2 and/or any other organic feedstock has been separately evolved as Microbial Electrosynthesis (MES) [108], whereas the enzymes involved-electrosynthesis has been termed as Enzymatic Electrosynthesis (EES). As a proof of concept, Nevin et al. [95] presented MES as a microbial catalysis of CO_2 reduction to multi-carbon organic compounds such as organic acids using electrical current at the cathode. Biocatalysts attached on the cathode (cathodic biofilms in case of microbial electrosynthesis) reduce the available terminal electron acceptor to produce value added products [86, 90, 108].

Bioelectrosynthesis is also prospected as an alternative strategy to capture electrical energy in covalent chemical bonds in the fields of microbiological and electrochemical engineering. Since the bioelectrosynthesis process uses electricity as the energy source, it can be coupled to renewable sources of energy (e.g. wind and solar). However, given the intermittent nature of renewable energy supply, this may require fast on/off switching or adjustments in capacity.

The major advantageous features of bioelectrosynthesis as compared to the classical electrochemical process are

(i)   the requirement of diminutive energy input for the biocatalysis,
(ii)   the adaptability of microbes for producing various commodities,
(iii)   Less expensive system and low operation cost,
(iv)   the reactions occurring at ambient conditions and no toxic substances involved,
(v)   the recyclability of the biocatalyst.

CO_2 molecule is thermodynamically and kinetically inert under normal conditions due its low energy level [51, 106] which means the conversion of CO_2 is non-spontaneous and slow. Consequently, a large energy input is required for its transformation. CO_2 conversions to higher carbon molecules occur in nature commonly via photosynthesis, in which sunlight provides the energy required for the dissociation of the CO_2 molecule.

CO_2 conversion technologies mimicking the natural photosynthesis are gaining attraction. However, the systems are limited in efficiency and discontinuity in operation (day/night variation). Instead, electrochemical CO_2 reduction using a metallic electrode/catalyst is an effective approach for the activation and transformation of the stable CO_2 molecule into fuels/chemicals such as methane, methanol, etc. [51, 54]. Proton-coupled multi-electron reductions of CO_2 are generally favorable because the products are thermodynamically stable. The major products of electrochemical reduction of CO_2 using aqueous electrolyte are carbon monoxide (CO), formic acid (HCOOH)/formate ($HCOO^-$), oxalic acid ($H_2C_2O_4$)/oxalate ($C_2O_4^{2-}$), formaldehyde (HCHO), methanol (CH_3OH), methane (CH_4), ethanol (CH_3CH_2OH) ethylene (CH_2CH_2) and others (Table 9.1). The electrochemical approach of CO_2 reduction is associated with high-energy requirement, low stability of the electrodes and catalyst, non-specificity for the product and high cathode cost [91].

Naturally found chemolithotrophic microorganisms have the ability to reduce CO_2 by utilizing inorganic electron donor such as H_2, a major energy carrier in biological CO_2 reduction systems. By integrating the electrochemical reaction with biological metabolism, CO_2 reduction can be driven with less energy input by using appropriate biological catalysts, either whole microbial cell [108] and/or enzymes [128]. Bioelectrosynthesis is proven less costly as compared to conventional electrosynthesis and it does not cause toxicity, and does not corrode and denature materials [82]. Reduction of CO_2 in BES is possible at the cathode using microbes or enzymes that can oxidize the cathode and use CO_2 as terminal electron acceptor. Electrosynthesis from CO_2 reduction becomes more product specific by incorporating biocatalysts on the cathode. The use of natural materials and neutral aqueous buffers close to ambient temperature in bioelectrosynthesis meets the sustainability criteria.

H_2 evolution via water electrolysis at the cathode (-0.41 V vs. SHE at biological conditions) occurs well beyond the CO_2 reduction potentials (Table 9.1). But due to the associated electrochemical overpotentials in CO_2 reduction, the potential has to go more negative than -0.41 V versus SHE. The produced H_2 can additionally serve as an electron donor for the CO_2 reduction in hydrogenotrophic methanogenic and acetogenic microorganisms. So far, H_2 mediated CO_2 reduction when using

Table 9.1 Possible CO_2 reduction half reactions in biological conditions with change in Gibb's energy and redox equilibrium potentials (Gibbs free energy of formation taken from Thauer et al. [131])

	Half reaction	No. of electron involved	$\Delta G_r^{O'}$ (kJ mol^{-1})	Standard reduction potential $E_r^{O'}$ (V vs. SHE* at pH = 7)
Formate	$CO_2 + 2e^- + H^+ \rightarrow HCOO^-$	2	83.2	−0.43
Carbon monoxide	$CO_2 + 2e^- + 2H^+ \rightarrow CO + H_2O$	2	99.77	−0.52
Oxalate	$CO_2 + 2e^- \rightarrow C_2O_4^{2-}$	2	114.678	−0.59
Formaldehyde	$CO_2 + 4e^- + 4H + \rightarrow CH_2O + H_2O$	4	186.12	−0.48
Methanol	$CO_2 + 6e^- + 6H^+ \rightarrow CH_3OH + H_2O$	6	221.01	−0.38
Methane	$CO_2 + 8e^- + 8\ H^+ \rightarrow CH4 + 2H_2O$	8	188.209	−0.24
Acetate	$2CO_2 + 8e^- + 7\ H + \rightarrow CH_3COO^- + 2H_2O$	8	224.04	−0.29
Ethanol	$2CO_2 + 12e^- + 12H^+ \rightarrow C_2H_5OH + 3H_2O$	12	373.87	−0.323
Ethylene	$2CO_2 + 12e^- + 12H^+ \rightarrow CH_2CH_2 + 4H_2O$	12	386.558	−0.33
Beta-hydroxy butyrate	$4CO_2 + 18e^- + 17\ H^+ \rightarrow CH_3CH(OH) CH_2COO^- + 5\ H_2O$	18	563.02	−0.324
Butyrate	$4CO_2 + 20\ e^- + 19\ H^+ \rightarrow CH_3CH_2CH_2COO^- + 6H_2O$	20	559.25	−0.29
Hydrogen	$2H^+ + 2e^- \rightarrow H_2$	2	79.74	−0.413

*V versus SHE, Volt versus Standard hydrogen electrode

biological catalysts seems to be the dominant mechanism in undefined mixed culture biocathodes, rather than direct electron transfer [18, 57]. Often, a mediator is required to transport electrons between the electrode and the whole cells dispersed through the electrolyte. In principle, the chemistry of whole cells could be less selective than that of a single enzyme [95], but the enzymatic reduction lasts only for a few hours, which reflected poor long-term stability.

9.2.1 Biocatalysts for CO_2 Reduction

A multitude of CO_2 fixation pathways and energy-acquiring systems have evolved in autotrophs that can be integrated as biocatalysts for the production of a wide range of biochemical and fuels. Microbial autotrophy is a natural and sustainable setting to convert CO_2 into biomass, biochemicals and biofuels by assimilating energy from light (photoautotrophy) or from reduced inorganic electron donors such as hydrogen, reduced metal ions, ammonia, formate and even direct electro-chemical energy (chemolithoautotrophy) [49]. The use of microorganisms and of enzymes involved in chemolithoautotrophy are separately discussed as biocatalysts for CO_2 reduction in the following sections.

9.2.1.1 Microbial Catalysts for Electrochemical CO_2 Reduction

The Wood–Ljungdahl, reverse tricarboxylic acid or hydroxypropionate cycles in chemolithoautotrophs are applicable for bioelectrosynthesis from CO_2. Via these pathways, microorganisms use inorganic electron donors such as H_2, formate, reduced metal ions for the reduction of CO_2. In electrochemical settings, the reduced inorganic electron donors can be generated at the cathode when electric energy is applied. In addition, various bacteria have evolved the ability to transfer/intake electrons extracellularly. Thus, the chemolithoautotrophs can be accomodated on electrodes to fix CO_2 efficiently with the input of electrical energy. The chemolithoautotrophic biofilm growing on electrodes may become electroactive so as to acquire the electrons from the cathode directly to produce organic products. Acetogens, methanogens and oxygen-reducing chemolithoautotrophs are mostly investigated in microbial electrosynthesis systems for the catalysis of CO_2 reduction as discussed below.

Acetogenic Microorganisms

Acetogens represent a group of several acetate producing taxonomically diverse gram-positive bacterial species such as *Acetobacterium, Sporomusa, Clostridium, Eubacterium, Syntrophococcus, Pentostreptococcus, Butyrobacterium*, etc. Acetogens are chemolithoautotrophs that can grow anaerobically using CO_2 and H_2 (electron donor) in natural environments using the Wood-Ljungdahl pathway, the most energy-efficient pathway known so far for the CO_2 fixation with H_2 [42]. The energy yield from the Wood-Ljungdahl pathway is low but the energy recovery in the extracellular organic end-products is high. Almost 95% of the electrons from H_2 oxidation and carbon from CO_2 is directed to the organic products instead of biomass production [42]. The pathway uniquely conserves energy (ATP) along with CO_2 fixation [25], conveying a form of anaerobic respiration.

Acetogens are promising biocatalysts for the electricity-driven biological CO_2 reduction. Acetogens are autotrophically grown on the cathode of the electrochemical system where they can acquire electrons from the cathode to reduce CO_2 to acetate or from the electrochemically generated H_2 or formate [21, 108]. For instance, Nevin et al. [95] showed production of acetate from CO_2 with a *Sporomusa ovata* biofilm at 85% coulombic efficiency when the graphite stick cathode was poised at -400 mV versus SHE. Several other acetogens such as *Clostridium ljungdahlii, Clostridium aceticum, Moorella thermoacetica, Sporomusa sphaeroides, Sporomusa silvacetica* have also shown to produce acetate from electrode-driven CO_2 fixation [96]. Some acetogens, namely *Clostridium species*, can also produce ethanol in addition to acetate under certain condition, and also 2,3-butanediol, lactate, butyrate and butanol in some instances [68, 115].

Enrichment of autotrophic electrotrophs from various natural sources in the biocathode is often used to select an effective biocatalyst. The selection for CO_2 reducers is based on the microbial ability to switch from heterotrophic to autotrophic metabolism. Mixed microbial communities from natural anaerobic sources have been used for CO_2 reduction in MES. The mixed cultures are enriched with acetogens by growing them in carbon dioxide/hydrogen gas mixtures followed by biofilm formation on the cathode at hydrogen evolving cathode potential with suppressed methanogenesis. The biocathode (suspended or biofilm) enriched from mixed cultures has shown stable and active performance in MES [23, 24]. Acetate production from CO_2 in MES has reached fairly high rates using enriched mixed cultures [57, 59]. A number of CO_2 reducing species from firmicute phylum were predominantly enriched in mixed cultures biocathodes. High-rate CO_2 reducing biocathodes developed from pond sediment and anaerobic wastewater sludge inoculum were dominated by *Acetoanaerobium* [58]. Bog sediment inoculum resulting in a CO_2 reducing biocathode contained *Trichococcus palustris* sp., *Oscillibacter* sp., and few other species [140]. In a number of other studies, the CO_2 reducing biocathode developed from wastewater inoculum was dominated by *Acetobacterium* sp. [89, 72, 88, 105].

Methanogens

Methanogenesis is a common natural anaerobic process. In hydrogenotrophic methanogenesis, archaea reduce CO_2 to methane using H_2 as the electron donor. *Methanobacterium thermoautotrophicum*, *Methanobacterium formicium* etc. are hydrogenotrophic methanogens widely reported for biomethane production [44, 130]. It was reported that an archaeal species *Methanobacterium palustre* sp. became predominant on a cathode polarized at -0.5 to -0.8 V (vs. SHE) generating methane from CO_2 [33]. Methanogens can directly intake electrons from the cathode at low potential or use electrochemically produced H_2 or a redox active mediator such as methyl red to acquire electrons [33, 86, 136, 137]. *Methanosarcina barkeri* oxidized H_2 coming from a platinum cathode to reduce CO_2 into methane [97]. It has also been suggested that a few methanogens might accept electrons from other organisms through biological or mineral electrical contacts [61, 81, 92], consistent with the potential for methanogens to make extracellular electrical contacts.

Aerobic Lithoautotrophs

Aerobic lithoautotrophs like *Cupriavidus necator* (formerly known as *Ralstonia eutropha*), are promising biocatalysts for CO_2 reduction in bioelectrosynthesis because of fast growth and the availability of genetic tools. The substrates for the lithoautotrophs are carbon dioxide from the waste streams as well as hydrogen and oxygen, which can be simultaneously obtained via splitting water by electrolysis. In electroautotrophic conditions, the bacteria oxidize electrochemically produced electron donors, such as H_2 or formate under aerobic conditions, while using the Calvin cycle for CO_2 fixation [99]. *C. necator* can naturally accumulate polyhydroxybutyrate (PHB), a bioplastic precursor, up to 70% of its biomass weight [99]. It has already been effectively tested in BES that generate H_2 [80, 132] or formate [76].

Bioelectrosynthesis of isobutanol and 3-methyl-1-butanol from CO_2 was reported using engineered *C. necator* in which electrochemically produced formate was providing the electrons and oxygen was the electron acceptor [76]. In addition, genetic engineering of heterologous production pathways into *C. necator* has led to the efficient production of compounds such as branched-chain alcohols [84] and alkanes [29].

Another aerobic lithoautotroph is *Acidithiobacillus ferrooxidans* that uses the Calvin cycle to fix CO_2 and can take up electrons directly from a cathode [31] or from Fe^{2+}, which can be produced electrochemically [129]. Genetic tools for *A. ferrooxidans* are being developed, which have recently enabled the production of isobutyric acid and heptadecane [63].

Besides hydrogen, electrochemically produced ammonia from nitrite can also serve as electron donor for the growth of *Nitrosomonas europaea*, which oxidizes ammonia back to nitrite with oxygen as electron acceptor and uses carbon dioxide as its sole carbon source to produce biomass [64]. Production of commodities from aerobic lithoautotrophs is inefficient because a high proportion of electrons is diverted to oxygen reduction to water and a large portion of carbon diverted to biomass. In studies with *N. europaea* in which biomass production was the goal, the energetic efficiency was less than 5% [64]. Additionally the reduction of oxygen at the cathode can also produce lethal reactive oxygen species [76].

9.2.1.2 Enzymes for Electrochemical CO_2 Reduction

Enzymes are remarkable biocatalysts with specific action. Carbonic anhydrase and a few dehydrogenases are the main candidates used as biocatalysts of CO_2 reduction [1, 123]. Carbonic anhydrase is used for capturing gaseous CO_2 in aqueous solution. Dehydrogenases are a sub-class of redox enzymes that relocate hydrogen or electrons from a donor to an acceptor. Carbon monoxide dehydrogenases (CODHs) and formate dehydrogenases (FDHs) are the most relevant enzymes considered for enzymatic electrochemical CO_2 reduction. These enzymes are involved in the reductive acetyl-CoA process, a CO_2 fixation metabolism in chemolithoautotrophic bacteria and methanogenic archaea to synthesize acetyl CoA from CO_2 [110]. Most of the CO_2 reducing dehydrogenases such as FDHs retain an electrochemically active cofactor, either Nicotinamide adenine dinucleotide-hydrogen (NADH) or Nicotinamide adenine dinucleotide phosphate-hydrogen (NADPH) which provides electrons or hydrogen in the reaction. Essentially, the role of the cofactor is to regenerate the enzyme to the active state by oxidation or reduction. The enzymatic CO_2 reduction studies are trending at the synthesis of CO, formate/formic acid, and methanol [85]. These products are generated using a single enzyme biocatalyst or by employing a cascaded multi-enzyme system in combination with CO_2 sequestration. An overview of studies on enzymatic electrosynthesis from CO_2 are provided in Table 9.2 and further details are discussed separately in following subsections.

Carbon Monoxide Dehydrogenase for CO Production

The conversion of CO_2 to CO by the enzyme CODH is observed in anaerobic bacteria and archaea such as *Moorella thermoacetica*, *Carboxydothermus hydrogenoformans*, and *Methanosarcina barkerii* [85]. The CODHs comprise [NiFe] active sites and readily catalyze the reversible transformation of CO_2 to CO [91]. The Ni center in CODH binds the CO_2 molecule and the Fe center stabilizes it to enable the reduction reaction [53]. The Ni containing CODH shows high activity as an electrocatalyst for CO_2 reduction. The first use of [NiFe] CODH as electrocatalyst for CO_2 reduction was in an in-vitro electrolysis system with methyl

Table 9.2 An overview of studies on enzymatic electrosynthesis from CO_2

Enzyme	Materials	Product	Catalytic efficiency/product yield	References
CODH *M. thermoacetica*	Glassy Carbon electrode (GCE), MV^{2+}	CO	$700 \ h^{-1}$	Shin [122]
(Ch)CODH	Pyrolytic graphitic edge (PGE)	CO	–	Parkin et al. [103]
W-FDH1 *S. fumaroxidans*	PGE electrode, buffer	HCOOH	$0.5 \times 10^3 \ s^{-1}$	Reda [111]
Mo containing FDH EcFDH-H *E. coli*	Graphite-epoxy electrode, MV^{2+}, MES,	HCOOH	$<1 \ s^{-1}$	Bassegoda et al. [26]
FDH	Graphite rod, NADH, Neutral red	HCOOH	−0.8 V SHE, 12.74%	Srikanth et al. [123]
TsFDH Thiobacillus	GCE, Pentamethyl cyclopentadienyl rhodium bipyridine [Cp*Rh (bpy)H₂O]+, NADH, photoanode-driven system	HCOOH	0.05% CO_2 to formate	Nam et al. [94]
FDH	GCE, Polydopamine nanoscale film, NADH, photoanode-driven system	HCOOH	0.06% conversion efficiency, $\sim$100% FE	Lee et al. [74]
CbsFDH	Cu foil electrode, NADH, [Cp*Rh(bpy) Cl]Cl mediator		−0.8 V SHE, 12% yield, 6.5×10^{-4} umol/min/mg	Kim [66]
W-containing FDH (FoDH1)	Gas-diffusion biocathode, 1,1'-trimethylene-2,2'-bipyridinium dibromide mediator	HCOOH	−0.8 V SHE, -17 ± 1 mA cm^{-2}	Sakai et al. [114]
Multi-enzyme heterodisulfide re-ductase supercomplex (Hdr-SC) of Methanococcus maripaludis	Graphite rod, DET		−0.6 V SHE CE 90%	Lienemann et al. [78]
FDH, Methanol dehydrogenase	MV^{2+}	CH_3OH	0.05 µmol h^{-1}	Kuwabata et al. [70]
FDH, FaldDH, alcohol dehydrogenase	GCE, neutral red, NADH, carbonic anhydrase, phosphate buffer	CH_3OH	−0.8 V SHE, 0.6 mmol g^{-1} h^{-1}	Addo et al. [1]

Fig. 9.2 CODH catalyzing the reduction of CO_2 to CO using two electrons (e−) from cathode

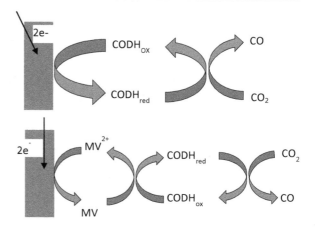

viologen (MV) as a mediator at pH = 6.3 and a faradaic efficiency close to 100% was reported with a turnover number of 700 h^{-1} at −0.57 V (vs. SHE) cathode potential [122]. The single step enzymatic CO_2 reduction by electrolysis at −0.51 V (vs. SHE) with [NiFe] CODH immobilized on pyrolytic graphite "edge" electrode displayed an increase in CO_2 reduction to CO compared to the bare enzyme [103]. Figure 9.2 illustrates the two mechanisms of CO_2 reduction to CO using the CODH enzyme.

Formate Dehydrogenase for Formate/Formic Acid Production

A typical redox enzyme from reductive acetyl-CoA pathways, FDH reduces CO_2 to formate with support of its cofactor NAD/NADH [9].

$$NADH + CO_2 \rightarrow HCOO^- + NAD^+$$

Mechanistically, NAD^+ is first reduced to NADH and then CO_2 is reduced by NADH. NADH serves as the electron and proton donor for reduction. Studies reported that CO_2 reduction by NADH-dependent FDHs undergoes a hydride (H$^-$) transfer mechanism [91]. A comparative study of FDH obtained from different sources showed that *Thiobacillus* sp. KNK65MA FDH (TsFDH) has the most superior CO_2 reducing capability with 5.8 times higher production of formate as compared to widely used *Candida boidinii* FDH [34].

A BES having FDH at the cathode with NADH as a cofactor reduced CO_2 to formate with a current efficiency of 12.74% at −0.8 V where NADH is electrochemically regenerated [123]. The use of TsFDH and the mediator, pentamethylcyclopentadienyl rhodium bipyridine [[Cp*Rh(bpy)H$_2$O]$^+$, Cp* = C$_5$Me$_5$, bpy = 2,2′-bipyridine] to regenerate NADH electrochemically in a photoanode-driven system, resulted in a steady formate production rate for more than 5 h with 0.05% CO_2 to formate conversion efficiency [94]. Electrochemically prepared Cu nanorods regenerated 62% of NADH when used for the immobilization of the enzyme and cofactor BES [65]. A biocathode with a polydopamine nanoscale film developed by electropolymerization uniquely embedded FDH and NADH for CO_2 electro-reduction and steadily produced formate for about 2 weeks with 0.042% conversion efficiency and $\sim 100\%$ faradaic efficiency in a photoanode-driven system [74]. So far, the CO_2 conversion efficiencies still remain low. The use of mediators instead of NADH cofactor is a way to make the CO_2 reducing system more feasible. With methyl viologen as a mediator, FDH coupled to a visible light-driven p-type Indium phosphide semiconductor at the cathode and externally polarized at -50 mV (vs. SHE) drove the photo-generated electrons towards methyl viologen reduction [104]. The system showed a current efficiency of 89% (formic acid specificity) over 4 h. Enzymatic electrocatalysis of CO_2 with immobilized FDH-viologen $CH_3V(CH_2)_9COOH$ (mediator) on indium titanium oxide cathode produced 29 mmol formic acid after 3 h [5]. A gas diffusion-type biocathode with tungsten-based FDH and a viologen 1,1'-trimethylene-2,2'-bipyridinium dibromide mediator for electron transfer produced high current density of 20 mA cm^2 at -0.8 V (vs. Ag/AgCl) for CO_2 reduction under a mild and quiescent condition [114].

Another idea gaining momentum is the direct injection of electrons into enzymes, independent of cofactors. This was evaluated for metal-dependent molybdenum-based or tungsten-based FDHs. Reda et al. [111] showed electrochemical reduction of CO_2 to formate assisted by FDH without NADH. High product specificity with faradaic efficiency $\sim 100\%$ was demonstrated when operated under mild condition with minute overpotential (110–210 mV). This is the first report of bioelectrochemical CO_2 reduction occurring at close to standard thermodynamic reduction potential (-0.4 V vs. Ag/AgCl) but the catalytic activity decreased with cycling due to enzyme denaturation. Likewise, Tungsten-containing FDH [111] and Molybdenum-containing FDH [26] were used as reversible electrocatalysts for interconversion of CO_2 and formate without NADH. Nevertheless, these FDHs were reported to be highly pH-sensitive and suffered slow turnover due to the difficulty of electrons and protons to reach the active site.

Immobilized Multiple Enzymes

A multi-enzymatic system was developed for CO_2 reduction to Formaldehyde (HCHO) by using FDH and Formaldehyde dehydrogenase FaldDH [121]. HCHO is the ultimate product that can be further selectively reduced to methanol (CH_3OH) by using alcohol dehydrogenase [1, 6].The multi-enzyme embedding approach from Addo et al. [1] is depicted in Fig. 9.3. In another case, Kuwabata et al. [70] presented an electrochemical approach for the conversion of CO_2 to methanol using two enzymes FDH and methanol dehydrogenase with methyl viologen as well as pyrroloquinolinequinone as supporting electron mediators. Electrodes with immobilized FDH onto it can be used efficiently for the electrocatalytic generation of higher alcohols such as butanol and the reduction of CO_2 to methanol [116, 118]. The co-immobilization of three dehydrogenases encapsulated in an alginate based matrix on a carbon felt was reported to obtain faradaic yields of 10% for methanol generation from electro-enzymatic CO_2 reduction [117].

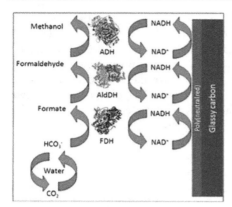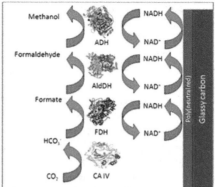

Fig. 9.3 Multi-enzyme embedding approach for the bioelectrosynthesis of methanol from CO_2 (Reproduced with permission from Addo et al. [1])

Immobilisation of enzymes on a suitable substrate can also enhance their stability and make them easier to handle in practical synthesis or chemical production. The multi-enzyme immobilized system for CO_2 reduction by the combination of carbonic anhydrase (CA), neutral red mediator and FDH increased the stability and CO_2 conversion efficiency at -0.55 V SHE in balance with the CO_2 solubility enhancement [124].

9.2.2 Interaction of Biocatalysts with Electrode for CO_2 Reduction

The biocatalyst-cathode interaction requires either the cells or enzyme attachment to the electrode or some substance mediate the connection for the electrons. Extracellular Electron Transfer (EET) may occur via two different mechanisms depending on the microorganism or enzyme involved: (i) mediator-free transfer (direct EET in microbial case) or (ii) mediator-dependent transfer (indirect EET) [82, 83]. Figure 9.4 displays the possible extracellular electron transfer mechanisms for electron uptake by microbes at cathode.

9.2.2.1 Mediator-Free Electron Transfer

The solid electrode is in physical connection with the microbial cells or redox active enzyme or with a cofactor in mediator-free electron transfer, in case of microbial biocatalyst, the term direct electron transfer (DET) is used. A number of CODH enzymes from autotrophic microorganisms have redox interaction via mediator-free electron transfer with the solid cathode to catalyse CO_2 reduction reactions [103]. However, these enzymes are difficult to employ on the solid electrode. They are often less stable and their chemistries are usually dependent on enzyme-bound cofactors, such as flavins or unbound free cofactors such as NAD^+, NADH,

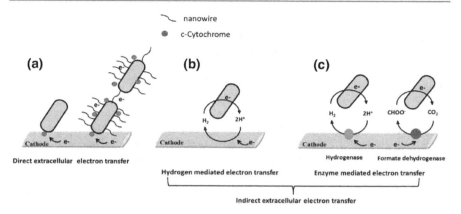

Fig. 9.4 Mechanism of extracellular electron transfer between the cathode and the microorganisms. cytochromes and/or nanowires involves for Direct electron transfer (**a**), and hydrogen mediation or enzyme mediation involves for Indirect electron transfer (**b–c**) Figure is adapted from Aryal et al. [11]

$NADP^+$, NADPH. In addition, the single step enzymatic electrosynthesis process is slow and the electron transfer between an enzyme and an electrode is generally poor. Because the redox center in the enzyme is usually far from the periphery, electrons have to jump a large distance during oxidation or reduction of the enzyme.

EET from cathode to microbe was hypothesised based on the electron transfer from microbe to anode. However; even though the inward and outward current may rely on the same redox chains, the electron transfer mechanisms seems to be different [112]. This is supported by the fact that some bacteria like *C. ljungdahlii* which do not have cytochromes were reported to be able to receive electrons from a poised electrode [96]. Undeniably, only a few has been known about DET in cathode compartment, however, understanding the molecular mechanism of electron transfer between a low redox surface and microbial cells is necessary for further development of bioelectrosynthesis.

Interestingly, in DET, extracellular conductive structures called nanowires [126] or redox moieties such as C-cytochrome [129] on the outer membrane of the microorganism are connected to the cathode for electron transfer. Self-assembled electrochemically active biofilms on the surface of cathodes are important to transfer the electrons for the CO_2 reduction. Electroautotrophs reported for being able to perform MES are mainly acetogenic bacteria [13, 95]. Indeed, these microorganisms are assumed to have high tendencies to form biofilm at the cathode surface. In fact, most of the operating condition so far applied either in pure or mixed culture driven MES, consisted in pre-establishing a biofilm. This step requires growing the bacteria with hydrogen for several days. This was suggested to promote the release of enzymes for instance hydrogenases and formate dehydrogenases due to the high cell lysis, which were reported to act as mediators in indirect EET [37, 113].

9.2.2.2 Mediated or Indirect Electron Transfer (MET or IET)

In case of enzymatic reduction, due to their low concentrations and large size, including that of the cofactors, the direct electron transfer from the electrode to the cofactor/enzyme is limited. In contrast to direct EET, no direct contact with the solid surface is required for indirect EET. The indirect EET requires diffusible extracellular compounds as intermediates that transport electrons between the electrode and the microorganism or enzyme. These diffusible shuttles are generally small redox molecules like formate, ammonia or Fe^{+2} or gases like oxygen O_2 or hydrogen H_2 [133]. Moreover endogenous soluble molecules secreted by the microorganism, like phenazine in *Pseudomonas*, were reported to act as diffusible shuttles for electron transfer from cathode to microbe. The mediator should be present at relatively high concentration and it should also have fast diffusion. Redox mediators are commonly single electron transferring reagents (e.g. ferrocene derivatives for oxidations or viologens for reductions) although two electrons mediators such as quinones or phenanthroline derivatives can also be used. The mediators used in enzymatic electrosynthesis help to increase the rate of regeneration of cofactor. The use of methyl viologen has been most commonly investigated for the regeneration of NADH and NADPH in enzymatic electrosynthesis from CO_2 using FDH.

Most of the MES studies reported H_2 mediated reduction of CO_2 as the applied cathode potentials were stimulating the electrochemical hydrogen evolution [18, 58]. In the microbial-biofilm matrix, enzymes or coenzymes secreted by the microbes were also suggested to act as electron shuttles in microbial electrosynthesis [37]. The mechanism of mediator-dependent EET is still unclear but is likely the dominant mechanism in microbial electrosynthesis.

9.2.3 Electrodes Material Development for CO_2 Reductions in Bioelectrosynthesis

The cathode electrode material should offer critical properties of high conductivity, excellent chemical stability, high mechanical strength, biocompatibility, high surface area, and low cost [141]. The electrode material is crucial for the performance by acting as an electron donor, providing biocompatibility to develop the electroactive biofilm and intrinsic conductivity for chemical synthesis from CO_2 supplied from different sources [11, 14, 16].

Carbon based electrode materials are the most widely used cathodes for BES due to their excellent properties of biocompatibility, electrical conductivity, low cost, and high chemical stability [11, 48, 75] Among the carbon electrodes, graphite has become the predominately applied commercial electrode material. Graphite electrodes are mainly used in the form of graphite block, graphite rods, and granular graphite in MES systems [11, 13]. Graphite has a plane sheet structure, and intrinsic advantages in aqueous solution, for example a wide electrochemical window, relative inertness, good electrical conductivity, low residual current, ease of modification, renewable surface, reproducibility, and very high biocompatibility [11].

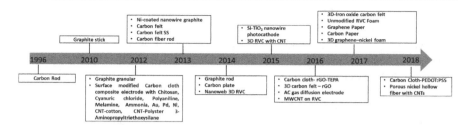

Fig. 9.5 Timeline representing the cathode material developments of for CO_2 reduction in MES systems. SS-Stainless Steel, Au-Gold, Pd-Palladium, Ni-Nickel, CNT-Carbon nanotube, Si-TiO$_2$, 3D-Three dimensional, RVC- Reticulated vitreous carbon, rGO-Reduced graphene oxide, TEPA-Tetraethylene pentamine, PEDOT:PSS poly(3,4-ethylenedioxythiophene):polystyrene, MWCNT-Multiwall carbon nanotube

The first proof of principle of CO_2 reduction was reported in MES with a graphite block electrode as electron donor [95]. Since then, commercially available plane electrode in particular graphite rod, graphite block, graphite stick, carbon cloth, carbon plate, activated carbon, gas diffusion activated carbon electrode, granular carbon, carbon fiber rod etc. are commonly applied. However, the application of a single graphite block has limitations to achieve higher productivity because of its low porosity and low surface area for microorganism adsorption. Therefore, commercially available two-dimensional carbon materials such as carbon plate, carbon cloth are also widely tested. Nevertheless, also 2D electrodes have limitations to the MES performance, such as low specific surface area and low catalytic activity for the generation of electron transfer mediators. Recently, commercially available three-dimensional (3D) electrode materials are therefore applied to advance MES productivity. Materials such as carbon felt, and carbon fiber rod electrode offer higher volume reactive surface area for the development of electrochemically active biofilm and minimize the mass transfer limitation compared to 2D electrodes. But there is still room for electrode material development and spatial surface modification for productivity enhancement (Fig. 9.5).

In a pioneering study, carbon cloth was modified with Chitosan, an amino and hydroxyl-group rich poly-saccharide. This allowed to increase the acetate production rate by 7.6 fold compared to unmodified carbon cloth electrode [141]. The authors reported the production was enhanced due to the better electron transfer owing to biocompatibility, better electrostatic interaction between the negatively charged bacterial and positively charged chitosan electrode and catalytic surface with larger pore sizes. Carbon cloth modified with conductive polymer poly (3, 4-ethylenedioxythiophene): polystyrene sulfonate (PEDOT:PSS) was also shown to improve acetate production [15]. Furthermore, a highly conductive 3D bioelectrode, NanoWeb-RVC, developed by coating multiwalled carbon nanotubes (MWCNT) on reticulated vitreous carbon (RVC), was demonstrated to have a high surface volume ratio optimal for bacterial attachment and mass transfer within the biofilm. Additionally, the same group developed electrode material by applying an electrophoretic deposition method to coat carbon nanotubes on RVC followed by chemical treatment. This resulted in highly conductive electrodes with 100% electron recovery comparing to 70% obtained with the NanoWeb-RVC [56, 57]. Graphene one of the best conductive materials, is also developed and tested for productivity enhancement of MES. Moreover, functionalized tetraethylene pentamine reducing graphene on a carbon cloth electrode was further enhanced by 11.8 fold in the acetate production rate compared to the carbon cloth electrode in microbial electrosynthesis [32]. Aryal et al. developed a 3D graphene carbon felt electrode and freestanding and flexible graphene papers cathodes and proved the excellency of graphene based electrode in MES due to better conductivity and enhanced surface area [10, 12]. These, research reported the superiority of cathode material for productivity enhancement. Also metal carbon complex electrodes were tested. Using a silicon-titanium oxide nanoparticles (Si-TiO_2) based photocathode, which harvested light as the sole electron source, up to 6 gL^{-1} acetate was produced with $S. ovata$ explained the important of cathode material [79]. Moreover Ni-coated nanowire graphite, iron oxide modified 3D carbon felt cathodes, 3D graphene Nickel foam form and porous Ni hollow fiber with CNTs have also shown further improvement of MES due to the characteristic of cathode materials [11].

The porous composite activated carbon gas diffusion electrode in particular VITO-CoRE™ electrode was tested in MES which provided an ideal three-phase interface (gas–liquid–solid) for the diffusion of CO_2 in liquid phase thereby improves the acetate production rate. VITO-CoRE™ electrode and gas diffusion electrodes (GDEs) consists of hydrophobic gas diffusion layer to enhanced the CO_2 diffusion, a current collector and catalytic layer to provide the active site to develop the microbial biofilm [20, 90, 120]. The mass transfer coefficient (k_La) for GDE was optimized and doubled than that of the traditional gas sparged system [20]. VITO-CoRE™ electrode with plastic inert support as the cathode further led to enhanced biofilm formation and better productivity MES due to high surface area [90].

9.3 Progress in Microbial Electrosynthesis

In recent years, MES has emerged as a promising bioprocess for multi-carbon chemical synthesis from CO_2 applying renewable electricity. Acetic acid has been the main targeted end-product from CO_2 since the technology was developed [95]. Recently, modified RVC with multi-walled CNTs (MWCNTs) resulted in acetate production rate of up to 1330 $gm^{-2} d^{-1}$, the highest to date based on the projected cathode surface area [58, 75] as shown in Table 9.3. Also, some product diversification was achieved through the improvement of MES system. So far, recent literature has reported the production of higher carbon chain organics like, isopropanol, butyric acid, and caproic acid from CO_2 [18, 43, 135]. Intensive investigations were performed on electron transfer, cathode modification, CO_2 supply, media modification and adaptive evolution of microbes for productivity enhancement [11].

Table 9.3 Overview of electrode materials and acetate production rates in microbial electrosynthesis from CO_2

Developed cathode materials	Microbial culture source	Applied potential (V vs. SHE)	Current density[a] (Am^{-2})	Acetate production		CE in acetate (%)	References
				Production Rate[a] $(gm^{-2}d^{-1})$	Achieved titer (gL^{-1})		
Carbon cloth + Chitosan[b]	S. ovata	−0.4	−0.47	13.51 ± 3.30	0.59[c]	86 ± 12	Zhang et al. [141]
Graphite Stick -Ni Nano wire	S. ovata	−0.4	−0.63	3.38	0.094	82 ± 14	Nie et al. [98]
Nanoweb 3D RVC	WWTP sludge	−0.85	−37	195 ± 30	1.2	70 ± 11	Jourdin et al. [56]
3D RVC with CNT	Enriched mix culture from WWTP sludge	−0.85	−102	685 ± 30	11	100 ± 4	Jourdin et al. [57]
MWCNT-RVC	same as above	−1.1	−200	1330	11	84 ± 2	Jourdin et al. [59]
Activated carbon VITO-CoRE™	Mix culture	−0.4	−0.165	9.49	4.1	29.91	Mohanakrishna [90]
Si-TiO2 nanowire photocathode	S. ovata	−0.595	−3.5	Ng	6	86 ± 9	Liu et al. [79]
Carbon Cloth- reduced graphene oxide tetraethylene pentamine (rGO-TEPA-CC)	Methanol adapted S. ovate	−0.69	−0.23	62.4 ± 26.64	1.88[d]	83 ± 3	Chen et al. [32]
3D-Graphene carbon felt composite	S. ovata	−0.69	−2.4	54.57 ± 1.7	1.4	86 ± 3	Aryal et al. [10]
Gas diffusion activated NORIT® carbon	Enriched Anaerobic sludge	−1	−20	36.6	2.89	35.46 ± 88	Bajracharya et al. [20]
3D Iron oxide modified carbon felt	S. ovata	−0.69	Ng	25.40[d]	1.8[d]	86 ± 9	Cui [36]
Graphene Paper	S. ovata	−0.69	−2.5	39.8	0.77	90.7	Aryal et al. [12]
3D Graphene Ni-Form	Mix culture from MFC	−0.85	−10.2	ng	5.46	70	Vassilev [135]
Carbon cloth PEDOT:PSS	S. ovata	−0.69	−3.2	59.5	1.77[c]	87.2	Aryal et al. [15]
Porous Ni-Hallow fiber with CNTs	S. ovata	−0.4	0.33	1.85	0.17[c]	83	(Bian, M. F. Alqahtani, et al., 2018)

[a]Calculation was done based on project surface area, [b]carbon cloth is base material for modification. [c]Approximate calculation from given production graph. TM: trade mark, CE-Coulombic efficiency, mfc-Microbial Fuel Cell, Ni-Nickel, CNT-carbon nanotubes, Si-TiO2-Silicon- titanium oxide, 3D-Three dimensional, RVC-reticulated vitreous carbon, ng-Not given, PEDOT:PSS poly(3,4-ethylenedioxythiophene):polystyrene sulfonate (PEDOT:PSS) polymer

9.4   Integrated Concepts for Bioelectrosynthesis Improvement

This section discusses a series of novel integrated concepts that have resulted in considerable improvements in bioelectrosynthesis from CO_2.

9.4.1   Gas Diffusion Biocathodes

High CO_2 dissolution/mass transfer rates are necessary for its easy availability in the biological reduction reactions. The use of gas diffusion electrodes (GDEs) is one of the efficient methods that can be employed for the gas to liquid mass transfer of CO_2. GDEs have been used in electrochemical cells and also in microbial fuel cells where the gas-phase reactant needs to react with liquid-phase reactants effectively at the same time interacting with the solid electrode [4, 100]. In VITO's proprietary GDEs (VITO CoRE), the combination of hydrophobic and hydrophilic micropores in the GDE creates a three-phase interface (gas–liquid–solid), which ensures an abundant availability of the gaseous reactants on the electrode surface [4]. This combination consists of a catalyst layer (porous activated carbon and Teflon binder) and a hydrophobic gas diffusion layer (GDL) (Fig. 9.6). GDEs have meanwhile also been used in bioelectrosynthesis research to reduce CO_2 to value added compounds [20, 69, 114]. It ensures sufficient availability of gaseous CO_2 directly to the electrochemically active biocatalysts residing on the electrode. Additionally, the size of the micropores in GDE biocathode can be fine-tuned to regulate the CO_2

Fig. 9.6 Schematic representation of VITO Core® gas diffusion electrode (Reproduced with permission from Alvarez-Gallego et al. [4])

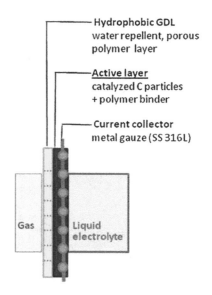

diffusion flux depending on the reaction rate. Attempts for the advancement and modification of biocathodes have shown some encouraging results for improvement of efficiency.

Apart from gas diffusion electrodes, a concept of conductive and porous nickel hollow fiber cathodes has been introduced recently for the bioelectrosynthesis from gaseous CO_2 and reported the improvement in the gas delivery and the rate of CO_2 reduction [3, 30]. The conductive and porous cathode architecture also supports the substrate-biocatalyst-electrode interfacial interaction for CO_2 reduction in similar manner as GDE. Katuri et al. [62] advocated the use of various materials modifications for such dual-functional electrocatalytic porous cathodes for CO_2 reduction process in bioelectrosynthesis.

9.4.2 Product Concentration and Separation

Bioelectrosynthesis from CO_2 has been reported to produce mainly CO and formic acid in enzymatic electrosynthesis or methane and acetate/acetic acid in MES. Despite the fact that a lot of effort is spent to optimize the processes, the product titers remain low (few g L^{-1} range), which brings down the economic viability of the process. Therefore, some attempts have been done to integrate separation techniques in the process which can also concentrate the product. For some products which are already inhibitory at very low titers, this also provides a means to reduce product inhibition.

The high solubility of short chained fatty acids complicates product separation from the liquid broth. One approach is to integrate membrane electrolysis. A pair of anion and cation exchange membrane can be placed between the anode and the cathode, providing an additional middle recovery/extraction chamber [46]. This does not only allow to transport acetate to the extraction chamber and separate it from the other products in the cathode compartment but it is also concentrated and simultaneously converted it into the acid form. Up to 13 gL^{-1} acetic acid in the extraction chamber was reported. Another approach to separate the acetate from the low concentration broth of MES was to use ion exchange resins for the sorption of acetate [22].

Further conversion of the short chain fatty acids produced from CO_2 reduction to medium chain fatty acids (MCFAs) also could be an approach for product extraction and separation As opposed to short chain fatty acids, the medium chain fatty acids (MCFAs) such as caproate (C6), caprylate (C8), caprate (C10), and laurate (C12) are more hydrophobic, have a lower solubility and easily phase separate from the aqueous medium. MCFAs have been recovered through (membrane-based) liquid-liquid extraction [27].

9.4.3   Product Diversification via Chain Elongation

Although acetate has been the main product of CO_2 reduction in MESs, its auto-trophic generation is not economically attractive due to its low commercial value. It was estimated that the production cost per ton of acetic acid from MES required be to lowered by more than 80% to compete with the industrial acetic acid production process [35]. Therefore, there is an interest to shift towards higher value chemicals. Some products with higher market values such as oxo-butyrate or alcohols, appear as minor by-products to acetic acid [43, 72, 96, 23]. However, several value-added products such as medium chain carboxylates, alcohols, bioplastics were reported as a result of further microbial conversion of the products obtained in MES from CO_2 [8, 23, 60, 107].

Bioelectrochemical chain elongation can for instance be applied to convert acetate with alcohols to medium chain fatty acids (up to C8) [2]. The production of fatty acids needs ethanol as an electron donor, which could be produced microbially using either the cathode or hydrogen as an electron donor [127]. The production of caproate with BES was investigated for the first time in 2013 using acetate without adding any electron mediator at a cathode potential of -0.9 V versus NHE where acetate was biologically up to 6.8 mM caproate and 3 mM butyrate [39]. This work studied the functionality of the cathode of a BES as an electron donor for the transformation of acetate into caproate. Continuous microbial production of caproate was also investigated using carbon felt electrodes [60]. The development of thick biofilm was observed on the electrode resulting in a current density of 14 kA m^{-3} with a caproate production rate that reached 0.95 g L^{-1} day. In other studies, butyrate was the main bioelectrochemical product (1.82 mM d^{-1}) from using CO_2 as a sole carbon source [43].

Also in continuous mode of operation using a mixed culture under an applied current of 3.1 Am^{-2}, butyrate was the main product, but also small amounts of propionate and caproate were produced as byproduct [109]. This study revealed that the solid-state electrode controlled the chain elongation reaction as an essential electron donor and determined the performance of the system. Ethanol is required as electron donor in chain elongation [119]. However, the electrons for the bio-electrochemical chain elongation process are available from the electric current and acetate. Possibly, the ethanol is produced as an intermediate via the reduction of acetate using electrons from cathode or using hydrogen. The main measured products of bioelectrochemical chain elongation were hydrogen, butyrate and caproate accounting an electron capture efficiency of 45% after 18 days of operation, in which the contribution of ethanol to the carbon conversion and cathodic electron was less than 1% [39].

Also a series of other multi-carbon compounds have been produced in various MES studies. Production rates were improved to 0.21 g L^{-1} d^{-1} for butyric acid [27], 0.18 g L^{-1} d^{-1} for ethanol [7, 77, 125], 0.157 g L^{-1} d^{-1} for isopropanol [8], 0.013 g L^{-1} d^{-1} for butanol/isobutanol [135] and 0.95 g L^{-1} d^{-1} for caproate [60, 135]. A two-stage direct solar-powered process was described for the production of acetate from CO_2 coupled with the biosynthesis of complex organic molecules from acetate, such as n-butanol, PHB biopolymer, and isoprenoid compounds using genetically engineered *Escherichia coli* in a separate vessel [79].

9.4.4 Bioanode-Biocathode Integration

The counter electrode chemistry is also an important aspect to consider in the development of bioelectrosynthesis from CO_2. Oxygen evolution is the common counter electrode reaction in aqueous solutions for bioelectrosynthesis from CO_2 and the consequent formation of protons should balance the pH throughout the cell. For water oxidation as a counter reaction, high energy input is required and the acid/base balance is difficult to maintain at both electrodes. The integration of a bioanode, degrading organic waste, to BES system presents an opportunity to supply the power demand in bioelectrosynthesis. A bioanode lowers the input of electricity to drive the cathode reaction in BES. With a biotic anode, the potential only fluctuates from 0 to 0.5 V versus SHE. In contrast, an abiotic anode potentially makes the power requirement significantly higher as the abiotic anode potential can largely fluctuate from 0.5 to 5 V versus SHE. As an alternative to a water-electrolysis anode, a bioanode MES has been proposed using a pure culture of *Desulfobulbus propionicus* to oxidize hydrogen sulfide as a source of electron [47].

Electrochemically active microorganisms maintain the anode potential by oxidizing the organic substrates in wastewater to generate energetic electrons which are transferred to the anode and subsequently flow to the biocathode, where they can be exploited for the electrosynthesis of value-added products. A bioanode-biocathode system was applied with small electrical input to treat artificial wastewater at the anode and reduce CO_2 at the cathode under different applied voltages up to 1.4 V [139]. Likewise, [77] employed the bioanode-biocathode system for bioelectrosynthesis from CO_2 in combination with reverse electrodialysis using the energy generated from salinity gradient [77]. Wastewaters from diverse sources can be treated and used for electricity production in BES. Mixed culture biofilm formation on the anode can be conveniently established and is also tolerant to environmental fluctuations.

9.4.5 Metabolic Engineering of Biocatalysts

Low CO_2 conversion rates and titers of cathodic products, limited end-product diversity and low product value are the main limitations for bioelectrosynthesis from CO_2. To overcome such hindrances, metabolic alteration of appropriate host microbes can be considered. Protein engineering techniques also allow the enzyme structure to be reformed to increase its activity as a catalyst for a specific reaction, or increase its tolerance to reaction conditions. The engineering of enzymes to meet the specific requirements of bioelectrosynthesis should be highlighted. Metabolic engineering of cathodic microbes offers an approach to introduce new pathways and optimize carbon assimilation and extracellular electron transports. It is important to identify the metabolic engineering targets for energy conservation and for electron transfer from electrodes to biocatalysts so as to reach higher rates for the bioelectrosynthesis from CO_2. A strain of *C. ljungdahlii* was modified by introducing the required genes on a plasmid to produce butanol from syngas [67]. In recent studies, metabolic engineering of a strain of *Clostridium ljungdahlii* was carried out to facilitate the production of butyrate from H_2 and CO_2 [134] and in another study, a strain of *Acetobacterium woodii* was genetically adapted to produce acetone [50]. The engineered strains of *Clostridium autoethanogenum* have already been employed in industrial scale applications for the production of ethanol from a mixture of H_2, CO_2 and CO [38]. The engineered microbes can be used in biocathodic application to produce specific high value chemicals from CO_2 reduction. On the other hand, defined co-cultures can be also used as biocatalyst to transform the low value short chained organic products from CO_2 reduction by acetogens into more valuable and energetic products using heterotrophs [52, 79].

9.4.6 Integration with Anaerobic Digestion

Bioelectrosyntheses have the potential to be integrated with anaerobic digestion (AD) technology. Biogas is produced during anaerobic digestion and can be used on-site for power generation but could also be more economically used as a transport fuel for buses, cars, etc. For the latter application, the CO_2 fraction needs to be removed to improve gas compression and storage and combustion efficiency. Therefore, an MES process to convert the CO_2 fraction to acetate could be suitable for such a biogas "cleaning" operation. This option could then be both environmentally and economically attractive for both applications (organics production and biogas utilization as fuel). Bioelectrosynthesis from CO_2 has also been proposed as a biogas upgrading technology with the conversion of the CO_2 fraction to methane [28]. With the latest development of bioelectrochemical chain elongation in bioelectrosynthesis from CO_2, high value MCFAs can also be produced from the CO_2 fraction of biogas in AD.

9.4.7 Integration with Syngas Fermentation

The use of solid electrodes is a simple way to electrochemically control the redox potential and regulate the products of fermentation process [55, 93]. Thus, by introducing bioelectrosynthesis process in syngas (a mixture of CO, CO_2 and H_2) fermentation, the cathode provides the electron source which can regulate the CO_2 reduction and higher product titer and conversion efficiency can be achieved. In another aspect, the conversion of exhaust CO_2 from the biomass gasification process could also be included in the integration of bioelectrosynthesis with syngas fermentation. Different solid waste materials in particular agriculture residue, forest residues can be used as feedstock for syngas production in biomass gasification. In such a way, the synthesis of chemicals and fuels from waste materials is possible by integrating bioelectrosynthesis with biomass gasification where syngas serve as carbon source and cathode as an electron source. In the similar fashion, the particulate free CO_2 exhaust generated from solid oxide fuel cell can also be directed to the cathode of an MES to generate high-value chemicals, such as methane or other mid-ranged organic compounds.

9.5 Techno-Economic Considerations for Upscaling of Bioelectrosynthesis from CO_2

The utilization of autotrophic microbes and/or enzymes from such microbes in the bioelectrosythesis process, is an attractive innovation that provides a sustainable and renewable energy powered systems for CO_2 reduction. This biocompatible and sustainable CO_2 conversion is also an indirect way of electricity storage in chemical form which is particularly attractive for industry because of the use of mild reaction conditions such as ambient temperature and pressure. However, bioelectrosynthesis has not yet reached industrial application.

High capital and operating/maintenance (O&M) costs are the key limitations in the industrialization of BES. Christodoulou and Velasquez-Orta [35] reported that capital and operating costs of acetate production in MES integrated with anaerobic fermentation is very high compared to production by pure chemical methods (methanol carbonylation and ethane oxidation). The low production yields and titers from the biological processes, electricity input and product recovery costs are the key factors that limit the commercial feasibility of BESs [35, 73, 89]. For the commercialization, production rates of >50 g m^{-2} h^{-1} acetic acid and >75 g m^{-2} h^{-1} formic acid with >50% energy efficiency are considered as economically feasible [40]. Pant et al. [101] considered that the Nafion 117 proton exchange membrane is the most expensive element in the BES reactor, when the cost was approximately \$1100 m^{-2}. Approaches to explore cheaper materials have been employed to diminish the capital costs. Several studies even tried to avoid using membranes by constructing membrane-less setups to lower the capital costs and alleviate pH gradients between anode and cathode [41, 45]. Electricity is another factor that comprises a prime cost element, which accounts for more than half of the total O&M costs. The use of renewable energy sources and surplus electricity could be an encouraging approach to reduce the O&M costs in BESs, but is not continuously available. Electrode durability was also reported to be key for commercial feasibility of MES [40]. Moreover, actual energy efficiencies are much lower than theoretical ones, which is another main challenges for the commercialization of this technology.

9.6 Future Outlook and Challenges

Bioelectrosynthesis from CO_2 offers the prospect to reuse CO_2 emissions as a feedstock and generate fuels and value-added chemicals from CO_2 and its derivatives. The technology has environmental advantages due to its sustainability, renewability and environmentally friendly qualities. The future potential of these systems can be associated to the framework of CO_2 biorefineries, the power-to-gas concept, or biogas upgrading, thus helping to step-up in the desired global transition from fossil fuel-based to electricity-based economy. Research has proven the feasibility of generating different value-added products from CO_2. However, BES still encounters numerous operational and technical issues. Some limiting factors include high capital and operating costs and low production yield, titers and rates. The knowledge gaps on the electroactivity of biocatalysts, the electrode materials, the reactor design and product recovery demand further research on the multidisciplinary aspects of CO_2 valorization. In addition, overcoming the drawbacks and optimizing and scaling of the process with possible integration with established processes are essential to understand under which conditions the technology is suited for commercialization. Further avenues to progress on bioelectrosynthesis from CO_2 are to use hybrid production systems comprising of autotrophs and heterotrophs or hybrids of autotrophic microorganisms with electrocatalysts or light-harvesting semiconductor materials. Optimization of bioelectrosynthesis process with possible integration with established processes are the exciting approaches for future development.

Acknowledgements Nabin Aryal is supported by a FutureGas project from Innovation Fund Denmark-Innovationfonden.

References

1. Addo PK et al (2011) Methanol production via bioelectrocatalytic reduction of carbon dioxide: role of carbonic anhydrase in improving electrode performance. Electrochem Solid-State Lett 14(4):E9–E13. https://doi.org/10.1149/1.3537463
2. Agler MT et al (2011) Waste to bioproduct conversion with undefined mixed cultures: the carboxylate platform. Trends Biotechnol 29(2):70–8 (Elsevier Ltd). https://doi.org/10.1016/j.tibtech.2010.11.006
3. Alqahtani MF et al (2018) Porous hollow fiber nickel electrodes for effective supply and reduction of carbon dioxide to methane through microbial electrosynthesis. Adv Funct Mater, 1804860 (Wiley). https://doi.org/10.1002/adfm.201804860
4. Alvarez-Gallego Y et al (2012) Development of gas diffusion electrodes for cogeneration of chemicals and electricity. Electrochim Acta 82:415–426. https://doi.org/10.1016/j.electacta.2012.06.096 (Pergamon-Elsevier Science Ltd)
5. Amao Y, Shuto N (2014) Formate dehydrogenase–viologen-immobilized electrode for CO_2 conversion, for development of an artificial photosynthesis system. Res Chem Intermed 40 (9):3267–3276. https://doi.org/10.1007/s11164-014-1832-1

6. Amao Y, Watanabe T (2007) Photochemical and enzymatic synthesis of methanol from formaldehyde with alcohol dehydrogenase from Saccharomyces cerevisiae and water-soluble zinc porphyrin. J Mol Catal B Enzym 44(1):27–31. https://doi.org/10.1016/J.MOLCATB.2006.08.001 (Elsevier)
7. Ammam F et al (2016) Effect of tungstate on acetate and ethanol production by the electrosynthetic bacterium *Sporomusa ovata*. Biotechnol Biofuels BioMed Central 9(1):1–10. https://doi.org/10.1186/s13068-016-0576-0
8. Arends JBA et al (2017) Continuous long-term electricity-driven bioproduction of carboxylates and isopropanol from CO_2 with a mixed microbial community. J CO2 Utilization 20:141–149. https://doi.org/10.1016/j.jcou.2017.04.014 (Elsevier)
9. Aresta M, Dibenedetto A, Quaranta E (2016) Enzymatic conversion of CO_2 (carboxylation reactions and reduction to energy-rich C1 molecules). In: Reaction mechanisms in carbon dioxide conversion. Springer, Heidelberg, pp 347–371. https://doi.org/10.1007/978-3-662-46831-9_9
10. Aryal N et al (2016) Enhanced microbial electrosynthesis with three-dimensional graphene functionalized cathodes fabricated via solvothermal synthesis. Electrochim Acta 217:117–122. https://doi.org/10.1016/j.electacta.2016.09.063 (Elsevier Ltd)
11. Aryal N, Ammam F et al (2017) An overview of cathode materials for microbial electrosynthesis of chemicals from carbon dioxide. Green Chem 19:5748–5760. https://doi.org/10.1039/C7GC01801K
12. Aryal N, Halder A et al (2017) Freestanding and flexible graphene papers as bioelectrochemical cathode for selective and efficient CO_2 conversion. Sci Rep 7(1):1–8. https://doi.org/10.1038/s41598-017-09841-7 (Springer, US)
13. Aryal N, Tremblay P-L et al (2017) Performance of different Sporomusa species for the microbial electrosynthesis of acetate from carbon dioxide. Bioresour Technol 233:184–190. https://doi.org/10.1016/j.biortech.2017.02.128 (Elsevier Ltd)
14. Aryal N, Kvist T et al (2018) An overview of microbial biogas enrichment. Bioresour Technol 264:359–369. https://doi.org/10.1016/J.BIORTECH.2018.06.013 (Elsevier)
15. Aryal N, Tremblay P-L, et al (2018) Highly conductive poly (3,4-ethylenedioxythiophene) polystyrene sulfonate polymer coated cathode for the microbial electrosynthesis of acetate from carbon dioxide. Front Energy Res, 18–20. https://doi.org/10.3389/fenrg.2018.00072
16. Aryal N, Kvist T (2018) Alternative of biogas injection into the Danish gas grid system—a study from demand perspective. ChemEngineering 2(3):43. https://doi.org/10.3390/chemengineering2030043
17. Baca M et al (2016) Microbial electrochemical systems with future perspectives using advanced nanomaterials and microfluidics. Adv Energy Mater 6(23):1600690. https://doi.org/10.1002/aenm.201600690 (Wiley-Blackwell)
18. Bajracharya S et al (2015) CO_2 reduction by mixed and pure cultures in microbial electrosynthesis using an assembly of graphite felt and stainless steel as a cathode. Bioresour Technol 15:14–24. https://doi.org/10.1016/j.biortech.2015.05.081 (Elsevier Ltd)
19. Bajracharya S, Sharma M et al (2016) An overview on emerging bioelectrochemical systems (BESs): technology for sustainable electricity, waste remediation, resource recovery, chemical production and beyond. Renew Energy 98:153–170. https://doi.org/10.1016/j.renene.2016.03.002
20. Bajracharya S, Vanbroekhoven K et al (2016) Application of gas diffusion biocathode in microbial electrosynthesis from carbon dioxide. Environ Sci Pollut Res 23(22):22292–22308. https://doi.org/10.1007/s11356-016-7196-x
21. Bajracharya S, Srikanth S et al (2017) Biotransformation of carbon dioxide in bioelectrochemical systems: state of the art and future prospects. J Power Sources 356. https://doi.org/10.1016/j.jpowsour.2017.04.024
22. Bajracharya S, van den Burg B et al (2017) In situ acetate separation in microbial electrosynthesis from CO_2 using ion-exchange resin. Electrochim Acta 237. https://doi.org/10.1016/j.electacta.2017.03.209

23. Bajracharya S, Yuliasni R et al (2017) Long-term operation of microbial electrosynthesis cell reducing CO_2 to multi-carbon chemicals with a mixed culture avoiding methanogenesis. Bioelectrochemistry 113:26–34. https://doi.org/10.1016/j.bioelechem.2016.09.001

24. Bajracharya S, Vanbroekhoven K et al (2017) Bioelectrochemical conversion of CO_2 to chemicals: CO_2 as a next generation feedstock for electricity-driven bioproduction in batch and continuous modes. Faraday Discuss 202:433–449. https://doi.org/10.1039/C7FD000 50B

25. Bar-Even A, Noor E, Milo R (2012) A survey of carbon fixation pathways through a quantitative lens. J Exp Bot 63(6):2325–2342. https://doi.org/10.1093/jxb/err417

26. Bassegoda A et al (2014) Reversible interconversion of CO_2 and formate by a molybdenum-containing formate dehydrogenase. J Am Chem Soc 136(44):15473–15476. https://doi.org/10.1021/ja508647u

27. Batlle-vilanova P et al (2017) Bioelectrochemistry microbial electrosynthesis of butyrate from carbon dioxide: production and extraction. 117:57–64. https://doi.org/10.1016/j.bioelechem.2017.06.004

28. Batlle-Vilanova P et al (2015) Deciphering the electron transfer mechanisms for biogas upgrading to biomethane within a mixed culture biocathode. RSC Adv. https://doi.org/10.1039/c5ra09039c

29. Bi C et al (2013) Development of a broad-host synthetic biology toolbox for *Ralstonia eutropha* and its application to engineering hydrocarbon biofuel production. Microbial Cell Fact 12(1):107. https://doi.org/10.1186/1475-2859-12-107

30. Bian B, Alqahtani MF et al (2018) Porous nickel hollow fiber cathodes coated with CNTs for efficient microbial electrosynthesis of acetate from CO_2 using *Sporomusa ovata*. J Mater Chem A 6:17201–17211. https://doi.org/10.1039/c8ta05322g (Royal Society of Chemistry)

31. Carbajosa S et al (2010) Electrochemical growth of *Acidithiobacillus ferrooxidans* on a graphite electrode for obtaining a biocathode for direct electrocatalytic reduction of oxygen. Biosens Bioelectron 26(2):877–880. https://doi.org/10.1016/J.BIOS.2010.07.037 (Elsevier)

32. Chen L et al (2016) Electrosynthesis of acetate from CO_2 by a highly structured biofilm assembled with reduced graphene oxide–tetraethylene pentamine. J Mater Chem A 4:8395–8401. https://doi.org/10.1039/C6TA02036D (Royal Society of Chemistry)

33. Cheng S et al (2009) Direct biological conversion of electrical current into methane by electromethanogenesis. Environ Sci Technol 43(10):3953–3958. Available at: http://www.ncbi.nlm.nih.gov/pubmed/19544913

34. Choe H et al (2014) Efficient CO_2-reducing activity of NAD-dependent formate dehydrogenase from thiobacillus sp. KNK65MA for formate production from CO_2 gas. PLoS ONE 9(7):1–10. https://doi.org/10.1371/journal.pone.0103111 (Public Library of Science)

35. Christodoulou X, Velasquez-Orta SB (2016) Microbial electrosynthesis and anaerobic fermentation: an economic evaluation for acetic acid production from CO_2 and CO. Environ Sci Technol 50(20). https://doi.org/10.1021/acs.est.6b02101

36. Cui M et al (2017) Three-dimensional hierarchical metal oxide-carbon electrode material for high efficient microbial electrosynthesis. Sustain Energy Fuels, 1–3. https://doi.org/10.1039/c7se00073a

37. Deutzmann J, Sahin M, Spormann A (2015) Extracellular enzymes facilitate electron uptake in biocorrosion and bioelectrosynthesis. mBio 6(2):1–8. https://doi.org/10.1128/mbio.00496-15.editor

38. Dürre P, Eikmanns BJ (2015) C1-carbon sources for chemical and fuel production by microbial gas fermentation. Curr Opin Biotechnol 35:63–72. https://doi.org/10.1016/J.COPBIO.2015.03.008 (Elsevier Current Trends)

39. Van Eerten-Jansen MCAA et al (2013) Bioelectrochemical production of caproate and caprylate from acetate by mixed cultures. ACS Sustain Chem Eng 1(5):513–518. https://doi.org/10.1021/sc300168z

40. ElMekawy A et al (2016) Technological advances in CO_2 conversion electro-biorefinery: a step towards commercialization. Bioresour Technol 215:357–370. https://doi.org/10.1016/j. biortech.2016.03.023

41. Escapa A et al (2015) Scaling-up of membraneless microbial electrolysis cells (MECs) for domestic wastewater treatment: Bottlenecks and limitations. Bioresour Technol 180:72–78. https://doi.org/10.1016/J.BIORTECH.2014.12.096 (Elsevier)

42. Fast AG, Papoutsakis ET (2012) 'Stoichiometric and energetic analyses of non-photosynthetic CO_2-fixation pathways to support synthetic biology strategies for production of fuels and chemicals. Curr Opin Chem Eng 1(4):380–395. https://doi.org/10. 1016/j.coche.2012.07.005 (Elsevier Ltd)

43. Ganigué R et al (2015) Microbial electrosynthesis of butyrate from carbon dioxide. Chem Commun 51:3235–3238. https://doi.org/10.1039/C4CC10121A

44. Geppert F et al (2016) Bioelectrochemical power-to-gas: state of the art and future perspectives. Trends Biotechnol 34(11):879–894. https://doi.org/10.1016/J.TIBTECH.2016. 08.010 (Elsevier Current Trends)

45. Giddings CGS et al (2015) Simplifying microbial electrosynthesis reactor design. Front Microbiol 6(MAY):1–6. https://doi.org/10.3389/fmicb.2015.00468

46. Gildemyn S et al (2015) Integrated production, extraction, and concentration of acetic acid from CO_2 through microbial electrosynthesis. Environ Sci Technol Lett 2(11):325–328. https://doi.org/10.1021/acs.estlett.5b00212 (American Chemical Society)

47. Gong Y et al (2013) Sulfide-driven microbial electrosynthesis Environ Sci Technol 47 (1):568–573. https://doi.org/10.1021/es303837j

48. Guo K et al (2015) Engineering electrodes for microbial electrocatalysis. Curr Opin Biotechnol 33:149–156. https://doi.org/10.1016/j.copbio.2015.02.014

49. Hawkins AS et al (2011) Extremely thermophilic routes to microbial electrofuels. ACS Catal 1(9):1043–1050. https://doi.org/10.1021/cs2003017 (American Chemical Society)

50. Hoffmeister S et al (2016) Acetone production with metabolically engineered strains of Acetobacterium woodii. Metab Eng 36:37–47. https://doi.org/10.1016/J.YMBEN.2016.03. 001 (Academic Press)

51. Hori Y (2008) Electrochemical CO_2 reduction on metal electrodes. In: Vayenas CG, White RE, Gamboa-Aldeco ME (eds) Modern aspects of electrochemistry. Springer, New York, pp 89–189

52. Hu P et al (2016) Integrated bioprocess for conversion of gaseous substrates to liquids. In: Proc Natl Acad Sci 113(14):3773 LP-3778. Available at: http://www.pnas.org/content/113/ 14/3773.abstract

53. Jeoung J-H, Dobbek H (2007) Carbon dioxide activation at the Ni,Fe-cluster of anaerobic carbon monoxide dehydrogenase. Science 318(5855):1461 LP-1464. Available at: http:// science.sciencemag.org/content/318/5855/1461.abstract

54. Jhong H-R, Ma S, Kenis PJ (2013) Electrochemical conversion of CO_2 to useful chemicals: current status, remaining challenges, and future opportunities. Curr Opin Chem Eng 2 (2):191–199. https://doi.org/10.1016/j.coche.2013.03.005 (Elsevier Ltd)

55. Jiang Y et al (2018) Electrochemical control of redox potential arrests methanogenesis and regulates products in mixed culture electro-fermentation. ACS Sustain Chem Eng 6 (7):8650–8658. https://doi.org/10.1021/acssuschemeng.8b00948 (American Chemical Society)

56. Jourdin L et al (2014) A novel carbon nanotube modified scaffold as an efficient biocathode material for improved microbial electrosynthesis. J Mater Chem A 2(32):13093–13102. https://doi.org/10.1039/C4TA03101F

57. Jourdin L et al (2015) High acetic acid production rate obtained by microbial electrosynthesis from carbon dioxide. Environ Sci Technol 49(22):13566–13574. https://doi.org/10. 1021/acs.est.5b03821

58. Jourdin L, Lu Y et al (2016) Biologically-induced hydrogen production drives high rate/high efficiency microbial electrosynthesis of acetate from carbon dioxide. ChemElectroChem 3 (4):581–591. https://doi.org/10.1002/celc.201500530

59. Jourdin L, Freguia S et al (2016) Bringing high-rate, CO_2-based microbial electrosynthesis closer to practical implementation through improved electrode design and operating conditions. Environ Sci Technol 50:1982–1989. https://doi.org/10.1021/acs.est.5b04431

60. Jourdin L et al (2018) Critical biofilm growth throughout unmodified carbon felts allows continuous bioelectrochemical chain elongation from CO_2 up to caproate at high current density. Front Energy Res 6:7. https://doi.org/10.3389/fenrg.2018.00007

61. Kato S, Hashimoto K, Watanabe K (2012) Methanogenesis facilitated by electric syntrophy via (semi)conductive iron-oxide minerals. Environ Microbiol 14(7):1646–1654. https://doi.org/10.1111/j.1462-2920.2011.02611.x

62. Katuri KP et al (2018) Dual-function electrocatalytic and macroporous hollow-fiber cathode for converting waste streams to valuable resources using microbial electrochemical systems. Adv Mater 30(26):1707072. https://doi.org/10.1002/adma.201707072

63. Kernan T et al (2015) Engineering the iron-oxidizing chemolithoautotroph *Acidithiobacillus ferrooxidans* for biochemical production. Biotechnol Bioeng 113(1):189–197. https://doi.org/10.1002/bit.25703 (Wiley-Blackwell)

64. Khunjar WO et al (2012) Biomass production from electricity using ammonia as an electron carrier in a reverse microbial fuel cell. PloS one 7(9):e44846. https://doi.org/10.1371/journal.pone.0044846

65. Kim S-H et al (2016) Electrochemical NADH regeneration and electroenzymatic CO_2 reduction on Cu nanorods/glassy carbon electrode prepared by cyclic deposition. Electrochim Acta 210:837–845. https://doi.org/10.1016/J.ELECTACTA.2016.06.007 (Pergamon)

66. Kim S et al (2014) Conversion of CO_2 to formate in an electroenzymatic cell using *Candida boidinii* formate dehydrogenase. J Mol Cataly B Enzym 102:9–15 (Elsevier B.V.). https://doi.org/10.1016/j.molcatb.2014.01.007

67. Köpke M et al (2010) *Clostridium ljungdahlii* represents a microbial production platform based on syngas. Proc Natl Acad Sci U S A 107(29):13087–13092. https://doi.org/10.1073/pnas.1004716107

68. Köpke M et al (2011) 2,3-Butanediol production by acetogenic bacteria, an alternative route to chemical synthesis, using industrial waste gas. Appl Environ Microbiol 77(15):5467–5475. https://doi.org/10.1128/aem.00355-11

69. Krieg T et al (2011) Gas diffusion electrode as novel reaction system for an electro-enzymatic process with chloroperoxidase. Green Chem 13(10):2686–2689. https://doi.org/10.1039/c1gc15391a

70. Kuwabata S, Tsuda R, Yoneyama H (1994) Electrochemical conversion of carbon dioxide to methanol with the assistance of formate dehydrogenase and methanol dehydrogenase as biocatalysts. J Am Chem Soc 116(12):5437–5443 (ACS Publications)

71. Laane C, Weyland A, Franssen M (1986) Bioelectrosynthesis of halogenated compounds using chloroperoxidase. Enzyme Microbial Technol 8(6):345–348. https://doi.org/10.1016/0141-0229(86)90133-X (Elsevier)

72. LaBelle EV et al (2014) Influence of acidic pH on hydrogen and acetate production by an electrosynthetic microbiome. PLoS ONE 9(10):e109935 (Public Library of Science). Available at: http://dx.doi.org/10.1371%252Fjournal.pone.0109935

73. LaBelle EV, May HD (2017) Energy efficiency and productivity enhancement of microbial electrosynthesis of acetate. Front Microbiol 8:756. https://doi.org/10.3389/fmicb.2017.00756

74. Lee SY et al (2016) Light-driven highly selective conversion of CO_2 to formate by electrosynthesized enzyme/cofactor thin film electrode. Adv Energy Mater 6(11):1502207. https://doi.org/10.1002/aenm.201502207

75. Lepage G et al (2014) Multifactorial evaluation of the electrochemical response of a microbial fuel cell. RSC Adv 4(45):23815–23825. https://doi.org/10.1039/C4RA03879G
76. Li H et al (2012) Integrated electromicrobial conversion of CO_2 to higher alcohols. Science 335(6076):1596 (New York, N.Y.). https://doi.org/10.1126/science.1217643
77. Li X, Angelidaki I, Zhang Y (2018) Salinity-gradient energy driven microbial electrosynthesis of value-added chemicals from CO_2 reduction. Water Res 142:396–404. https://doi.org/10.1016/J.WATRES.2018.06.013 (Pergamon)
78. Lienemann M et al (2018) Mediator-free enzymatic electrosynthesis of formate by the Methanococcus maripaludis heterodisulfide reductase supercomplex. Bioresour Technol 254:278–283. https://doi.org/10.1016/J.BIORTECH.2018.01.036 (Elsevier)
79. Liu C et al (2015) Nanowire–bacteria hybrids for unassisted solar carbon dioxide fixation to value-added chemicals. Nano Lett 15(5):3634–3639. https://doi.org/10.1021/acs.nanolett.5b01254 (American Chemical Society)
80. Liu C, Ziesack M, Silver PA (2016) Water splitting—biosynthetic system with CO_2 reduction efficiencies exceeding photosynthesis. Science 352(6290):1210–1213. https://doi.org/10.1126/science.aaf5039
81. Liu F et al (2012) Promoting direct interspecies electron transfer with activated carbon. Energy Environ Sci 5(10):8982–8989. https://doi.org/10.1039/C2EE22459C (The Royal Society of Chemistry)
82. Lovley DR (2011) Powering microbes with electricity: direct electron transfer from electrodes to microbes. Environ Microbiol Rep 3(1):27–35. https://doi.org/10.1111/j.1758-2229.2010.00211.x
83. Lovley DR, Nevin KP (2013) Electrobiocommodities: powering microbial production of fuels and commodity chemicals from carbon dioxide with electricity. Curr Opin Biotechnol 24(3):385–390. https://doi.org/10.1016/J.COPBIO.2013.02.012 (Elsevier Current Trends)
84. Lu J et al (2012) Studies on the production of branched-chain alcohols in engineered Ralstonia eutropha. Appl Microbiol Biotechnol 96(1):283–297. https://doi.org/10.1007/s00253-012-4320-9
85. Majumdar P et al (2018) Enzymatic Electrocatalysis of CO_2 reduction. In: Wandelt K (ed) Encyclopedia of interfacial chemistry: surface science and electrochemistry. Elsevier Inc. https://doi.org/10.1016/b978-0-12-409547-2.13353-0
86. Marshall CW et al (2012) Electrosynthesis of commodity chemicals by an autotrophic microbial community. Appl Environ Microbiol 78(23):8412–8420. https://doi.org/10.1128/aem.02401-12
87. Marshall CW et al (2013) Long-term operation of microbial electrosynthesis systems improves acetate production by autotrophic microbiomes. Environ Sci Technol 47(11):6023–6029. https://doi.org/10.1021/es400341b
88. Marshall CW et al (2017) Metabolic reconstruction and modeling microbial electrosynthesis. Sci Rep 7(1):1–12. https://doi.org/10.1038/s41598-017-08877-z (Springer, US)
89. Marshall CW, LaBelle EV, May HD (2013) Production of fuels and chemicals from waste by microbiomes. Curr Opin Biotechnol 24(3):391–397. https://doi.org/10.1016/J.COPBIO.2013.03.016 (Elsevier Current Trends)
90. Mohanakrishna G et al (2015) An enriched electroactive homoacetogenic biocathode for the microbial electrosynthesis of acetate through carbon dioxide reduction. Faraday Discuss. https://doi.org/10.1039/c5fd00041f
91. Mondal B et al (2015) Bio-inspired mechanistic insights into CO_2 reduction. Curr Opin Chem Biol 25:103–109. https://doi.org/10.1016/J.CBPA.2014.12.022 (Elsevier Current Trends)
92. Morita M et al (2011) Potential for direct interspecies electron transfer in methanogenic wastewater digester aggregates. In: Casadevall A (ed) mBio, vol 2, issue 4. Available at: http://mbio.asm.org/content/2/4/e00159-11.abstract

93. Moscoviz R et al (2016) Electro-fermentation: how to drive fermentation using electrochemical systems. Trends Biotechnol 34(11):856–865. https://doi.org/10.1016/J. TIBTECH.2016.04.009 (Elsevier Current Trends)
94. Nam DH et al (2016) Enzymatic photosynthesis of formate from carbon dioxide coupled with highly efficient photoelectrochemical regeneration of nicotinamide cofactors. Green Chem 18(22):5989–5993. https://doi.org/10.1039/C6GC02110G (The Royal Society of Chemistry)
95. Nevin KP et al (2010) Microbial electrosynthesis: feeding microbes electricity to convert carbon dioxide and water to multicarbon extracellular organic. mBio 1(2):e00103-10-. https://doi.org/10.1128/mbio.00103-10.editor
96. Nevin KP et al (2011) Electrosynthesis of organic compounds from carbon dioxide is catalyzed by a diversity of acetogenic microorganisms. Appl Environ Microbiol 77(9):2882–2886. https://doi.org/10.1128/aem.02642-10
97. Nichols EM et al (2015) Hybrid bioinorganic approach to solar-to-chemical conversion. Proc Natl Acad Sci 112(37):11461 LP-11466. Available at: http://www.pnas.org/content/112/37/11461.abstract
98. Nie H et al (2013) Improved cathode for high efficient microbial-catalyzed reduction in microbial electrosynthesis cells. Phys Chem Chem Phys PCCP 15(34):14290–14294. https://doi.org/10.1039/c3cp52697f
99. Nybo SE et al (2015) Metabolic engineering in chemolithoautotrophic hosts for the production of fuels and chemicals. Metab Eng 30:105–120. https://doi.org/10.1016/J. YMBEN.2015.04.008 (Academic Press)
100. Pant D et al (2010) Use of novel permeable membrane and air cathodes in acetate microbial fuel cells. Electrochim Acta 55(26):7710–7716. https://doi.org/10.1016/j.electacta.2009.11. 086 (Pergamon-Elsevier Science Ltd)
101. Pant D et al (2011) An introduction to the life cycle assessment (LCA) of bioelectrochemical systems (BES) for sustainable energy and product generation: relevance and key aspects. Renew Sustain Energy Rev 15(2):1305–1313. https://doi.org/10.1016/j.rser.2010.10.005 (Elsevier Ltd)
102. Pant D et al (2012) Bioelectrochemical systems (BES) for sustainable energy production and product recovery from organic wastes and industrial wastewaters. RSC Adv 2(4):1248. https://doi.org/10.1039/c1ra00839k (Royal Soc Chemistry)
103. Parkin A et al (2007) Rapid and efficient electrocatalytic CO_2/CO interconversions by carboxydothermus hydrogenoformans CO dehydrogenase i on an electrode. J Am Chem Soc 129(34):10328–10329. https://doi.org/10.1021/ja073643o (American Chemical Society)
104. Parkinson BA, Weaver PF (1984) Photoelectrochemical pumping of enzymatic CO_2 reduction. Nature 309:148 (Nature Publishing Group). Available at: http://dx.doi.org/10. 1038/309148a0
105. Patil SA et al (2015) Selective enrichment establishes a stable performing community for microbial electrosynthesis of acetate from CO_2. Environ Sci Technol 49(14):8833–8843. https://doi.org/10.1021/es506149d
106. Pearson RJ et al (2012) Energy storage via carbon-neutral fuels made from CO_2, water, and renewable energy. Proc IEEE 100(2):440–460. https://doi.org/10.1109/JPROC.2011. 2168369
107. Pepè Sciarria T et al (2018) Bio-electrorecycling of carbon dioxide into bioplastics. Green Chem 20(17):4058–4066. https://doi.org/10.1039/C8GC01771A (The Royal Society of Chemistry)
108. Rabaey K, Rozendal RA (2010) Microbial electrosynthesis—revisiting the electrical route for microbial production.pdf. Nat Rev Microbiol 8(10):706–716. https://doi.org/10.1038/nrmicro2422 (Nature Publishing Group)
109. Raes SMT et al (2016) Continuous long-term bioelectrochemical chain elongation to butyrate. ChemElectroChem 4(2):386–395. https://doi.org/10.1002/celc.201600587

110. Ragsdale SW, Pierce E (2008) Acetogenesis and the Wood-Ljungdahl pathway of CO_2 fixatio. Biochim Biophys Acta Proteins Proteomics 1784(12):1873–1898. https://doi.org/10.1016/J.BBAPAP.2008.08.012 (Elsevier)

111. Reda T et al (2008) Reversible interconversion of carbon dioxide and formate by an electroactive enzyme. Proc Natl Acad Sci U S A 105(31):10654–10658. https://doi.org/10.1073/pnas.0801290105

112. Rosenbaum M et al (2011) Cathodes as electron donors for microbial metabolism: Which extracellular electron transfer mechanisms are involved? Bioresour Technol 102(1):324–333. https://doi.org/10.1016/j.biortech.2010.07.008 (Elsevier Ltd)

113. Ross DE et al (2011) Towards electrosynthesis in Shewanella: energetics of reversing the Mtr pathway for reductive metabolism. PLoS ONE 6(2). https://doi.org/10.1371/journal.pone.0016649

114. Sakai K et al (2016) Efficient bioelectrocatalytic CO_2 reduction on gas-diffusion-type biocathode with tungsten-containing formate dehydrogenase. Electrochem Commun 73:85–88. https://doi.org/10.1016/J.ELECOM.2016.11.008 (Elsevier)

115. Schiel-Bengelsdorf B, Dürre P (2012) Pathway engineering and synthetic biology using acetogens. FEBS Lett 586(15):2191–2198. https://doi.org/10.1016/j.febslet.2012.04.043 (Federation of European Biochemical Societies)

116. Schlager S et al (2015) Direct electrochemical addressing of immobilized alcohol dehydrogenase for the heterogeneous bioelectrocatalytic reduction of butyraldehyde to butanol. ChemCatChem 7(6):967–971. https://doi.org/10.1002/cctc.201402932 (Wiley-Blackwell)

117. Schlager S, Haberbauer M et al (2016) Bio-electrocatalytic application of microorganisms for carbon dioxide reduction to methane. Chemsuschem 10(1):226–233. https://doi.org/10.1002/cssc.201600963 (Wiley-Blackwell)

118. Schlager S, Dumitru LM et al (2016) Electrochemical reduction of carbon dioxide to methanol by direct injection of electrons into immobilized enzymes on a modified electrode. Chemsuschem 9(6):631–635. https://doi.org/10.1002/cssc.201501496 (Wiley-Blackwell)

119. Seedorf H et al (2008) The genome of *Clostridium kluyveri*, a strict anaerobe with unique metabolic features. Proc Natl Acad Sci 105(6):2128 LP-2133. Available at: http://www.pnas.org/content/105/6/2128.abstract

120. Sharma M et al (2013) Bioelectrocatalyzed reduction of acetic and butyric acids via direct electron transfer using a mixed culture of sulfate-reducers drives electrosynthesis of alcohols and acetone. Chem Commun 49(58):6495–6497 (Cambridge, England). https://doi.org/10.1039/c3cc42570c

121. Shi J et al (2012) 'Constructing spatially separated multienzyme system through bioadhesion-assisted bio-inspired mineralization for efficient carbon dioxide conversion. Bioresour Technol 118:359–366. https://doi.org/10.1016/J.BIORTECH.2012.04.099 (Elsevier)

122. Shin W et al (2003) Highly selective electrocatalytic conversion of CO_2 to CO at −0.57 V (NHE) by carbon monoxide dehydrogenase from moorella thermoacetica. J Am Chem Soc 125(48):14688–14689 (American Chemical Society). https://doi.org/10.1021/ja037370i

123. Srikanth S et al (2014) Enzymatic electrosynthesis of formate through CO_2 sequestration/reduction in a bioelectrochemical system (BES). Bioresour Technol 165:350–354. https://doi.org/10.1016/j.biortech.2014.01.129 (Elsevier Ltd)

124. Srikanth S et al (2017) Enzymatic electrosynthesis of formic acid through carbon dioxide reduction in a bioelectrochemical system: effect of immobilization and carbonic anhydrase addition. ChemPhysChem 18(22):3174–3181. https://doi.org/10.1002/cphc.201700017

125. Srikanth S et al (2018) Electro-biocatalytic conversion of carbon dioxide to alcohols using gas diffusion electrode. Bioresour Technol 265(February):45–51. https://doi.org/10.1016/j.biortech.2018.02.058 (Elsevier)

126. Steidl R, Lampa-Pastirk S, Reguera G (2016) Mechanistic stratification in electroactive biofilms of Geobacter sulfurreducens mediated by pilus nanowires. Nat Comm (submitted). https://doi.org/10.1038/ncomms12217

127. Steinbusch KJJ et al (2011) Biological formation of caproate and caprylate from acetate: fuel and chemical production from low grade biomass. Energy Environ Sci 4(1):216–224. https://doi.org/10.1039/C0EE00282H

128. Sultana S et al (2016) A review of harvesting clean fuels from enzymatic CO_2 reduction. RSC Adv 6(50):44170–44194. https://doi.org/10.1039/c6ra05472b (Royal Society of Chemistry)

129. Sydow A et al (2014) Electroactive bacteria-molecular mechanisms and genetic tools. Appl Microbiol Biotechnol, 8481–8495. https://doi.org/10.1007/s00253-014-6005-z

130. Thauer RK et al (2008) Methanogenic archaea: ecologically relevant differences in energy conservation. Nat Rev Microbiol 6:579 (Nature Publishing Group). Available at: http://dx.doi.org/10.1038/nrmicro1931

131. Thauer RK, Jungermann K, Decker K (1977) Energy conservation in chemotrophic anaerobic bacteria. Bacteriol Rev 41(3):809. Available at: http://www.ncbi.nlm.nih.gov/pubmed/16350228

132. Torella JP et al (2015) Efficient solar-to-fuels production from a hybrid microbial–water-splitting catalyst system. Proc Natl Acad Sci 112(12):201503606–201503607. https://doi.org/10.1073/pnas.1503606112

133. Tremblay P-L, Angenent LT, Zhang T (2017) Extracellular electron uptake: among autotrophs and mediated by surfaces. Trends Biotechnol 35(4):360–371. https://doi.org/10.1016/J.TIBTECH.2016.10.004 (Elsevier Current Trends)

134. Ueki T et al (2014) Converting carbon dioxide to butyrate with an engineered strain of *Clostridium ljungdahlii*. mBio 5(5). https://doi.org/10.1128/mbio.01636-14

135. Vassilev I et al (2018) Microbial electrosynthesis of isobutyric, butyric, caproic acids, and corresponding alcohols from carbon dioxide, pp 4–12. https://doi.org/10.1021/acssuschemeng.8b00739

136. Villano M et al (2010) Bioelectrochemical reduction of CO(2) to CH(4) via direct and indirect extracellular electron transfer by a hydrogenophilic methanogenic culture. Bioresour Technol 101(9):3085–3090 (Elsevier Ltd). https://doi.org/10.1016/j.biortech.2009.12.077

137. Villano M et al (2011) Electrochemically assisted methane production in a biofilm reactor. J Power Sources 196(22):9467–9472 (Elsevier B.V.). https://doi.org/10.1016/j.jpowsour.2011.07.016

138. Wang H, Ren ZJ (2013) A comprehensive review of microbial electrochemical systems as a platform technology. Biotechnol Adv 31(8):1796–1807. https://doi.org/10.1016/J.BIOTECHADV.2013.10.001 (Elsevier)

139. Xiang Y et al (2017) High-efficient acetate production from carbon dioxide using a bioanode microbial electrosynthesis system with bipolar membrane. Bioresour Technol 233:227–235. https://doi.org/10.1016/j.biortech.2017.02.104 (Elsevier Ltd)

140. Zaybak Z et al (2013) Enhanced start-up of anaerobic facultatively autotrophic biocathodes in bioelectrochemical systems. J Biotechnol 168:478–485 (Elsevier B.V.). https://doi.org/10.1016/j.jbiotec.2013.10.001

141. Zhang T et al (2013) Improved cathode materials for microbial electrosynthesis. Energy Environ Sci 6(1):217. https://doi.org/10.1039/c2ee23350a

Enhanced Biological Fixation of CO_2 Using Microorganisms

10

Fuyu Gong, Huawei Zhu, Jie Zhou, Tongxin Zhao, Lu Xiao, Yanping Zhang and Yin Li

Abstract

Microbial fixation of carbon dioxide (CO_2), represented by photosynthesis, is an important link of the global carbon cycle. It provides the majority of organic chemicals and energy for human consumption. With the great development and application of fossil resources in recent years, more and more CO_2 has been released into the atmosphere, and the greenhouse effect is looming. Therefore,

F. Gong · H. Zhu · J. Zhou · T. Zhao · L. Xiao · Y. Zhang · Y. Li (✉)
CAS Key Laboratory of Microbial Physiological and Metabolic Engineering, State Key
Laboratory of Microbial Resources, Institute of Microbiology, Chinese Academy of Sciences,
Beijing 100101, China
e-mail: yli@im.ac.cn

F. Gong
Institute of Process Engineering, Chinese Academy of Sciences, Beijing 100190, China
e-mail: gongfuyu1987@126.com

H. Zhu · T. Zhao · L. Xiao
University of Chinese Academy of Sciences, Beijing 100049, China

H. Zhu
e-mail: zhuhuawei0901@163.com

J. Zhou
e-mail: jiezhouw@im.ac.cn

T. Zhao
e-mail: txzhao2014@sina.com

L. Xiao
e-mail: 276115320@qq.com

Y. Zhang
e-mail: zhangyp@im.ac.cn

© Springer Nature Switzerland AG 2019
M. Aresta et al. (eds.), *An Economy Based on Carbon Dioxide and Water*,
https://doi.org/10.1007/978-3-030-15868-2_10

359

more efficient carbon fixation processes are urgently needed. In view of this, the microbial conversion of exhaust CO_2 into valuable fuels and chemicals based on an efficient CO_2 fixation pathway is very promising. With the rapid development of systems biology, more and more insights into the natural carbon fixation processes have become available. Many attempts have been made to enhance the biological fixation of CO_2, by engineering the key carbon fixation enzymes, introducing natural carbon fixation pathways into heterotrophs, redesigning novel carbon fixation pathways, and even developing novel energy supply patterns. In this review, we summarize the great achievements made in recent years, and discuss the main challenges as well as future perspectives on the biological fixation of CO_2.

It is the key step of the global carbon cycle to convert inorganic carbon into organic chemicals. Organic chemicals and energy currently consumed by humans in the world are mainly derived from organic carbon fixed from CO_2. Energy and the environment are two major issues that are closely related to human life. The release of tremendous and increasing amounts of CO_2 into the atmosphere has attracted worldwide attention due to the greenhouse effect. Recycling the abundant waste CO_2 directly into fuels or chemicals is a potential approach to provide a sustainable energy supply and to reducing the greenhouse effect.

Carbon atoms in common fuels and chemicals are in lower states, while those in CO_2 molecules are in the highest oxidation state. Significant energy input is thus required to convert inorganic CO_2 to organic compounds. Autotrophs fix CO_2 through evolved carbon fixation pathways, with light or inorganic chemical as the energy source. Plants can utilize light to fix atmospheric CO_2 through the well-known process of photosynthesis and the Calvin cycle, while CO_2 fixation by microorganisms proceeds through many different carbon fixation pathways and modules that utilize different energy sources. Furthermore, autotrophs convert CO_2 into a variety of products, including C2, C3, C4 and longer chain carbohydrates. Recently, there have been numerous studies on autotrophic microorganisms with CO_2 fixation capacity. Such microorganisms can be reconstructed through metabolic engineering to produce target chemicals from CO_2. In the past 5 years, great progress has been made in this area. Ethanol, butanol, lactic acid, acetone, isobutyraldehyde, isoprene, oil, and other fuels and chemicals can be synthesized from CO_2 using engineered autotrophic microorganisms, demonstrating the great potential of biological carbon sequestration [1, 4, 8, 18, 44, 43, 82]. However the specific carbon fixation efficiency of microorganisms is relatively low [23]. For example, the carbon fixation rate of cyanobacteria, which is 1 to 5 mg/L/h, cannot meet the industrial demands, which amount to 1–10 g/L/h [2]. In addition, another bottleneck of carbon fixation is limited energy availability. Natural autotrophs generally include two types, photo- and chemoautotrophs. Photoautotrophs fix carbon dioxide, with light as the energy source, while the energy source of chemoautotrophs is reduced compounds such as hydrogen, sulfite, nitrite and other inorganic compounds [58]. Although here is still is a long way to their industrial applications, the natural carbon fixation processes also provide a variety of carbon-fixing enzymes and pathways, which opens multiple possibilities for artificial engineering.

In recent years, with greater understanding of natural carbon fixation pathways coupled with the rapid development of synthetic biology techniques, research concepts related to biological carbon fixation have been greatly expanded. Carbon fixation is divided into two parts, energy supply modules and carbon fixation pathways. The two parts can be redesigned and freely combined to develop new synthetic carbon fixation processes, which is expected to result in higher efficiencies than what can be achieved with their natural counterparts. This article focuses on reviewing the advances in microbial carbon fixation in recent years, including the research and exploration of natural carbon fixation processes, introducing natural carbon fixation pathways into heterotrophs for CO_2 sequestration, designing and constructing novel synthetic CO_2-assimilation pathways, as well as the characterization and engineering of energy supply patterns for carbon fixation. Finally, challenges and future prospects of biological carbon fixation are discussed.

10.1 Exploring Natural Carbon Fixation Processes

Six natural carbon fixation pathways have been discovered to date (Table 10.1, Fig. 10.1): the Calvin cycle, reductive tricarboxylic acid cycle (rTCA cycle), the Wood–Ljungdahl pathway (WL pathway), 3-hydroxypropionate bicycle (3HP bicycle), dicarboxylate/4-hydroxybutyrate cycle (di-4HB cycle), and 3-hydroxypropionate-4-hydroxybutyrate cycle (3HP-4HB cycle) [10, 14, 20, 33, 67, 72]. The Calvin cycle, the 3HP bicycle, and the 3HP-4HB cycle function well under aerobic conditions, while the rTCA cycle, the WL pathway, and the di-4HB cycle contain certain oxygen-sensitive enzymes, and are therefore considered to only function anaerobically [26]. Interestingly, although the six natural carbon fixation pathways generate different metabolites from CO_2, they consume the same amount of NAD(P)H (0.5 mol/mol) to reduce C atoms to the univalent, but they require different amounts of ATP.

The Calvin cycle is the most widespread carbon fixation pathway, and the most thoroughly studied. More than 90% of CO_2 in nature is fixed by plants, algae and microorganisms using this carbon fixation cycle. It was discovered in the 1940s by Melvin Ellis Calvin, who won the 1961 Nobel Prize in Chemistry for his discovery [15, 13]. The essence of this cycle is that three molecules of CO_2 are converted into one molecule of glyceraldehyde 3-phosphate at the expense of nine molecules of ATP and six molecules of NADPH. Interestingly, this cycle has the highest energy consumption among the six naturally occurring carbon fixation pathways. Because all crop biomass comes from the Calvin cycle, the efficiency of this carbon fixation cycle is closely related to food production, and is therefore of utmost concern to scientists.

Table 10.1 Comparison of the six known natural carbon fixation pathways

Pathway	Reaction	Valence state of C atoms	Energy input(s)	ATP and NAD(P)H/ univalent state of C atoms	Energy sources(s)	References
Calvin cycle	$3CO_2 \rightarrow$ Glyceraldehyde-3-phosphate	$3(+4) \rightarrow 3(0)$	9 ATP, 6 NAD (P)H	0.75, 0.5	Light	[14]
rTCA cycle	$2CO_2 \rightarrow$ Acetyl-CoA	$2(+4) \rightarrow 2(0)$	2 ATP, 4 NAD (P)H	0.25, 0.5	Light and sulfur	[20]
WL pathway	$2CO_2 \rightarrow$ Acetyl-CoA	$2(+4) \rightarrow 2(0)$	1 ATP, 4 NAD (P)H	0.125, 0.5	Hydrogen	[67]
3HP bicycle	$3HCO_3^- \rightarrow$ Pyruvate	$3(+4) \rightarrow 3(0.67)$	5 ATP, 5 NAD (P)H	0.5, 0.5	Light	[71]
di-4HB cycle	$1CO_2 + 1HCO_3^- \rightarrow$ Acetyl-CoA	$2(+4) \rightarrow 2(0)$	3 ATP, 4 NAD (P)H	0.375, 0.5	Hydrogen and sulfur	[10]
3HP-4HB cycle	$2HCO_3^- \rightarrow$ Acetyl-CoA	$2(+4) \rightarrow 2(0)$	4 ATP, 4 NAD (P)H	0.5, 0.5	Hydrogen and sulfur	[33]

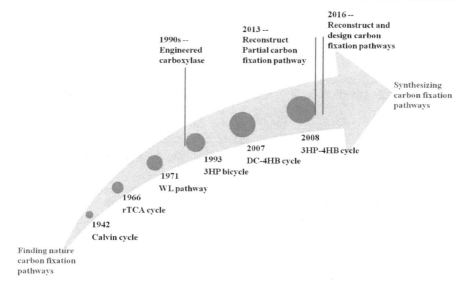

Fig. 10.1 The time axis of discovery of natural carbon fixation pathways and the development of synthetic carbon fixation pathways

The reductive TCA cycle exists in photosynthetic green sulfur bacteria and some other anaerobic bacteria. The total reaction of this cycle consumes two molecules of carbon dioxide to synthesize one molecule of acetyl coenzyme A by utilizing two molecules of ATP and four molecules of NAD(P)H [9, 37].

The Wood–Ljungdahl pathway was discovered in the 1970s by Harland G. Wood and Lars G. Ljungdahl [19]. This pathway is mainly present in acetate-producing anaerobes and is driven by hydrogen. It is the only non-cyclic pathway. Moreover, it is the shortest and consumes the least energy of the six natural carbon fixation pathways. This pathway has two branches and essentially converts two molecules of carbon dioxide (or one molecule of carbon dioxide and one of carbon monoxide) into one molecule of acetyl coenzyme A, utilizing one molecule of ATP and four molecules of NAD(P)H. It is also called the anaerobic acetyl coenzyme A pathway [21].

The 3-hydroxypropionate bicycle exists in photosynthetic green nonsulfur bacteria, and is driven by light [32, 71]. This cycle contains 16 enzymatic reaction steps with 13 enzymes. The inorganic carbon source of this pathway is the bicarbonate ion. Thus, each cycle involves the input of three molecules of HCO$_3^-$ and utilizes five molecules of ATP and five of NAD(P)H, with an output of one molecule of pyruvate.

The 3-hydroxypropionate-4-hydroxybutyrate cycle was discovered in archaea in 2007 [10]. This pathway is driven by elemental sulfur and hydrogen. Thus, each cycle involves the input of two molecules of HCO$_3^-$ to synthesize one molecule of acetyl coenzyme A, thereby consuming four molecules of ATP and four of NAD(P)H.

Discovered in 2008, the dicarboxylate/4-hydroxybutyrate cycle is a strictly anaerobic pathway and is also present in archaea [32]. In contrast with other carbon fixation pathways, its substrates are one molecule of CO_2 and one molecule of HCO_3^-. Owing to its strong carbon-fixation capacity, the doubling time of the autotrophic archaeon *Ignicoccus hospitalis*, which utilizes this pathway, is only 1 h under optimal growth conditions, which approximates that of heterotrophic organisms.

Unfortunately, the pathways other than the Calvin cycle are poorly understood, especially those from anaerobes. The anaerobic carbon fixation enzymes, such as CO dehydrogenate/acetyl-CoA synthase, 2-oxoglutarate synthase, and pyruvate synthase, should therefore be further characterized.

10.2   Characterization of Photosynthetic Pathways and Enhancing CO_2 Fixation in the Calvin Cycle

Photosynthetic carbon sequestration is the most important chemical process on Earth. It is an important link in the global carbon and oxygen cycles. Almost all organic matter on Earth comes from photosynthetic carbon sequestration. Photosynthetic carbon sequestration is a biochemical process in which photosynthetic organisms convert solar energy into chemical energy through photosynthesis, and then use the thus obtained energy to reduce CO_2 and water to organic matter, most of them releasing oxygen in the process. Photosynthetic organisms use solar energy to convert CO_2 into carbohydrates through photosynthesis, which provides the material and energy basis for practically the entire living system of the planet. Therefore, the direct conversion of CO_2 into organic matter utilizing solar energy through photosynthesis can not only be used to implement the conversion of solar energy into useful forms, but also the resource utilization of CO_2. It is a potentially ideal way to alleviate the global energy crisis and convert the greenhouse gas CO_2 from a waste product into a resource. Consequently, it has become the focus of current international scientific and technological competition and cooperation.

Cyanobacteria, as photosynthetic autotrophic prokaryotes, are important primary producers on Earth. Twenty to thirty percent of the earth's organic carbon comes from photosynthetic carbon sequestration by cyanobacteria [75]. Although cyanobacteria are prokaryotes, their photosynthetic mechanism is very similar to that of higher plants, since chloroplasts evolved from endosymbiotic cyanobacteria. Both photosystem I and photosystem II carry out oxygen-releasing photosynthesis. In addition, cyanobacteria, as prokaryotes, have simple cell structure and are easy to genetically manipulate. Moreover, their growth rate is much higher than that of higher plants. Some cyanobacteria can produce one generation in two hours [35]. Therefore, cyanobacteria are ideal organisms for studying and reforming photosynthetic carbon sequestration [57, 73, 74].

Currently, CO_2 bioconversion technology in cyanobacteria mainly uses metabolic engineering to modify natural photosynthetic organisms to synthesize specific organic chemicals. With the development of molecular biology and metabolic engineering methods for cyanobacteria, important progress has been made in the production of green fuels and chemicals using these organisms (Table 10.2). In 2011, James Liao's team transferred the butanol biosynthesis pathway from *Clostridium acetobutylicum* to *Synechococcus elongatus* PCC 7942, and for the first time, photosynthetic production of 1-butanol was achieved, demonstrating the feasibility of the attractive prospect of converting carbon dioxide into biofuels [5]. In 2011, our team expressed CoA transferase (CtfAB) and acetoacetate decarboxylase (ADC) from *Clostridium acetobutylicum* in *Synechocystis* sp. 6803 and blocked the synthesis of PHB and acetic acid. It was the first time that acetone, an important chemical, was synthesized in Cyanobacteria [82]. In 2012, Lv Xuefeng's team introduced pyruvate decarboxylase (PDC) and type-II alcohol dehydrogenase from *Zymomonas mobilis* into *Synechocystis* sp. 6803 to achieve ethanol synthesis. On this basis, NADPH-dependent alcohol dehydrogenase was replaced by NADH-dependent alcohol dehydrogenase and the competing PHB metabolic pathway was knocked out. A mutant with strong ethanol production capacity was obtained. After 24 days of cultivation, the ethanol production reached 5.5 g/L and the synthesis rate reached 212 mg/L/day, which was at the leading international level [24]. Through a series of efforts, CO_2 can be directly converted to sucrose in cyanobacteria, yielding a concentration of 2 g/L within 24 h, with the highest synthesis efficiency of 83 mg/L/h [49, 70].

In summary, through the joint efforts of many researchers from around the world, nearly 30 chemicals have been biosynthesized from CO_2 in cyanobacteria [65], including ethanol, lactic acid, isoprene, butanol, hydrogen, acetone, isopropanol and 3-hydroxybutyric acid. These advances give us a glimpse of the potential of the bioconversion of CO_2 in cyanobacteria. However, the yield of the target chemicals is basically at the level of mg/L (mmol/L), which is nearly 100 times lower than what is required for industrial production.

As the rate-limiting enzyme of the Calvin cycle, RuBisCO has long been the primary engineering target. Different strategies to improve the carbon fixation efficiency have been tested, including replacing the native RuBisCO with homologs from other sources, constructing a hybrid RuBisCO, and activity-directed selection for more efficient RuBisCO mutants [12, 25, 34, 62].

Recently, researchers also attempted to introduce a part of Calvin cycle into heterotrophic hosts using the key enzymes for carbon fixation, which enables the host to re-assimilate CO_2 at the expense of energy derived from external carbohydrates. Two key enzymes of the Calvin cycle have been introduced into both *Escherichia coli* and *Saccharomyces cerevisiae* to enable CO_2 cycling and to increase the ethanol yield [28, 31, 48, 83]. Moreover, researchers tried to reconstruct the complete Calvin cycle in heterotrophs. In 2016, Antonovsky et al. evolved a fully functional Calvin cycle in *E. coli*—the first successful horizontal transfer of a full natural carbon fixation pathway [3].

Table 10.2 Recent progress in the design and engineering of carbon fixation pathways by the optimization of natural pathways and synthetic biology strategies (expanded from [29])

Pathway	Source organism	Engineered organism	Strategy	Result	References
CBB	*Clostridium acetobutylicum*	Cyanobacteria	Transferring the butanol biosynthesis pathway from *Clostridium acetobutylicum*	Photosynthesis of 1-butanol in *Synechococcus elongatus* PCC 7942	[43]
CBB	*Clostridium acetobutylicum*	Cyanobacteria	Expression of CoA transferase (CtfAB) and acetoacetate decarboxylase (ADC) from *Clostridium acetobutylicum*	Acetone, an important chemical, was synthesized in cyanobacteria	[82]
CBB	*Zymomonas mobilis*	Cyanobacteria	Introduction of pyruvate decarboxylase (PDC) and type-II alcohol dehydrogenase from *Zymomonas mobilis* into *Synechocystis* sp. 6803	Ethanol production reached 5.5 g/L and the synthesis rate reached 212 mg/L/day	[24]
CBB	Plants	Rice *Chlamydomonas*	Construction of a hybrid RuBisCO from different large and small subunits of RuBisCO	The enzymatic properties of the hybrid RuBisCO combined those of the source enzymes	[25, 34]
CBB	Cyanobacteria	*E. coli*	Activity-directed selection for RuBisCO	Increased the specific carboxylation activity of RuBisCO by 85%	[12]
CBB	Plants	Tobacco	Overexpression of sedoheptulose-1-7 bisphosphatase	Improved photosynthetic carbon gain and yield	[62]
CBB	Cyanobacteria	Yeast	Introduction of a partial cyanobacterial Calvin cycle	Reduced glycerol formation in batch cultures by 60% and increased the ethanol yield from galactose by 8%	[31]
CBB	Cyanobacteria	*E. coli*	Introduction of a partial cyanobacterial Calvin cycle	Recycled CO_2 in an engineered *E. coli*, which fixed 67 mg CO_2 per mole of arabinose L^{-1} h^{-1}	[83]
CBB	Cyanobacteria	*E. coli*	Introduction of a partial cyanobacterial Calvin cycle and carbon concentrating mechanisms	A specific rate of 22.5 mg CO_2 g DCW^{-1} h^{-1} in *E. coli*, which is comparable with natural autotrophs	[28]

(continued)

Table 10.2 (continued)

Pathway	Source organism	Engineered organism	Strategy	Result	References
CBB	Cyanobacteria	E. coli	A fully functioning Calvin cycle was evolved in E. coli	Horizontal transfer of a natural carbon fixation pathway	[3]
WL pathway	Methylobacterium extorquens	E. coli	Insertion of a partial WL pathway and glycine cleavage/synthase system into E. coli	Biosynthesis of glycine and serine from formate and CO₂ in E. coli	[77]
WL pathway	–	E. coli	Carbon fixation by combining biological- and electrocatalysis	Conversion of two formate and one CO₂ to one pyruvate in E. coli	[78]
WL pathway	–	E. coli	Reconstructed tetrahydrofolate (THF) cycle and reverse glycine cleavage (gcv) pathway	The engineered strain was able to produce pyruvate from formate and CO₂	[5]
3HP-bicycle	Chloroflexus aurantiacus	E. coli	Divided the 3HP-bicycle from Chloroflexus aurantiacus into four sub-pathways and expressed them separately in E. coli	Horizontal transfer of a part of the natural carbon fixation pathway	[52]
3HP-4HB	Metallosphaera sedula	Pyrococcus furiosus	Introduction of a partial 3HP-4HB cycle from Metallosphaera sedula	Production of 0.6 mM 3-hydroxypropionate from CO₂ by Pyrococcus furiosus	[36]
	All available genomes	In vitro	Computationally obtained a series of synthetic CO₂-fixation pathways	Provides more choices for constructing new synthetic carbon fixation pathways	[6]
CETCH cycle	Some natural organisms	In vitro	Use of an efficient carboxylase to create a synthetic CO₂ fixation pathway	A carbon fixation rate comparable to that of the Calvin cycle	[68]
MCG pathway	Some natural organisms	In vitro E. coli	Design and synthesis of an artificial malyl-CoA-glycerate (MCG) pathway	Improved carbon fixation efficiency in photosynthetic organisms	[79]

10.3 Applying the Wood–Ljungdahl Pathway for CO_2 Fixation

The WL pathway is divided into the "eastern" branch and the "western" branch [61]. In the "eastern" branch, one molecule of CO_2 is reduction using six electrons to form a methyl group, while in the "western" branch, the other CO_2 molecule is reduction to carbon monoxide, followed by the formation of acetyl-CoA via the condensation of the bound methyl group with CO and coenzyme A (CoA). Acetogens use the whole Wood–Ljungdahl pathway for carbon fixation to synthesize cell components and also to produce many chemicals, such as 2,3-Butanediol, acetate, butyrate, an ethanol, using $CO_2/CO/H_2$ as substrates [66].

The WL pathway is the most efficient pathway for the synthesis of acetyl-CoA among the six naturally occurring carbon-fixation pathways [21]. However, the investigations of this promising pathway for synthetic carbon fixation took many years and required a lot of effort. Due to a series of strictly anaerobic complex enzymes, including methyltransferase, corrinoid protein, and CO dehydrogenase/acetyl-CoA synthase in the "western" branch, it is difficult to functionally express the WL pathway in heterotrophic host. Nevertheless, the gene cluster encoding the Wood–Ljungdahl pathway in *Clostridium ljungdahlii* was successfully introduced into *Clostridium acetobutylicum* [22] in 2018 for the first time, only to find that the carbonyl carbon from acetyl-CoA is exchanged with headspace CO.

In the last few years, scientists paid more attention to the "eastern" branch and combined this branch with other pathways in *E. coli*, using CO_2 or formate to synthesize amino acids or intermediate metabolites (Table 10.2). Bar-Even's group integrated a partial "eastern" branch (namely the C1 metabolic pathway) with the glycine cleavage system to produce glycine [77]. They also integrated it with threonine biosynthesis into a new cycle to produce acetyl-CoA [78]. Similarly, Kondo's group combined the C1 metabolic pathway with the glycine cleavage system to produce pyruvate through serine deamination. In order to evaluate if it is possible to produce pyruvate from C1 resources, Sang Yup Lee's group used ^{13}C-labeled formate and CO_2 to trace intermediate metabolites and confirmed that the new combination was able to sustain slight cell growth during fermentation after glucose depletion [5].

In summary, it is difficult to introduce the "western" branch into heterologous hosts, but the new combination strategies including a partial "eastern" branch provided new insights for engineering CO_2 fixation into heterotrophs.

10.4 Introducing Other Natural Carbon Fixation Pathways into Heterotrophs for CO_2 Sequestration

The redesign of natural carbon fixation pathways a research area focusing on the reconstruction of carbon fixation pathways in heterotrophs. However, the natural carbon fixation pathways other than the Calvin cycle and WL pathways are poorly understood, and there are only two related publications known to us (Table 10.2). In 2013, Mattozzi et al. expressed four sub-pathways of the 3-hydroxypropionate bicycle from *Chloroflexus aurantiacus* in *Escherichia coli* [52]. This work proved that each smaller sub-pathway is functional, providing evidence for horizontal gene transfer in bacterial evolution and a foundation for further metabolic engineering. In the same year, Keller et al. expressed a part of the 3-hydroxypropionate-4-hydroxybutyrate cycle from the archaeon *Metallosphaera sedula* (whose optimal growth temperature is 73 °C) in the archaeon *Pyrococcus furiosus* (optimal growth temperature 100 °C) [36]. This engineered strain can synthesize a valuable industrial chemical, 3-hydroxypropionic acid, from carbon dioxide using hydrogen.

These two studies have successfully reconstructed a heterologous carbon fixation side-pathway, which can utilize CO_2 as one of the substrates. However, the engineered strains could not grow using only CO_2. Moreover, the introduced side-pathway does not participate in the central metabolic network of the engineered host, which is why efficient carbon fixation cannot be associated with cell growth.

10.5 Designing and Constructing Novel Synthetic CO_2 Pathways

Autotrophs have some disadvantages for carbon fixation, such as slow cell growth, inefficient protein expression and low chemical productivity. Therefore, many attempts have been made to overcome these shortcomings by introducing natural carbon fixation pathways into heterotrophic hosts, or even creating novel non-natural carbon fixation pathways. Excitingly, non-natural carbon fixation pathways provide many more possibilities. However, designing non-natural carbon fixation pathways is a great challenge. In 2010, Bar-Even et al. assessed more than 5000 types of naturally occurring enzymes, and computationally obtained a series of new combinations, which provides more choices for synthetic CO_2-fixation pathways. They compared the synthetic carbon fixation pathways with natural pathways, from the perspectives of dynamics, energy efficiency, and topology. Compared to the Calvin cycle, the better candidate pathways including phosphoenolpyruvate carboxylase for CO_2 fixation would have a 2 or 3-fold higher carbon fixation rate [6]. However, many obstacles to achieving these cycles remain, including the activity, stability, localization and regulation, as well as the expression levels of pathway enzymes, and others. However, research findings provide a wealth of information on the design of new, more efficient artificial carbon fixation pathways.

Recently, some exciting pathways for CO_2 fixation have been designed and created. Schwander et al. constructed a new synthetic CO_2 fixation pathway in vitro, namely the crotonyl-CoA/ethylmalonyl-CoA/hydroxybutyryl-CoA (CETCH) cycle. In this pathway, 12 enzymatic reactions can achieve an unprecedented CO_2 fixation efficiency, due to the use of the highly efficient enoyl-CoA carboxylases/reductases (ECRs) as the starting point for designing a pathway. The carbon fixation rate of

CETCH was then comparable to that of the natural Calvin cycle [27]. Furthermore, new breakthroughs are expected by a stepwise optimization in this pathway. Moreover, to overcome the limit of acetyl-CoA synthesis by the Calvin cycle in *Synechococcus elongatus* PCC7942, a synthetic malyl-CoA-glycerate (MCG) pathway was designed and introduced. It can convert a C3 metabolite to two acetyl-CoA by fixation of one additional CO_2 equivalent. As expected, the intracellular acetyl-CoA pool and CO_2 fixation were increased about 2-fold [79]. Based on these successful studies, designing new efficient carbon fixation pathways will be easier in the future.

10.6 Characterizing and Engineering Energy Supply Patterns for Carbon Fixation

As mentioned above, a robust energy supply module is necessary for CO_2 fixation. In nature, solar energy is the most widely available energy source, with a yearly input into the earth's atmosphere of approximately 2,200,000 EJ [45]. This is much higher than the world energy use every year. Photoautotrophs can use light as the energy source. They absorb light and split water to produce ATP and NADPH, which provides the necessary energy for CO_2 fixation [81]. Unfortunately, the intercepted energy is usually restricted to the visible region of the spectrum (400–700 nm), which means that photoautotrophs can capture only $\sim 50\%$ of the input solar energy [11]. To broaden the same absorption spectrum, photosystem variants or even bacteriochlorophylls from some species of photosynthetic organisms can be replaced or combined. Recently, very specific photosystems I and II with absorption spectra further into far-red light were found in *Chroococcidiopsis thermalis*. Applying these specific photosystems is expected to improve the light utilization efficiency [59]. It provides new elements for constructing synthetic photosystems. Notably, there is also a relatively simple photosystem in marine picoplankton that uses a transmembrane proteorhodopsin as a light-driven proton pump, which absorbs light and thereby produces ATP [16, 51]. However, since no NAD(P)H is produced by this photosystem, it should be combined with an NAD(P)H regeneration system for CO_2 fixation.

In addition to the limits of the absorption spectrum, there are also many restrictions that lower the energy efficiency of photosystems. It was reported that 60% of the initially captured energy is lost during the electron transfer between photosystems I and II [76]. Many efforts have been mad to improve the photosynthetic energy transfer efficiency. In view of the photo-absorption process, light-gathering antenna systems play a crucial role. They carry specialized pigments (typically several hundred) that collect solar energy and transfer it to the photosynthetic reaction center. Because many pigment molecules are recruited, the harvested light rapidly exceeds the capacity of the photosynthetic apparatus and results in nonphotochemical quenching. By minimizing the size of the phycobilisome light-harvesting antenna, Kirst et al. increased the mass productivity of an engineered cyanobacterium by 57% [38]. This intervention not only mitigates energy losses, but also allows a greater transmittance of light into deep layers under water. Moreover, in our group, an additional NADPH consumption module was introduced into cyanobacteria to balance the ATP and NADPH produced from photosynthesis. It significantly increased the photosynthetic efficiency, and the light saturation point of the engineered cyanobacterium was 2-fold higher than that of the wild-type strain [81]. In addition, chemical photosensitizers have been introduced to utilize light energy for microbes. By decorating target cells with cadmium sulfide nanoparticles, artificial photosynthesis of acetic acid from CO_2 was achieved [63].

In addition to sunlight, there are some reduced chemicals (such as hydrogen and reduced sulfur) that can be also used as energy sources by autotrophs through various energy modules or electron transfer chains. It has been found that some hydrogenases from special species of microorganisms can uptake hydrogen and produce ATP and/or NADH [30]. The soluble hydrogenase (SH) from *Ralstonia eutropha* is a typical one, which can generate both ATP and NADH from hydrogen. Since no NADH-regenerating hydrogenase was found in *E. coli* to date [46, 64], Lamont et al. introduced the SH from *R. eutropha* into *E. coli* and demonstrated that intracellular NADH was increased in the engineered strain [42]. It brought a great potential to reconstruct a chemoautotrophic *E. coli* that utilizes hydrogen for CO_2 fixation. Notably, while the SH from *R. eutropha* catalyzes hydrogen uptake, it can produce hydrogen with a much higher rate (100-fold) [17]. It is expected to discover novel hydrogenases with high specificity of uptaking hydrogen rather than releasing hydrogen.

High-energy inorganic substances, such as H_2S, S, Fe, H_2, NH_3, etc., can be used as energy sources for carbon fixation by chemoautotrophs. Some chemoautotrophic organisms can achieve rapid growth. For example, the doubling time of *Thiomicrospira crunogena* with sulfur as an energy source is only 1.6 h, which indicates that the utilization efficiency of chemical energy can reach very high levels. It is also convenient to add inorganic substances in large-scale industrial fermentation. At present, there is no complete energy-metabolic network model for chemical autotrophs. However, in the case of sulfur as energy source, the key metabolic steps have been resolved.

The metabolism of sulfur in chemical autotrophs proceeds mainly via oxidation to species containing lower valence states such as H_2S, S, $S_2O_3^-$ to SO_4^{2-}, and generates electrons and ATP in the process to supply cells. Depending on the different enzyme clusters, it can be divided into the Sox-dependent pathway, the Sox-independent pathway, and the archaeal pathway [53].

The Sox-dependent pathway of alphaproteobacteria requires the cytochrome C oxidoreductase Sor and total of 15 genes comprising a single sox gene cluster. Seven genes, *soxXYZABCD*, code for proteins essential for sulfur oxidation in vitro. There are four sulfur oxidases clusters, SoxYZ, SoxXA, SoxB, and SoxCD. The role of SoxYZ is that of a covalently sulfur-binding and sulfur compound chelating protein. SoxXA delivers electrons to external cytochromes and/or high-potential iron-sulfur protein complexes. SoxB, a dimanganese-containing protein, act as the sulfate thiol esterase component. SoxCD is a sulfate dehydrogenase (SoxC: binding a molybdenum cofactor, SoxD: di-heme cytochrome C), which catalyzes the step in which the SoxYZ-SS-complex is oxidized to release electrons, and under further hydrolysis by SoxB, the SoxYZ-SS-complex returns to the original state [80].

The Sox-independent pathway of gammaproteobacteria and anaerobic phototrophs encompasses the sulfite-dissimilating reductase Dsr, the sulfur-quinone oxidoreductase Sqr, 5'-adenosine sulfate reductase, APS reductase, and 3'-adenosine sulphate sulfatase [54]. In the third, archaeal pathway of *Acidianus ambivalens*, the enzymes involved include sulfur oxygenase reductase SOR, sulfur-quinone oxidoreductase SQR, sulfite:acceptor oxidoreductase SAOR, thiosulfate-quinone oxidoreductase TQO, tetrathionate hydrolase TetH, adenylyl-sulfate APS, and adenylyl-sulfate phosphate adenosyltransferase APAT [39].

In addition to hydrogen and sulfur, microbes use electricity as the energy source for carbon fixation is another choice. Many efforts have been made to achieve microbial electrosynthesis [60]. Acetogens and methanogens acquire electrons directly from a cathode to reduce CO_2 to the final product acetate and methane [7, 41, 55, 56], However, the mechanism of direct electron transfer is unclear, and the engineering of non-electroactive microbes is limited [69]. Another strategy required electron shuttles or mediators, which are then transported into the microbes to power carbon fixation, electron shuttles or mediators included hydrogen, formate, flavin, Fe^{2+} and ammonia on the cathode [50]. In 2012, James Liao's reported that *Ralstonia eutropha* H16 used electricity as the sole energy source assimilate CO to produce higher alcohols, with formate as a bioelectrochemical intermediate [47]. The understanding of extracellular electron transfer mechanism also needs more effort in the future, it is advisable to use electricity for CO_2 fixation in alternative ways.

10.7 Challenges of Biological Carbon Fixation

The material conversion rate of microorganisms using CO_2 as raw material is about 2–20 mg/L/h, which is 100 times lower than that of bio-manufacturing using glucose as raw material, which reaches 2–4 g/L/h. Hence, the utilization rate of CO_2 by microorganisms is far from meeting the requirements of industrial application. It should also be acknowledged that improvements of the carbon fixation efficiency still face some core challenges.

As summarized in our recent review [29], the main challenge is the low reaction activity of key carbon fixation enzymes. It leads to a low productivity of organic compounds from CO_2 and an inevitable long production period. Though many efforts have been made to discover new CO_2 fixation enzymes based on the rapid development of genome-, metagenome- and microbiome projects, no breakthrough has been obtained to date. From the increasingly more abundant omics data, novel carbon-fixing enzymes are expected. Furthermore, the screening of natural efficient carboxylases or the de novo design of synthetic enzymes are approaches to address this question.

Adiitionally, the designing ability of carbon fixation pathways is also a main challenge. The natural carbon fixation pathways are the result of a billion years' evolution process, and were beneficial for the survival and growth of hosts in a constantly changing, sub-optimal environment, rather than for the efficient production of target chemicals under stable bioreactor conditions. With the development of synthetic biology, exploring new natural carbon fixation pathways and designing novel carbon fixation pathways may fundamentally solve this problem.

In addition, most efforts were focused on the carbon fixation enzymes and pathways, but were not combined with energy supplying modules. As described above, due to the low energy state of CO_2, significant energy input is required to synthesize organic compounds from inorganic CO_2. Solar energy in its raw form is very cheap, but the light utilization efficiency of current systems is too low to provide sufficient energy for CO_2 fixation. Researchers have developed strategies to increase the efficiency of solar energy conversion. Firstly, artificial photovoltaic power generation has been able to increase the efficiency of solar energy utilization from between 0.1 and 2.5 to 16.2%. The design and implementation of a branch of photosynthetic electron transport by reforming and optimizing the pathway of electron transfer is another approach for avoiding the loss of energy in the light-to-electron conversion process. Secondly, an NADPH consumption module can be introduced to break the inherent NADPH balance of cells and enhance the light reaction through the effective coupling of light and dark reactions. The basic principle of ATP coupling with NADPH in the photoreaction is a potential solution for the problem that the ATP produced by the photoreaction cannot meet the energy requirements of carbon fixation in the dark reaction. Thirdly, it is possible to modify and utilize the photorespiration pathway to synthesize target products, which reduces the 30% loss of photosynthetically fixed carbon to photorespiration [40]. Moreover, increasing the range of biologically available wavelengths can also improve the efficiency of light utilization. Finally, various energy supply patterns for direct carbon fixation, such as electric energy and hydrogen, are expected to meet the need of the carbon fixation processes in industrial bioreactors.

10.8 Perspectives

As summarized above, with the rapid development of biotechnology, more and more insights into the natural carbon fixation pathways are becoming available. In recent years, many new strategies have been developed for constructing biological carbon fixation in engineered strains. Among the numerous unrelenting efforts, redesigning the carbon fixation pathways based on novel and efficient enzymes has brought more than a glimmer of hope. It seems a promising strategy to obtain enhanced biological fixation of CO_2. Hopefully, novel and efficient carbon fixation enzymes and pathways will be created based on the diverse natural enzymes in the near future. Additionally, the increasing attention paid to intracellular bio-energy and electron transfer processes makes it possible to construct efficient energy

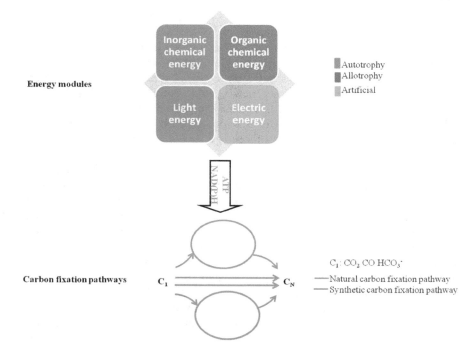

Fig. 10.2 The different energy modules that can be used to provide ATP and NAD(P)H for the carbon fixation pathways

modules in target hosts, which is expected to provide the required ATP and NAD (P)H for CO_2 fixation (Fig. 10.2). More importantly, the combination of different energy sources, in addition to solar irradiation, provides new potential beyond the most widespread photosynthesis process. Assembling different modules of carbon fixation and novel energy supply patterns is feasible today. It is promising to meet the economic requirements of the carbon fixation processes.

Acknowledgements This work was supported by the Key Research Program of the Chinese Academy of Sciences (ZDRW-ZS-2016-3), and the National Natural Science Foundation of China (31470231, 31670048, and 31700047). Yin Li was supported by the Hundreds of Talents Program of the Chinese Academy of Sciences, and Yanping Zhang was supported by the Youth Innovation Promotion Association of the Chinese Academy of Sciences (No. 2014076).

References

1. Angermayr SA, Paszota M, Hellingwerf KJ (2012) Engineering a cyanobacterial cell factory for production of lactic acid. Appl Environ Microbiol 78:7098–7106
2. Angermayr SA, Rovira AG, Hellingwerf KJ (2015) Metabolic engineering of cyanobacteria for the synthesis of commodity products. Trends Biotechnol 33:352–361

3. Antonovsky N, Gleizer S, Noor E et al (2016) Sugar synthesis from CO_2 in *Escherichia coli*. Cell 166:115–125
4. Atsumi S, Higashide W, Liao JC (2009) Direct photosynthetic recycling of carbon dioxide to isobutyraldehyde. Nat Biotechnol 27:1177–1180
5. Bang J, Lee SY (2018) Assimilation of formic acid and CO_2 by engineered Escherichia coli equipped with reconstructed one-carbon assimilation pathways. Proc Natl Acad Sci USA 115: E9271–E9279
6. Bar-Even A, Noor E, Lewis NE et al (2010) Design and analysis of synthetic carbon fixation pathways. Proc Natl Acad Sci USA 107:8889–8894
7. Beese-Vasbender PF, Grote JP, Garrelfs J et al (2015) Selective microbial electrosynthesis of methane by a pure culture of a marine lithoautotrophic archaeon. Bioelectrochemistry 102:50–55
8. Bentley FK, Melis A (2012) Diffusion-based process for carbon dioxide uptake and isoprene emission in gaseous/aqueous two-phase photobioreactors by photosynthetic microorganisms. Biotechnol Bioeng 109:100–109
9. Berg IA (2011) Ecological aspects of the distribution of different autotrophic CO_2 fixation pathways. Appl Environ Microbiol 77:1925–1936
10. Berg IA, Kockelkorn D, Buckel W et al (2007) A 3-hydroxypropionate/4-hydroxybutyrate autotrophic carbon dioxide assimilation pathway in archaea. Science 318:1782–1786
11. Blankenship Robert E et al (2011) Comparing photosynthetic and photovoltaic efficiencies and recognizing the potential for improvement. Science 332:805–809
12. Cai Z, Liu G, Zhang J, et al (2014) Development of an activity-directed selection system enabled significant improvement of the carboxylation efficiency of Rubisco. Protein cell:1–11
13. Calvin M (1949) The path of carbon in photosynthesis. J Chem Educ 26:639
14. Calvin M, Benson AA (1948) The path of carbon in photosynthesis. Science 107:476–480
15. Calvin M, Massini P (1952) The path of carbon in photosynthesis. Cell Mol Life Sci 8:445–457
16. Claassens NJ, Volpers M, dos Santos V et al (2013) Potential of proton-pumping rhodopsins: engineering photosystems into microorganisms. Trends Biotechnol 31:633–642
17. Darensbourg MY, Lyon EJ, Smee JJ (2000) The bio-organometallic chemistry of active site iron in hydrogenases. Coord Chem Rev 206:533–561
18. Dexter J, Fu PC (2009) Metabolic engineering of cyanobacteria for ethanol production. Energy Environ Sci 2:857–864
19. Drake HL (1994) Acetogenesis, acetogenic bacteria, and the acetyl-CoA "Wood/Ljungdahl" pathway: past and current perspectives. Acetogene:3–60
20. Evans MCW, Buchanan BB, Arnon DI (1966) A new ferredoxin-dependent carbon reduction cycle in a photosynthetic bacterium. Proc Natl Acad Sci USA 55:928–934
21. Fast AG, Papoutsakis ET (2012) Stoichiometric and energetic analyses of non-photosynthetic CO_2-fixation pathways to support synthetic biology strategies for production of fuels and chemicals. Curr Opin Chem Eng 1:380–395
22. Fast AG, Papoutsakis E T (2018) Functional expression of the Clostridium ljungdahlii Acetyl-Coenzyme A Synthase in clostridium acetobutylicum as demonstrated by a novel in vivo CO exchange activity en route to heterologous installation of a functional Wood-Ljungdahl pathway. Appl Environ. Microbiol 84
23. Field CB, Behrenfeld MJ, Randerson JT et al (1998) Primary production of the biosphere: Integrating terrestrial and oceanic components. Science 281:237–240
24. Gao ZX, Zhao H, Li ZM et al (2012) Photosynthetic production of ethanol from carbon dioxide in genetically engineered cyanobacteria. Energy Environ Sci 5:9857–9865
25. Genkov T, Meyer M, Griffiths H et al (2010) Functional hybrid rubisco enzymes with plant small subunits and algal large subunits: engineered rbcS cDNA for expression in chlamydomonas. J Biol Chem 285:19833–19841
26. Gong FY, Cai Z, Li Y (2016) Synthetic biology for CO_2 fixation. Sci China-Life Sci 59:1106–1114

27. Gong FY, Li Y (2016) Fixing carbon, unnaturally. Science 354:830–831
28. Gong F, Liu G, Zhai X et al (2015) Quantitative analysis of an engineered CO_2-fixing *Escherichia coli* reveals great potential of heterotrophic CO_2 fixation. Biotechnol Biofuels 8:1–10
29. Gong F, Zhu H, Zhang Y et al (2018) Biological carbon fixation: from natural to synthetic. J CO_2 Util 28:221–227
30. Greening C, Cook GM (2014) Integration of hydrogenase expression and hydrogen sensing in bacterial cell physiology. Curr Opin Microbiol 18:30–38
31. Guadalupe-Medina V, Wisselink HW, Luttik MAH et al (2013) Carbon dioxide fixation by Calvin-Cycle enzymes improves ethanol yield in yeast. Biotechnol Biofuels 6:1–12
32. Herter S, Farfsing J, Gad'On N et al (2001) Autotrophic CO_2 fixation by *Chloroflexus aurantiacus*: study of glyoxylate formation and assimilation via the 3-hydroxypropionate cycle. J Bacteriol 183:4305–4316
33. Huber H, Gallenberger M, Jahn U et al (2008) A dicarboxylate/4-hydroxybutyrate autotrophic carbon assimilation cycle in the hyperthermophilic Archaeum Ignicoccus hospitalis. Proc Natl Acad Sci USA 105:7851–7856
34. Ishikawa C, Hatanaka T, Misoo S et al (2011) Functional incorporation of sorghum small subunit increases the catalytic turnover rate of Rubisco in transgenic rice. Plant Physiol 156:1603–1611
35. Kang RJ, Zhou WQ, Cai ZL et al (2000) Photoautotrophic cultivation of *Synechococcus* sp. PCC7002 in photobioreactor. Sheng Wu Gong Cheng Xue Bao 16:618–622
36. Keller MW, Schut GJ, Lipscomb GL et al (2013) Exploiting microbial hyperthermophilicity to produce an industrial chemical, using hydrogen and carbon dioxide. Proc Natl Acad Sci USA 110:5840–5845
37. Kim BW, Chang HN, Kim IK et al (1992) Growth kinetics of the photosynthetic bacterium *Chlorobium thiosulfatophilum* in a fed-batch reactor. Biotech Bioeng 40:583–592
38. Kirst HFC, Melis A (2014) Maximizing photosynthetic efficiency and culture productivity in cyanobacteria upon minimizing the phycobilisome light-harvesting antenna size. Biochim Biophys Acta-Bioenerg 10:1653–1664
39. Kletzin A, Urich T, Muller F et al (2004) Dissimilatory oxidation and reduction of elemental sulfur in thermophilic archaea. J Bioenerg Biomem 36:77–91
40. Kozaki A, Takeba G (1996) Photorespiration protects C3 plants from photooxidation. Nature 384:557–560
41. Kracke F, Vassilev I, Kromer JO (2015) Microbial electron transport and energy conservation —The foundation for optimizing bioelectrochemical systems. Front Microbiol 6:575
42. Lamont CM, Sargent F (2017) Design and characterisation of synthetic operons for biohydrogen technology. Arch Microbiol 199:495–503
43. Lan EI, Liao JC (2011) Metabolic engineering of cyanobacteria for 1-butanol production from carbon dioxide. Metab Eng 13:353–363
44. Lan EI, Liao JC (2012) ATP drives direct photosynthetic production of 1-butanol in cyanobacteria. Proc Natl Acad Sci USA 109:6018–6023
45. Larkum AWD (2010) Limitations and prospects of natural photosynthesis for bioenergy production. Curr Opin Biotech 21:271–276
46. Laurinavichene TV, Tsygankov AA (2001) H_2 consumption by *Escherichia coli* coupled via hydrogenase 1 or hydrogenase 2 to different terminal electron acceptors. FEMS Microbiol Lett 202:121–124
47. Li H, Opgenorth PH, Wernick DG et al (2012) Integrated electromicrobial conversion of CO_2 to higher alcohols. Science 335:1596
48. Li YJ, Wang MM, Chen YW et al (2017) Engineered yeast with a CO_2-fixation pathway to improve the bio-ethanol production from xylose-mixed sugars. Sci Rep 7:1–9
49. Lou WJ, Tan XM, Song K et al (2018) A specific single nucleotide polymorphism in the ATP synthase gene significantly improves environmental stress tolerance of *Synechococcus elongatus* PCC 7942. Appl Environ Microbiol 84

50. Lovley DR, Nevin KP (2013) Electrobiocommodities: powering microbial production of fuels and commodity chemicals from carbon dioxide with electricity. Curr Opin Biotechnol 24:385–390

51. Martinez A, Bradley AS, Waldbauer JR et al (2007) Proteorhodopsin photosystem gene expression enables photophosphorylation in a heterologous host. Proc Natl Acad Sci USA 104:5590–5595

52. Mattozzi M, Ziesack M, Voges MJ et al (2013) Expression of the sub-pathways of the *chloroflexus aurantiacus* 3-hydroxypropionate carbon fixation bicycle in *E. coli*: toward horizontal transfer of autotrophic growth. Metab Eng

53. Nakagawa S, Takai K (2008) Deep-sea vent chemoautotrophs: diversity, biochemistry and ecological significance. FEMS Microbiol. Eco. 65:1–14

54. Nakagawa S, Takaki Y, Shimamura S et al (2007) Deep-sea vent epsilon-proteobacterial genomes provide insights into emergence of pathogens. Proc Natl Acad Sci USA 104:12146–12150

55. Nevin KP, Hensley SA, Franks AE et al (2011) Electrosynthesis of organic compounds from carbon dioxide is catalyzed by a diversity of acetogenic microorganisms. Appl Environ Microbiol 77:2882–2886

56. Nevin KP, Woodard TL, Franks AE et al (2010) Microbial electrosynthesis: feeding microbes electricity to convert carbon dioxide and water to multicarbon extracellular organic compounds. mBio 1:e00103–00110

57. Niederholtmeyer H, Wolfstadter BT, Savage DF et al (2010) Engineering cyanobacteria to synthesize and export hydrophilic products. Appl Environ Microbiol 76:3462–3466

58. Nybo SE, Khan NE, Woolston BM et al (2015) Metabolic engineering in chemolithoautotrophic hosts for the production of fuels and chemicals. Metab Eng 30:105–120

59. Nürnberg DJ, Morton J, Stefano S, et al (2018) Photochemistry beyond the red limit in chlorophyll f-containing photosystems. Science 360:1210–1213

60. Rabaey K, Rozendal RA (2010) Microbial electrosynthesis-revisiting the electrical route for microbial production. Nat Rev Microbiol 8:706–716

61. Ragsdale SW (1997) The Eastern and Western branches of the Wood/Ljungdahl pathway: how the East and West were won. BioFactors 6:3–11

62. Rosenthal DM, Locke AM, Khozaei M et al (2011) Over-expressing the C3 photosynthesis cycle enzyme Sedoheptulose-1-7 Bisphosphatase improves photosynthetic carbon gain and yield under fully open air CO$_2$ fumigation (FACE). BMC Plant Biol 11:123

63. Sakimoto KK, Wong AB, Yang PD (2016) Self-photosensitization of nonphotosynthetic bacteria for solar-to-chemical production. Science 351:74–77

64. Sargent F (2016) The model NiFe-hydrogenases of *Escherichia coli*. Adv Microb Physiol 68:433–507

65. Savakis P, Hellingwerf KJ (2015) Engineering cyanobacteria for direct biofuel production from CO$_2$. Curr Opin Biotechnol 33:8–14

66. Schiel-Bengelsdorf B, Durre P (2012) Pathway engineering and synthetic biology using acetogens. FEBS Lett 586:2191–2198

67. Schulman M, Wood HG, Ljungdahl LG, et al (1972) Total synthesis of acetate from CO$_2$ V. determination by mass analysis of the different types of acetate formed from ^{13}CO$_2$ by heterotrophic bacteria. J Bacteriol 109:633–644

68. Schwander T, von Borzyskowski LS, Burgener S et al (2016) A synthetic pathway for the fixation of carbon dioxide in vitro. Science 354:900–904

69. Shi L, Dong HL, Reguera G et al (2016) Extracellular electron transfer mechanisms between microorganisms and minerals. Nat Rev Microbiol 14:651–662

70. Song K, Tan X, Liang Y et al (2016) The potential of *Synechococcus elongatus* UTEX 2973 for sugar feedstock production. Appl Microbiol Biotechnol 100:7865–7875

71. Strauss G, Fuchs G (1993) Enzymes of a novel autotrophic CO, fixation pathway in the phototrophic bacterium *Chloroflexus aurantiacus*, the 3-hydroxypropionate cycle. Euro J Biochem 215:633–643

72. Strauss G, Fuchs G (1993) Enzymes of a novel autotrophic CO_2 fixation pathway in the phototrophic bacterium *Chloroflexus-aurantiacus*, the 3-hydroxypropionate cycle. Eur J Biochem 215:633–643
73. Varman AM, Yu Y, You L et al (2013) Photoautotrophic production of D-lactic acid in an engineered cyanobacterium. Microb Cell Fact 12
74. Wang B, Wang J, Zhang W et al (2012) Application of synthetic biology in cyanobacteria and algae. Front Microbiol 3
75. Waterbury JB, Watson SW, Guillard RRL et al (1979) Widespread occurrence of a unicellular, marine, planktonic, cyanobacterium. Nature 277:293–294
76. Work VH, D'Adamo S, Radakovits R et al (2012) Improving photosynthesis and metabolic networks for the competitive production of phototroph-derived biofuels. Curr Opin Biotechnol 23:290–297
77. Yishai O, Bouzon M, Doring V et al (2018) In vivo assimilation of one-carbon via a synthetic reductive glycine pathway in *Escherichia coli*. ACS Synth, Biol
78. Yishai O, Goldbach L, Tenenboim H et al (2017) Engineered assimilation of exogenous and endogenous formate in *Escherichia coli*. ACS Synth Biol 6:1722–1731
79. Yu H, Li X, Duchoud F et al (2008) Augmenting the Calvin-Benson-Bassham cycle by a synthetic malyl-CoA-glycerate carbon fixation pathway. Nat Commun 2018:9
80. Zander U, Faust A, Klink BU et al (2011) Structural basis for the oxidation of protein-bound sulfur by the sulfur cycle molybdohemo-enzyme sulfane dehydrogenase SoxCD. J Bio Chem 286:8349–8360
81. Zhou J, Zhang F, Meng H et al (2016) Introducing extra NADPH consumption ability significantly increases the photosynthetic efficiency and biomass production of cyanobacteria. Metab Eng 38:217–227
82. Zhou J, Zhang HF, Zhang YP et al (2012) Designing and creating a modularized synthetic pathway in cyanobacterium *Synechocystis* enables production of acetone from carbon dioxide. Metab Eng 14:394–400
83. Zhuang ZY, Li SY (2013) Rubisco-based engineered *Escherichia coli* for in situ carbon dioxide recycling. Bioresour Technol 150:79–88

Enhanced Fixation of CO$_2$ in Land and Aquatic Biomass

Angela Dibenedetto

Abstract

Biomass, either terrestrial or aquatic, can efficiently fix CO$_2$ from a variety of sources, such as the atmosphere, power plant and industrial exhaust gases, and soluble (hydrogen)carbonate salts thanks to the enzyme Ribulose-1,5-bisphosphate carboxylase/oxygenase (RuBisCO). In addition to the carboxylation of ribulose that fixes CO$_2$ into glucose (*ca.* 60%), used by the biomass as a source of energy and for building cellulose, RuBisCO also oxidizes the substrate (*ca.* 40%). Attempts have been made to engineer the enzyme for an enhanced carboxylation. Besides, efficient light-using organisms are under deep investigation. In particular, enhanced bio-fixation using microalgae has recently become an attractive approach to CO$_2$ capture and C-recycling with a benefit derived from downstream utilization and application of the resulting microalgal biomass. This is a paradigmatic example of use of water and CO$_2$ for stepping from the linear- to the circular-C-economy. It is of importance to select appropriate microalgal species that have a high growth rate, high CO$_2$ fixation ability into valuable components, resistance to contaminants, low operation cost, and are easy to harvest and process. Strategies for the enhanced production of bioproducts and biofuels from microalgae, based on the manipulation of the strain physiology by controlling light, nutrient and other environmental conditions, which determine an efficient carbon conversion, are underway since long time. They are discussed in this Chapter together with an assessment of the value of the algal strain *P. Tricornutum*.

A. Dibenedetto (✉)
Department of Chemistry and CIRCC, University of Bari "Aldo Moro", Bari, Italy
e-mail: angela.dibenedetto@uniba.it

© Springer Nature Switzerland AG 2019
M. Aresta et al. (eds.), *An Economy Based on Carbon Dioxide and Water*,
https://doi.org/10.1007/978-3-030-15868-2_11

11.1 Introduction

Photosynthesis uses CO_2 and water for building a myriad of organics. Aquatic and terrestrial biomass absorb about 25% of carbon dioxide emissions during the Calvin cycle (a step of the photosynthesis process) where the enzyme Ribulose-1,5-bisphosphate carboxylase/oxygenase (RuBisCO) (Fig. 11.1) catalyzes the reaction that turns ribulose and CO_2 into glucose, used by the biomass as a source of energy and for building structural materials such as cellulose.

RuBisCO, one of the most abundant enzymes in the biosphere, is a key enzyme in the global carbon cycle. It is defined a "slow enzyme", in fact the turnover frequency of an average RuBisCO is only between 1 and 10 s^{-1} [1], representing a limiting factor in photosynthetic CO_2-fixation under optimal conditions. In particular, the enzyme RuBisCO catalyzes the addition of gaseous carbon dioxide to ribulose-1,5-bisphosphate (**1**) (Scheme 11.1). The reaction gives two molecules of glycerate-3-phosphate (**3**). RuBisCO is also an error-prone enzyme (Scheme 11.1) (error rate: 20–40%) [2–4], since it can capture oxygen instead of CO_2, slowing down the process of carbon dioxide fixation and producing 2-phosphoglycolate (2PG) (**4**) besides 3-phosphoglycerate (PGA) (**3**) (photorespiration).

PGA is the normal product of carboxylation, and productively enters the Calvin cycle. Phosphoglycolate, instead, inhibits certain enzymes involved in photosynthetic carbon fixation: hence, it is often said to be an 'inhibitor of photosynthesis'. Improving the selectivity of RuBisCO towards carboxylation would improve the overall CO_2 fixation rate, reducing its accumulation in the atmosphere. The majority of biomass on our planet (85%) are C_3 plants, which have no special features to combat photorespiration (oxygen uptake): the Calvin cycle occurs into a single compartment as depicted in Fig. 11.2.

Conversely, C_4 plants minimize photorespiration by separating initial CO_2 fixation and the Calvin cycle in space, performing these steps in different cell types; Crassulacean Acid Metabolism (CAM) plants minimize photorespiration and save water by separating these steps in time, between night and day (Fig. 11.3).

Fig. 11.1 The four subunits of Ribulose-1,5-bisphosphate carboxylase/oxygenase (RuBisCO) enzyme

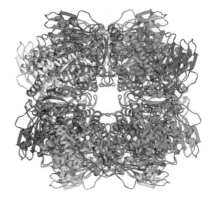

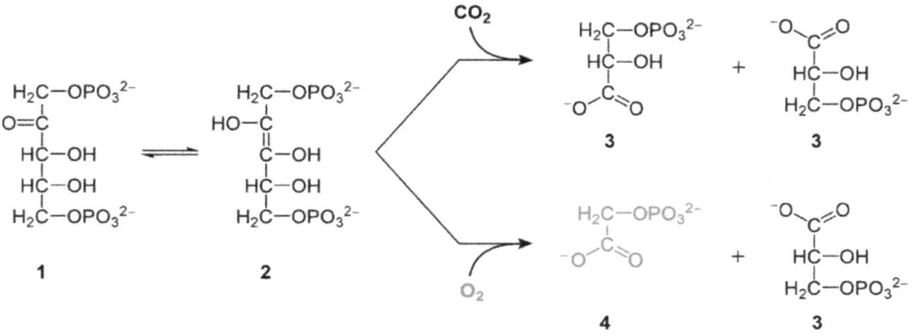

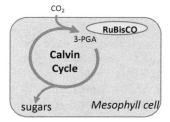

Scheme 11.1 Carboxylation of ribulose-1,5-bisphosphate (**1**) to glycerate 3-phosphate (**3**) and its oxygenation to afford (**3**) and 2-phosphoglycolate (**4**): both reactions are catalyzed by RuBisCO

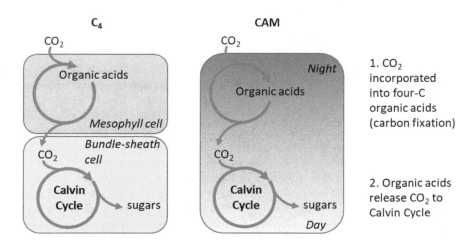

Fig. 11.2 The C_3 pathway

Fig. 11.3 The C_4 and CAM pathways

In C_3 plants, RuBisCO has a low catalytic activity, operates below its $K_m(CO_2)$ and is inhibited by O_2. In order to stimulate C_3 photosynthesis and reduce the photorespiration, it is possible to increase the CO_2/O_2 ratio. The effect of CO_2 enrichment is particularly evident with C_3 plants, but not with C_4 species [5]. Very often, an acclimation effect is observed with a reduction of the photosynthesis process (due to the reduction of the activity of RuBisCO).

Several efforts have been done to increase the biological CO_2-fixation [6]. Recently, faster RuBisCO with higher reaction rate and specificity [7–9] or transplanting natural CO_2-fixation pathways into non-autotrophic organisms, such as E. coli [10], have been realized.

Genetic manipulation of RuBisCO to increase its specificity for CO_2 would theoretically increase $A_{(max)}$ and improve the photosynthetic rate [11]. An essential step in RuBisCO activity is the carbamation of a lysine present close to the active site [12]. The carbamation process depends on the: (i) concentration of both CO_2 and Mg^{2+}; (ii) absence from the non-carbamated sites of certain phosphorylated compounds and particularly RuBP; (iii) activity of an enzyme called RuBisCO activase [13].

The chloroplast transformation processes have been of fundamental importance for the development of RuBisCO bioengineering [14]. In particular, it was necessary to understand the regulation of the chloroplast gene [15] and to make a catalytic screening of the RuBisCOs to select the most reactive forms [16–20]. To improve the assimilation of CO_2 two strategies have been considered, namely: (i) the implementation of CO_2 concentrating mechanism (CCM) for C_3 plants, and (ii) the generation of alternative metabolic pathways to bypass oxydation [21–24].

So far, the potential of RuBisCO itself has not been increased, but other approaches have been developed to create designed CO_2-fixation pathways. The objective is to create artificial CO_2-fixation routes with no kinetic or thermodynamic limitations. Studies report that the synthetic pathways could have significant quantitative advantages over their natural counterparts, such as an increased overall kinetic rate [25].

The drawback of such artificial systems is their realization: as a matter of fact, there is a gap between theory and realization of experiments in natural processes with living organisms.

A synthetic pathway to fix CO_2 in vitro has been proposed by Schwander et al. [26] who used a mix of seventeen enzymes derived from different microorganisms. Such enzyme-system seems more efficient (5 time more active) than the natural pathway and represents a possible route of enhanced artificial fixation of CO_2 for biotechnological applications. Further insights and studies are required for optimization. The transfer of such enhanced fixation ability into high plants, due to their complexity, represents a great challenge. Photosynthetic microorganisms such as microalgae may be more responsive to such modification, especially if mono-cellular organisms are considered. On the other hand, microalgae may respond to physical stress and modify their biochemical pathways. In following paragraphs attention will be devoted to the enhanced fixation of CO_2 into microalgae (and, in part, macroalgae) as they are more suitable than higher plants to biochemical pathways manipulation (even not considering genetic engineering not accepted in several countries).

11.2 Generalities About the Production and Nature of Aquatic Biomass

Compared with higher plants, microalgae have a greater ability to fix CO_2 in a variety of products, such as fine chemicals, proteins, fuels and materials of industrial interest [27]. They are characterized by the presence of various carbon-concentration mechanisms (CCMs) that allow to increase the local CO_2 concentration around RuBisCO [28]. In microalgae, CO_2 for reaching RuBisCO has to cross the cell wall, cell membrane, cytoplasm, chloroplast membrane, stroma and extracellular boundary layer [29] and this means that the most important limiting factors for CO_2 fixation are the resistance to CO_2 transportation and diffusion. In most cases CO_2 is up-taken by microalgae in form of gas, but a few cases exist where microalgae strains can assimilate HCO_3^- which is then converted to CO_2 by *Carbonic Anhydrase* (CA) [30–32]. Concerning macroalgae, it is very interesting that their photosynthesis is saturated at different levels of carbon dioxide, ranging from 500 to 2000 ppm [33], that means that with carbon dioxide concentration five times higher than the atmospheric one, under the correct light conditions and nutrient supply, macroalgae may grow with a better performance than that they show in natural environments [34]. The growth rate of algae is dependent on the temperature and the season (higher growth rate in summer and lower growth rate in winter).

CO_2 can be supplied into the algal suspension in the form of fine bubbles. In general, in this way, a lot of CO_2 is lost to the atmosphere and only 13–20% of CO_2 is effectively used. A different method to supply CO_2 is the gas exchanger which consists of a plastic frame which is covered by transparent sheeting and immersed in the medium. CO_2 is fed into the unit and the exchanger is floated on the water surface. Also if this is a most effective method, it presents the drawback that CO_2 should be very concentrated and pure.

The production of large quantities of biomass also requires large volumes of CO_2. A total of 1.8–2.2 t of CO_2 is needed to produce 1 t of dry algal biomass [35]. The CO_2 fixation efficiency for microalgae is about 5–8 times higher than for terrestrial biomass [36–38]: best for superior land plants is around 1.8–2%, while for microalgae ranges between 6–8% in nature and 8–12% in photobioreactors (PBR). In recent times, the use of microalgae as biodiesel source has been claimed. Based on the figures reported above, one can calculate that 1.3 billion tons of CO_2 would be required for the production of 0.4 billion m^3 of biodiesel to supply the European transportation market. The European Union produces about 6 billion tons of CO_2 per year [39]. The production of biodiesel from microalgae would go some way towards relieving one fourth of the CO_2 excess, a quite interesting figure.

Another important factor is water supply. In fact, it has been estimated that only for photosynthesis, one kg of produced biomass uses *ca.* 0.75 L of water [35]. Therefore, considering a content of 40% of lipid in the algal biomass, 1.9 L of water per liter of biodiesel would be required. The issue that will be discussed later in this Chapter is: is the production of biodiesel from microalgae economically sustainable?

A parameter of interest is the fact that microalgae can be adapted to endure the post-combustion power plants flue gas which typically contains 4–14% (or even more, depending on the fuel used) of CO_2 together with possibly toxic compounds such as NOx and SOy, at high flow rate and even high temperature (80–120 °C or higher) [40]. It has been reported [41] that high concentration of CO_2 may have an "anesthetic" effect on microalgal cells and, in general, algae growth. *Anabaena variabilis* grows quite normally with CO_2 concentration in the range 4–13%, but when the concentration of CO_2 is risen to 18% or higher the growth is completely inhibited. This is most probably due to the decreased pH of the medium at CO_2 concentration higher than 18%. On the contrary, some algal strains (very limited number) are tolerant to high concentration of CO_2. This is the case of the strain *Chlorella ZY-1* [42] that shows its maximum growth rate with CO_2 concentration around 15%. Interestingly enough, the growth rate and cell density remain very high also if the CO_2 concentration reaches 30 and 50%. When the concentration of CO_2 is 70% the growth rate decreases but the algae concentration still reaches 0.776 g/L after 6 days cultivation.

The use of a microalgal consortium may be considered of interest as it may have high potential for both CO_2 reduction and bio-oil production. In particular, it has been shown that a microalgal consortium (where the most abundant species were *Acutodesmus* (*Scenedesmus*) sp., *A. dimorphus* (Turpin) Tsarenko and *Scenedesmus obliquus* (Turpin) Kützing) [43] cultivated with 30% CO_2 was able to produce 21.1 mg L^{-1} day^{-1} of biomass and 4.8 mg L^{-1} day^{-1} of lipid (27.6% of dry weight, dw) with an ability to fix CO_2 of 0.0271 g CO_2 L^{-1} day^{-1} (higher than in the ambient air conditions). By using the same microalgal consortium cultivated under exhaust gas (19% CO_2) from a power generator supplied with biogas from chicken manure, the achieved biomass and lipid productivity were 25.82 and 5.2 mg L^{-1} day^{-1} (16.96% of dw), respectively.

11.3 Products From Microalgae

Assimilated CO_2 or carbon substances are potentially metabolized to a variety of bioproducts some of which can be used as biofuels, either directly or after processing. Bioproducts are widely used in food, pharmaceutical and feeding industries as well as in energy area. Microalgae are characterized by a large entropy of products, and this may represent a limitation to their use because of the economic and energetic cost of isolation of a single product. Carotenoids and fatty acids from microalgae have got increasing attention and their potential health benefits are being revealed [44–46].

In addition, microalgae have been for long time pointed as excellent source of biodiesel, bioethanol and biogas because of their higher photosynthetic efficiency and less land demand than higher plants [36].

In general, one may divide the products obtained from algae in two main categories:

– fuel products that include bio-oil, biodiesel, ethanol, methane and hydrogen.
– non-fuel products that include specialty chemicals, food and feed, nutraceuticals, personal care products, natural pigments and other novel products.

The former are low price and large market, the latter have a smaller market but consistently higher value.

11.3.1   Fuel Products

Microalgae are a potential source of renewable energy, and they can be converted into energy vectors such as biofuels, oil and biogas [47] or even hydrogen. Since microalgae have high water content (90%) they are often dried before use and not all biomass energy conversion processes can be applied. By using thermochemical processes, oil and gas can be produced, and by using biochemical processes, ethanol, biodiesel and bio-hydrogen can be produced [47]. The energetic value of algae as such is quite low (LHV ranging from 10 to 22 MJ/kg_{dw} depending on the content of lipids and carbohydrates) and the transportation cost quite high. It is more convenient to process them and produce a biofuel that can be transported. Among fuels produced from algae, biodiesel has attracted much attention, but algae cannot be evaluated only on the basis of their triglicerides (TG) or in general lipids content: all kinds of fuels that can be produced (e.g. ethanol, hydrogen gas or biogas) must be taken into consideration. This means that the direct combustion or applying pyrolysis is not the most convenient approach.

The quality and composition of the algal biomass determined with a careful analysis plan will suggest the best option for the biofuel to be produced. A biomass rich in lipids (30–70% dry weight) will be suitable for the production of biooil and biodiesel (hydrogenation maybe required, depending on the concentration of polyunsaturated fatty acids). Usually lipids are converted into Fatty Acid Methyl Esters (FAMEs, vide infra) better suited than lipids to be used in Diesel engines. A biomass rich in sugars will be better suited for the production of bioethanol. The anaerobic fermentation of the residual biomass after extraction of sugars, proteins, organics, etc. will produce biogas.

Several companies are engaged in producing fuels from macroalgae (Table 11.1). Because of the low lipid level, production of biofuels from macroalgae is expected to depend on conversion of carbohydrate feedstocks, rather than extraction of energy-rich oils that can be processed to biodiesel or

Table 11.1 Companies currently engaged in macroalgae to fuels production

Company	Activity
Algenol Biofuels	Production of ethanol
Solazyme	Oily genetically engineering algal strains
	Direct fermentation of biomass
Seaweed Energy Solutions	Production of biogas and bioethanol
Green Gold Algae and Seaweed Sciences, Inc	Production of ethanol
Butamax Advance Fuels-Dupont-BioArchitecture Lab-Statoil	Production of ethanol and butanol
Seambiotic Ltd	Production of ω-3 fatty acids
Blue Sun Energy	Production of jet fuel
Holmfjord AS	Production of biofuel
Oil Fox	Production of biofuel

hydrocarbons. Many companies have been for long time involved in algal biodiesel research and some major players have claimed large scale algal biodiesel production since the year 2010. Among these are Algenol Biofuels, Solix Biofuels, Sapphire Energy, Solazyme, and Seambiotic. Nevertheless, the production of biodiesel only from microalgae is not economically affordable (vide infra).

Algal biomass thermally processed to produce bio-oil also produce highly acidic compounds that make the oil not suited for a direct use as fuel. Lipids contain more than a single type of fatty-acid (FA), with variable number of unsaturations that make unlike the use of FAMEs as fuel. A biodiesel with good combustion properties should have maximum one unsaturated bond in the molecule [48]. Hydrogenation can be used to reduce the unsaturation number and produce a better-quality fuel with obvious increase of the production cost. It has been found that the degree of unsaturation may increase with the concentration of CO_2 bubbled in the pond/PBR [49, 50].

11.3.1.1 Production of Biodiesel

Different extraction methods can be used to isolate the lipid fraction from algae, some of them are most suited for micro- some others for macro-algae. They can be categorized as: (i) expeller/oil press (very suited for macroalgae, less for microalgae), (ii) extraction with solvent, (iii) extraction under supercritical conditions, (iv) extraction with the assistance of microwave and/or ultrasound techniques [51, 52]. After extraction the lipid fraction, constituted by neutral lipids (cytoplasm), glycolipids and phospholipids (cell membrane), is reacted with short-chain alcohols (methanol or ethanol) under suitable reaction conditions (transesterification process) to obtain the alkyl esters that can be used as biodiesel in existing engines. Usually, methanol is used and the products are said Fatty Acid Methyl Esters (FAMEs). The transesterification process is a catalyst-driven reaction and either acid or basic catalysts can be used, as discussed below. Alkaline catalysis is commonly used essentially because it is characterized by faster kinetics and lower cost than the acid catalysis. An important aspect to consider is the presence of free fatty acids (FFAs) and in general of water that can hydrolyze lipids.

In optimal conditions, the transesterification products are a mixture of fatty acids alkyl esters and glycerol [53, 54] (11.1). The reaction occurs into three different steps in which di- and mono-glycerides are formed as intermediates [54]. Although the stoichiometry of reaction requires a ratio triglyceride:alcohol of 1:3 mol/mol, transesterification is carried out with an excess of alcohol in order to increase the conversion yield and simplify the glycerol separation work-up.

Several research groups have studied the transesterification process by using as catalysts sulfonic [55] and sulfuric acids [54, 56, 57] that produce alkyl esters in high yield, but at long reaction times (20–50 h) at temperatures in the range 60–120 °C [58], conditions under which corrosion may occur. Moreover, the use of acid catalysts may favor the formation of free carboxylic acids [59], which, in principle, should be esterified in the reaction medium.

$$
\begin{array}{l}
\text{triglyceride} \xrightarrow[]{\text{H}_3\text{C—OH}} \longrightarrow \text{di-glyceride} \xrightarrow[]{\text{H}_3\text{C—OH}} \longrightarrow \text{mono-glyceride} \xrightarrow[]{\text{H}_3\text{C—OH}} \longrightarrow \text{glycerol} \\
\text{fatty acid methyl esters FAMEs}
\end{array}
$$

$$
(11.1)
$$

Conversely, alkaline catalysts are less corrosive and promote the transesterification reaction at higher rate [54]. A large number of industrial processes are based on the use of alkaline catalysts such as: sodium or potassium hydroxide, carbonates or alkoxides [60–62]. Nevertheless, basic catalysts present some drawbacks as they require refined oil (water free and with a low content of free fatty acids in order to avoid losses due to the formation of soaps) [63, 64] and produce large amounts of wastewater at the end of the process.

Such negativities are not in line with the production of sustainable biodiesel where the starting material should have low price and the conversion process should be based on simple technologies [65]. In order to fulfil such stringent requisites, our Research Group has developed since some years tunable mixed-oxides that work as catalysts in absence of water [66, 67] and produce a mixture of fatty acids alkyl esters and water-free glycerol [68]. The latter can be easily separated from the reaction medium and used for further reactions [69, 70].

The transesterification process can be also catalyzed by enzymes, a process that needs deep investigation before can match requisites for industrial application and can compete with chemical catalysis.

The attractiveness of biodiesel is its low environmental impact as it does not contain sulphur and does not emit SO_2. Moreover, both carbon monoxide and carbon dioxide emissions are reduced [71] with respect to fossil-diesel. Palm oil contains mainly saturated FAs and the biodiesel produced from it has found large acceptance in several countries.

It is also worth to mention that from aquatic biomass it is possible to obtain secondary metabolites with properties very close to fossil gasoline [72]. Terpenes (oligomers- or polymers- of isoprene) can be used directly in current gasoline engines without important modification [73]. As a matter of fact, terpenes are not used as gasoline substitute but find a much more remunerative utilization as cosmetics, pharmaceuticals and food additives [74, 75].

Interestingly, biodiesel can substitute the diesel fuels having similar properties [76–78].

11.3.1.2 Production of Bio-alcohol

Ethanol and other alcohols can be produced starting from algal biomass through a fermentation process. Aquatic biomass contains simple sugars and starch that can be easily fermented representing an alternative to the corn- or land plants cellulose-route. The field of third generation biofuels is a leading research theme today. The main algal strains used to obtain bio-ethanol are microalgae *Chlorella, Chlamydomonas, Scenedesmus, Spirulina*, which are rich of starch and glycogen [79–90] (Table 11.2). In the same way, macro-algae can be used to obtain bio-ethanol by converting their storage material into fermentable sugars [91, 92].

Recently, algal biomass genetically engineered to produce bio-ethanol has gained attention: indeed, genetic engineering has been and is successfully used in algae modification [93].

11.3.1.3 Production of Biomethane

Due to composition of cells, microalgae are a suitable substrate for anaerobic digestion, which gives biogas, a mixture of CH_4 and CO_2 with a ratio CH_4/CO_2 that depends on the feed and the operative conditions [94–96]. Depending on the microorganisms that drive the anaerobic digestion and the micro-algae used, different kinds of reactors with different characteristics have been designed [97]. Several studies have been published on the anaerobic digestion of green algae, cyanobacteria, euglenophyceans, diatoms and macroalgae [98].

Table 11.2 Starch content of some micro-algae

Algal strain	Starch (g/dry weight)/%	References
Clorella	12–37	[83–85]
Chlamydomonas r.	17–53	[83, 84, 86, 87]
Scenedesmus	7–24	[85, 88]
Spirulina	37.3–56.1	[89, 90]

Aquatic biomass may produce also biomethane with higher potential with respect to terrestrial biomass because of the absence of lignin and non-fermentable constituents and the higher productivity of algal biomass with respect to terrestrial ones. It has been found that from *Gracilaria sp.* and *Macrocystis* (kelp) is possible to obtain bio-methane with a yield (0.28–0.4 m^3 kg^{-1}) [99, 100] not far from the methane yield from organic waste (0.54 m^3 kg^{-1}) [101]. Nevertheless, bio-methanation of microalgae is not a standing alone process (vide infra).

11.3.2 Non-fuel Products

As reported in Table 11.3 different types of useful substances can be extracted from algal biomass.

All products listed in Table 11.3 have a high value and their recovery can increase the value of the aquatic biomass if biomass is grown for commercial purposes. If biomass is intended to reduce CO$_2$ emission, then a careful LCA analysis is necessary.

11.3.2.1 Polyunsaturated Fatty Acids

Consumer demand for omega-3 products will continue to grow briskly over 2020 and will influence the activities of manufacturers and marketers worldwide in supplying omega-3 products across various categories. It is well known that fish such as tuna, salmon, mackerel, herring as well as sardines are used as common sources of long-chain polyunsaturated fatty acids (PUFAs). Safety issues have been raised (EPA-US discourage pregnant women, infants and nursing mothers to eat shellfish and fish because of high Hg contamination) [102] about using these sources because of the possible accumulation of toxins in fish [103]. Table 11.4 reports the amount of mercury and omega-3 fatty acids assumed when eating common seafood products. Moreover, persistent typical fishy smell, unpleasant taste and poor stability to oxidation [104–107] of fish-derived PUFAs make algal omega-3 preferred.

However, micro-algae have a great potential as source of PUFAs [109, 110]. PUFAs of particular interest are γ-linolenic acid (GLA), arachidonic acid (AA),

Table 11.3 Main components of algae

Pigments/carotenoids	β-carotene, astaxanthin, lutein, zeaxanthin, canthaxanthin, chlorophyll, phycocyanin, phycoerythrin, fucoxanthin
Polyunsaturaed fatty acids (PUFAs)	DHA(C22:6), EPA(C20:5), ARA(C20:4), GAL(C18:3)
Antioxidants	Catalases, polyphenols, superoxide dismutase, tocopherols
Vitamins	A, B1, B6, B12, C, E, biotin, riboflavin, nicotinic acid, pantothenate, folic acid
Other	Antifungal, antimicrobial and antiviral agents, toxins, amino acids, proteins, sterols, MAAs for light protection, inorganics, amines

Table 11.4 Levels of mercury and omega-3 fatty acids in common seafood dishes [108]

Seafood	Average mercury level (ppm)	Omega-3 fatty acid content (g/3 oz. serving)
Canned tuna (light)	0.12	0.26–0.73
Shrimp	<0.01	0.27
Salmon	0.01	0.68–1.83
Scallops	0.05	0.17
Catfish	0.05	0.15–0.20
Lobster	0.31	0.07–0.41
Grouper	0.55	0.21
Halibut	0.26	0.40–1.00

Table 11.5 Companies that commercialize DHA oil extracted from micro-algae

Company	Algal species	Applications
Martek (USA)	*Crypthecodinium cohnii*	Infant milk, feed
OmegaTech (USA)	*Schizochytrium* sp.	Health food, feed
Bio-Marine (USA)	*Schizochytrium* sp.	Feed
Advanced BioNutrition	*Schizochytrium* sp.	Feed
Nutrinova (Germany)	*Ulkenia*	Health food

eicosapentaenoic acid (EPA) and docosahexaenoic acid (DHA) which extraction from microalgae is less expensive that the extraction from fish. Micro-algae rich in DHA include autotrophic (*Isochrysis, Pavlova*) and heterotrophic (*Cryptheco-dinium, Schizochytrium, Thraustochytrium, Auratiochytrium*) microorganisms. Several companies (Table 11.5) commercialize DHA oil extracted from micro-algae, that seems to be competitive even with that produced from fungal species.

11.3.2.2 Pigments

According to their structure, algal pigments can be classified as: (i) chlorophylls *a* and *b* (chlorins), (ii) chlorophyll *c* (porphyrins), (iii) phycobilipigments (open tetra-pyrrols), and (iv) carotenoids (poly-isoprenoids with terminal cyclohexane rings; carotenes and xanthophylls) (Table 11.6). The distribution of pigments is characteristic of each algal species. Cyanobacteria, for example, are characterized by the presence of chlorophyll *a*, phycobilins, carotenoids such as ε-carotene, lycopene, γ-carotene and β-carotene, and xanthophylls such as astaxanthin, canthaxanthin, β-cryptoxanthin, echinenone, myxoxanthophyll and oscillaxanthin [111]. Green microalgae (e.g. *Dunaliella*) and Cyanobacteria (e.g. *Spirulina*) [112, 113] are mostly used for the production of carotenoids (β-, ε-, α carotenes, xanthophylls) and phycobilins. Carotenoids are known for their antioxidant power [114] and anti-cancer activity [115, 116] and are used in food and beverage

Table 11.6 Structure of selected pigments

Chlorophyll *a* and *b*	Chlorophyll *c*
Chlorophyll a Chlorophyll b	
Phycobilipigments (open tetra-pyrrols)	Carotenoids

industries as colorants or supplement for human food and animal feed [80, 114]. In general the extraction yield is low if natural strains are used (only 0.1–2%), but if algal biomass is cultivated under best growth conditions the amount of carotenoids may rise up to 14% (*Dunaliella* is the most suitable strain for the production of β-carotene) [83, 117–120]. DSM N.V. and BASF SE are the major producers of β-carotene in a global market of over 432 MUS$ in 2015 with a forecast to exceed 500 MUS$ in 2023, with a focus in Europe (France and Germany) and Israel.

Natural β-carotene was first obtained around 1980 with the commercial production of *Dunaliella salina* in Australia. In 1994, natural β-carotene was extracted in small scale from algae and its cost was prohibitive with respect to that of synthetic one. With the increase of the number and size of plants producing β-carotene-rich *Dunaliella* the situation changed [118] and this sector is now economically viable [117, 121].

Xanthophylls derived from algae find utilization in several fields at commercial level: as colorant [86] and in medicine and food sector [122, 123]. Among xanthophylls, astaxanthin is mainly used as nutraceutical and anti-oxidant, with a major use as salmon feed. The annual worldwide aquaculture market of this pigment was estimated to be worth about 200 MUS$ in 2015 with an average price of 2500 US$/kg [124]. The best natural producer is *Haematococcus pluvialis* (up to 3% weight), a freshwater alga that normally grows in shallow natural basins. It requires a two-stage culture process: in the first the optimal growing conditions are set, in the second astaxanthin is produced, under intense light and nutrient poor conditions [83]. The price of the natural product is not at all competitive with that of synthetic forms [83, 119]. Several pilot plants have been built in order to make an economic evaluation of the astaxanthin production. Considering a theoretical plant with a production capacity of about 900 kg/y, a cost of astaxanthin from *Haematococcus* of about 718 $/kg [125] has been estimated. The global astaxanthin market size was estimated at US$ 555.4 million in 2016 with a share of 58% of natural product and a foreseen rise to 770 MUS$ by 2024.

The extraction process of astaxanthin is based on the use of supercritical CO_2. The manufacturer Valensa International started off as a supercritical CO_2 fluid extraction (SCFX) facility specializing in US Plus Saw Palmetto and in 2002 acquired La Haye Laboratories (producer of astaxanthin since 1986) along with patent called Deep Extract™ [126]. Valensa subsequently perfected the extraction and stabilization of astaxanthin becoming the first company to obtain European Novel Foods approval for their Zanthin® brand astaxanthin.

In vitro, these structures can be obtained through double Wittig condensation reacting a symmetrical C10-dialdehyde with two equivalent of C15-phosphonium salt (11.2) [127, 128].

$$(11.2)$$

The main producers of synthetic astaxanthin are BASF (Ludwigshafen, Germany) and Hoffman–La Roche (Basel, Switzerland) [129–131]. Differently from natural systems where only isomers 3S and 3'S are obtained [132], in vitro a mixture of different stereoisomers [(3S, 3'S), (3R, 3'S), (3S, 3'R), (3R, 3'R), in a 1:2:2:1 ratio is produced.

In order to estimate if natural astaxanthin can be produced at lower cost than synthetic astaxanthin, a pilot plant with two large scale outdoor photobioreactors and a raceway pond were built and operated for 2 years to produce astaxanthin from *Haematococcus*. The developed processes were scaled up to a hypothetical plant with a production capacity about 900 kg astaxanthin per year, and the economic assessment of the process was made. The analysis of data suggests that the production cost of astaxanthin and microalgae biomass can be as low as 718 US$/kg and 18 US$/kg, respectively. Such results were very encouraging because the optimized estimated cost was lower than that of chemically synthesized astaxanthin [133]. In an independent study, Nguyen [134] considered the economical, environmental, and societal impacts of astaxanthin production via chemical synthesis (Wittig reaction), yeast fermentation (*Phaffia rhodozyma*), and algal induction (*H. pluvialis*). Yeasts and microalgae produce not only astaxanthin but also other products. This comparative study confirms that the chemical synthesis is the cheapest at 40 US$/kg of astaxanthin, followed by yeast at 140 US$/kg and algae at 164 US$/kg. Land usage is about the same for yeast and algae, with chemical synthesis requiring about less than half the space. The most energy intensive was the yeast process, with algae being 20% less, and the synthetic process requiring only 1/10th with respect to yeast. Nevertheless, astaxanthin is produced at commercial level in Hawaii, India and Israel by Algatech from *Haematococcus* for pharmaceutical application [83, 135, 136].

Phycocyanins can be obtained from aquatic biomass, in particular from red algae and *Cyanobacteria,* and are used for cosmetics, food industries and biomedical applications [111, 137].

11.3.2.3 Proteins

A variety of proteins are present in algae and are located in different part of cells: for example, they are in the cellular wall, or are enzymes, eventually are linked to pigments and carbohydrates. Green algae are rich of glycoproteins and hydrox-yproline [138]. Some other strains contain protein levels very similar to meat or eggs. The content of protein in some macroalgae is around 40% dw, essentially they are rich in aspartic and glutamic acids [139]. The distribution depends on the season and temperature.

Algal proteins are characterized by several properties. For example pepsin-digests from *Porphyra yezoensis* have anti-mutagenic, blood sugar reducing, calcium precipitation inhibition, cholesterol lowering, antioxidant, and improved hepatic function activities [140]. They have also antibiotic, cytotoxic, anti-inflammatory, anti-adhesion, anti-HIV, reduction to pain sensitivity, human platelet aggregation inhibition and anti-cancer, antimicrobial, anti-viral and anti-hypertensive properties [141]. Proteins derived from algae are used essentially in human nutrition.

11.3.2.4 New Materials

Modern synthetic biology and genomics may greatly contribute to produce new chemicals and materials from algae with parallel increase of the biomass productivity. For example, the inclusion of selected bacterial genes into an alga may generate a new species able to produce selectively new components/materials (with high yield) that can be extracted and used to produce biomaterials (bioplastics) with different properties (biodegradability, compostability, sustainability, high processability, good mechanical properties), even meeting the required economic standards. Moreover, the new material, for effect of the engineering modification, can bear new functional groups able to interact with other reactive moieties or to form materials with very well defined properties as thermal stability or brittleness. In this way it is possible to produce 100% bio-copolymers using exclusively biomass-derived monomers. Alternatively, both monomers derived from fossil fuels and those derived from biomass can be copolymerized. Different kinds of polymers have been produced such as collagen, gelatine, alginates, casein, elastin essentially used for medical applications.

Currently, the market of bioplastics represents a tiny niche, 3.1 Mt/y or 1% of the global market (about 335 Mt/y) of plastics [142], but it is continuously growing with a forecast of about 12.62 Mt/y in 2023. Starting from algae it is possible to obtain a variety of bio-plastics such as: (i) hybrid plastics (mixing denatured algae biomass with conventional plastics derived from fossil-C); (ii) cellulose-based plastics where cellulosic algal strains are used as feedstock; (iii) poly-lactic acid

(PLA) where algal biomass produce lactic acid by microbial fermentation which is then polymerized to produce PLA; (iv) bio-poly-ethene where algal biomass is fermented to produce bio-ethanol used to produce bio-ethene. In order to exploit at commercial level such bioplastics, a techno-economic analysis is required.

Alginate

Alginate, a polysaccharide distributed widely in the cell walls of some algal strains, has been used as natural biopolymer as gelling agent and colloidal stabilizer in the food and beverage industry. It has also a strong potential in the area of drug delivery. Alginate polymers (Fig. 11.4) are extracted mainly from *Laminaria hyperborean, Ascophyllum nodosum,* and *Macrocysis pyrifera* [143]. They are characterized by having matrices with very high biodegradability as they can be depolymerized in very mild conditions [144].

Chitins and chitosan

Chitin and chitosan are biopolymers extracted form algae and other biosystems with several properties such as anti-bacterial, anti-fungi and anti-viral. They are non-toxic and non-allergenic. They are characterized by soft fibers with good breathability, absorbency, smoothness and used in food, cosmetics, biomedical and pharmaceutical applications. Chitin is the second most important natural polymer in the world.

Chitosan, a linear polysaccharide composed of randomly distributed β-(1 $\rightarrow$ 4)-linked D-glucosamine (deacetylated unit) and *N*-acetyl-D-glucosamine (acetylated unit), is obtained by chitin via de-acetylation (Fig. 11.5). It can be used as support

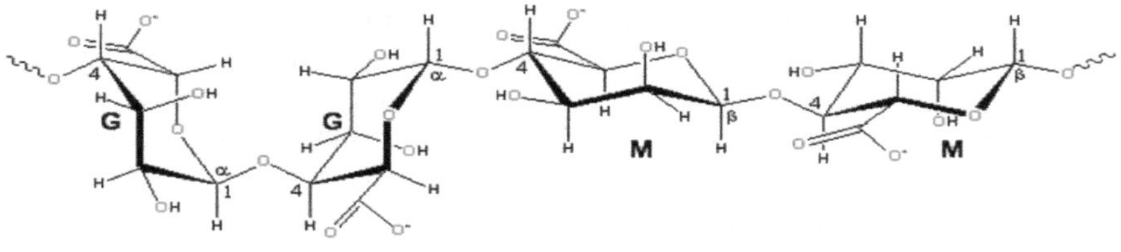

Fig. 11.4 Chemical structure of alginate

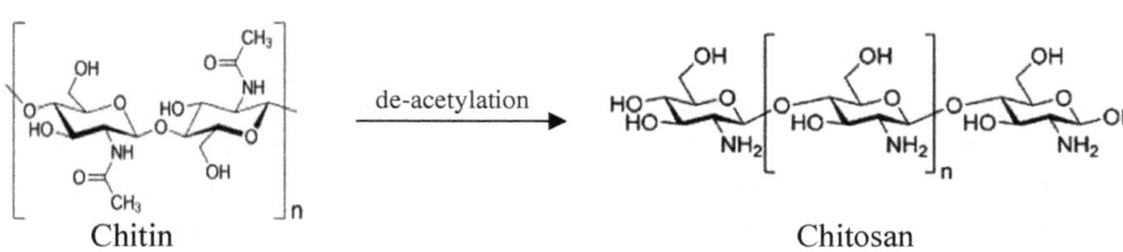

Fig. 11.5 Chitosan formation

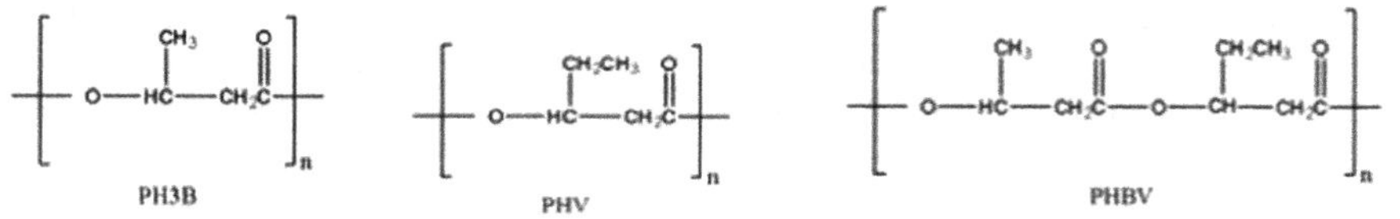

Fig. 11.6 PH3B: poly-3-hydroxybutyrate, PHV: poly-3-hydroxyvalerate, PHBV: Poly (3-hydroxybutyrate-*co*-3-hydroxyvalerate)

matrix, as gel, or in the form of beads also to cover alginate and to adsorb antibiotics. Chitosan can be used in food and beverage industry, as dehydrating agents in fruit juice or to reduce the acid content of the coffee drinks [145].

Poly-hydroxo-alkanoates

Poly-hydroxo-alkanoates (PHAs, Fig. 11.6) are a kind of polymers extracted from algae and other microorganisms.

They can be made by a single monomer such as hydroxobutyrric acid that forms polyhydroxobutirrates-PH3B consisting of 1000–30,000 hydroxy fatty acid monomers. The PH3B is a polyester with thermoplastic properties, with excellent biodegradability used as bioplastic for various industrial applications. Today, the highest levels of PHB synthesis in plants with fertile offspring are obtained in the plastid of *Nicotiana tabacum* resulting in up to 18% PHB of cellular dry weight. In general, the commercial production of PHAs is a high-cost process if heterotrophic species of bacteria are used. Alternatively, algae can be used as they are able to accumulate PHA. For example, *Cyanobacterium synechococcus elongates* cultivated under stress (nitrogen starvation) yields almost 18% of PHA by using 1% sucrose as carbon source [146]. Moreover, has been demonstrated [147] that reduction of the cost of the PHA production is possible by introducing the bacterial PHB pathway into the cytosolic compartment of the diatom *Phaeodactylum tricornutum*.

11.4 Commercial Production of Microalgae

Mass cultivation of microalgae (in particular *Chlorella*) has been studied during the 1950s [148] essentially in Japan as human food or animal feed [149, 150]. As an example, Cyanotech Co. [151] and DIC Co. [152] (formerly Dainippon Ink and Chemicals Inc.) have produced *Arthrospira platensis* for human nutrition. As said above, microalgae have been utilized as source of polyunsaturated fatty acids (PUFA, e.g. g-linolenic acid, arachidonic acid, eicosapentaenoic acid, and docosahexaenoic acid) and pigments (e.g. β-carotene, astaxanthin, and phycoery-thrin) [153].

Recently, several start-ups have been established with the focus to cultivate microalgae for biofuel production [154–156]. When comparing the production of algae for fuel production with that for production of added value products, one has to consider that fuels from algae must match two key requisites: cost must be comparable or lower than that of fossil-C derived fuels, and the CO_2 emission per unit of fuel produced must be quasi zero. Both are high barriers to exploitation of algae for only fuels production. To make the production of biofuel economically feasible a large production scale is required with a full exploitation of the potential of algal biomass. As a matter of fact fuels are the less valuable products that can be made from algae. However, main companies have diversified their business to high-value molecule production. For example, Terravia, previously known as Solazyme, has developed an innovative approach to use algal biomass (microalgae in particular) with the aim to make the production process very efficient under different aspects such as cost, scale up, time, sustainability. Terravia grows microalgae by using stainless-steel reactors, which offer advantages in terms of control, purity and consistency for the fermentation process while allowing to grow microalgae independently by the season and place.

Since the 1970s, sizeable research and development programs have addressed the production of microalgae and their use as a source of fuel [157–160]. These programs discovered many of the high-lipid microalgae strains and developed systems for algae cultivation. For example, the Japanese research program "Biological CO_2 Fixation and Utilization" addressed photobioreactor design for microalgae cultivation in the 1990s [158]. The cost of algal biodiesel cannot stand the price of fossil-diesel. For this reason, it is necessary to fractionate the algal biomass into different products and analyze the total value of the biomass. The best approach to algal biomass utilization is the application of the "biorefinery" concept that may average production costs and incomes from a variety of products.

11.5 Economic Evaluation of Microalgae

The production cost of aquatic biomass and the cost of the extracted products are influenced by a number of factors, such as: (i) *Location of the growing site*; (ii) *Nature of the container structure where algae are grown*; (iii) *Use of light and nutrients*; (iv) *Harvesting*; (v) *Drying*; (vi) *Processing*.

Either open ponds of photobioreactors can be used. When algae are grown in natural conditions they are able to receive light from sun, absorb CO_2 from the air and nutrients from the aquatic habitats. However, when algal biomass is grown under artificial conditions it is necessary to replicate and improve the natural conditions. Concerning light, growing algae for commercial use under natural conditions has the great advantage that light is free (sun), also if it is limited by day/night cycles and seasonability. At pilot scale stages, light can be supplied by fluorescent lamps [161] and it can be modulated for its wavelength that can be adapted to the algal strain and the target product. Supplying artificial light may assure a non-stop production, but it has a cost.

The requirement of N- and P-compounds and micronutrients is another key requisite. Considering the average content of nitrogen (7%) and phosphorus (1%) present in algal biomass and the EU consumption of biodiesel, it has been estimated that if the EU biodiesel were all derived from aquatic biomass the amount of nitrogen and phosphorus required would be about twice the amount available on the market as fertilizers in Europe [162]. The use of residual nutrients or the recycling of nutrients represents a way to a sustainable production of biodiesel from algae. In particular, the use of wastewater represents a good option: municipal wastewater, water from fishery ponds, wastewater from biogas plants and some industrial processes exempt from contaminants may have a high content of nitrogen and phosphorus. They need to be sanitized in order to avoid contamination of the aquatic biomass. Their use as growing medium allows the recovery and utilization of nutrients together with the remediation of wastewater that can be, at the end, reused or discharged in natural basins. Noteworthy, if nutrients are added to water, the biomass produced cannot be considered as a source of "zero C-emission fuel" because a large amount of carbon dioxide is emitted for the production of N- and P-nutrients. Therefore, the use of wastewater rich in N and P compounds brings significant economic benefits to algal growth in both ponds and photobioreactors.

Another key parameter is the use of water, particularly critical in temperate regions, which are poor of water while present ideal climatic conditions for growing algae. Besides the physiological amount discussed above, water is required in several other operations and can be recycled (cooling water in closed system, large amounts of water used as growing medium) or is lost (evaporation process that occurs in open ponds). To reduce the amount of fresh water, algae can be grown in marine water, even highly saline. Interestingly, because of the possibility of recycling water, algal biomass requires much less water for growing than land biomass (watering causes loss of water in soil): 10,000 L of water are required to obtain 1 L of biodiesel from crops [163].

After production, algal biomass needs to be harvested, the lipids and other products extracted, and the remaining cell components recovered. Harvesting costs depend on the biomass. For microalgae the cost is much higher than for macroalgae. Due to the size of the microalgae cells, centrifugation is often used as a preferred harvesting method. However, because the biomass concentration is generally low (<3 g/L), centrifugation of diluted streams requires a large capacity of the centrifuge, which makes the process much energy-demanding and expensive. Flocculation, followed by sedimentation and flotation, before centrifugation or filtration will substantially reduce the volume to treat and, thus, harvesting costs and energy requirements. Harvested biomass must be dried before processing. Drying can be carried out using a variety of cheap techniques alternative to electric heating, such as solar power, use of residual industrial heat and other cheap sources of thermal energy. Energy consumption in this step can deeply influence the cost of production of algae and algae-sourced products as demonstrated by a LCA study [164].

After harvesting and drying, the extraction procedure starts. Most microalgae strains are, in general, relatively small and have a thick cell wall. To break the cells different systems can be used (e.g., mechanical, chemical, and physical). This step may affect the functionality of some compounds like proteins present in the algal cell so that particular care must be put in selecting the most suited technology according to the target products. After disruption, oil can be extracted with organic solvents or with more environmentally benign, but more expensive, solvents (e.g., supercritical CO_2) even coupled to microwave and ultrasounds. Recently, the extraction in water-amine-CO_2 has been attempted in order to avoid drying of the biomass [165]. The scientific community has to develop new methodologies (mild cell disruption, extraction, and separation) that retain the functionality of the different cell components (e.g., proteins, carbohydrates, ω-3 fatty acids, pigments, and vitamins).

Fractionation of microalgae is an approach for the sequential recovery of classes of compounds (proteins first, then lipids followed by carbohydrates) maintaining the functionality of each of them.

Several efforts have been done to make an economic evaluation of the use of microalgae as source of biofuels and chemicals. As an example, Davis et al. [166] reported the cost for capital (CAPEX) and operational (OPEX) investment comparing PBR and open ponds showing that economics are driven more by CAPEX than by OPEX for both open pond and PBR.

Considering the data reported in the literature [166–168], one can see that the installation of PBRs is more expensive than open ponds even considering the various types of open pond (raceway, circular, etc.) and their characteristics (water depth, mixing system, operational days). The cost of PBR varies according to several parameters: kind of tubes, length, degassing systems, cooling systems, etc. A key difference between the two techniques is the amount of water (PBR use only 30% of the amount used in open ponds). PBR can represent an optimal solution in water-limited areas.

Considering all costs, it seems that for open pond CAPEX are smoothly distributed across the whole system, while for PBR the CAPEX depend mainly on the equipment ($\sim 80\%$) [166–168].

On the basis of literature data, the minimum selling price for crude TAGs (triacylglycerides) or hydrocarbon fuels ("Diesel,") (Fig. 11.7), assuming a production of 38,000 m^3 of crude algae oil per year [166–168], was calculated to range between 8 and 20 US$/gal (2.1–5.3 US$/L): too high with respect to fossil-C derived fuels.

In this scenario, for PBR the capital costs (depreciation) are of main importance contributing significantly to the overall cost of production. This theoretical study suggests that even on this relatively large scale of production, algae-based fuels cannot compete with petroleum-based fuels in the foreseeable future [72].

The market price of algal biomass and its components is not straightforward to calculate because it depends on several factors such as the geographic area where the biomass is grown, the market demand, the nature of products and required purity. In the literature, algal biomass is reported to have, on large-scale,

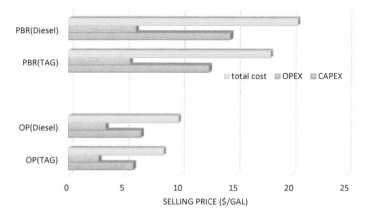

Fig. 11.7 Minimum selling price required to achieve 10% rate for algal TAG and Diesel production. OP = open ponds, PBR = photobioreactors

a production cost that ranges between 0.50 and 6 €/kg$_{dw}$ [169, 170]. In particular, it has been reported that the production cost of algal biomass may vary depending on the cultivation method (open ponds, horizontal tubular photobioreactors and flat panel photobioreactors) from 4 to 6 € per kg$_{dw}$ [72]. Key factors are the light supply, the mixing, the type and cost of cultivation medium, the cost of nutrients. Considering these factors and optimizing the productivity, it is possible to reduce the cost down to €0.68 per kg$_{dw}$ [171].

Wijffels et al. [172] have defined that application of the biorefinery concept may make algae-based biofuel economically viable [169]. In order to make an evaluation of the global algal biomass based on its components a value has been attached to each component on the basis of its use. The lipid fraction used as a feedstock for the energy or chemical industry was estimated to have a value of *ca.* 2 €/kg while the general market price for biodiesel production was set at 0.50 €/kg [172]. The protein fraction used as food or feed supplement was given a value of 5 and 0.75 €/kg, respectively. Carbohydrates (sugars), used as chemical building blocks were priced at €1/kg. PUFAs are also valuable products that have a quite high value and cost due to the complex processing for isolation. Taking into consideration the above reference market prices, an economic evaluation of *T. obliquus* and *P. tricornutum* cultivated in the local Southern Italy climatic conditions was carried out in our laboratory [173]. Between the two, the latter was much more suited for minimizing production costs and we have carried a full evaluation of products obtainable from it after fractionation and deep analytical characterization of each fraction. The value of 1.1 €/kg$_{dw}$ as non-fully optimized value for growing microalgae under the conditions used by Buono et al. was considered [174].

In our study we considered that microalgae are multitask and should be fully exploited. So, for example, saturated fatty acids (SFAs), such as stearic and palmitic acids, can be used not only for the production of biodiesel, but also for the production of detergents, soaps, and cosmetics such as shampoos and shaving cream, or even, as reported by Lai et al. [175], or as plasticizers for zein sheets. According to the use, a quite different income is foreseeable: an interesting exercise for optimizing the income is a correct allocation of each fraction to a given use. The use of microalgae as source of fuels is the less remunerative one can imagine, this is now clear. Moreover, the production of fuels has to match two must: have a cost comparable to, if not lower than, fossil fuels and reduce the CO_2-emission. As already said, the former objective is difficult to be reached, the latter is even more complex due to the energy required for growing, harvesting, processing microalgae. The production of chemicals has similar economic constraints as it must reduce both the cost and emissions (less waste and lower energy input) with respect to synthetic products: this is not always verified, as discussed above. The separation of various valuable components may make economic the growth and use of algal biomass even for the partial production of selected fuels. The same fraction could be diversified in use for increasing the income. For example, mono-unsaturated fatty acids (MUFAs) can be used for making biodiesel or can be converted into mono-carboxylic and di-carboxylic acids [176] (Scheme 11.2). Specifically, oleic acid oxidative cleavage affords di-carboxylic acids such as azelaic acid (AA) and mono-carboxylic acids such as pelargonic acid (PA), both of considerable industrial importance [72, 176].

The former finds application in the production of polymers such as: polyamides, polyesters, plasticizers (used for packaging, electronics, textiles and automotive [177]), pharmaceuticals, lubricants, hydraulic fluids. Pelargonic acid is used for the preparation of lubricating oil, alkydresin and perfumes [178], or can find application as bioagrochemical, namely as herbicide. Di-unsaturated fatty acids (DUFAs) such as linoleic acid can be selectively hydrogenated into either MUFAs or even SFAs used in the cosmetic and pharmaceutical industry [179–181].

The fractionation of biomass, as said above, produces, besides lipids, proteins and carbohydrates. The former could find application as animal feed and human food supplement (more complex treatment and higher production cost), since nutritional and toxicological evaluations have demonstrated the suitability of microalgae biomass is such applications [182].

Algal carbohydrates are mainly composed of various kinds of polysaccharides such as starch, cellulose/hemicelluloses: noteworthy, they do not contain lignin making the hydrolysis of cellulose less problematic [183] than with land plants. Conventionally, starch and glucose are used for bioethanol production, while

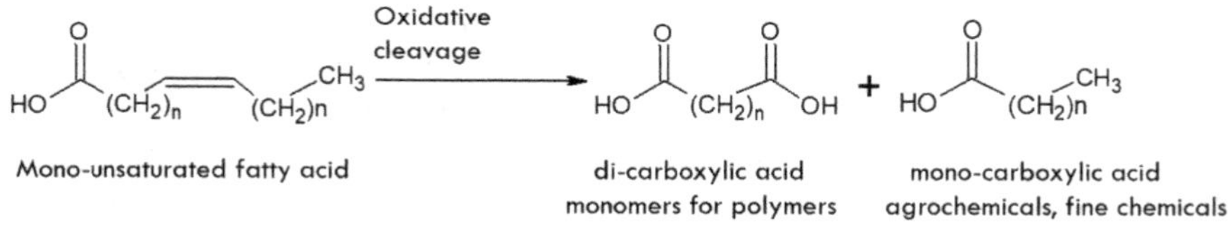

Scheme 11.2 Oxidative cleavage of mono-unsaturated fatty acids (MUFAs)

polysaccharides find many applications in food, textiles, cosmetics, as emulsifiers, lubricants, and clinical drugs [184].

Noteworthy, glucose can be converted into 5-HMF (Scheme 11.3) that plays a key role as platform molecule. The production of the latter molecule is a two-step process: isomerization of glucose into fructose, which is base catalyzed, and dehydration of fructose, an acid catalyzed process.

Bottlenecks in such process is the formation of polymeric humins, which are formed especially when the reaction is carried out in water. Such solids, in addition to cause loss of the starting reagent reducing the yield of the process, can affect the catalyst activity as they can deposit on its surface (when heterogeneous catalysts are used) and deactivate it [185]. In order to improve the catalyst life and yield of production of 5-HMF, organic solvents can be used as extracting agent of the target product from water [186]. Their correct choice plays a key role, as they increase the efficiency of the catalysts and their recyclability [186]. 5-HMF is a platform molecule from which several monomeric compounds can be derived (Scheme 11.4). 5-HMF bears two moieties, the alcoholic and the aldehydic, which can undergo oxidation and reduction processes. To perform in a selective way the oxidation of the former or the latter moiety without touching the other is a challenging task. Each of the oxidation products shown in Scheme 11.4 has a specific industrial application, as fine chemical, intermediate or monomer for polymers. A key issue in such conversion reaction is the choice of the catalyst, oxidant and reaction medium. Our target is to use cheap and abundant catalysts and oxygen or

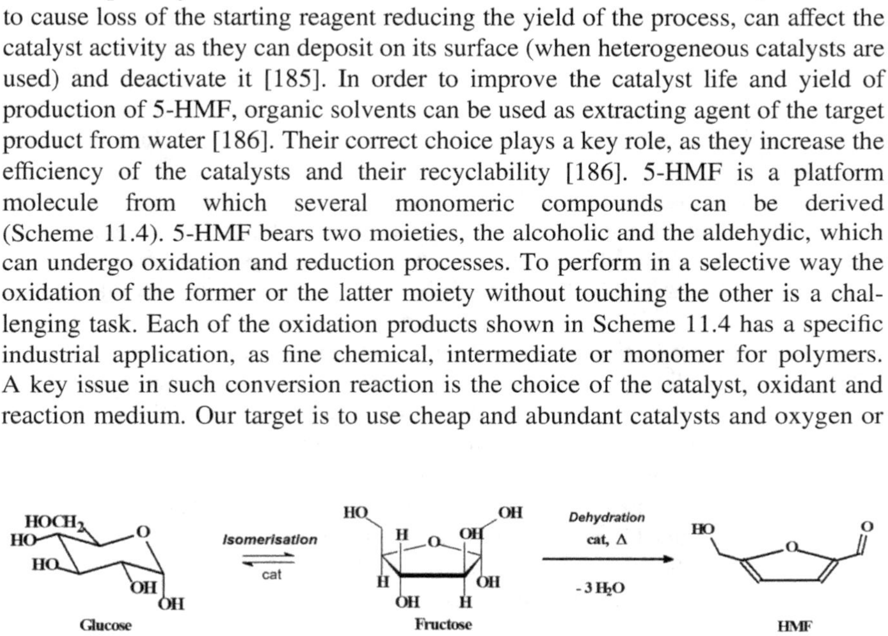

Scheme 11.3 Isomerization of glucose into fructose and its dehydration

Scheme 11.4 Products of side-chain oxidation of 5-HMF

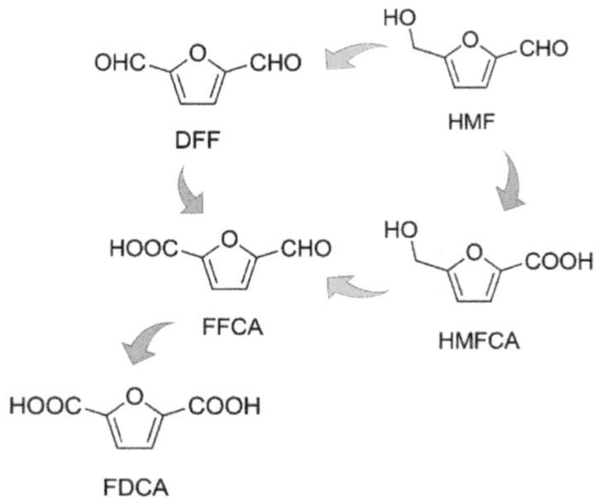

air as oxidant, working in water [187–191]. Under such conditions, we have developed catalysts that convert quite selectively HMF into FFCA [187], DFF [188, 189], FDCA [190] or even cleave the ring to afford oxalic (OA) and succinic acid (SA) [191]. Each derivative is produced with a yield and selectivity close to 100%, by using a customized catalyst, using oxygen as oxidant in water.

5-HMF can be also converted into levulinic acid (LA) and formic acid (FA) that find several applications, including the use as precursor of biofuels. The latter conversion requires hydrogen that must be cheap and derived from a non fossil source, if the reduction of CO_2 emission is targeted.

Noticeably, both microalgal strains *T. obliquus* and *P. tricornutum* mentioned above are rich in PUFAs such as EPA (eicosapentaenoic acid, C20:5) and ALA (α-linolenic acid, C18:3), which have a high market value [72] due to their important role for cancer prevention and cardiac protection and find medicinal applications.

After such general introduction, let us go back to the target microalga *P. tricornutum* and draw some conclusions. The profit with pond cultivation may be higher than when using PBR technology, due to lower capital and possibly operational costs. Nevertheless, because of the strong influence of external factors (temperature, photoperiod, CO_2 transfer rate, evaporation loss, contamination), waving biomass productivities were observed during the cultivation in open pond [174], which made the cultivation in PBRs (indoor or outdoor) preferred. *P. tricornutum* seemed to be a more suitable strain than *T. obliquus* from an economic point of view, since its cultivation in indoor photobioreactors under standard conditions produced a larger amount of biomass [174] than *T. obliquus* with an interesting amount of fractions (lipids and carbohydrates) that could be transformed into valuable products through total fractionation.

As mentioned above, a factor that influences the microalgae production cost is the price of nutrients and micronutrients that can be recovered from wastewater. However, in order to increase the profit we did grow microalgae using sanitized-wastewater (municipal or from a biogas process) as medium. Sanitation is necessary for eliminating potential contaminants.

In order to quantify the value of the biomass produced in our laboratory, the work of Norsker et al. [171] was taken as reference. In particular, after fractionation a complete analysis of each fraction was carried out, with great emphasis on lipids and the determination of the abundance of SFAs, MUFAs, DUFAs and PUFAs and their chain length in order to decide the best use of each fraction. We found that the composition of the algal biomass was: lipid (40–45%), protein (~50%) and carbohydrates (10–12%) [173, 174]. The economic value of the products extracted from *P. tricornutum* was determined on the basis of their use. For example, the value of lipids was set in the range 2–0.50 €/kg according to the fact that they were considered for use as raw materials for the chemical industry (we allocated 25% of the total to this use) or as biodiesel (75%). Similarly, the value of proteins fraction varied from 5 €/kg (if they were used as food) to 0.75 €/kg (if they were used as feed, 50% allocation), while carbohydrate fraction had an estimated value of 1 €/kg.

Benefits were calculated from wastewater treatment. A value of 0.14 and 0.1 €/kg$_{dw}$ of algae produced for removal of nitrogen compounds and phosphate, respectively, was estimated. Moreover, oxygen produced by algae was estimated at 0.16 €/kg$_{dw}$.

Considering all products obtained the calculated value of the biomass was 1.65 €/kg of algae with respect to a production cost for microalgae growth and processing that varied from 1.1 €/kg (non optimized value) to 0.68 (optimized) €/kg [173, 174], or 0.40 €/kg [171]. The fractionation of biomass and the optimized use of various components affords an economic benefit. Conversely, if from algae only biodiesel was produced a value of 0.37 €/kg was obtained with a net loss.

Recently, Benemann and others have assessed the production of biodiesel and biogas from microalgae and have demonstrated that such single option is not economically feasible [192]. Therefore, algal biomass cultivation for biodiesel production alone is not feasible from the economic point of view, while the application of the biorefinery concept may represent a chance to the use of aquatic biomass [173, 193].

Actually, the use of microalgae to produce biodiesel is very limited with respect to terrestrial biomass. Microalgae are essentially used for the production of carotenoids, ω-3 fatty acids and in general for high added value products used in nutraceutical and in food sectors which global market is around €1.25 billion with an average market value for microalgae of €250/kg$_{dw}$ biomass much higher than the production cost [194]. The production of biodiesel is much more conveniently done by using terrestrial oleaginous crops: the world production of palm oil is estimated around 40 million tons with an average market price of ∼0.50 €/kg [195].

11.6 Conclusions

Aquatic biomass is the most effective example that it would be possible to exploit, under the due circumstances and conditions, an *economy based on water and CO$_2$*. Algae can be used as source of fine chemicals, inorganics, proteins, monomers for polymers, liquid (biodiesel, alcohols) and gaseous (methane, H$_2$) fuels and can greatly contribute to the sustainability of the chemical and polymer industry and produce some fuels for the transport sector. The potential is quite high, supposed that waters rich of nutrients are used (process water from biogas plants, waters from fisheries, industrial waters, municipal sanitized waters, marine water, and so on) as growing medium for reducing the cost and impact of growing microalgae. Due to the high cost of growing and harvesting/drying/processing, it is essential that all components of the biomass are used at their best. However, the application of the Biorefinery concept is essential for algal biomass may deliver their best economic value.

References

1. http://brenda-enzymes.org
2. Sharkey TD (1988) Estimating the rate of photorespiration in leaves. Physiol Plant 73:147–152
3. Sage RF, Sage TL, Kocacinar F (2012) Photorespiration and the evolution of C4 photosynthesis. Annu Rev Plant Biol 63:19–47
4. Walker BJ, VanLoocke A, Bernacchi CJ, Ort DR (2016) The costs of photorespiration to food production now and in the future. Annu Rev Plant Biol 67:107–129
5. Bowes G (1991) Growth at elevated CO_2: photosynthetic responses mediated through Rubisco. Plant, Cell Environ 14(8):795–806
6. Erb TJ, Zarzycki J (2016) Biochemical and synthetic biology approaches to improve photosynthetic CO_2-fixation. J Curr Opin Chem Biol 34:72–79
7. Lin MT, Occhialini A, Andralojc PJ, Parry MAJ, Hanson MR (2014) A faster Rubisco with potential to increase photosynthesis in crops. Nature 513(7519):547–550
8. Greene DN, Whitney SM, Matsumura I (2007) Artificially evolved *Synechococcus PCC6301* Rubisco variants exhibit improvements in folding and catalytic efficiency. Biochem J 404(3):517–524
9. Kreel NE, Tabita FR (2015) Serine 363 of a Hydrophobic Region of Archaeal Ribulose 1,5-bisphosphate carboxylase/oxygenase from *Archaeoglobus fulgidus* and *Thermococcus kodakaraensis* affects CO_2/O_2 substrate specificity and oxygen sensitivity. PLoS One 10(9): e0138351, 1–25
10. Mattozzi MD, Ziesack M, Voges MJ, Silver PA, Way JC (2013) Expression of the sub-pathways of the *Chloroflexus aurantiacus* 3-hydroxypropionate carbon fixation bicycle in *E. coli*: toward horizontal transfer of autotrophic growth. Metab Eng 16:130–139
11. Reynolds MP, van Ginkel M, Ribaut JM (2000) Avenues for genetic modification of radiation use efficiency in wheat. J Exp Bot 51:459–473
12. Lorimer GH, Miziorko HM (1980) Carbamate formation on the ϵ-amino group of a lysyl residue as the basis for the activation of ribulosebisphosphate carboxylase by CO_2 and Mg^{2+}. Biochemistry 19:5321–5324
13. Portis AR (1992) Regulation of ribulose 1,5-bisphosphate carboxylase/oxygenase activity. Annu Rev Plant Physiol Plant Mol Biol 43:415–437
14. Bock R (2014) Genetic engineering of the chloroplast: novel tools and new applications. Curr Opin Biotechnol 26:7–13
15. Germain A, Hotto AM, Barkan A, Stern DB (2013) RNA processing and decay in plastids. Wiley Interdisc Rev: RNA 4:295–316
16. Whitney SM, Andrews TJ (2001) The gene for the ribulose-1,5-bisphosphate carboxylase/oxygenase (RuBisCO) small subunit relocated to the plastid genome of tobacco directs the synthesis of small subunits that assemble into RuBisCO. Plant Cell 13:193–205
17. Whitney SM, Andrews TJ (2001) Plastome-encoded bacterial ribulose-1,5-bisphosphate carboxylase/oxygenase (RuBisCO) supports photosynthesis and growth in tobacco. Proc Natl Acad Sci USA 98:14738–14743
18. Galmes J, Kapralov MV, Andralojc PJ, Conesa MA, Keys AJ, Parry MA, Flexas J (2014) Expanding knowledge of the RuBisCO kinetics variability in plant species: environmental and evolutionary trends. Plant, Cell Environ 37:1989–2001
19. Orr D, Alcantara A, Kapralov MV, Andralojc J, Carmo-Silva E, Parry MA (2016) Surveying RuBisCO diversity and temperature response to improve crop photosynthetic efficiency. Plant Physiol 172:702–717
20. Sharwood RE, Ghannoum O, Whitney SM (2016) Prospects for improving CO_2 fixation in C_3 crops through understanding C_4 RuBisCO biogenesis and catalytic diversity. Curr Opin Plant Biol 31:135–142
21. von Caemmerer S, Quick WP, Furbank RT (2012) The development of C4 rice: current progress and future challenges. Science 336:1671–1672

22. Price GD, Howitt SM (2014) Plant science: towards turbocharged photosynthesis. Nature 513:497–498
23. Hanson MR, Lin MT, Carmo-Silva AE, Parry MA (2016) Towards engineering carboxysomes into C3 plants. Plant Journal 87:38–50
24. Long BM, Rae BD, Rolland V, Forster B, Price GD (2016) Cyanobacterial CO concentrating mechanism components: function and prospects for plant metabolic engineering. Curr Opin Plant Biol 31:1–8
25. Bar-Even A, Noor E, Lewis NE, Milo R (2010) Design and analysis of synthetic carbon fixation pathways. Proc Natl Acad Sci USA 107(19):8889–8894
26. Schwander T, von Borzyskowski LS, Burgener S, Cortina NS, Erb TJ (2016) A synthetic pathway for the fixation of carbon dioxide in vitro. Science 354(6314):900–904
27. Dibenedetto A (2011) The potential of aquatic biomass for CO₂-enhanced fixation and energy production. GHG 1(1):58–71
28. Raven JA (2010) Inorganic carbon acquisition by eukaryotic algae: four current questions. Photosynth Res 106:123–134
29. Calvin M (1989) Forty years of photosynthesis and related activities. Photosynth research 21:3–16
30. Giordano M, Beardall J, Raven JA (2005) CO₂ concentrating mechanisms in algae: mechanisms, environmental modulation, and evolution. Annu Rev Plant Biol 56:99–131
31. Colman B, Huertas IE, Bhatti S, Dason JS (2002) The diversity of inorganic carbon acquisition mechanisms in eukaryotic microalgae. Funct Plant Biol 29:261–270
32. Wang B, Li Y, Wu N, Lan C (2008) CO₂ bio-mitigation using microalgae. Appl Microb Biotechnol 79(5):707–718
33. Brown DL, Tregunna EB (1967) Inhibition of respiration during photosynthesis by some algae. Can J Bot 45:1135–1143
34. Aresta M, Alabiso G, Cecere E, Carone M, Dibenedetto A, Petrocelli A (2005) VIII conference on carbon dioxide utilization, Oslo, Book of abstracts, 56, 20–23
35. Kliphuis AM, de Winter L, Vejrazka C, Martens DE, Janssen M, Wijffels RH (2010) Photosynthetic efficiency of *Chlorella sorokiniana* in a turbulently mixed short light-path photobioreactor. Biotechnol Prog 26(3):687–696
36. Chisti Y (2007) Biodiesel from microalgae. Biotechnol Adv 25(3):294–306
37. Costa JAV, Linde GA, Atala DIP, Mibielli GM, Krüger RT (2000) Modelling of growth conditions for cyanobacterium Spirulina platensin in microcosms. World J Microb Biotecnol 16(1):15–18
38. Dibenedetto A, Colucci A (2015) In: Aresta M, Dibenedetto A, Dumeignil F (eds) Biorefineries: an introduction. Berlin/Boston, Walter de Gruyter GmbH & Co KG, pp 57–77
39. European Environment Agency, Greenhouse Gas Emission Trends and Projections in Europe (2018) EEA report 16. EEA, Copenhagen, Denmark. ISSN 1977-8449
40. Zhao B, Su Y (2014) Process effect of microalgal-carbon dioxide fixation and biomass production: a review. Renew Sustain Energy Rev 31(1):121–132
41. Yoon JH, Sim SJ, Kim M-S, Park TH (2002) High cell density culture of *Anabaena variabilis* using repeated injections of carbon dioxide for the production of hydrogen. Int J Hydrogen Energy 27:1265–1270
42. Yue L, Chen W (2005) Isolation and determination of cultural characteristics of a new highly CO₂ tolerant fresh water microalgae. Energy Convers Manag 46:1868–1876
43. Boonma S, Chaiklangmuang S, Chaiwongsar S, Pekkoh J, Pumas C, Ungsethaphand T, Tongsiri S, Peerapornpisal Y (2015) Enhanced carbon dioxide fixation and bio-oil production of a microalgal consortium. Clean Soil Air Water 43(5):761–766
44. Liu XJ, Luo QX, Rakariyatham K, Cao Y, Goulette T, Liu X, Xiao H (2016) Antioxidation and anti-ageing activities of different stereoisomeric astaxanthin in vitro and in vivo. J Funct Foods 25:50–61
45. Gong M, Bassi A (2016) Carotenoids from microalgae: a review of recent developments. Biotechnol Adv 34:1396–1412

46. Santos-Sanchez NF, Valadez-Blanco R, Hernandez-Carlos B, Torres-Arino A, Guadarrama-Mendoza PC, Salas-Coronado R (2016) Lipids rich in omega-3 polyunsaturated fatty acids from microalgae. Appl Microbiol Biotechnol 100:8667–8684
47. Amin S (2009) Review on biofuel oil and gas production processes from microalgae. Energy Convers Manag 50:1834–1840
48. Renaud SM, Luong-Van JT (2006) Seasonal variation in the chemical composition of tropical australian marine macroalgae. J Appl Phycol 18:381–387
49. Fu FX, Warner ME, Zhang Y, Feng Y, Hutchins DA (2007) Effects of increased temperature and CO_2 on photosynthesis, growth, and elemental ratios in marine *synechococcus* and *prochlorococcus*(cyanobacteria). J Phycol 43(3):485–496
50. Andersen T, Andersen F (2006) Effects of CO_2 concentration on growth of filamentous algae and *Littorella uniflora* in a Danish softwater lake. Aquat Bot 84:267–271
51. Harun R, Singh M, Forde GM, Danquah MK (2010) Bioprocess engineering of microalgae to produce a variety of consumer products. Renew Sustain Energy Rev 14(2010):1037–1047
52. Aresta M, Dibenedetto A, Carone M, Colonna T, Fragale C (2005) Production of biodiesel from macroalgae by supercritical CO_2 extraction and thermochemical liquefaction. Env Chem Lett 3(3):136–139
53. Wright HJ, Segur JB, Clark HV, Coburn SK, Langdon EE, DuPuis EN (1944) A report on ester interchange. Oil Soap 21:145–148
54. Freedman B, Butterfield RO, Pryde EH (1986) Transesterification kinetics of soybean oil 1. J Am Oil Chem Soc 63:1375–1380
55. Stern R, Hillion G (1990) Purification of esters. Eur Pat Appl EP 356317
56. Harrington KJ, D'Arcy-Evans C (1985) Transesterification in situ of sunflower seed oil. Ind Eng Chem Prod Res Dev 24:314–318
57. Graille J, Lozano P, Pioch D, Geneste P (1986) Essais d'alcoolyse d'huiles végétales avec des catalyseurs naturels pour la production de carburants diesels. Oleagineux 41:457–464
58. Freedman B, Pryde EH, Mounts TL (1984) Variables affecting the yields of fatty esters from transesterified vegetable oils. J Am Oil Chem Soc 61:1638–1643
59. Schuchardt U, Sercheli R, Vargas RM (1998) Transesterification of vegetable oils: a review. J Braz Chem Soc 9(1):199–210
60. Sivasamy A, Cheah KY, Fornasiero P, Kemausuor F, Zinoviev S, Miertus S (2009) Catalytic applications in the production of biodiesel from vegetable oils. Chemsuschem 2(4):278–300
61. Helwani Z, Othman MR, Aziz N, Kim J, Fernando WJN (2009) Solid heterogeneous catalysts for transesterification of triglycerides with methanol: a review. Appl Catal A 363:1–10
62. Demirbas AH, Demirbas I (2007) Importance of rural bioenergy for developing countries. Energy Convers Manag 48:2386–2398
63. Knothe G, Van Gerpen J, Krahl J (2005) The biodiesel handbook. AOCS Press, Champaign, IL. ISBN: 9781893997622
64. Sharma YC, Singh B, Upadhyay SN (2008) Advancements in development and characterization of biodiesel: a review. Fuel 87:2355–2373
65. Demirbas A (2008) Biofuels sources, biofuel policy, biofuel economy and global biofuel projections. Energy Convers Manag 49:2106–2116
66. (a) Aresta M, Dibenedetto A, Pastore C (2004) Group 5 (V, Nb and Ta) element-alkoxides as catalysts in the trans-esterification of ethylene-carbonate with methanol, ethanol and allyl alcohol. In: Studies on surface sciences and catalysis (Carbon Dioxide Utilization for Global Sustainability), 153, 221. (b) Dibenedetto A, Aresta M, Angelini A, Ethiraj J, Aresta BM (2012) Synthesis, characterization, and use of Nb V/Ce IV-mixed oxides in the direct carboxylation of ethanol by using pervaporation membranes for water removal. Chem A Eur J 18(33):10524–10534
67. Angelini A, Dibenedetto A, Fasciano S, Aresta M (2017) Synthesis of di-*n*-butyl carbonate from *n*-butanol: comparison of the direct carboxylation with butanolysis of urea by using recyclable heterogeneous catalysts. Catal Today 281:371–378

68. Dibenedetto A, Angelini A, Colucci A, di Bitonto L, Pastore C, Aresta BM, Giannini C, Comparelli R (2016) Tunable mixed oxides: efficient agents for the simultaneous trans-esterification of lipids and esterification of free fatty acids from bio-oils for the effective production of FAMEs. Int J Renew Energy Biofuels. Article ID 204112. https://doi.org/10.5171/2016.204112

69. Dibenedetto A, Angelini A, Aresta M, Ethiraj J, Fragale C, Nocito F (2011) Converting wastes into added value products: from glycerol to glycerol carbonate, glycidol and epichlorohydrin using environmentally friendly synthetic routes. Tetrahedron 67:1308–1313

70. Dibenedetto A, Nocito F, Papai I, Angelini A, Mancuso R, Aresta M (2013) Catalytic synthesis of hydroxymetyl-2-oxazolidinones from glycerol or glycerol carbonate and urea. Chemsuschem 6(2):345–352

71. Ben-Amotz AW, Polle JE, Subba DV, Rao DV (eds) (2008) The alga *Dunaliella*: biodiversity, phisiology, genomics and biotechnology. Science Publ

72. Aresta M, Dibenedetto A (2019) Beyond fractionation in the utilization of microalgal components. In: Pires JCM and Goncalves ALC (eds), Bioenergy with carbon capture and storage. ISBN 9780128162293, Elsevier Publ

73. Sialve B, Bernet N, Bernard O (2009) Anaerobic digestion of microalgae as a necessary step to make microalgal biodiesel sustainable. Biotechnol Adv 27(4):409–416

74. Hill RA, Connolly JD (2018) Triterpenoids. Nat Prod Rep 35:1294–1329

75. Tarkowská D, Strnad M (2018) Isoprenoid-derived plant signaling molecules: biosynthesis and biological importance. Planta 1–16

76. Kumar V, Nanda M, Joshi HC, Singh A, Sharma S, Verma M (2018) Production of biodiesel and bioethanol using algal biomass harvested from fresh water river. Renew Energy 116:606–612

77. Knothe G, Dunn RO, Bagby MO (1997) Biodiesel: the use of vegetable oils and their derivatives as alternative diesel fuels. In: Fuels and chemicals from biomass, Chapter 10, pp 172–208, ACS symposium series, vol 666. ISBN 13: 9780841235083

78. de Almeida VF, García-Moreno PJ, Guadix A, Guadix EM (2015) Biodiesel production from mixtures of waste fish oil, palm oil and waste frying oil: optimization of fuel properties. Fuel Process Technol 133:152–160

79. Huntley M, Redalje DG (2007) CO_2 mitigation and renewable oil from photosynthetic microbes: a new appraisal. Mitigat Adapt Strat Global Change 12:573–608

80. Rosenberg JN, Oyler GA, Wilkinson L, Betenbaugh MJ (2008) A green light for engineered algae: redirecting metabolism to fuel a biotechnology revolution. Curr Opin Biotechnol 19 (5):430–436

81. Sheehan J, Dunahay T, Benemann J, Roessler PA (1998) A look back at the U.S. DOE aquatic species program: biodiesel from algae. NREL close out report

82. John RP, Anisha GS, Nampoothiri KM, Pandey A (2011) Micro and macroalgal biomass: a renewable source for bioethanol. Biores Technol 102(1):186–193

83. Spolaore P, Joannis-Cassan C, Duran E, Isambert A (2006) Review: commercial application of microalgae. J Biosci Bioeng 101:87–96

84. Hirano A, Ueda R, Hirayama S, Ogushi Y (1997) CO_2 fixation and ethanol production with microalgal photosynthesis and intracellular anaerobic fermentation. Energy 22:137–142

85. Rodjaroen S, Juntawong N, Mahakhant A, Miyamoto K (2007) High Biomass production and starch accumulation in native green algal strains and cyanobacterial strains of Thailand. Kasetsart J Nat Sci 41:570–575

86. Kim MS, Baek JS, Yun YS, Sim SJ, Park S, Kim SC (2006) Hydrogen production from *Chlamydomonas reinhardtii* biomass using a two-step conversion process: anaerobic conversion and photosynthetic fermentation. Int J Hydrogen Energy 31:812–816

87. Anandraj A, White S, Mutanda T (2019) Photosystem I fluorescence as a physiological indicator of hydrogen production in *Chlamydomonas reinhardtii*. Biores Technol 273:313–319

88. Harun R, Danquah MK, Forde GM (2010) Microalgal biomass as a fermentation feedstock for bioethanol production. J Chem Technol Biotechnol 85:199–203

89. Rafiqul IM, Hassan A, Sulebele G, Orosco CA, Roustaian P, Jalal KCA (2003) Salt stress culture of blue-green algae *Spirulina fusiformis*. Pakistan J Biol Sci 6:648–650

90. Shirnalli GG, Kaushik MS, Kumar A, Abraham G, Singh PK (2018) Isolation and characterization of high protein and phycocyanin producing mutants of *Arthrospira platensis*. J Basic Microbiol 58(2):162–171

91. Ueda R, Hirayama S, Sugata K, Nakayama H (1996) Process for the production of ethanol from microalgae. US Patent 5578472

92. Sivaramakrishnan R, Incharoensakdi A (2018) Utilization of microalgae feedstock for concomitant production of bioethanol and biodiesel. Fuel 217:458–466

93. de Farias Silva C E, Bertucco A (2016) Bioethanol from microalgae and cyanobacteria: a review and technological outlook. Process Biochem 51(11):1833–1842

94. Demirbas A (2001) Biomass resource facilities and biomass conversion processing for fuels and chemicals. Energy Convers Manag 42:1357–1378

95. Nigam PS, Singh A (2011) Production of liquid biofuels from renewable resources. Prog Energy Combust Sci 37(1):52–68

96. Jain MS, Kalamdhad AS (2018) A review on management of *Hydrilla verticillata* and its utilization as potential nitrogen-rich biomass for compost or biogas production. Bioresour Technol Rep 1:69–78

97. Vergara-Fernández A, Vargas G, Alarcón N, Velasco A (2008) Evaluation of marine algae as a source of biogas in a two-stage anaerobic reactor system. Biomass Bioenerg 32(4): 338–344

98. González-González LM, Correa DF, Ryan S, Jensen PD, Pratt S, Schenk PM (2018) Integrated biodiesel and biogas production from microalgae: towards a sustainable closed loop through nutrient recycling. Renew Sustain Energy Rev 82:1137–1148

99. Bird KT, Chynoweth DP, Jerger DE (1990) Effects of marine algal proximate composition on methane yields. J Appl Phycol 2:207–213

100. Lenhart K, Klintzsch T, Langer G, Nehrke G, Bunge M, Schnell S, Keppler F (2016) Evidence for methane production by marine algae (*Emiliana huxleyi*) and its implication for the methane paradox in oxic waters. Biogeosciences 13:3163–3174

101. Chynoweth DP, Turick CE, Owens JM, Jerger DE, Peck MW (1993) Biochemical methane potential of biomass and waste feedstocks. Biomass Bioenerg 5:95–111

102. https://www.epa.gov/fish-tech/2017-epa-fda-advice-about-eating-fish-and-shellfish

103. Apt KE, Behrens PW (1999) Commercial developments in microalgal biotechnology. J Phycol 35:215–226

104. Certik M, Shimizu S (1999) Biosynthesis and regulation of microbial polyunsaturated fatty acid production. J Biosci Bioeng 87:1–14

105. Luiten EEM, Akkerman I, Koulman A, Kamermans P, Reith H, Barbosa MJ, Sipkema D, Wijffels RH (2003) Realizing the promises of marine biotechnology. Biomol Eng 20:429–439

106. Abril R, Garrett J, Zeller SG, Sander WJ, Mast RW (2003) Safety assessment of DHA-rich microalgae from *Schizochytrium sp.* Part V: target animal safety/toxicity study in growing swine. Regul Toxicol Pharm 37:73–82

107. Wu BCP, Stephen D, Morgenthaler GE, Jones DV (2015) US Patent application no. 14/505, 427

108. AHA (American Heart Association). Fish 101 (2014) http://www.heart.org/HEARTORG/GettingHealthy/NutritionCenter/Fish-101_UCM_305986_Article.jsp

109. Jiang Y, Chen F, Liang SZ (1999) Production potential of docosahexaenoic acid by the heterotrophic marine dinoflagellate *Crypthecodinium cohnii*. Process Biochem 34:633–637

110. Puri M (2017) Algal biotechnology for pursuing omega-3 fatty acid (bioactive) production. Microbiology Australia 38(2):85–88

111. Prasanna RA, Sood A, Jaiswal P, Nayak S, Gupta V, Chaudhary V, Joshi M, Natarajan C (2010) Rediscovering cyanobacteria as valuable sources of bioactive compounds. Appl Biochem Microbiol 46:119–134
112. Chaneva G, Urnadzhieva S, Minkova K, Lukavsky J (2007) Effect of light and temperature on the cyanobacterium *Arthronema africanum*—a prospective phycobiliprotein producing strain. J Appl Phycol 19:537–544
113. Prasanna RA, Sood A, Suresh S, Nayak S, Kaushik BD (2007) Potentials and applications of algal pigments in biology and industry. Acta Bot Hung 49:131–156
114. Sachindra NM, Sato E, Maeda H, Hosokawa M, Niwano Y, Kohno M, Miyashita K (2007) Radical scavenging and singlet oxygen quenching activity of marine carotenoid fucoxanthin and its metabolites. J Agri Food Chem 55:8516–8522
115. Hosokawa M, Kudo M, Maeda H, Kohno H, Tanaka T, Miyashita K (2004) Fucoxanthin induces apoptosis and enhances the antiproliferative effect of the PPARg ligand, troglitazone, on colon cancer cells. Biochim Biophys Acta 1675:113–119
116. Bolhassani A (2015) Cancer chemoprevention by natural carotenoids as an efficient strategy. Anti-Cancer Agents Med Chem (Formerly Current Medicinal Chemistry-Anti-Cancer Agents) 15(8):1026–1031
117. Singh S, Kate BN, Banerjee UC (2005) Bioactive compounds from cyanobacteria and microalgae: an overview. Crit Rev Biotechnol 25:73–95
118. Becker EW (1994) Microalgae. Biotechnology and microbiology. Cambridge University Press, Cambridge. ISBN 978-0-521-06113
119. Wijffels RH (2007) Potential of sponges and microalgae for marine biotechnology. Trends Biotechnol 26(1):26–31
120. Cognis (2008) Cognis Launches Betatene_10% WDP. Cognis [Online] Cognis, 22 10 2008. http://www.cognis.com/company/Press?and?Media/Press?Releases/2008/221008_ EN_NHa.htm
121. Chisti Y (2006) Microalgae as sustainable cell factories. Environ Eng Manag J (EEMJ) 5(3)
122. Hussein G, Sankawa U, Goto H, Matsumoto K, Watanabe H (2006) Astaxanthin, a carotenoid with potential in human health and nutrition. J Nat Prod 69:443–449
123. Vilchez C, Forjan E, Cuaresma M, Bedmar F, Garbayo I, Vega JM (2011) Marine carotenoids: biological functions and commercial applications. Mar Drugs 9:319–333
124. Hejazi MA, Wijffels RH (2004) Milking of microalgae. Trends Biotechnol 22:189–194
125. Li J, Zhu D, Niu J, Shen S, Wang G (2011) An economic assessment of astaxanthin production by large scale cultivation of *Haematococcus pluvialis*. Biotechnol Adv 29 (6):568–574
126. Tso MOM, Lam T-T (1996) Method of retarding and ameliorating central nervous system and eye damage. US Patent 5527533. The method was called Deep Extract™
127. Paust J (1996) Carotenoids, vol 2: Synthesis. In: Britton G, Liaaen-Jensen S, Pfander H (eds), Chap 3, Part VII, pp 259–292. Birkhäuser, Basel
128. Ernst H (1996) Carotenoids, vol 2: Synthesis. In: Britton G, Liaaen-Jensen S, Pfander H (eds), Chap 2, Part III, pp 79–102. Birkhäuser, Basel
129. Olaizola M (2003) Commercial development of microalgal biotechnology: from the test tube to the marketplace. Biomol Eng 20:459–466
130. Widmer E, Zell R, Broger EA, Crameri Y, Wagner HP, Dinkel J, Schlageter M, Lukác T (1981) Technische Verfahren zur Synthese von Carotinoiden und verwandten Verbindungen aus 6-Oxo-isophoron. II. Ein neues Konzept für die Synthese von (3RS, 3′ RS)-Astaxanthin. Helv Chim Acta 64:2436–2446
131. Ernst H, Dobler W, Paust J, Rheude U (1994) BASF, Europ. Pat. 633 258
132. Higuera-Ciapara I, Felix-Valenzuela L, Goycoolea FM (2006) Astaxanthin: a review of its chemistry and applications. Crit Rev Food Sci Nutr 46:185–196
133. Li J, Zhu D, Niu J, Shen S, Wang G (2011) An economic assessment of astaxanthin production by large scale cultivation of *Haematococcus pluvialis*. Biotechnol Adv 29 (6):568–574

134. Nguyen KD (2013) Astaxanthin: a comparative case of synthetic vs. natural production. Chemical and Biomolecular Engineering Publications and other works. http://trace. tennessee.edu/utk_chembiopubs/94/
135. Algatech (2004) Astaxanthin—the algatech story [Online]. Algatech. http://www.algatech. com/
136. Borowitzka MA (2006) Biotechnological & environmental applications of microalgae. Biotechnological & Environmental Applications of Microalgae. http://www.bsb.murdoch. edu.au/groups/beam/BEAM-Appl0.html
137. Eriksen N (2008) Production of phycocyanin—a pigment with applications in biology, biotechnology, foods and medicine. Appl Microbiol Biotechnol 80:1–14
138. Sekar S, Chandramohan M (2008) Phycobiliproteins as a commodity: trends in applied research, patents and commercialization. J Appl Phycol 20:113–136
139. Harnedy P, FitzGerald RJ (2011) Review of bioactive protein, peptides and amino acids from macroalgae. J Phycol 47:218–232
140. Gotelli IB, Cleland R (1968) Differences in the occurrence and distribution of hydroxyproline-proteins among the algae. Am J Bot 55:907–914
141. Fleurence J (1999) Seaweed proteins: biochemical, nutritional aspects and potential uses. Trends Food Sci Technol 10:25–28
142. https://www.european-bioplastics.org/market
143. Smidsrod O, Skjak-Braek G (1990) Alginate as immobilization matrix for cells. TIBTECH 8:71–78
144. Gombotz WR, Wee SF (1998) Protein release from alginate matrices. Adv Drug Rev 31:267–285
145. Shahidi F, Abuzaytoun R (2005) Chitin, Chitosan, and co-products: chemistry, production, applications, and health effects. Adv Food Nutr. Res 49:93–135. https://www.sciencedirect. com/science/journal/10434526/49/supp/C
146. Mendhulkar VD, Shetye LA (2017) Synthesis of biodegradable polymer polyhydroxyalka- noate (PHA) in cyanobacteria *Synechococcus elongates* under mixotrophic nitrogen- and phosphate-mediated stress conditions. Ind Biotechnol 13(2):85–93
147. Hempel F, Bozarth AS, Lindenkamp N, Klingl A, Zauner S, Linne U, Steinbüchel A, Maier UG (2011) Microalgae as bioreactors for bioplastic production. Microb Cell Fact 10:81–86
148. Burlew JS (1953) Algal culture: from laboratory to pilot plant. Carnegie Inst., Washington Publ.
149. Tamiya H, Iwamura T, Shibata K, Hase E, Nihei T (1953) Correlation between photosynthesis and light-independent metabolism in the growth of Chlorella. Biochim Biophys Acta 12:23–40
150. Benson AA (2005) Hiroshi Tamiya. Biogr Mem 86:335–353
151. Cysewski GR (1994) (Cyanotech Corporation): ocean-chill drying of microalgae and microalgal products. US5276977 A
152. Sakakibara M, Fukuda Y, Sekiya A, Nishihashi H, Hirahashi T (2008) Process for treating spirulina. US7326558 B2, Dainippon Ink And Chemicals, Inc.
153. Ayalon O (2014) Astaxanthin derivatives for heat stress prevention and treatment. WO2014057493 A1, Algatechnologies Ltd.
154. Franklin S, Somanchi A, Espina K, Rudenko G, Chua P (2011) Recombinant microalgae cells producing novel oils. US7935515 B2, Solazyme, Inc.
155. Franklin S, Somanchi A, Espina K, Rudenko G, Chua P (2017) Production of tailored oils in heterotrophic microorganisms. EP3098321 A3, TerraVia Holdings, Inc.
156. Roussis SG, Cranford RJ (2014) Compositions of matter comprising extracted algae oil. US20140249338 A1, Sapphire Energy, Inc.
157. Ryan C (2009), In: Hartley A (ed) The promise of algae biofuels. Cultivating clean energy. NRDC Report

158. Usui N, Ikenouchi M (1997) The biological CO_2 fixation and utilization project by RITE(1). Highly-effective photobioreactor system. Energy Convers Manag 38:S487–S492
159. Sapphire Energy Inc (2012) Sapphire Energy Announces $ 144 Million Series C Funding. Press release
160. Solazyme Inc (2016) Solazyme focuses its breakthrough algae platform to redefine the future of food. Press release
161. Muller-Feuga A, Le Gue´des R, Herve´ A, Durand P. (1998) Comparison of artificial light photobioreactors and other production systems using *Porphyridium cruentum*. J Appl Phycol 10(1):83–90
162. van Egmond K, Bresser T, Bouwman L (2002) The European nitrogen case. Ambio 31 (2):72–78
163. de Fraiture C, Giordano M, Liao Y (2008) Biofuels and implications for agricultural water use: blue impacts of green energy. Water Policy 10(Supplement 1):67–81
164. Jez S, Spinelli D, Fierro A, Dibenedetto A, Aresta M, Busi E, Basosi R (2017) Comparative life cycle assessment study on environmental impact of oil production from micro-algae and terrestrial oilseed crops. Biores Technol 239:266–275
165. Boyd AR, Champagne P, McGinn PJ, MacDougall KM, Melanson JE, Jessop PG (2012) Switchable hydrophilicity solvents for lipid extraction from microalgae for biofuel production. Biores Technol 118:628–632
166. Davis R, Aden A, Pienkos PT (2011) Techno-economic analysis of autotrophic microalgae for fuel production. Appl Energ 88:3524–3531
167. Hess SK, Lepetit B, Kroth PG, Mecking S (2018) Production of chemicals from microalgae lipids—status and perspectives. Eur J Lipid Sci Technol 120:1700152
168. Cheirsilp B, Torpee S (2012) Enhanced growth and lipid production of microalgae under mixotrophic culture condition: effect of light intensity, glucose concentration and fed-batch cultivation. Bioresour Technol 110:510–516
169. FAO (2010) Designing viable algal bioenergy co-production concepts. In: Algae-based biofuels applications and co-products. n° 44, Roma, FAO
170. Aresta M, Dibenedetto A, He LN (2013) Analysis of demand for captured CO_2 and products from CO_2 conversion. TCGR report
171. Norsker N-H, Barbosa MJ, Vermuë MH, Wijffels RH (2011) Microalgal production—a close look at the economics. Biotechnol Adv 29(1):24–27
172. Wijffels RH, Barbosa MJ, Eppink MHM (2010) Microalgae for the production of bulk chemicals and biofuels. Biofuels Bioprod Bioref 4:287–295
173. Dibenedetto A, Colucci A, Aresta M (2016) The need to implement an efficient biomass fractionation and full utilization based on the concept of "biorefinery" for a viable economic utilization of microalgae. Env Sci Pollut Res 23:22274–22283
174. Buono S, Colucci A, Angelini A, Langelotti AL, Massa M, Martello A, Fogliano V, Dibenedetto A (2016) Productivity and biochemical composition of *Tetradesmus obliquus* and *Pheodactylum tricornutum*: effect of different cultivation approaches. J Appl Phycol 28 (6):3179–3192
175. Lai H-M, Padua GW, Wei LS (1997) Properties and microstructure of zein sheets plasticized with palmitic and stearic acids. Cereal Chem 74(1):83–90
176. Aresta M, Dibenedetto A, Cornacchia D (2017) Mixed oxides for the oxidative cleavage of lipids using oxygen to afford mono- and di-carboxylic acids WO2017202955A1
177. Köckritz A, Martin A (2011) Synthesis of azelaic acid from vegetable oil-based feedstocks. Eur J Lipid Sci Technol 113:83–89
178. Janz A, Köckritz A, Habil MA (2011) Producing mono- and dicarboxylic acids, useful in pharmaceutical and plastic industries, comprises oxidatively splitting oxidized derivatives of vegetable oil or fat with molecular oxygen or air using gold-containing catalyst and solvent. Patent DE 102010002603 A1. Br
179. Brandhorst M, Dubois J-L (2015) Method for cleaving unsaturated fatty chains. US Patent 9035079

180. Dibenedetto A, Nocito F in preparation
181. Mäki-Arvela P, Kuusisto J, Sevilla EM, Simakova I, Mikkola J-P, Myllyoja J, Salmi T, Murzin DY (2008) Catalytic hydrogenation of linoleic acid to stearic acid over different Pd- and Ru-supported catalysts. Appl Cat A 345(2):201–212
182. Yaakob Z, Ali E, Zainal A, Mohamad M, Takriff MS (2014) An overview: biomolecules from microalgae for animal feed and aquaculture. J Biol Res 19, 21(1):6–16
183. Li J, Liu Y, Cheng JJ, Mos M, Daroch M (2015) Biological potential of microalgae in China for biorefinery-based production of biofuels and high value compound. New Biotechnol 32 (6):588–596
184. Yen H-W, Hu I-C, Chen C-Y, Ho S-H, Lee D-J, Chango J-S (2013) Microalgae-based biorefinery-From biofuels to natural products. Bioresour Technol 135:166–174
185. Dibenedetto A, Aresta M, Pastore C, di Bitonto L, Angelini A, Quaranta E (2015) Conversion of fructose into 5-HMF: a study on the behaviour of heterogeneous Ce-based catalysts and their stability in aqueous media under mild conditions. RSC Adv 5:26941–26948
186. Dibenedetto A, Aresta M, di Bitonto L, Pastore C (2016) Organic carbonates: efficient extraction solvents for the synthesis of 5-HMF in aqueous media witn Ce-phosphates as catalysts. ChemSusChem 9:118–125
187. Ventura M, Aresta M, Dibenedetto A (2016) Selective Aerobic Oxidation of 5-(Hydrox- ymethyl)furfural to 5-Formyl-2-furancarboxylic acid in water. ChemSusChem 9(10):1096– 1100
188. Dibenedetto A, Ventura M, Lobefaro F, de Giglio E, Distaso M, Nocito F (2018) Selective aerobic oxidation of 5-(hydroxymethyl) furfural to 2, 5-diformylfuran or 2-formyl-5- furancarboxylic acid in water using MgO·CeO$_2$ mixed oxides as catalysts. ChemSusChem 11(8):1305–1315
189. Nocito F, Ventura M, Aresta M, Dibenedetto A (2018) Selective oxidation of 5- (Hydroxymethyl)furfural to DFF using water as solvent and oxygen as oxidant with earth-crust-abundant mixed oxides. ACS Omega 3(12):18724–18729
190. Ventura M, Nocito F, de Giglio E, Cometa S, Altomare A, Dibenedetto A (2018) Tunable mixed oxides based on CeO$_2$ for the selective aerobic oxidation of 5-(hydroxymethyl) furfural to FDCA in water. Green Chem 20:3921–3926
191. Dibenedetto A, Ventura M, Williamson D, Lobefaro F, Jones MD, Mattia D, Nocito F, Aresta M (2018) Sustainable synthesis of oxalic (and succinic) acid via aerobic oxidation of C6 polyols by using M@ CNT/NCNT (M = Fe, V) based catalysts in mild conditions. ChemSusChem 11(6):1073–1081
192. Lundquist TJ, Woertz IC, Quinn NWT, Benemann JR (2010) A realistic technology and engineering assessment of algae biofuel production. Energy Biosci Instit 1–178
193. Aresta M, Dibenedetto A, Dumeignil F (Eds) (2015) Biorefineries: an introduction Walter de Gruyter. GmbH & Co KG, Berlin/Boston. ISBN 978-3-11-033158-5
194. Pulz O, Gross W (2004) Valuable products from biotechnology of microalgae. Appl Microbiol Biotechnol 65:635–648
195. Food and Agriculture Organization of the United Nations. http://faostat.fao.org

Technoenergetic and Economic Analysis of CO$_2$ Conversion

Suraj Vasudevan, Shilpi Aggarwal, Shamsuzzaman Farooq,
Iftekhar A. Karimi and Michael C. G. Quah

Abstract

Mere improvements in energy efficiency and development of alternative energy sources may not be sufficient and timely to reverse the continuing rise of the CO$_2$ emissions before it crosses dangerous levels. Given the mixed feelings on the geological sequestration of captured CO$_2$ and the scale of worldwide CO$_2$ emissions, the idea of utilizing CO$_2$ to produce fuels and chemicals is receiving increasing attention as a potential long-term solution to this problem. The source of hydrogen is vital for producing fuels and chemicals from CO$_2$. We consider both renewable (i.e., solar) and nonrenewable (i.e., fossil fuels) sources of hydrogen and identify several fuels and chemicals that can be produced from CO$_2$ while meeting the hard constraint of net zero CO$_2$ emission. Taking a small, geologically disadvantaged, and developed city-state of Singapore as an example, we analyze and compare thermodynamically feasible production of fuels/chemicals, whose global demands can make a significant dent in CO$_2$ emissions. We also identify the hydrogen source and the cost at which it will make economic sense under various carbon tax regimes.

12.1 Introduction

With increasing focus on the tackling of the CO$_2$ emissions and resulting global warming problem, the pertinent question today is what next after the capture of CO$_2$. Two very promising options are sequestration and utilization. For large

S. Vasudevan (✉) · S. Aggarwal · S. Farooq · I. A. Karimi · M. C. G. Quah
Department of Chemical & Biomolecular Engineering, National University of Singapore,
4 Engineering Drive 4, Singapore 117585, Singapore
e-mail: chesura@nus.edu.sg

© Springer Nature Switzerland AG 2019
M. Aresta et al. (eds.), *An Economy Based on Carbon Dioxide and Water*,
https://doi.org/10.1007/978-3-030-15868-2_12

countries, sequestration is a feasible option due to the advantages of geography. On the other hand, for smaller countries like Singapore, geographical limitations will mean difficulties in the location of sequestration/storage sites. To add to this, heavy industrialization mainly in the refining and petrochemical sectors has and will lead to a big jump in the CO_2 emissions. In this context, the relatively smaller amount of emissions in countries like Singapore and Hong Kong, compared to larger countries is a blessing in disguise as it offers the scope to look into the utilization of captured CO_2 to produce important fuels/chemicals.

Carbon capture and utilization (CCU) as an option to tackle the CO_2 emissions problem has received less attention compared to carbon capture and storage/sequestration (CCS). Moreover, CCU schemes are not widely commercial as CO_2 being a stable molecule means any conversion to chemicals and materials constitutes 'moving up' the thermodynamic ladder where additional energy and materials are needed. We first look at some of the latest works on CCU (up till the year 2017). In general, there is a lack of well-proven technologic routes—most routes are either at conceptual or bench-scale stages, with a handful in demonstration stage.

Most CCU works focus on one particular route such as *methanol* [11, 21, 28, 33, 37, 44, 48, 56, 59, 62], *formic acid* [23, 45], *methane* [5, 12, 18, 31], and *dimethyl ether* [8, 49]. The works of Cheah et al. [10], ElMekawy et al. [19], Oh and Martin [41], Wilson et al. [57], Yadav et al. [60], Dibenedetto et al. [16], Bajracharya et al. [7] and Goyal et al. [24] focus on *biotechnology applications* (such as biofuels and microalgae). Konig et al. [27] and Kuo and Wu [29] discuss the *syngas route to produce liquid hydrocarbons*. Some of the other common routes discussed are *electrochemical conversion* [32, 35, 58], *enhanced oil recovery* [20, 38, 52], and *carbonates* [1, 4, 53].

Dry reforming of methane has been the focus in several recent works such as Horváth et al. [25], Huang et al. [26], Gao et al. [22], Yao et al. [61], Wang et al. [54], Oemar et al. [40] and Wang et al. [55]. A few works look at a *combination* of different routes/products such as those of Rebecca Khoo et al. [47]—polymers, star-shaped molecules and nanocarbons; Meylan et al. [36]—solar fuels, mineralization, polymer synthesis, biological utilization; Otto et al. [43]—bulk chemicals such as formic acid, oxalic acid, formaldehyde, methanol, urea, dimethyl ether, and fine chemicals such as methyl urethane; Li et al. [30]—microalgae ponds, accelerated carbonation using alkaline solid wastes; Chiuta et al. [13]—power-to-methane and power-to-syngas; Matzen and Demirel [34] and da Silva et al. [15]—methanol and dimethyl ether; Dutta et al. [17]—dimethyl ether, ethylene, formic acid, ethylene carbonate, urea, etc.; Choi et al. [14]—direct CO_2-Fischer Tropsch synthesis with solar hydrogen to liquid transport fuels.

Detailed reviews and insights on the current status of CCU is available in Aresta et al. [3], Aresta [2] and Naims [39].

The current work proposes a framework for CCU with several possible CO_2 utilization options and applies the results to a realistic scenario. The results of a preliminary study on a small and developed city-state Singapore are presented. Two important issues when evaluating various utilization options are whether the selected route is profitable and whether it is able to make a real dent in the CO_2 emissions. While economics would favor high value-added products, their relatively low volumes will not be able to make a significant dent in the emissions. Hence, products with high global demands need to be considered to bring about a substantial reduction in CO_2 emissions. However, the low costs of such products will make the economics and carbon tax more critical. The current work focusses on such products—and analyzes their ability to make a dent in CO_2 emissions as well as their economics.

The following analysis encompasses a range of fuels/chemicals that could be produced preferably with renewable energy and hydrogen from renewable sources. We minimize the energy penalty needed for these 'thermodynamically uphill' reactions with solar hydrogen; this will also minimize the emissions needed for their production. Given the constraints of energy and emissions, the focus is limited to C1–C4 chemistries. The proposed overall CCU scheme is presented in Sect. 12.2. The alternative CCU routes considered in this study are described in Sect. 12.3. The more specific Singapore case study results are presented in Sect. 12.4, followed by concluding remarks in Sect. 12.5.

12.2 Overall Scheme for CCU

Figure 12.1 gives a broad overview of the proposed CCU. The stationary CO_2 emissions from the power plants and refining/chemical industry can be directed to utilization to produce transport fuels, chemical intermediates and commodity chemicals. Carbon recycle is crucial in CO_2 mitigation via utilization, that is, the fuels/chemicals produced from CO_2 can offset part of their current production through the traditional route. This will reduce the consumption of crude oil and the resulting CO_2 emissions.

12.2.1 Sources of Hydrogen

For a highly stable molecule like CO_2, conversion will require renewable energy inputs, novel catalysts and processes, and high-energy reactants. One apparent example of a high-energy reactant is hydrogen. CO_2 transformation through hydrogen depends on the source and cost of the latter. Since the broad aim is to reduce CO_2 emissions, it would be desirable to employ a renewable source of hydrogen. One possibility is to produce hydrogen by solar splitting of water (as shown in Fig. 12.2). However, electrolysis of water is an energy-intensive process. Its power consumption at 100% efficiency is 39.4 kWh/kg of hydrogen [42]. At the highest achievable efficiency of 73% [42], the actual power consumption for electrolysis of water is at least 53 kWh/kg of hydrogen.

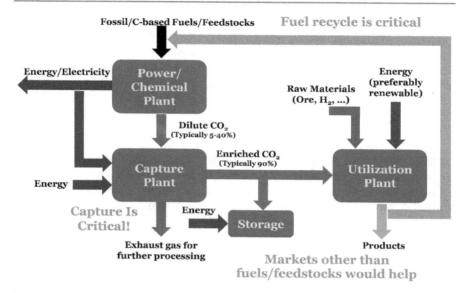

Fig. 12.1 A schematic of the carbon capture and sequestration/utilization process

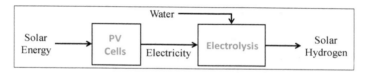

Fig. 12.2 Hydrogen from solar splitting of water

Fossil fuels can also be used as the source of hydrogen. However, combustion of fossil fuels will emit additional CO_2. In order to achieve a zero CO_2 emission system, the fossil fuel must supply additional hydrogen for utilizing CO_2 emissions after consuming its own carbon. Thus, not all fossil fuels are suitable. For example, coal is not feasible due to the non-availability of extra hydrogen to utilize the CO_2 emitted by other processes, as shown by the following reactions:

$$CH + 0.5O_2 \rightarrow CO + 0.5H_2$$

$$CO + H_2O \rightarrow CO_2 + H_2$$

In fact, even for methane, which has the highest hydrogen-to-carbon ratio among the fossil fuels, the possible chemicals under zero emission condition are limited as can be seen in Sect. 12.3.2 later.

12.3 Potential Fuels/Products from CCU with Zero Emissions

12.3.1 Fuels/Products from Solar Hydrogen

There are numerous chemicals that can be produced by reacting CO_2 with solar hydrogen. In this study, a few possibilities are considered: methanol, dimethyl ether (DME), ethylene, natural gas, and formic acid. Table 12.1 lists the stoichiometric reactions that represent the conversion of CO_2 to the above chemicals, together with their heats of reaction.

In order to evaluate the various alternatives, we need to get an idea of the CO_2 mitigation efficiency of each of them. We propose to compute the CO_2 avoidance for a given utilization option using the following equation:

$$[\text{Stoichiometric } CO_2 \text{ reduction} + CO_2 \text{ avoided, if any}]$$
$$- [CO_2 \text{ emissions from the capture of } CO_2]$$
$$- [CO_2 \text{ emissions from energy input, if necessary}]$$
$$- [CO_2 \text{ emissions from solar photo voltaics (PV) for hydrogen production}]$$

$$(12.1)$$

The components in the above equation are now considered one-by-one.

1. The stoichiometric CO_2 reduction follows straight from the reaction stoichiometry given in Table 12.1.
2. The next component is the emissions that can be avoided by producing the chemicals from solar hydrogen instead of producing the same via conventional processes. These amounts will vary for each chemical. In the case of ethylene, CO_2 avoided is the amount that would otherwise be emitted during the production of ethylene via naphtha cracking. For methanol, CO_2 avoided is the difference between the emissions from gasoline and methanol engines for the same energy output. Methanol is assumed to be blended with gasoline and used as M85 fuel (85 vol% methanol and 15 vol% gasoline). A fuel equivalent factor

Table 12.1 Chemical conversion of CO_2 to selected products

Chemical products	Stoichiometric reactions	Heat of reaction, ΔH (MJ/kmol)
Methanol	$CO_2 + 3H_2 \rightarrow CH_3OH + H_2O$	−49
DME	$2CO_2 + 6H_2 \rightarrow CH_3OCH_3 + 3H_2O$	−61
Ethylene	$2CO_2 + 6H_2 \rightarrow CH_3CH_2OH + 3H_2O$ followed by $CH_3CH_2OH \rightarrow CH_2CH_2 + H_2O$	−40 (overall)
Methane	$CO_2 + 4H_2 \rightarrow CH_4 + 2H_2O$	−160
Formic acid	$CO_2 + H_2 \rightarrow HCOOH$	15

is used to estimate the amount of gasoline that would give the same mileage as M85. A similar procedure is also applied to DME with respect to diesel. The emission factors for gasoline, diesel, M85, and DME are obtained from the literature (www.eia.gov). For methane and formic acid, there is no CO_2 avoided component.

3. Next, it is assumed that the capture energy is provided in the form of electricity. Hence, its CO_2 emissions are calculated using the appropriate grid emissions factor. Furthermore, any energy released from the reaction is used to partially offset the capture energy. An energy transfer efficiency of 75% is assumed for these calculations.

4. As for the energy input component, the only chemical in Table 12.1 that is affected is formic acid as the reaction is endothermic. CO_2 emitted from supplying this energy is computed by assuming the form of this energy to be electricity. This gives a conservative estimate (i.e., upper limit) of the additional CO_2 emissions to be subtracted to get the net CO_2 reduction.

5. Finally, CO_2 emitted in hydrogen production via solar PV is calculated using a CO_2 emissions factor of 0.032 t/MWh of power generation [51].

Having discussed the assumptions made to calculate the net CO_2 avoidance from the various utilization options, the economics aspect is next considered. For this, the break-even hydrogen price (BEHP, in $ per kg) is computed such that it satisfies the following equation:

$$
\begin{aligned}
[\text{market price of the product} &+ \text{carbon tax}] \\
= [\text{carbon capture and} &\text{ concentration (CCC) cost} \\
&+ \text{fixed cost of producing hydrogen} \\
&+ \text{energy price for producing hydrogen (BEHP)} \\
&+ \text{fixed cost of producing the chemical} \\
&+ \text{energy cost (if any) for producing the chemical}]
\end{aligned} \tag{12.2}
$$

At a given rate of carbon tax, a higher BEHP for a particular chemical would mean that the economic feasibility of producing that chemical is better. The commercial prices for the various products are obtained from the literature and listed in Table 12.2. All calculations in Sect. 12.4 have been performed for the year 2015–2016. CO_2 capture cost factor is assumed to be $100/t CO_2. This is the upper limit of the range of costs presented in the IEA CCS Roadmap (www.iea.org). It is used to obtain a more attainable estimate of the capture cost. Capital cost for producing hydrogen from electrolysis is assumed to be 11% of the total hydrogen cost [42]. It is assumed that the capital cost factor for the production of various fuels/chemicals through CO_2 utilization is the same as that for the production of methanol from natural gas, which was equal to $130 (per annual t) for the year 1987 [6]. Based on CEPCI of 323.8 for 1987 and 556.8 for 2015, the calculated typical capital cost factor for the year 2010 is $223.5 (per annual t). Since all the chemicals

Table 12.2 Market prices of various chemicals/reactants considered in this study

Chemical	Formic acid	Dimethyl ether	Ethylene	Methanol	Methane	Oxygen	Water
Reference	[45]	Based on China market price in 2016	Chemical market reporter February 2016	[28]	http:// www.eia. gov (2016 price)	[45]	Public Utilities Board (PUB), Singapore
Price ($/kg)	0.715	0.36	0.568	0.44	0.185	0.06	0.47×10^{-3}

considered are produced through CO_2 utilization in a way similar to methanol production through syngas, the capital cost factor for each of the chemicals is considered to be the same. As for the energy cost, the only chemical affected is formic acid. As mentioned earlier, it is assumed that energy is supplied in the form of electricity, and thus the cost of this electricity is computed.

12.3.2 Products from Fossil-Fuel Hydrogen

As mentioned earlier in Sect. 12.2.1, when methane is used as the source of hydrogen, the CO_2 utilization options are limited. Only low-energy chemicals with a maximum hydrogen-to-carbon ratio of two can be produced. Accordingly, formic acid, formaldehyde, and acetic acid are considered. These bulk chemicals have wide-ranging uses. Formic acid has its uses in the tanning, agriculture, pharmaceuticals, and food industries, while formaldehyde finds use in the construction, furniture and automotive industries. Acetic acid has its uses in the production of chemical compounds like vinyl acetate monomer and acetic anhydride.

For all the calculations in this section, a scenario of zero net CO_2 emission is assumed. This means that the CO_2 emitted from the additional electricity required for post-combustion CCC is also utilized.

The productions of formic acid, formaldehyde, and acetic acid from CO_2 and methane have reactions common to Gas-to-Liquid technology (GTL)—an interesting route to synthesize important chemicals from H_2 and CO, that is, syngas [46]. The ratio of H_2 and CO required in GTL varies according to the specific chemical being synthesized. However, one factor that is common for all products is the generation of syngas being the most expensive and energy-intensive step. Syngas is traditionally produced by steam reforming (SR)—a technology that has high energy requirements and capital costs. A widely recognized alternative to SR is autothermal reforming (ATR), due to certain advantages like relative compactness, lower capital costs, etc. In ATR, the hydrocarbons undergo complete conversion to a mixture of H_2 and CO in a single reactor. Partial oxidation (POX) of the hydrocarbon feed provides the heat required for the endothermic reforming reactions—this makes the process auto-thermal. The reactions are summarized in Fig. 12.3.

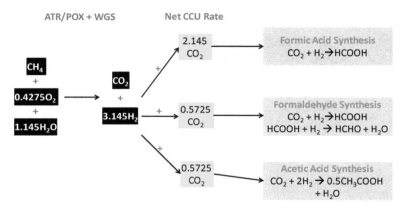

Fig. 12.3 Zero CO_2 emission products with fossil-fuel hydrogen

The basis for GTL through natural gas to produce low-energy chemicals is as follows:

$$ATR: 4CH_4 + O_2 + 2H_2O \rightarrow 10H_2 + 4CO, \quad \Delta H = 340 \text{ MJ/kmol}$$

$$POX: CH_4 + 0.5O_2 \rightarrow CO + 2H_2, \quad \Delta H = -36 \text{ MJ/kmol}$$

In order to make the above system thermo-neutral or slightly exothermic, for every mol of CH_4, 0.71 mol has to go to POX and 0.29 mol to ATR. Therefore,

$$ATR : 0.29CH_4 + 0.0725O_2 + 0.145H_2O \rightarrow 0.725H_2 + 0.29CO$$
$$POX: 0.71CH_4 + 0.355O_2 \rightarrow 0.71CO + 1.42H_2$$
$$\text{Overall: } CH_4 + 0.4275O_2 + 0.145H_2O \rightarrow CO + 2.145H_2O, \quad \Delta H = -1 \text{ MJ/kmol}$$

When combined with the following water-gas shift (WGS) reaction:

$$CO + H_2O \rightarrow CO_2 + H_2, \quad \Delta H = -41 \text{ MJ/kmol}$$

The overall reaction is:

$$CH_4 + 0.4275O_2 + 1.145H_2O \rightarrow CO_2 + 3.145H_2, \quad \Delta H = -42 \text{ MJ/kmol}$$

It can be seen that 1 mol of CO_2 is generated per mole of CH_4 oxidized. Now, with the above overall reaction, the individual chemicals are next considered one-by-one.

12.3.2.1   Formic Acid

The reaction scheme is:

$$CH_4 + 0.4275O_2 + 1.145H_2O \rightarrow CO_2 + 3.145H_2, \quad \Delta H = -42\,MJ/kmol$$
$$CO_2 + H_2 \rightarrow HCOOH, \quad \Delta H = 15\,MJ/kmol$$

From the stoichiometry for production of one mole of formic acid (1 mol H$_2$ reacting with one mole of CO$_2$), 2.145 extra moles of H$_2$ are available for the utilization of CO$_2$ from the capture process. Consequently, the overall reaction is:

$$CH_4 + 0.4275O_2 + 1.145H_2O + 2.145CO_2 \rightarrow 3.145HCOOH,$$
$$\Delta H = 4.8\,MJ/kmol$$

12.3.2.2   Formaldehyde

The reaction scheme is:

$$CH_4 + 0.4275O_2 + 1.145H_2O \rightarrow CO_2 + 3.145H_2, \quad \Delta H = -42\,MJ/kmol$$

$$CO_2 + H_2 \rightarrow HCOOH, \quad \Delta H = 15\,MJ/kmol$$

$$HCOOH + H_2 \rightarrow HCHO + H_2O, \quad \Delta H = 21\,MJ/kmol$$

From the stoichiometry of formaldehyde production (2 mol H$_2$ reacting with one mole of CO$_2$), 1.145 extra moles of H$_2$ are available for CO$_2$ utilization. Consequently, the overall reaction is:

$$CH_4 + 0.4275O_2 + 0.5725CO_2 \rightarrow 1.5725HCHO + 0.4275H_2O,$$
$$\Delta H = 15\,MJ/kmol$$

12.3.2.3   Acetic Acid

The reaction scheme is:

$$CH_4 + 0.4275O_2 + 1.145H_2O \rightarrow CO_2 + 3.145H_2, \quad \Delta H = -42\,MJ/kmol$$

$$CO_2 + 2H_2 \rightarrow 0.5CH_3COOH + H_2O, \quad \Delta H = -66\,MJ/kmol$$

Based on the stoichiometry for the production of acetic acid (2 mol H$_2$ reacting with 1 mol of CO$_2$), 1.145 extra moles H$_2$ is available for CO$_2$ utilization. Consequently, the overall reaction is:

$$CH_4 + 0.4275O_2 + 0.5725CO_2 \rightarrow 0.78125CH_3COOH + 0.4275H_2O,$$
$$\Delta H = -150\,MJ/kmol$$

Table 12.3 Stoichiometric production amounts and reactant consumptions for CO_2 utilization using hydrogen from methane under zero CO_2 emission scenarios

Chemical	Amount produced		Reactant consumption					
			Methane		Oxygen		Water	
	mpm	tpt	mpm	tpt	mpm	Tpt	mpm	tpt
Formic acid	1.466	1.533	0.466	0.169	0.199	0.145	0.534	0.218
Formaldehyde	2.747	1.873	1.747	0.635	0.747	0.543	−0.747	−0.306
Acetic acid	1.367	1.864	1.747	0.635	0.747	0.543	−0.747	−0.306

Note (1) mpm = mol per mol CO_2, tpt = tonne per tonne CO_2
(2) Negative consumption means net production of reagent

The stoichiometric consumption factors (mol/mol CO_2 and t/t CO_2) for the above overall reactions are summarized in Table 12.3.

12.3.3 Break-Even Production Cost

To evaluate the economics of this route, the break-even production cost is computed under different carbon tax scenarios from the following equation:

$$[\text{Market price of product} + \text{carbon tax}]$$
$$= [\text{cost of methane/oxygen/water} + \text{CCC cost} \qquad (12.3)$$
$$+ \text{production cost of the product (both capital and energy)}]$$

The above equation is very similar to (12.2) presented in Sect. 12.3.1. The basic idea is to determine at what production costs, one can afford these products at the current market prices under various carbon tax scenarios. These production costs should be the target that future R&D could strive to achieve.

12.4 Case Study: Singapore

The utilization scenarios presented in Sect. 12.3 are now applied to the case of Singapore, a developed small island nation which is geographically disadvantaged for any storage/sequestration to be considered. In this context, utilization is a potential option that Singapore could consider. Utilization options comprising of state-of-the-art reactions to convert CO_2 into valuable products like fuels, value-added chemicals, building, and construction materials, via organic, inorganic, or biological pathways become crucial for Singapore. The markets for many of these chemicals are small, the CO$_2$ emissions are also small—this makes utilization advantageous for Singapore. The reactions/routes presented in Sects. 12.3.1 and 12.3.2 are considered in the Singapore context, and the results are evaluated in terms of the CO$_2$ reduction capacity and economic viability.

12.4.1 CO$_2$ Emissions Scenario

Before performing any utilization calculations, the emissions scenario in Singapore must be studied. The sector-wise CO$_2$ emissions profile of Singapore for the year 2015 is considered. These annual emissions (presented in Fig. 12.4) have been determined and updated from a separate detailed study [9], which is not elaborated here for the sake of brevity. To summarize, the power plants, refineries, and petrochemical plants (i.e., ethylene) constitute Singapore's capturable emissions. This amounts to 38.86 mtpa (million tonnes per annum), and this forms the basis for all the utilization calculations. Hydrogen and ethylene glycol sectors are excluded as they emit CO$_2$ at 100% concentration, which is assumed to be already utilized in various forms.

12.4.2 Scenarios for Solar Hydrogen

Solar energy represents the most attractive and feasible renewable energy source for Singapore as energy sources such as wind and tidal are minimal, and nuclear power is debatable. Solar energy could either be used as an energy source for CO$_2$ conversion or to directly substitute fossil fuel-based electricity generation. This work focuses on the former. The evaluation of the feasibility of hydrogen-based routes for CO$_2$ utilization first requires an estimation of the availability of the solar

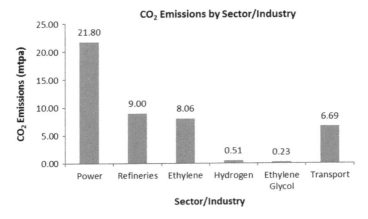

Fig. 12.4 CO$_2$ emissions from various sectors in Singapore for the year 2015

hydrogen. A good initial estimate would be to calculate the maximum possible solar power that can be generated in Singapore, given the limitations of area available for installing PV cells, which directly convert solar energy into electrical energy.

A recent study [50] has estimated the annual solar power potential for Singapore to be 7 TWh by the year 2050 under the baseline progress scenario, and 15 TWh under the accelerated development scenario (based on novel concepts for wafer-based solar cells). The former is used for a more conservative estimate; the results presented in Table 12.4 will be approximately doubled for the latter scenario. The solar power potential of 7 TWh divided by the power needed for electrolysis (see Sect. 12.2.1) gives a maximum annual hydrogen production of 0.132 mtpa. Based on the solar PV emissions factor given in Sect. 12.3.1, generation of this amount of solar power will emit 0.224 mtpa CO_2, which is just around 0.5% of the total emissions of 46.29 mtpa (38.86 mtpa capturable from power, refineries and ethylene sectors, 6.74 mtpa from uncapturable mobile sources and 0.7 mtpa assumed to be already utilized) for Singapore. Thus, 0.132 mtpa of hydrogen is used as the basis for the following utilization calculations.

Using the analysis presented in Sect. 12.3.1, the calculated stoichiometric CO_2 reduction for each fuel/chemical is given in Table 12.4. Note that the current price of electricity in Singapore (i.e., $ 0.1562/kWh taken from www.ema.gov.sg) is used to calculate the capture energy cost. Table 12.4 also shows the amount of each fuel/chemical that can be produced annually together with the normalized values. The normalization is done with respect to the annual solar power potential of Singapore.

It should be noted that the production of each fuel/chemical in Table 12.4 corresponds to fully utilizing the solar energy potential for Singapore to produce solar hydrogen. Hence the annual production amount shown is valid for producing that fuel/chemical only and they are not additive. The results indicate that formic acid is the best product in terms of the net CO_2 reduction, followed by DME and ethylene. However, looking at the current world demand/production of formic acid given later in Table 12.6, it is clear that the amount will far exceed its total world demand.

Table 12.4 Reduction in CO_2 emissions using options with solar hydrogen

Fuel/Chemical	Amount produced		Stoichiometric CO_2 consumption	CO_2 Avoidance	
	t/GWh	mtpa[a]	mtpa	mtpa	%
Formic acid	434.0	3.04	2.91	2.10	4.5
Dimethyl ether	72.3	0.51	0.97	1.69	3.6
Ethylene	44.0	0.31	0.97	1.24	2.6
Methanol	100.6	0.70	0.97	0.77	1.6
Methane	37.7	0.26	0.73	0.50	1.1

[a]Total amount produced from 7 TWh of solar electricity

Next, the BEHP's for the different CO_2 utilization options at different carbon tax rates are presented in Table 12.5. It can be concluded that the best option would be to produce formic acid as its production can be achieved at the highest possible hydrogen price of $7.44/kg without the imposition of any carbon tax. This can be favorably compared with a solar hydrogen price of $6.63/kg assuming solar electricity price at $0.125/kWh (http://www.eia.gov) in the current scenario. The only other feasible option with zero carbon tax is methanol. The negative price of hydrogen for methane and DME indicates that these options cannot be pursued without the imposition of any carbon tax. However, even with a carbon tax of $200/t CO_2, methane production by this route is feasible only if hydrogen is available almost free (see Table 12.5).

12.4.3 Scenarios for Hydrogen from Fossil Fuels

Using the baseline capturable emissions of 38.86 mtpa (see Sect. 12.4.1), the total production amounts of formic acid, formaldehyde, and acetic acid through the zero CO_2 emissions fossil-fuel route are computed by multiplying with the corresponding stoichiometric factors from Table 12.3. The results are given in Table 12.6; for comparison of scales, the actual world production/demands for these chemicals obtained from literature is also included in Table 12.6. Next, the break-even production costs for the different options at different carbon tax rates are computed based on (12.3) and presented in Table 12.7; the current product prices are also shown for comparison. Here the production cost of the product constitutes capital and operating costs excluding the cost of raw materials. The costs of methane/oxygen/water are computed using the current cost/price data in Table 12.2 and the stoichiometric consumption factors in Table 12.3.

From Table 12.6, it is clear that the complete utilization of even Singapore's relatively small CO_2 emissions will in fact flood the markets for all the product options. Taking for instance formic acid, its production from Singapore's CO_2 would be 67.7 mtpa—nearly 100 times its world demand! However, formic acid seems to yield the most favorable economics of the three chemicals, as it has the largest affordable cost of production (see Table 12.7). As expected, greater carbon tax makes it even more favorable; but the market size challenge remains formidable. The results presented above are also sensitive to periodic changes in demands and costs.

Table 12.5 BEHP's ($/kg) under various carbon tax ($/t) scenarios

	Carbon tax ($/t)			
	0	25	100	200
Methanol	0.39	0.51	0.90	1.42
Dimethyl ether	−0.19	0.10	0.95	2.09
Ethylene	0.06	0.27	0.90	1.73
Methane	−0.56	−0.47	−0.22	0.12
Formic acid	7.44	7.80	8.86	10.27

Table 12.6 World demand of chemicals considered for production from CO_2 using fossil–fuel hydrogen, and their production amounts, under zero net CO_2 emission scenario

Chemical	World demand/production (mtpa)	Production to achieve zero net CO_2 emission	
		Total production[a] (mtpa)	Production rate (tpt CO_2)[b]
Formic acid	0.72	67.7	1.53
Formaldehyde	23.0	82.7	1.87
Acetic acid	10.7	82.7	1.87

[a]Based on 44.13 mtpa of total CO_2 emissions from power, refinery and chemical plants including the additional CO_2 emissions from the CCC processes
[b]tpt = tonne chemical per tonne CO_2

Table 12.7 Break-even production costs ($/t) for various chemicals from CO_2 utilization using fossil-fuel hydrogen under zero emission scenario for different carbon tax ($/t) rates

	Carbon tax ($/t)				Current product price ($/t)
	0	25	100	200	
Formic acid	622.1	637	681.4	740.7	715 [45]
Formaldehyde	202.8	214.9	251.3	299.8	354 (based on latest China market price)
Acetic acid	566.6	578.7	615.1	663.6	700 (Chemical Market Reporter, March 2015)

12.5 Conclusions

A preliminary assessment of the potential for CCU in a small but developed country like Singapore has been presented in this study. Two possible routes for CCU to produce fuels and petrochemicals are with solar or fossil-fuel hydrogen. With solar hydrogen, the main determinant is the price of solar hydrogen. On the other hand, with the use of fossil-fuel hydrogen, the key concern is the limitation arising from the need of additional hydrogen to utilize the additional CO_2; this limits both the choice of fossil fuel as well as the possible chemicals that can be produced under zero emission condition.

In general, a major limitation with CCU is the imbalance between the large amounts of CO_2 emissions that need to be handled vis-à-vis the small demands for the CCU products. The results presented in this work indicate that the only global markets of comparable size, which can make a real dent in the CO_2 emissions, are liquid fuels (e.g., methanol, ethanol, hydrocarbons, etc.).

The utilization of CO_2 for fuels and value-added chemical production is slowly receiving attention in the global research landscape. Even in countries where geological utilization and/or sequestration are an apparently viable short-term option, its long term consequences and risks are still being investigated. Hence, the importance of CCU to produce fuels and chemicals is not necessarily unique to Singapore, but offers widespread scope for investigation and application. We have presented some preliminary results including the effect of carbon tax for the various scenarios, which will guide the cost target for solar hydrogen and production of the fuels/chemicals considered for CCU to become a reality.

Acknowledgements (1) This work is an extension of a project carried out by an NUS team for the Energy Technology Roadmapping exercise, partially supported by NCCS/NRF. (2) The authors would like to thank Professor Rakesh Agrawal (Purdue University) for the insightful discussions during the course of the roadmap development. (3) We dedicate this work to the late Professor Michael C.G. Quah who contributed immensely to this project. His dynamism, intellect, sense of humor, and laughter are greatly missed to this day.

References

1. Angelini A, Dibenedetto A, Fasciano S, Aresta M (2017) Synthesis of di-n-butyl carbonate from n-butanol: comparison of the direct carboxylation with butanolysis of urea by using recyclable heterogeneous catalysts. Catal Today 281:371–378
2. Aresta M (2016) ICCDU and JCOU: two different entities, one common goal. J CO$_2$ Util. 15:3–5
3. Aresta M, Dibenedetto A, Angelini A (2013) The changing paradigm in CO$_2$ utilization. J CO$_2$ Util 3–4:65–73
4. Aresta M, Dibenedetto A, Dutta A (2017) Energy issues in the utilization of CO$_2$ in the synthesis of chemicals: the case of the direct carboxylation of alcohols to dialkyl-carbonates. Catal Today 281:345–351
5. Ashok J, Ang M, Kawi S (2017) Enhanced activity of CO$_2$ methanation over Ni/CeO$_2$–ZrO$_2$ catalysts: influence of preparation methods. Catal Today 281:304–311
6. Baasel WD (1990) Preliminary chemical engineering plant design, 2nd edn. Van Nostrand Reinhold, New York
7. Bajracharya S, Vanbroekhoven K, Buisman CJ, Pant D, Strik DP (2016) Application of gas diffusion biocathode in microbial electrosynthesis from carbon dioxide. Environ Sci Pollut Res 23:22292–22308
8. Bonura G, Cannilla C, Frusteri L, Mezzapica A, Frusteri F (2017) DME production by CO$_2$ hydrogenation: key factors affecting the behavior of CuZnZr/ferrierite catalysts. Catal Today 281:337–344
9. Carbon Capture and Storage/Utilization—Singapore Perspectives (2014) Technology Roadmap, National Climate Change Secretariat, Prime Minister's Office Singapore, 2014
10. Cheah WY, Ling TC, Juan JC, Lee DJ, Chang JS, Show PL (2016) Biorefineries of carbon dioxide: from carbon capture and storage (CCS) to bioenergies production. Biores Technol 215:346–356
11. Chen Q, Lv M, Tang Z, Wang H, Wei W, Sun Y (2016) Opportunities of integrated systems with CO$_2$ utilization technologies for green fuel & chemicals production in a carbon-constrained society. J CO$_2$ Util 14:1–9
12. Chen X, Su X, Duan H, Liang B, Huang Y, Zhang T (2017) Catalytic performance of the Pt/TiO$_2$ catalysts in reverse water gas shift reaction: controlled product selectivity and a mechanism study. Catal Today 281:312–318
13. Chiuta S, Engelbrecht N, Human G, Bessarabov DG (2016) Techno-economic assessment of power-to-methane and power-to-syngas business models for sustainable carbon dioxide utilization in coal-to-liquid facilities. J CO$_2$ Util 16:399–411

14. Choi YC, Jang YJ, Park H, Kim WY, Lee YH, Choi SH, Lee JS (2017) Carbon dioxide Fischer-Tropsch synthesis: a new path to carbon-neutral fuels. Appl Catal B 202:605–610

15. da Silva RJ, Pimentel AF, Monteiro RS, Mota CJ (2016) Synthesis of methanol and dimethyl ether from the CO_2 hydrogenation over Cu-ZnO supported on Al_2O_3 and Nb_2O_5. J CO_2 Util 15:83–88

16. Dibenedetto A, Colucci A, Aresta M (2016) The need to implement an efficient biomass fractionation and full utilization based on the concept of "biorefinery" for a viable economic utilization of microalgae. Environ Sci Pollut Res 23:22274–22283

17. Dutta A, Farooq S, Karimi IA, Khan SA (2017) Assessing the potential of CO_2 utilization with an integrated framework for producing power and chemicals. J CO_2 Util 19:49–57

18. Duyar MS, Wang S, Arellano-Treviño MA, Farrauto RJ (2016) CO_2 utilization with a novel dual function material (DFM) for capture and catalytic conversion to synthetic natural gas: an update. J CO_2 Util 15:65–71

19. ElMekawy A, Hegab HM, Mohanakrishna G, Elbaz AF, Bulut M, Pant D (2016) Technological advances in CO_2 conversion electro-biorefinery: a step toward commercialization. Biores Technol 215:357–370

20. Fukai I, Mishra S, Moody MA (2016) Economic analysis of CO_2-enhanced oil recovery in Ohio: implications for carbon capture, utilization, and storage in the Appalachian Basin region. Int J Greenhouse Gas Control 52:357–377

21. Gai S, Yu J, Yu H, Eagle J, Zhao H, Lucas J, Doroodchi E, Moghtaderi B (2016) Process simulation of a near-zero-carbon-emission power plant using CO_2 as the renewable energy storage medium. Int J Greenhouse Gas Control 47:240–249

22. Gao X, Tan Z, Hidajat K, Kawi S (2017) Highly reactive Ni–Co/SiO_2 bimetallic catalyst via complexation with oleylamine/oleic acid organic pair for dry reforming of methane. Catal Today 281:250–258

23. Georgopoulou C, Jain S, Agarwal A, Rode E, Dimopoulos G, Sridhar N, Kakalis N (2016) On the modelling of multidisciplinary electrochemical systems with application on the electrochemical conversion of CO_2 to formate/formic acid. Comp Chem Eng 93:160–170

24. Goyal N, Zhou Z, Karimi IA (2016) Metabolic processes of *Methanococcus maripaludis* and potential applications. Microb Cell Fact 15:107

25. Horváth É, Baán K, Varga E, Oszkó A, Vágó Á, Töro M, Erdohelyi A (2017) Dry reforming of CH_4 on Co/Al_2O_3 catalysts reduced at different temperatures. Catal Today 281:233–240

26. Huang X, Ji C, Wang C, Xiao F, Zhao N, Sun N, Wei W, Sun Y (2017) Ordered mesoporous CoO–NiO–Al_2O_3 bimetallic catalysts with dual confinement effects for CO_2 reforming of CH_4. Catal Today 281:241–249

27. Konig DH, Baucks N, Dietrich RU, Worner A (2015) Simulation and evaluation of a process concept for the generation of synthetic fuel from CO_2 and H_2. Energy 91:833–841

28. Kourkoumpas DS, Papadimou E, Atsonios K, Karellas S, Grammelis P, Karellas E (2016) Implementation of the power to Methanol concept by using CO_2 from lignite power plants: techno-economic investigation. Int J Hydrogen Energy 41:16674–16687

29. Kuo PC, Wu W (2016) Thermodynamic analysis of a combined heat and power system with CO_2 utilization based on co-gasification of biomass and coal. Chem Eng Sci 142:201–214

30. Li P, Pan SY, Pei S, Yupo Lin J, Chiang PC (2016) Challenges and perspectives on carbon fixation and utilization technologies: an overview. Aerosol Air Qual Res 16:1327–1344

31. Liang B, Duan H, Su X, Chen X, Huang Y, Chen X, Delgado JJ, Zhang T (2017) Promoting role of potassium in the reverse water gas shift reaction on Pt/mullite catalyst. Catal Today 281:319–326

32. Lu X, Dennis Leung YC, Wang H, Maroto-Valer MM, Xuan J (2016) A pH-differential dual-electrolyte microfluidic electrochemical cells for CO_2 utilization. Renewable Energy 95:277–285

33. Luu MT, Milani D, Abbas A (2016) Analysis of CO_2 utilization for methanol synthesis integrated with enhanced gas recovery. J Clean Prod 112:3540–3554

34. Matzen M, Demirel Y (2016) Methanol and dimethyl ether from renewable hydrogen and carbon dioxide: Alternative fuels production and life-cycle assessment. J Clean Prod 139:1068–1077
35. Merino-Garcia I, Alvarez-Guerra E, Albo J, Irabien A (2016) Electrochemical membrane reactors for the utilisation of carbon dioxide. Chem Eng J 305:104–120
36. Meylan FD, Moreau V, Erkman S (2015) CO₂ utilization in the perspective of industrial ecology, an overview. J CO₂ Util 12:101–108
37. Mondal K, Sasmal S, Badgandi S, Chowdhury DR, Nair V (2016) Dry reforming of methane to syngas: a potential alternative process for value added chemicals—a technoeconomic perspective. Environ Sci Pollut Res 23:22267–22273
38. Morales Mora MA, Vergara CP, Leiva MA, Delgadillo SAM, Rosa-Domínguez ER (2016) Life cycle assessment of carbon capture and utilization from ammonia process in Mexico. J Environ Manage 183:998–1008
39. Naims H (2016) Economics of carbon dioxide capture and utilization—a supply and demand perspective. Environ Sci Pollut Res 23:22226–22241
40. Oemar U, Hidajat K, Kawi S (2017) High catalytic stability of Pd–Ni/Y₂O₃ formed by interfacial Cl for oxy-CO₂ reforming of CH₄. Catal Today 281:276–294
41. Oh ST, Martin A (2016) Thermodynamic efficiency of carbon capture and utilisation in anaerobic batch digestion process. J CO₂ Util 16:182–193
42. Olah GA, Goeppert A, Surya Prakash GK (2009) Chemical recycling of carbon dioxide to methanol and dimethyl ether: from greenhouse gas to renewable, environmentally carbon neutral fuels and synthetic hydrocarbons. J Org Chem 74:487–498
43. Otto A, Grube T, Schiebahn S, Stolten D (2015) Closing the loop: captured CO₂ as a feedstock in the chemical industry. Energy Environ Sci 8:3283–3297
44. Pérez-Fortes M, Schöneberger JC, Boulamanti A, Tzimas E (2016) Methanol synthesis using captured CO₂ as raw material: techno-economic and environmental assessment. Appl Energy 161:718–732
45. Pérez-Fortes M, Schöneberger JC, Boulamanti A, Harrison G, Tzimas E (2016) Formic acid synthesis using CO₂ as raw material: techno-economic and environmental evaluation and market potential. Int J Hydrogen Energy 41:16444–16462
46. Piña J, Borio DO (2006) Modeling and simulation of an autothermal reformer. Latin Am Appl Res 36:289–294
47. Rebecca Khoo SH, Luo HK, Braunstein P, Andy Hor TS (2015) Transformation of CO₂ to value-added materials. J Mol Eng Mater 3(1540007):12
48. Roh K, Lee JH, Gani R (2016) A methodological framework for the development of feasible CO₂ conversion processes. Int J Greenhouse Gas Control 47:250–265
49. Schakel W, Oreggioni G, Singh B, Strømman A, Ramírez A (2016) Assessing the techno-environmental performance of CO₂ utilization via dry reforming of methane for the production of dimethyl ether. J CO₂ Util 16:138–149
50. Solar Photovoltaic (PV) Roadmap for Singapore (2014) (A summary) Solar Energy Research Institute of Singapore (SERIS)
51. Sovacool BJ (2008) Energy Policy 36:2940–2953
52. Tapia JFD, Lee JY, Raymond Ooi EH, Dominic Foo CY, Raymond Tan R (2016) Optimal CO₂ allocation and scheduling in enhanced oil recovery (EOR) operations. Appl Energy 184:337–345
53. Wang L, Ammar M, He P, Li Y, Cao Y, Li F, Han X, Li H (2017) The efficient synthesis of diethyl carbonate via coupling reaction from propylene oxide, CO₂ and ethanol over binary PVEImBr/MgO catalyst. Catal Today 281:360–370
54. Wang C, Sun N, Zhao N, Wei W, Zhao Y (2017) Template-free preparation of bimetallic mesoporous Ni–Co–CaO–ZrO₂ catalysts and their synergetic effect in dry reforming of methane. Catal Today 281:268–275

55. Wang F, Xu L, Yang J, Zhang J, Zhang L, Li H, Zhao Y, Li HX, Wu K, Xu GQ (2017) Enhanced catalytic performance of Ir catalysts supported on ceria-based solid solutions for methane dry reforming reaction. Catal Today 281:295–303

56. Wiesberg IL, de Medeiros JL, Alves RMB, Coutinho PLA, Araújo OQF (2016) Carbon dioxide management by chemical conversion to methanol: hydrogenation and Bi-reforming. Energy Convers Manag 125:320–335

57. Wilson MH, Mohler DT, Groppo JG, Grubbs T, Kesner S, Frazar EM, Shea A, Crofcheck C, Crocker M (2016) Capture and recycle of industrial CO$_2$ emissions using microalgae. Appl Petrochem Res 6:279–293

58. Wu K, Birgersson E, Kim B, Kenis Paul JA, Karimi IA (2015) Modeling and experimental validation of electrochemical reduction of CO$_2$ to CO in a microfluidic cell. J Electrochem Soc 162:F23–F32

59. Xiao S, Zhang Y, Gao P, Zhong L, Li X, Zhang Z, Wang H, Wei W, Sun Y (2017) Highly efficient Cu-based catalysts via hydrotalcite-like precursors for CO$_2$ hydrogenation to methanol. Catal Today 281:327–336

60. Yadav A, Choudhary P, Atri N, Teir S, Mutnuri S (2016) Pilot project at Hazira, India, for capture of carbon dioxide and its biofixation using microalgae. Environ Sci Pollut Res 23:22284–22291

61. Yao L, Wang Y, Shi J, Xu H, Shen W, Hu C (2017) The influence of reduction temperature on the performance of ZrO$_x$/Ni-MnO$_x$/SiO$_2$ catalyst for low-temperature CO$_2$ reforming of methane. Catal Today 281:259–267

62. Zhang Y, Zhong L, Wang H, Gao P, Li X, Xiao S, Ding G, Wei W, Sun Y (2016) Catalytic performance of spray-dried Cu/ZnO/Al$_2$O$_3$/ZrO$_2$ catalysts for slurry methanol synthesis from CO$_2$ hydrogenation. J CO$_2$ Util 15:72–82

Perspective Look on CCU Large-Scale Exploitation

13

Michele Aresta

Abstract

CCU is a step towards a *Cyclic Carbon Economy*, with different effects over diverse time-scale. In this chapter, the potential benefits of CCU are summarized with a perspective look on the amount of used/avoided CO_2 in the short, medium, long term and on the conditions that must be fulfilled for an extensive utilization of such abundant source of carbon.

It is now clear that CCS and CCU are not two faces of the same medal. They are two different approaches to "CO_2-problem" solving, with quite diverse effects and roles. Differences, pros and cons have been presented in Chap. 1. Summarizing, the key difference is that while *CCS* continues the *Linear Carbon Economy*, *CCU* is a step towards a *Cyclic Carbon Economy*, with different effects over diverse time-scale perspective.

CCU can contribute to the reduction of CO_2 emission through "*innovation*" that means more efficient processes with lower consumption of raw-materials and direct reduction of the emission even in a fossil-C based energy scheme. The use of CO_2 promotes raw materials diversification that brings to natural resources saving. However, CCU will cause to last longer the limited fossil-C resources of our planet, while CCS expands the extraction and use of fossil-C by 25–50% depending on the operative conditions and logistics (distance of the disposal site from the source, purity of shipped-disposed CO_2, composition of the original gas-stream, etc).

It is clear that neither CCU nor CCS (alone or combined) can solve the "CO_2 problem" in a fossil-C based energy system. Interestingly, both technologies can take profit by an extensive use of "perennial energy" (SWGH). It has been said in previous Chapters that the way to an extended conversion of CO_2 is the use of SWGH. If such sources are also used for the separation, shipping, housing of CO_2,

M. Aresta (✉)
IC²R srl, Lab H124, Tecnopolis, via Casamassima km 3, 70010 Valenzano, Italy
e-mail: michele.aresta@uniba.it

© Springer Nature Switzerland AG 2019
M. Aresta et al. (eds.), *An Economy Based on Carbon Dioxide and Water*,
https://doi.org/10.1007/978-3-030-15868-2_13

CCS would take advantage as CO_2-disposal would not be done at expenses of fossil-C and natural resources would last for longer time.

The use of perennial energy sources for delivering energy to human activities is a key factor for our future. SWGH direct exploitation, without implying CO_2 conversion, is preferred by many people because it makes a better use of energy. Such direct approach is possible in a number of cases, but not for all human activities and not everywhere. It is a matter of economics and logistics. The direct injection of PV- or wind-generated electrons into the electricity network avoids the inefficiency (*ca*. 30% of energy loss) of the chemical conversion of CO_2, but it is impossible to use directly the electricity generated by perennial energies for all activities. Avio, navy and road transport sectors cannot use PV at an extended scale and still for several decades fuels will be necessary. As a matter of fact, intensive energy applications can only partially be defossilized and human life cannot be decarbonized. Electricity production for Industries, megalopolies, electrified transport will require fossil-C, the most dense energy carriers, still for long time. Human life will need C-based goods forever.

The intelligence of scientists, technologists, policy-makers must be focused on designing a network of options that optimizes the use of all forms of energy with the target of making available for the longest time and in the cleanest contest the existing resources while alleviating the environmental burden within the shortest time. There is not a single technology that can solve the CO_2-problem. Such complex problem needs an integrated solution: "how can we use all available options for shaping a best future"?

Turning to CCU, this Book has presented a variety of applications, based on CO_2 combined with energy rich substrates or water, giving the perception of the permeation of each of them in our life. Here an attempt will be done to size the amount of carbon dioxide that can be used and avoided, supposed that all options described in previous Chapters are exploited at their best.

It is worth to recall that "used" and "avoided" are not synonym. The former term refers to the amount of CO_2 used in a synthesis (dictated by the stoichiometry), the latter to the amount of CO_2 not emitted (with respect to existing conventional processes) while using CO_2 in syntheses. In an average case, for an "innovative" procedure based on CO_2 carried out in most effective conditions, the ratio "avoided/used" ranges around 2.8. This means that per each tonne of used CO_2, 2.8 are avoided.

The analysis will be carried out per class of compounds. Table 13.1 presents the actual and perspective (2030) use of CO_2 in the synthesis of some classes of chemicals.

It is clear that right now major contributes come from three sources (bold in the Table): urea, inorganic carbonates and methanol. The former is by far the largest and oldest application for CO_2 conversion.

Most of CO_2 used in the urea manufacture is recovered from the synthesis of ammonia, co-reagent with CO_2 in the synthesis of urea, a very old example of industrial CO_2-utilization and C-recycling. Technologies for the synthesis of polycarbonates are also at the demo level. Inorganic carbonates can be made through well- known procedures, but using recovered CO_2.

Table 13.1 Uses of CO_2 in the synthesis of chemicals (>1Mt/y)

Compound	Formula $C_{oxidation\ state}$		Market 2017 Mt/y	CO_2 Used Mt/y	Market 2030 Mt/y	CO_2 used Mt/y
Urea	**$(H_2N)_2CO$**	**+4**	**180**	**132**	**210**	**154**
Carbonates linear	$OC(OR)_2$	+4	>2	<0.1	10	5
Carbonates cyclic		+4		<0.1		
Polycarbonates	**$-[OC(O)$ $OOCH_2CHR]-n$**	**+4**	**5**	**<0.1**	**9–10**	**2–3**
Carbamates	$RHN–COOR$	+4	>6	1	11	ca. 4
Acrylates	$CH_2=CHCOOH$	+3	5	1.5	8	5
Formic acid	HCO_2H	+2	1	0.9	>10	>9
Inorganic carbonates	**M_2CO_3 $M'CO_3$**	**+4**	**$CaCO_3$ 250**	**70**	**400**	**100**
Methanol	**CH_3OH**	**−2**	**80**	**10**	**100**	**>28**
Total				**207**		**>332**
Ethene	$CH_2=CH_2$	−2	160	0	223 (ca. 3.5%/y growth)	ca. 140
Propene	$CH_3CH=CH_2$	−2	80	0	108 (ca. 3.5%/y growth)	ca. 65
Ethene glycol	$HOCH_2CH_2OH$	−1	30	0	45 (ca. 5%/y growth)	ca. 64
Grand total				207		ca. 600

The synthesis of methanol from CO_2 and non-fossil H_2 is in as expanding phase (demo plant), but not yet at the industrial full exploitation level. Noteworthy, some CO_2 is already used in the conventional production of methanol from Syngas.

Methodologies for the synthesis of other chemicals are still at a low TRL (3-5), but bottlenecks are known as well as what is needed to boost their exploitation. Attention demand the last three chemicals listed in Table 13.1: ethene, propene and etheneglycol. Today such chemicals are totally derived from fossil-C, but routes are known (electrochemical) that may produce such chemicals from CO_2. Assuming that a prudential share of 20% of such chemicals will be produced from CO_2 in 2030, considering a growth rate of their market equal to 3.5% per year (https://www.vci.de/langfassungen-pdf/vci-analyis-on-the-future-of-basic-chemicals-production-in-germany.pdf) an interesting amount of ca. 270 Mt/y of used CO_2 can be calculated for the three chemicals. However, such routes could be exploited if the correct investment in R&I is done.

However, one can foresee that the CO_2 utilization in the synthesis of chemicals can rise from actual ca. 200 Mt to ca. 350–600 Mt by 2030. This may appear not to be a great contribution to CC control. If we look at the *avoided* CO_2 derived from the implementation of innovative technologies based on CO_2, then the amount of avoided CO_2 can be as high as 1–1.7 Gt/y. Furthermore, an additional major contribution to CO_2 reduction in this sector could come from process innovation in the synthesis of ammonia and more precisely in the production of H_2, avoiding coal (13.1) or methane (13.2) reforming that causes large emission of CO_2(WGS). Innovation in synthetic methodologies is expected to significantly contribute to the reduction of CO_2 emission.

$$C + 2H_2O \rightarrow 2H_2 + CO_2 \qquad (13.1)$$

$$CH_4 + 2H_2O \rightarrow 4H_2 + CO_2 \qquad (13.2)$$

A recent LCA study on the synthesis of C1 molecules [A Sternberg, CM Jens and A Bardow (2017) Life cycle assessment of CO_2-based C1-chemicals. Green Chem 19:2244–2259] from CO_2 has shown that the benefit is maximized for formic acid that has the smaller market (600–800 kt/y) with respect to CO, formaldehyde and methanol. One can question if such application may have an impact on CC or not. A figure that quantifies the real benefit is the global reduction of CO_2 that is given by the product of the avoided CO_2 times the amount of the chemical produced. Any contribution of the size of 1 Mt/y can be worth to be considered in making an overall balance.

Growing aquatic biomass (microalgae) for the production of chemicals would greatly contribute to CO_2 fixation (ca. 1.8–2.2 t_{CO_2} per t of dry biomass) and conversion into fine chemicals (not shown in the Table) and fuels, with interesting emission reduction. In fact, if one considers the E-factor ($t_{waste\ produced}/t_{useful\ product}$) realizes that fine chemicals are strong waste producers ($E = 10–100$) and CO_2 emitters, as they produce mostly organic waste that usually gives off CO_2 as ultimate C-form. The production of chemicals from CO_2 will, thus, be beneficial to making more sustainable the chemical and polymer industry by reducing emissions and waste (as discussed in Chap. 1) and will have a positive impact on CC not only for the amount of CO_2 used, but for all reductions linked to innovative syntheses.

Noteworthy, it is the conversion of CO_2 into fuels that attracts attention as can be responsible for a major CO_2 conversion and cut of CO_2 emissions. In fact, fuels have a market that is some 15 times larger than chemicals.

From what has been said in previous Chapters, it is not realistic to imagine that all used fuels are synthetically made from CO_2, nor that the energy sector is totally based on synthetic fuels. However, it is conceivable that fuels derived from CO_2 hydrogenation are made for some specific transport sectors such as avio, navy, urban. Such application will grow with growing of PV instalments and with decreasing of the price of PV-electrons necessary for water-electrolysis. Most likely, the CO_2 hydrogenation will slowly grow until 2030–2040 and will be exploited at its best after such deadline. Such forecast is in line with the fact that installed PV power will most likely be maximized by 2040 (>3000 GW installed with respect to ca. 300 today) with concurrent lower costs. By then the price of PV-H_2 should be comparable to that of reforming-H_2 and enough facilities could be installed for electrochemical production of H_2 may reach the required level for large-scale exploitation. A bottleneck for a full exploitation of PV-H_2 is the

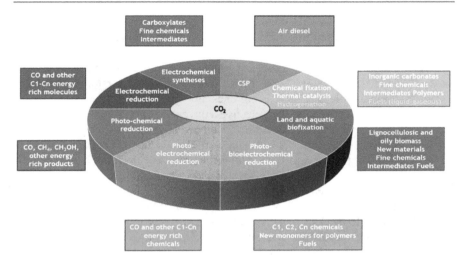

Scheme 13.1 CCU: Technologies and products. With the exception of a few routes to chemicals (bluish frame), all others require water

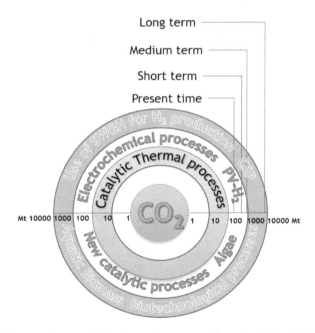

Scheme 13.2 Perspective use of CO_2 that can be converted into chemicals and fuels. Short term: 3–5 y; Medium term: 6–8 y; Long term: 15+ y

availability of large scale and durable electrolysers. If one wishes to exploit the production of H_2 *via* electrolysis of water, the actual capacity should be multiplied by 10 or 20.

New technologies are also necessary, such as high-pressure electrolysis (for making H_2 under pressure ready for distribution) and high temperature (solid-state) electrolysis, for complementing existing ones.

Concurrently, catalysts for the selective conversion of CO_2 into liquid fuels (HCn), methanol and methane should be developed so that all such options will be available allowing large scale fuels production. The use of Concentrators of Solar Power (CSP) may make usable the simultaneous recovery from the atmosphere of CO_2 and H_2O, with thermal-catalysed water and CO_2 splitting with production of "Syngas" that would generate "air diesel".

By integrating catalysis and biotechnology a consistent bunch of technologies should be available by 2040 for the conversion of a few Gt/y of CO_2 into useful products, avoiding consistent amount of fossil-C extraction and use. All such options are summarized in Scheme 13.1.

A study carried out by the Catalyst Group has shown that combining the potential of all technologies it will be possible by 2040 avoid some 7–9 Gt_{CO2}/y, a figure that agrees with data from other studies: this is a very interesting target that will contribute to CC control (Scheme 13.2).

As discussed in Chap. 1, the implementation of value-chains may largely improve the utilization of CO_2 and maximize the benefits.

CPSIA information can be obtained
at www.ICGtesting.com
Printed in the USA
LVHW010014220719
624811LV00004B/122/P

9 783030 158675